Phosphoinositides I: Enzymes of Synthesis and Degradation

SUBCELLULAR BIOCHEMISTRY

SERIES EDITOR

J. ROBIN HARRIS, University of Mainz, Mainz, Germany

ASSISTANT EDITOR

P.J. QUINN, King's College London, London, U.K.

Recent Volumes in this Series

Volume 33 **Bacterial Invasion into Eukaryotic Cells**
Tobias A. Oelschlaeger and Jorg Hacker
Volume 34 **Fusion of Biological Membranes and Related Problems**
Edited by Herwig Hilderson and Stefan Fuller
Volume 35 **Enzyme-Catalyzed Electron and Radical Transfer**
Andreas Holzenburg and Nigel S. Scrutton
Volume 36 **Phospholipid Metabolism in Apoptosis**
Edited by Peter J. Quinn and Valerian E. Kagan
Volume 37 **Membrane Dynamics and Domains**
Edited by P.J. Quinn
Volume 38 **Alzheimer's Disease: Cellular and Molecular Aspects of Amyloid beta**
Edited by R. Harris and F. Fahrenholz
Volume 39 **Biology of Inositols and Phosphoinositides**
Edited by A. Lahiri Majumder and B.B. Biswas
Volume 40 **Reviews and Protocols in DT40 Research**
Edited by Jean-Marie Buerstedde and Shunichi Takeda
Volume 41 **Chromatin and Disease**
Edited by Tapas K. Kundu and Dipak Dasgupta
Volume 42 **Inflammation in the Pathogenesis of Chronic Diseases**
Edited by Randall E. Harris
Volume 43 **Subcellular Proteomics**
Edited by Eric Bertrand and Michel Faupel
Volume 44 **Peroxiredoxin Systems**
Edited by Leopold Flohd J. Robin Harris
Volume 45 **Calcium Signalling and Disease**
Edited by Ernesto Carafoli and Marisa Brini
Volume 46 **Creatine and Creatine Kinase in Health and Disease**
Edited by Gajja S. Salomons and Markus Wyss
Volume 47 **Molecular Mechanisms of Parasite Invasion**
Edited by Barbara A. Burleigh and Dominique Soldati-Favre
Volume 48 **The Coronin Family of Proteins**
Edited by Christoph S. Clemen, Ludwig Eichinger and Vasily Rybakin
Volume 49 **Lipids in Health and Disease**
Edited by Peter J. Quinn and Xiaoyuan Wang
Volume 50 **Genome Stability and Human Diseases**
Edited by Heinz Peter Nasheuer
Volume 51 **Cholesterol Binding and Cholesterol Transport Proteins**
Edited by J. Robin Harris
Volume 52 **A Handbook of Transcription Factors**
Edited by T.R. Hughes

For further volumes:
http://www.springer.com/series/6515

Tamas Balla • Matthias Wymann • John D. York
Editors

Phosphoinositides I: Enzymes of Synthesis and Degradation

Editors
Dr. Tamas Balla
National Institutes of Health
NICHD
Bethesda, MD
USA

Dr. Matthias Wymann
University of Basel
Cancer- and Immunobiology
Basel, Switzerland

Dr. John D. York
Duke University Medical Center
Pharmacology and Cancer
Biology
Durham, NC
USA

ISBN 978-94-007-3011-3   e-ISBN 978-94-007-3012-0
DOI 10.1007/978-94-007-3012-0
Springer Dordrecht Heidelberg London New York

Library of Congress Control Number: 2012931684

© Springer Science+Business Media B.V. 2012
No part of this work may be reproduced, stored in a retrieval system, or transmitted in any form or by any means, electronic, mechanical, photocopying, microfilming, recording or otherwise, without written permission from the Publisher, with the exception of any material supplied specifically for the purpose of being entered and executed on a computer system, for exclusive use by the purchaser of the work.

Printed on acid-free paper

Springer is part of Springer Science+Business Media (www.springer.com)

Preface

When I was approached to shape a book about phosphoinositide signaling, I first felt honored and humbled. On second thought, this appeared to be an impossible task. Phosphoinositides have grown from being just a curious lipid fraction isolated from bovine brain, showing increased radioactive metabolic labeling during intense stimulation protocols, to become the focus of immense interest as key regulatory molecules that penetrate every aspect of eukaryotic biology. The expansion of this field in the last three decades has been enormous: it turned from a basic science exercise of a devoted few to highly translatable science relevant to a large number of human diseases (isn't this the nature of good basic science?). These include cancer, metabolic-, immuno- and neurodegenerative disorders, to name just a few. Reviewing the large number of enzymes that convert phosphoinositides would fill a book—let alone the diverse biological processes in which phosphoinositides play key regulatory roles. Given the interest, a collection of up-to-date reviews compiled in a book is clearly warranted, which was enough to sway me to accept this assignment. As one editor is unable to handle this enormous task, I was delighted when Matthias Wymann and John York were kind enough to join me in this ambitious effort.

When thinking about potential authors, the obvious choice would have been to approach the people whose contributions have been crucial to push and elevate this field to the level it is today. Bob Michell, prophetically placed phosphoinositides in the center of signal transduction in a 1975 Biochem. Biophys. Acta review (Michell 1975), Michael Berridge had a key role in linking phosphoinositides and Ca^{2+} signaling and whose fascinating reviews have inspired many of us (Berridge and Irvine 1984). Robin Irvine, whose group found that $InsP_3$ was a mixture of two isomers, the active $Ins(1,4,5)P_3$ and an inactive $Ins(1,3,4)P_3$, and who described the tetrakisphosphate pathway (Irvine et al. 1986), and who always challenges us with most provocative ideas. Philip Majerus, who has insisted on the importance of inositide phosphatases (Majerus et al. 1999) very early on. The group of Lewis Cantley, with the discovery of PI 3-kinase activities and the mapping of downstream effectors (Whitman et al. 1988; Franke et al. 1997), or the Waterfield lab where the first PI 3-kinase catalytic subunit was isolated and cloned (Otsu et al. 1991; Hiles et al. 1992). Peter Downes, who recognized the translational value of phosphoinositide research. Jeremy Thorner and Scott Emr, whose work in baker's yeast still forms the foundation of our understanding of the role of inositol lipids in trafficking (Strahl and Thorner 2007) or Pietro De Camilli, whose group documented the central role of inositides in brain and synaptic biology (Cremona et al. 1999). There are many others who made valuable or even greater contributions to phosphoinositide research. The above list reflects my bias, as these researchers had the largest impact on my thinking and the directions of my work. Research is, however, a constantly evolving process and we (now Matthias and John being involved) wanted to involve contributions of scientists who represent a second or third wave of researchers infected with the interest in phosphoinositides. We made an effort to recruit authors who have been trainees of these founding laboratories. With this selection our goal was to sample the view of the current and future generation. By selecting their trainees, we feel that we pay tribute to the "Founding Fathers", and show that the research they put in motion is alive and continues with fresh ideas, new ambitions and a translational and therapeutic value.

Phosphoinositide research in the 1980s went hand in hand with research on Ca^{2+} signaling pursued in "non-excitable" cells and was also marked with the discovery of the family of protein kinase C enzymes, regulated by diacylglycerol, one of the products of phosphoinositide-specific phospholipase C enzymes. These areas of research developed and expanded to form their own fields, and could not be discussed here in detail—even though they are linked historically to the development of phosphoinositide signaling. The enormous work of the groups of Yasutomi Nishizuka on protein kinase C, and Katsuhiko Mikoshiba on cloning and characterizing the Ins(1,4,5)P_3 receptors are prime examples of these achievements. Although we could not cover all these areas, we included a chapter on Ca^{2+} signaling via the Ins(1,4,5)P_3 receptor by Colin Taylor, a trainee of the Michael Berridge's lab, where important links between Ca^{2+} release and Ins(1,4,5)P_3 receptor signaling were discovered. We also decided to allocate some space to inositol phosphates, the soluble counterparts of some of the phosphoinositides. These molecules for long had been viewed only as the metabolic products of the second messenger Ins(1,4,5)P_3 but recently gained significant prominence as regulators of important physiological processes. With the discovery of the highly phosphorylated and pyrophosphorylated inositols and the enzymes that produce them, it became clear that this system represents a whole new regulatory paradigm with exciting new developments.

Finally, it was a difficult dilemma whether to include a Chapter on the early history of phosphoinositides. We decided against it for a number of reasons. First, the really interesting history is traced back to studies that preceded the landmark 1975 Bob Michell review and included the work of the Hokins (1987), Bernard Agranoff (2009) and other pioneers of phosphoinositide research. Nobody could tell these early developments better than Bob Michell in his several recollections (Michell 1995) or Robin Irvine who commemorated the 20 years of Ins(1,4,5)P_3 and the period leading to its discovery (Irvine 2003). We encourage the young readers to go back and read these recollections, as they show several examples of how seemingly uninspiring observations formed the beginning of something that became huge as it unfolded. What came after these landmark discoveries is so overwhelming that each one of us has own views and subjective memories and stories to tell on some aspects of it. As Editors we felt that our views should not be elevated above others on these historical aspects, and leave it to the authors of the individual Chapters to elucidate the diversity in this respect. The only exception is a Chapter on the history of PI 3-kinases by Alex Toker that we felt deserves special emphasis as it had the most transforming impact on the field since the late 1980s.

One needs to understand that selection of authors is a subjective process and does not always reflect on who contributed the most in a selected field. However, we are confident that proper credit is given in the individual Chapters to each groups and individuals whose work has moved this field forward. It should also be understood that a field that generates over 10,000 entries in PubMed with each keyword that relates to phosphoinositides cannot be covered without missing some aspects that could be important. However, we trust that this collection will be found useful for both the experts and the novices.

References

Agranoff BW (2009) Turtles all the way: Reflections on myo-Inositol. *J Biol Chem* 284(32):21121–21126

Berridge MJ, Irvine RF (1984) Inositol trisphosphate, a novel second messenger in cellular signal transduction. *Nature* 312:315–321

Cremona O, et al (1999) Essential role of phosphoinositide metabolism in synaptic vesicle recycling. *Cell* 99:179–188

Franke TF, Kaplan DR, Cantley LC, Toker A (1997) Direct regulation of the Akt protooncogene product by PI3,4P2. *Science* 275:665–668

Hiles ID et al (1992) Phosphatidylinositol 3-kinase: structure and expression of the 110 kDa catalytic subunit. *Cell* 70:419–429

Hokin LE (1987) The road to the phosphoinositide-generated second messengers. *Trends Pharmacol Sci* 8:53–56

Irvine RF (2003) 20 years of Ins(1,4,5)P3, and 40 years before. *Nat Rev Mol Cell Biol* 4(7):586–590

Irvine RF, Letcher AJ, Heslop JP, Berridge MJ (1986) The inositol tris/tetrakis phosphate pathway—demonstration of inositol (1,4,5)trisphosphate-3-kinase activity in mammalian tissues. *Nature* 320:631–634

Majerus PW, Kisseleva MV, Norris FA (1999) The role of phosphatases in inositol signaling reactions. *J. Biol Chem* 274:10669–10672

Michell B (1995) Early steps along the road to inositol-lipid-based signalling. *Trends Biochem Sci* 20(8):326–329

Michell RH (1975) Inositol phospholipids and cell surface receptor function. *Biochim Biophys Acta* 415:81–147

Otsu M et al (1991) Characterization of two 85 kDa proteins that associate with receptor tyrosine kinases, middle-T/pp60c-src complexes, and PI3-kinase. *Cell* 65:91–104

Strahl T, Thorner J (2007) Synthesis and function of membrane phosphoinositides in budding yeast, Saccharomyces cerevisiae. *Biochim Biophys Acta* 1771(3):353–404

Whitman M, Downes CP, Keeler M, Keller T, Cantley L (1988) Type-I phosphatidylinositol kinase makes a novel inositol phospholipid, phosphatidylinositol-3-phosphate. *Nature* 332:644–646

Contents

1. **The Phosphatidylinositol 4-Kinases: Don't Call it a Comeback** 1
   Shane Minogue and Mark G. Waugh

2. **PIP Kinases from the Cell Membrane to the Nucleus** 25
   Mark Schramp, Andrew Hedman, Weimin Li, Xiaojun Tan
   and Richard Anderson

3. **The Phospholipase C Isozymes and Their Regulation** 61
   Aurelie Gresset, John Sondek and T. Kendall Harden

4. **Phosphoinositide 3-Kinases—A Historical Perspective** 95
   Alex Toker

5. **PI3Ks—Drug Targets in Inflammation and Cancer** 111
   Matthias Wymann

6. **Phosphoinositide 3-Kinases in Health and Disease** 183
   Alessandra Ghigo, Fulvio Morello, Alessia Perino and Emilio Hirsch

7. **Phosphoinositide Phosphatases: Just as Important as the Kinases** ... 215
   Jennifer M. Dyson, Clare G. Fedele, Elizabeth M. Davies,
   Jelena Becanovic and Christina A. Mitchell

8. **The PTEN and Myotubularin Phosphoinositide 3-Phosphatases:
   Linking Lipid Signalling to Human Disease** 281
   Elizabeth M. Davies, David A. Sheffield, Priyanka Tibarewal,
   Clare G. Fedele, Christina A. Mitchell and Nicholas R. Leslie

Glossary 337

Index 343

Abbreviations

AD	Alzheimer's disease
AMPK	5'-AMP-activated protein kinase
ALL	Acute lymphocytic leukemia
ALS	Amyotrophic lateral sclerosis
AML	Acute myeloblastic leukemia
ARNO	Arf nucleotide binding site opener
ASK1	Apoptosis signal-regulating kinase 1
ATM	Ataxia telangiectasia mutated
ATX	Arabidopsis trithorax 1
Bad	Bcl-XL/Bcl-2-associated death promoter
BAFF	B cell activation factor of the TNF family
BCR	B cell receptor
Bcr/Abl	Break point cluster region/Abelson kinase fusion protein
Btk	Bruton's tyrosine kinase
c-Kit	Stem cell growth factor receptor
CAD	Caspase activated DNase
CCR(L)	C-C chemokine receptor (ligand) type
CDK	Cyclin-dependent kinase
CDKN2A	Cyclin-dependent kinase inhibitor 2A
CERT	Ceramide transfer protein
CIN85	Cbl-interactin protein of 85kD (also Ruk (regulator of ubiquitous kinase), SETA (SH3 domain-containing gene expressed in tumorigenic astrocytes))
CML	Chronic myeloid leukemia
CMT	Charcot-Marie-Tooth
COPI/II	Coatomer protein complex I/II
CXCR(L)	C-X-C chemokine receptor (ligand) type
DAAX	Death domain-associated protein
DAG	Diacylglycerol
DGK	Diacylglycerol kinase
DH	Dbl-homology
DMSO	Dimethyl sulfoxide

DNA-PK$_{cs}$	DNA-dependent protein kinase, catalytic subunit
DOCK2	Dedicator of cytokinesis 2
Dpm1	Dolichol phosphate mannosyltransferase
EGF(R)	Epidermal growth factor (receptor)
eEF1A	Eukaryotic elongation factor 1A
eIF4E	Elongation initiation factor 4E
EMT	Epithelial-to-mesenchymal transition
EnaC	Epithelial sodium channel
ER	Estrogen receptor, or endoplasmic reticulum
ErbB1	Epidermal growth factor receptor
ERM	Ezrin/radexin/moesin
FAK	Focal adhesion kinase
FAPP1	Phosphoinositol 4-phosphate adaptor protein 1
FAPP2	Phosphoinositol 4-phosphate adaptor protein 2
FcεRI	High affinity receptor for Fc fragment of IgE
FOXO	Forkhead transcription factor, class O
FYVE	Fab1, YOTB, Vac1, EEA-1 homology
G6P	Glucose-6-phosphatase
Gab	Grb2-associated binder
GAP	GTPase-activating protein
GEF	Guanine nucleotide exchange factor
GFP	Green fluorescent protein
GIST	Gastrointestinal stromal tumors
GK	Glucokinase
GLUT4	Glucose transporter type 4
GM-CSF	Granulocyte and macrophage colony stimulating factor
GPCR	G protein-coupled receptors
GRK2	G protein-coupled receptor kinase 2 (also βARK1 (adrenergic receptor kinase 1)
Grp1	General receptor for phosphoinositides
GSK-3	Glycogen synthase kinase-3
GST-2xFYVE	Glutathione S-transferase-tagged to tandem FYVE domains
HAUSP	Herpesvirus-associated ubiquitin-specific protease
Hdac2	Histone deacetylase 2
HSCs	Hematopoietic stem cells
IκBK	IκB kinase
ING2	Inhibitor of growth protein 2
Inpp5e/INPP5E	72 kDa inositol polyphosphate 5-phosphatase
Ins	*Myo*-inositol
IGF1(R)	Insulin-like growth factor (receptor)
ILK	Integrin-linked kinase
Ins(1,4)P_2	Inositol 1,4-bisphosphate
Ins(1,4,5)P_3	Inositol 1,4,5-trisphosphate; also used InsP$_3$
IPMK	Inositol polyphosphate multikinase
IRS	Insulin receptor substrate

Abbreviations

ITAM	Immunoreceptor tyrosine-based activation motif
ITIM	Immunoreceptor tyrosine-based inhibitory motif
JAK	Janus-activated kinase
JNK	Jun N-terminal Kinase
Kv1.3	Voltage-gated K^+ channel
LAT	Linker for activation of T cells
LOH	Loss of heterozygosity
LSCs	Leukemic stem cells
LTP	Long term potentiation
MAPK	Mitogen-activated protein kinase
MAPKAP-2	Mitogen-activated protein kinase-activated kinase 2
M-CSF	Macrophage colony-stimulating factor
MDM2	Murine double minute 2
MDS	Myelodysplastic syndrome
MEFs	Mouse embryonic fibroblasts
miRNA	Microrna
MPP(+)	1-methyl-4-phenylpyridinium iodide
MSN	Medium sized spiny projection neurons
MTM	Myotubularin
MTMR	Myotubularin related
mTOR	Mammalian target of rapamycin, see also TOR
MVB	Multivesicular body
MVP	Major vault protein
Nedd4	Neural-precursor-cell-expressed developmentally down-regulated 4
NFκB	Nuclear factor κB
NLS	Nuclear localization signal
NMDA(R)	N-methyl-D-aspartate (receptor)
NOS3/eNOS	NO-synthase 3
NTAL	Non-T cell activation linker, also named LAB (Linker of activation for B cells) or LAT2
OSBP	Oxysterol binding protein
OCRL	Oculocerebrorenal syndrome of Lowe
OGD	Oxygen–glucose deprivation
PAO	Phenylarsine oxide
PCAF	p300/CBP-associated factor
PDE	Phosphodiesterase
PDGF(R)	Platelet-derived growth factor (receptor)
PDZ	Post synaptic density protein, Drosophila disc large tumor suppressor, zonula occludens-1 protein
PDK1	Phosphoinositide-dependent kinase 1
PEPCK	Phosphoenolpyruvate carboxy kinase
PEST	Proline, glutamic acid, serine, threonine
PH	Plecktrin-homology
PHD	Plant homeodomain
PH-GRAM	Pleckstrin homology glucosyltransferase Rab-like GTPase activator

PHTS	PTEN hamartoma tumor syndrome
PI3K	Phosphoinositide 3-kinase; catalytic subunits of class I PI3K are referred to as p110α, p110β, p110γ and p110δ
PI3Kc	PI3K catalytic domain
PI3Kr	PI3K regulatory subunit
PI4K	Phosphatidylinositol 4-kinase
PI4KII	Type II phosphatidylinositol 4-kinase
PI4KIII	Type III phosphatidylinositol 4-kinase
PICS	Pten-loss–induced cellular senescence
PID	Phosphoinositide interacting domain
PIKE	PI-3-kinase enhancer
PIKK	Phosphoinositide 3-kinase-related kinase
PIP4K	Phosphatidylinositol 5-phosphate 4-kinase (also called type II PIP kinase)
PIP5K	Phosphatidylinositol 4-phosphate 5-kinase (also called type I PIP kinase)
PIPP	Proline-rich inositol polyphosphate 5-phosphatase
PIX	PAK-associated guanine nucleotide exchange factor
PKA	Protein kinase A
PKB/Akt	Protein kinase B, also called Akt after the transforming kinase encoded by the AKT8 retrovirus
PKC	Protein kinase C
PLC	Phospholipase C
PLD	Phospholipase D
PM	Plasma membrane
PML	Promyelocytic leukemia protein
PPI	Polyphosphoinositide
pRB	Retinoblastoma protein
PRD	Proline-rich domain
P-Rex	PtdIns(3,4,5)P_3-dependent Rac exchanger
PSD95	Post synaptic density protein 95
PtdIns	Phosphatidylinositol
PtdIns4P	Phosphatidylinositol 4-phosphate; short PIP
PtdIns3P	Phosphatidylinositol 3-phosphate; PIP should not be used here
PtdIns5P	Phosphatidylinositol 5-phosphate; PIP should not be used here
PtdIns(4,5)P_2	Phosphatidylinositol 4,5-bisphosphate; short PIP$_2$
PtdIns(3,4)P_2	Phosphatidylinositol 3,4-bisphosphate; the abbreviation PIP$_2$ should not be used here.
PtdIns(3,5)P_2	Phosphatidylinositol 3,5-bisphosphate; the abbreviation PIP$_2$ should not be used here.
PtdIns(3,4,5)P_3	Phosphatidylinositol 3,4,5-trisphosphate; short PIP$_3$
PtdOH	Phosphatidic acid (also used PA)
PTEN	Phosphatase and Tensin homolog deleted on chromosome Ten, [also MMAC (mutated in multiple advanced cancers), TEP1 (TGF-β-regulated and epithelial cell enriched phosphatase 1)]

Abbreviations

PX	Phox-homology
RAN	Ras-related nuclear protein
RID	Rac-induced recruitment domain
RNAi	Ribonucleic acid interference
ROS	Reactive oxygen species
RSK	Ribosomal S6 kinase
R-SMAD	Receptor regulated SMAD
RTK	Receptor tyrosine kinase
Rb2	Retinoblastoma-related gene p130^{Rb2}
RYR1	Type 1 ryanodine receptor
Sac	Suppressor of actin
SCIP	Sac domain-containing inositol phosphatases
SCV	*Salmonella*-containing vacuole
SGK	Serum- and glucocorticoid-induced protein kinase
SH2	Src homology 2
SHIP	SH2 domain-containing inositol 5'-phosphatase
SID	Set interacting domain
siRNA	Short-interfering RNA
SKICH	SKIP carboxy homology
SKIP	Skeletal muscle and kidney enriched inositol phosphatase
SNP	Single nucleotide polymorphism
SSC	Squamos cell carcinoma
Star-PAP	Poly(A) polymerase
Syk	Spleen tyrosine kinase, member of the Src tyrosine kinase family
TAC	Transverse aortic constriction
TDLU	Terminal ductal lobuloalveolar units
TGFβ	Transforming growth factor β
Tiam	T-lymphoma invasion and metastasis inducing protein
TNF	Tumour necrosis factor
TopoIIα	Topoisomerase IIα
TOR	Target of rapamycin (also called FRAP or mTOR)
TPIP	TPTE and PTEN homologous inositol lipid phosphatase
TPTE	Trans-membrane phosphatase with tensin homology
TRAPs	Transmembrane adapter proteins, link immune-receptors to downstream signaling cascades. Examples: LAT, NTAL/LAB
TSC	Tuberous sclerosis complex
UTR	Untranslated region
Vps34p	Vacuolar protein sorting mutant 34 protein
WASP	Wiskott Aldrich Syndrome protein
Wm	Wortmannin
WT	Wild type

Chapter 1
The Phosphatidylinositol 4-Kinases: Don't Call it a Comeback

Shane Minogue and Mark G. Waugh

Abstract Phosphatidylinositol 4-phosphate (PtdIns4P) is a quantitatively minor membrane phospholipid which is the precursor of PtdIns(4,5)P_2 in the classical agonist-regulated phospholipase C signalling pathway. However, PtdIns4P also governs the recruitment and function of numerous trafficking molecules, principally in the Golgi complex. The majority of phosphoinositides (PIs) phosphorylated at the D4 position of the inositol headgroup are derived from PtdIns4P and play roles in a diverse array of fundamental cellular processes including secretion, cell migration, apoptosis and mitogenesis; therefore, PtdIns4P biosynthesis can be regarded as key point of regulation in many PI-dependent processes.

Two structurally distinct sequence families, the type II and type III PtdIns 4-kinases, are responsible for PtdIns4P synthesis in eukaryotic organisms. These important proteins are differentially expressed, localised and regulated by distinct mechanisms, indicating that the enzymes perform non-redundant roles in trafficking and signalling. In recent years, major advances have been made in our understanding of PtdIns4K biology and here we summarise current knowledge of PtdIns4K structure, function and regulation.

Keywords Golgi complex · Membrane traffic · Phosphatidylinositol 4-phosphate · Signalling

1.1 Introduction

The phosphatidylinositol kinases (PtdInsKs) of eukaryotes have a long history. Their existence was originally inferred by the discovery of their enzymatic products, the phosphoinositides (PIs), whose synthesis from phosphatidylinositol (PtdIns) in the exocrine pancreas could be induced with cholinergic agonists (Hokin and

---

S. Minogue (✉) · M. G. Waugh
Centre for Molecular Cell Biology, Department of Inflammation, Division of Medicine, University College London, Rowland Hill Street, Hampstead, London NW3 2PF, United Kingdom
e-mail: s.minogue@medsch.ucl.ac.uk

Table 1.1 Mammalian PI4K encoding genes

Locus	Common name	SwissProt/ Protein (UniProt)	Accession nr. Hs	Interacts with	Reference MIM	Gene Map (Hs/Mm)
PIK4CA	PI4K230, PI4K55	PI4KA_human	NM_058004 NM_002650		(Wong and Cantley 1994) MIM 600286	22q11.21 16 A3
PIK4CB	PI4K92	PI4KB_human	NM_002651		(Meyers and Cantley 1997) MIM 602758	1q21 3 F2.1
PI4K2A	PI4KIIα	PI4K2A human	NM_018425	AP-3	Barylko et al. [2001]; Minogue et al. [2001]; Craige et al. [2008]	10q24 19 C3; 19 47.0 cm
PI4K2B	PI4KIIβ	PI4K2B human	NM_018323	Rac1, Hsp90	Minogue et al. [2001]; Balla et al. [2002]; Wei et al. [2002]; Jung et al. [2008]	4p15.2 5 C1;5

Hokin 1953). The enzymes themselves were first identified in membrane fractions and tissue extracts as activities that could transfer the γ-phosphate of radiolabelled ATP to PtdIns (Michell et al. 1967, Harwood and Hawthorne 1969). It soon became apparent that biochemically distinct activities could be detected in these extracts, and these were subsequently named the type I, II and III phosphoinositide kinases (Endemann et al. 1987; Whitman et al. 1987). The type I PI kinase was later shown to phosphorylate the D3 position of the inositol headgroup and these enzymes are now better known as the phosphoinositide 3-kinases (PI3Ks). The type II and III enzymes both phosphorylate at the D4 position but exhibit very different biochemical properties for example, the type II isoforms are inhibited by adenosine but not PI3K inhibitors and have K_{mATP} in the micromolar range; meanwhile, the type III isoforms are larger proteins, insensitive to adenosine and inhibited by PI3K inhibitors, albeit at higher concentrations than PI3Ks (Pike 1992; Balla and Balla 2006; Hunyady et al. 1983). For many years these biochemical characteristics were the only means of distinguishing PtdIns4K activities in experiments and furthermore, results were sometimes confounded by the presence of trace amounts of the highly active type II activity (Yamakawa et al. 1991). Final and unambiguous identification of the isoforms had to wait until the proteins were characterised at the molecular level (Table 1.1) and this was finally made possible during the genomics age of the 1990s. The cloning of the type II (PtdIns4KIIα and PtdIns4KIIβ) and type III isoforms (PtdIns4KIIIα and PtdIns4KIIIβ) revealed that they belong to divergent sequence families with the PtdIns4KIIIs showing a greater degree of similarity to the PI3K/protein kinase superfamily. Thus the biochemical differences, particularly the relative sensitivity to PI3K inhibitors, were finally explained and sequence specific reagents, along with some inhibitors, became available to study PtdIns4K functions. These have contributed to a resurgence of interest in these key activities and the last few years have seen the

Fig. 1.1 Metabolism of D4 phosphorylated phosphoinositides. The glycerophospholipid phosphatidylinositol (PtdIns) is a component of membranes in all eukaryotic cells. It is made up of a hydrophobic diacyglycerol backbone (R1 and R2, are predominantly stearoyl and arachidonyl in mammalian cells) linked to a polar inositol headgroup. The headgroup can be phosphorylated at the D3, D4 and D5 positions by various kinase activities to generate eight higher phosphoinositides (PIs); PIs phosphorylated by PtdIns 4-kinases (PtdIns4Ks) are shown. Reverse (dephosphorylation) reactions catalysed by specific PI phosphatases are shown

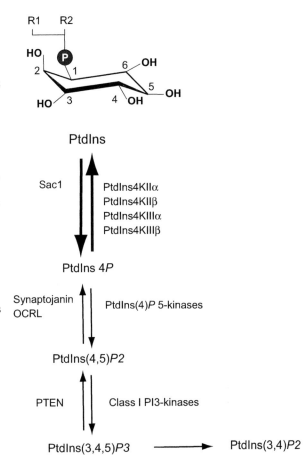

emergence of numerous new and unexpected functions for the PtdIns4Ks, the first PtdInsK activities to be identified all those years ago.

In order to understand PtdIns4K function it is often necessary to study the enzymatic output of individual isoforms, that is, not just PtdIns4P but also the D4 metabolic derivatives whose levels may change in response to modulation of PtdIns4K activity. Due regard should also be paid to dephosphorylation reactions by PI phosphatases since together these activities determine the flux of intermediates through PI pathways (Fig. 1.1). Herein lie the central technical problems in PI research: firstly, different PI species are not evenly distributed between organelles and biochemical methods that rely on total PI analysis are unable to distinguish separate pools of PIs which, although chemically identical, may not have equivalent functions. Secondly, the steady-state subcellular localisation of PI kinase protein is not a reliable indicator of its locus of activity. The study of PtdInsK activity is therefore increasingly reliant on protein domains which bind PIs *in vitro* and such reagents have been widely used as overexpressed GFP chimeras to report PtdIns4K activity in cultured cells (Balla et al. 2000b; Szentpetery et al. 2009). However, whilst useful, these tools are sometimes of dubious specificity *in vivo* and their overexpression may sequester PIs preventing further metabolism or interaction with effector proteins (Balla et al. 2000b). Thus the cell models we rely on may be prone to observer effects where the experimental system can generate complex and uninterpretable phenomena. Fortunately other strategies are available, for instance, using recombinant PI-binding domains to indirectly stain fixed cells to report steady-state levels of PIs (Watt et al. 2002; Hammond et al. 2009). Nevertheless, our inability to accurately image specific pools of PIs and unambiguously attribute results to specific PI kinase activities *in vivo* remains a significant technical limitation. A similar caveat applies to the use of RNAi as a method of studying loss of function. Although convenient, RNAi only reveals phenotypes after 48–72 h of knockdown—a relatively long period of time during which a cell may respond by upregulating other PtdInsK activities or compensating by making previously separate pools of PIs available. This last problem may be circumvented by the use of cell-permeable drugs able to specifically inhibit PI kinases but there are currently few such compounds available. A separate approach, which acutely targets individual PI pools by inducing the heterodimerisation of a specific PI phosphatase with organelle markers has been employed to selectively hydrolyse PIs (Varnai and Balla 2006). No perfect strategy yet exists but used together, approaches such as these have revealed a great deal about PtdIns4K function in cells.

As with many conserved PI kinases, our understanding of PtdIns4K function has been hugely aided by genetic studies in the yeast *S. cerevisiae* and, despite the vast differences between yeast and multicellular eukaryotic biology, these studies have often led the way in elucidating gene function in higher organisms. This fact is perhaps best illustrated by Pik1, the yeast orthologue of mammalian PtdIns4KIIIβ, which was the first PtdIns4K gene identified in any organism and which shares many functional features with its mammalian counterpart. Since several extensive and authoritative reviews have been published recently (Balla and Balla 2006; Strahl and Thorner 2007), this chapter will concentrate mainly on the function of PtdIns4Ks in animal cells and genetic models. However, we will discuss those functions that overlap with or provide particular insight into the relevant mammalian orthologues.

1.2 The Type II PtdIns 4-Kinases

1.2.1 *PtdIns4KII Gene Family and Domain Structure*

Sequences homologous to PtdIns4KII exist in all metazoan genomes and all vertebrates appear to contain genes encoding the two closely related alpha and beta isoforms. Our analysis indicates that most invertebrate genomes, with the exception of *C. elegans*, contain just a single PtdIns4KII gene. Lower eukaryotic organisms such as the yeasts also contain just one gene and it is difficult to determine whether these are more closely related to the mammalian alpha or beta isoforms.

The mammalian alpha and beta PtdIns4KIIs are both membrane proteins of ~55 kDa and the human proteins share 68% identity in their conserved catalytic

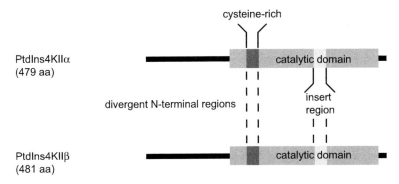

Fig. 1.2 The mammalian type II PtdIns 4-Kinases. A schematic illustration of conserved features in the sequences of mammalian PtdIns4KIIs. The PtdIns4KII isoforms share homology in the conserved catalytic core with the exception of an insert region of varying length in different species. The pre-eminent conserved feature is the cysteine-rich region in the kinase domain which mediates membrane targeting and probably places the enzyme in close proximity to membrane-bound PtdIns substrate. The N-termini are highly dissimilar: whereas PtdIns4KIIα is proline-rich and amphiphillic in character, the N-terminus of PtdIns4KIIβ is highly acidic

domains. No X-ray structures currently exist and the proteins have only weak homology to the PI3K/protein kinase superfamily of phosphotransferase enzymes. Nevertheless, local regions of homology corresponding to recognisable protein kinase subdomains are present (Minogue et al. 2001) and the requirement for invariant residues within these conserved sequences has been tested by site-directed mutagenesis (Barylko et al. 2002). The conserved catalytic core is separated by a non-conserved insert region which varies in length between species (Balla and Balla 2006) (illustrated in Fig. 1.2).

An interesting feature of the otherwise highly related type II PtdIns4K isoforms is that they contain divergent amino termini: PtdIns4KIIα has a proline-rich region (Balla and Balla 2006) followed by a stretch of hydrophobic and charged residues, which are predicted to adopt the structure of an amphipathic helix and may mediate membrane insertion (Barylko et al. 2009); meanwhile PtdIns4KIIβ contains a highly acidic amino terminus. Both PtdIns4KII isoforms contain a conserved CCPCC motif within the predicted kinase domain which undergoes palmitoylation (Barylko et al. 2001). This motif is present across species but acylation may be less important in lower eukaryotes such as the yeasts where the cysteine residues are conservatively substituted by a hydrophobic residue (Barylko et al. 2009).

1.2.2 PtdIns4KII Expression, Localisation and Regulation

The PtdIns4KIIs are widely expressed in human tissues with PtdIns4KIIα somewhat enriched in brain and kidney (Minogue et al. 2001; Balla et al. 2002); meanwhile, PtdIns4KIIβ is relatively enriched in liver and cell lines of haematopoietic origin (Balla et al. 2002). In cultured cells, both isoforms are membrane proteins (Waugh

Table 1.2 The subcellular localisation and functions of PtdIns4K isoforms

PtdIns4K isoform	Subcellular localisation	Function
PtdIns4KIIα	TGN	Recruits adaptors AP-1 and GGA TGN-PM transport
	Late endosome	Interacts with AP-3; regulates degradation of EGFR
	PM	Not known
	ER	Not known
PtdIns4KIIβ	PM	Recruited in response to PDGF and V12Rac
	Endosomes	Not known
PtdIns4KIIIα	ER	Resupply of PM PtdIns4*P*
	Nucleus	Not known
	Golgi	Recruitment of GBF1
PtdIns4KIIIβ	TGN	Golgi-to-PM traffic
	Nucleus	Not known

et al. 2003b) which localise to the *trans*-Golgi network (TGN), endosomes and plasma membrane (PM) (Balla et al. 2002; Wang et al. 2003; Salazar et al. 2005; Minogue et al. 2006) (Table 1.2).

Both PtdIns4KII isoforms can undergo palmitoylation in the conserved CCPCC motif and this appears to be an important, though not exclusive, determinant of membrane localisation and probably therefore, access to PtdIns substrate. Although PtdIns4KIIα lacks identifiable transmembrane sequences it is tightly membrane associated, effectively behaving as an integral membrane protein. Deletion of $_{174}$CCPCC$_{178}$ (human numbering will be used throughout) prevents the incorporation of radiolabelled palmitate and is sufficient to render the enzyme extractable from membranes in sodium carbonate buffers at pH 10–11, but not 1 M NaCl (Barylko et al. 2001; Barylko et al. 2009). This indicates that the deacyl form retains the ability to interact tightly with membranes, albeit in a peripheral manner. Deletion of $_{174}$CCPCC$_{178}$ also dramatically reduces kinase activity. Since the CCPCC motif lies within the conserved catalytic domain, this could be due to gross effects on kinase structure. However, this does not seem to be the case because inhibition of acylation with 2-bromopalmitate or by mutation of all the cysteine residues to serine has identical effects on activity, membrane binding and subcellular localisation (Barylko et al. 2009).

All four cysteine residues in PtdIns4KIIα are capable of accepting [^{3}H]-palmitate in metabolic labelling experiments (Barylko et al. 2009). Site-directed mutagenesis indicates that the first two are preferentially palmitoylated and better able to support integral membrane binding and activity (Barylko et al. 2009). This observation has interesting implications for the regulation of PtdIns4KII enzymes of lower eukaryotes, where the first two cysteines are substituted with hydrophobic residues. The sole PtdIns4KII of *S. cerevisiae* Lsb6 is, however, palmitoylated outside of this conserved motif on a C-terminal cysteine residue (Roth et al. 2006) which may contribute to membrane binding.

Palmitoylation is a potentially reversible modification but as yet, no compelling evidence has been presented to suggest that dynamic palmitoylation regulates

PtdIns4KIIα. Unlike all other PI kinases, PtdIns4KIIα is constitutively membrane-associated, and therefore the mechanism responsible for the observed active pool in membrane fractions (Waugh et al. 2003b) probably occurs within the membrane itself. Indeed, good evidence exists for an intrinsic mechanism of regulation because PtdIns4KIIα activity is extremely sensitive to membrane environment and, in particular, intra-membrane cholesterol concentration. Targeting to cholesterol-containing membranes alters the lateral diffusion kinetics of PtdIns4KIIα and there is strong evidence that a high activity and slowly diffusing pool of PtdIns4KIIα associates with lipid-raft like domains of the TGN (Minogue et al. 2010). Moreover the size of the mobile pool of PtdIns4KIIα is affected by TGN membrane connectivity, indicating that the enzyme's biophysical properties are affected by the overall membrane architecture (Minogue et al. 2010). A further level of cholesterol-dependent control may result from PtdIns4KIIα being highly responsive to oxysterol-binding protein (OSBP)-generated sterol gradients on Golgi and post-Golgi membranes (Banerji et al. 2010). Therefore, PtdIns4P generation by PtdIns4KIIα is a function of sterol-based membrane ordering and topological organisation. In future work, it may be interesting to consider the degree to which PtdIns4*P* and sterol concentrations synergistically determine both the functional and morphological identity of PtdIns4*P* and cholesterol-rich membranes such as those of the TGN.

PtdIns4KIIβ also contains the conserved CCPCC motif but in contrast to PtdIns4KIIα, the enzyme is partially cytosolic. The membrane-bound fraction is palmitoylated and displays ~20-fold higher activity (Jung et al. 2008) suggesting that the enzyme is dynamically acylated resulting in membrane targeting. PtdIns4KIIβ is recruited to the PM in response to PDGF or overexpression of constitutively active Rac; this appears to require the C-terminal 160 amino acids of PtdIns4KIIβ in COS cells and is accompanied by activation of kinase activity (Jung et al. 2008). The enzyme is also phosphorylated *in vivo* on a serine residue in the N-terminus but the significance of this modification is unclear since mutation at this site does not affect kinase activity, palmitate incorporation or membrane localisation (Jung et al. 2008).

Phosphopeptides derived from PtdIns4KIIα have been identified in several proteomic screens and indicate potential for regulation by phosphorylation (Olsen et al. 2006; Villen et al. 2007). In one such study PtdIns4KIIα phosphorylation was agonist-dependent (Olsen et al. 2006), reminiscent of earlier work showing the co-immunoprecipitation of PtdIns4KII activity with activated EGF receptors in a phosphoprotein-containing complex (Cochet et al. 1991; Kauffmann-Zeh et al. 1994). No endogenous protein regulators of PtdIns4KIIα activity have yet been described, however, along with some other interfacial enzymes such as PLD2 (Chahdi et al. 2003), the activity of PtdIns4KIIα can be greatly stimulated by the wasp venom peptide mastoparan and related amphipathic peptides (Waugh et al. 2006). The physiological significance of these biochemical observations have yet to be determined; however, it is noteworthy that the cationic phospholipid phosphatidylcholine has also been shown to be a potent *in vitro* activator of PtdIns4KII activity (Olsson et al. 1993), inferring that alterations to membrane charge may augment PtdIns4*P* synthesis *in vivo*.

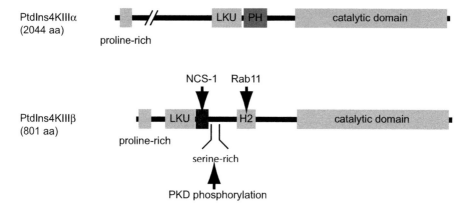

Fig. 1.3 The mammalian type III PtdIns 4-kinases. A schematic illustration of conserved features in the sequences of mammalian PtdIns4KIIIs. The catalytic domains of the PtdIns4KIIIα and PtdIns4KIIIb display minimal homology to the type II PtdIns4Ks and are instead more similar to the PI3K family of enzymes. This similarity includes the presence of the conserved lipid kinase unique (LKU) domain. PtdIns4KIIIα contains an N-terminal proline-rich domain and PH domain. Non catalytic domains in PtdIns4KIIIb include a proline-rich sequence, the LKU, a region of homology (H2) shared with other PtdIns4KIIIb orthologues including the *S. cerevisae* Pik1p, regions implicated in the binding of NCS-1 and Rab11 and a serine-rich region which is phosphorylated by PKD

1.3 The Type III PtdIns 4-Kinases

1.3.1 Gene Family and Domain Structure

PtdIns4KIIIβ and PtdIns4KIIIα were first identified in *S. cerevisae* as the essential gene PIK1 (Flanagan et al. 1993) and STT4, whose mutation confers sensitivity to the protein kinase C (PKC) inhibitor staurosporine (Yoshida et al. 1994). Both PtdIns4KIIIα and -β sequences are present in all metazoan genomes.

The type III PtdIns4Ks are larger enzymes whose catalytic domains display strong sequence similarities to the PI3Ks. Like the PI3Ks and the more closely related serine/threonine kinases TOR, ATM and DNA-PK, they also contain the so-called lipid kinase unique (LKU) domain which is predicted to adopt a helical structure (Balla and Balla 2006) and which, in the yeast PtdIns4KIIIβ orthologue at least, is essential for kinase activity (Strahl and Thorner 2007) (Fig. 1.3). In PtdIns4KIIIα the LKU region is closely juxtaposed with a PH domain (Wong and Cantley 1994) and although this PH domain has poor homology to other such domains, the isolated region from a plant PtdIns4KIIIα binds PtdIns4*P* in lipid overlay assays (Stevenson et al. 1998) and may contribute to product inhibition (Stevenson-Paulik et al. 2003). Binding sites for NCS-1 and Rab11 have been identified in PtdIns4KIIIβ (discussed in Sect. 4.2.4) close to a homology region (Hom2) conserved in other PtdIns4KIIIβ orthologues. This region, situated between the LKU and catalytic domain, also contains a serine-rich domain that is phosphorylated by protein kinase D (Sect. 4.2.4). A splice variant exists which extends this serine-rich region by 15 amino acids (Balla and Balla 2006). This insert can be phosphorylated by casein kinase II *in vitro* but the modification does not lead to substantial activation of kinase activity (S. Minogue, unpublished data).

The high degree of conservation between the PtdIns4KIII and PI3K proteins confers sensitivity to inhibitors such as wortmannin and LY294002, with *in vitro* IC_{50} values that are approximately 20–100 fold higher than those for class I PI3Ks. Whilst this property has proved useful in distinguishing between type II and type III activities, it is not an effective way to discriminate between PtdIns4KIII activities. However, the recent characterisation of PIK93, a cell-permeable inhibitor relatively selective for the PtdIns4KIIIβ isoform (Knight et al. 2006; Balla et al. 2008) offers a means of selectively inhibiting PtdIns4KIIIβ function.

1.3.2 PtdIns4KIIIα Expression, Localisation and Regulation

PtdIns4KIIIα is widely expressed in tissues but is particularly enriched in brain (Wong et al. 1997; Gehrmann et al. 1999; Zolyomi et al. 2000). PtdIns4KIIIα predominantly localises to the ER in mammalian cells (Wong et al. 1997) but it has also been detected in the pericentriolar Golgi region (Nakagawa et al. 1996). Many studies have been forced to use overexpressed protein to determine the pattern of subcellular distribution because of problems detecting endogenous PtdIns4KIIIα and this may affect the observed localisation of the protein. However, immuno-electron microscopy of endogenous PtdIns4KIIIα in the rat CNS localised the protein to the ER and atypical membranes surrounding mitochondria, and multivesicular bodies (Balla et al. 2000a). It is tempting to speculate that these membranes represent some form of inter-organelle junction (Levine and Rabouille 2005) that permits the communication of PtdIns4 or PtdIns4KIIIα protein between closely juxtaposed compartments. Endogenous PtdIns4KIIIα has also been detected in the nucleolus (Kakuk et al. 2006).

1.3.3 PtdIns4KIIIβ Expression, Localisation and Regulation

PtdIns4KIIIβ is ubiquitously expressed in mammalian tissues (Balla et al. 1997) and localises to *cis*- and *trans*-regions of the Golgi complex (Wong et al. 1997; Godi et al. 1999). Golgi localisation is brefeldin A-sensitive and therefore likely to depend on Arf (discussed in Sect. 4.2.4). PtdIns4KIIIβ binds Rab11 in a region between residues 401–506 (Fig. 1.3) and it appears that this interaction, which is conserved in yeast, is required to recruit Rab11 to Golgi membranes rather than PtdIns4KIIIβ (de Graaf et al. 2004). Like PtdIns4KIIIα, PtdIns4KIIIβ has also been detected in the nucleus (de Graaf et al. 2002). The mechanism of nuclear localisation has not been determined but it is notable that the yeast orthologue Pik1 was identified as a component of the nuclear pore complex (Garcia-Bustos et al. 1994).

1.4 Cellular Functions of PtdIns4Ks

1.4.1 PtdIns4Ks in Signalling Pathways

PtdIns4P is the immediate precursor of PtdIns(4,5)P_2, the major *in vivo* substrate for receptor-linked phospholipase C (PLC) and PI3K enzymes. Both PIs have been found at the PM where they can be detected with monoclonal antibodies and PH domains (Hammond et al. 2006, 2009); meanwhile, PtdIns4KIIα, PtdIns4KIIβ and possibly PtdIns4KIIIα all localise to the PM (Table 1.2). Historically, type II PtdIns4K activity was linked to agonist-regulated pathways because of its association with activated EGFR (Cochet et al. 1991; Kauffmann-Zeh et al. 1994). However, these studies were performed prior to the availability of sequence-specific reagents and therefore we cannot rule out the contribution of PtdIns4KIIβ in these results or the possibility that PtdIns4KIIα associates with the receptors in an endosomal trafficking complex (Minogue et al. 2006).

1.4.1.1 PtdIns4KIIs in Signalling Pathways

Although there is evidence for regulated membrane association and kinase activation, relatively little is known about PtdIns4KIIβ function. Partial recruitment in response to PDGF and constitutively active Rac expression is intriguing (Wei et al. 2002) and suggests that the isoform may be acutely regulated by receptor activation and GTP-Rac, although details of the mechanism remain elusive. Rac is a well known signalling molecule regulating cell adhesion and migration (Heasman and Ridley 2008) and one report links PtdIns4KIIβ to hepatocellular cancer cell migration: PtdIns4KIIβ co-immunoprecipitates with overexpressed tetraspannin CD81 from HepG2 lysates and the proteins partially colocalise by immunofluorescence microscopy. The mechanism linking PtdIns4KIIβ and CD81 is as yet unclear but loss of CD81 expression in hepatocellular carcinoma correlates with poorly differentiated and metastatic tumours (Mazzocca et al. 2008).

An interesting development has been the recent finding that PtdIns4KIIα is a component of the canonical Wnt pathway (Pan et al. 2008). The Wnts (from wingless, the *Drosophila* mutation) are a large and conserved family of hydrophobic ligands which control an array of developmental and oncogenic processes (Kikuchi et al. 2009). In the canonical β-catenin-dependent Wnt pathway in the absence of ligand, β-catenin is phosphorylated by GSK3β and CKIα and is subsequently ubiquitylated and targeted for degradation in the proteosome, thereby regulating levels of β-catenin. When Wnt ligands bind to their cognate receptors LRP5 or 6, β-catenin phosphorylation and subsequent proteosomal degradation is suppressed, leading to increased cytosolic levels. β-catenin is then able to translocate to the nucleus where it stimulates expression of a number of genes. PtdIns4KIIα was initially identified in a siRNA library screen aimed at identifying proteins involved in the Wnt3a-dependent phosphorylation of LRP6 and therefore the β-catenin pathway to the nucleus (Pan et al. 2008). PtdIns4P 5-kinases were also identified in this screen suggesting that together the kinases form a PtdIns(4,5)P_2 biosynthetic complex. Supporting this is the finding that exogenous PtdIns(4,5)P_2 stimulates Wnt3a-dependent LRP5/6 phosphorylation *in vitro* and that reduction of endogenous PtdIns(4,5)P_2 levels using a 5-phosphatase has the opposite effect (Pan et al. 2008). Wnt3a also stimulates the production of PtdIns4P and PtdIns(4,5)P_2 but this is abolished in HEK293 cells treated with siRNA to knockdown PtdIns4KIIα and PtdIns4P 5KIβ (Pan et al. 2008; Qin et al. 2009).

Wnt3a probably regulates PtdIns4KIIα and PtdIns4P 5KIβ activities by directly interacting with the scaffold protein dishevelled (Dvl), a critical adaptor protein in Wnt signalling. Dvl co-immunoprecipitates with PtdIns4KIIα and PtdIns4P 5KIβ in a tripartite complex when the proteins are overexpressed in HEK293 cells and the activity of both enzymes is activated by Dvl binding *in vitro* (Qin et al. 2009). In the case of bacterially expressed PtdIns4KIIα protein at least, this interaction does not appear to involve the enzyme's divergent N-terminal region (Qin et al. 2009).

1.4.1.2 PtdIns4KIIIs in Signalling Pathways

Despite the association of PtdIns4KII with activated receptors and involvement in Wnt signalling, far stronger evidence supports a role for the type III PtdIns4Ks in agonist-regulated turnover of PIs, indeed the wortmannin sensitivity of angiotensin II (AngII) and PDGF-sensitive pools was identified long ago (Nakanishi et al. 1995). Evidence of a signalling role for PtdIns4KIIIα also exists from *S. cerevisiae* where the orthologue Stt4p localises to the PM and participates in the yeast non-PLC-linked PKC-MAPK cascade (Audhya et al. 2000). More recent work has sought to identify the PtdIns4K isoform responsible for maintaining the PM pool of PtdIns4P for PtdIns(4,5)P_2 biosynthesis, using a combination of pharmacological inhibitors and siRNA targeting of specific PtdIns4K isoforms (Balla et al. 2008a). Use of the PH domain from OSH2 fused to GFP and the PLCδ1 PH domain fused to mRFP allowed simultaneous monitoring of PM PtdIns4P and PtdIns(4,5)P_2, respectively. The PM localisation of both fusion proteins was sensitive to wortmannin as were the sustained hydrolysis of PM PtdIns(4,5)P_2, changes in levels of total [^{32}P]-labelled PtdIns4P, PtdIns(4,5)P_2, [^{3}H]-InsP$_3$ and Ca^{2+} in AngII-stimulated cells (Balla et al. 2008a). Both PtdIns4KIII isoforms are sensitive to wortmannin but the availability of the new inhibitor PIK93 (Knight et al. 2006; Balla et al. 2008b), relatively selective for PtdIns4KIIIβ, allowed investigators to rule out this isoform in the response to AngII stimulation (Balla et al. 2008b).

This work clearly implicates PtdIns4KIIIα as the isoform responsible for maintenance of the G-protein coupled agonist-responsive pool of PtdIns4P. Curiously however, endogenous PtdIns4KIIIα is present on ER and Golgi membranes but cannot be detected at the PM. This means that either only a very small amount of PtdIns4KIIIα is necessary for PtdIns4P supply to the PLC signalling pathway, or that PtdIns4P is rapidly transported from the ER to replenish that consumed at the PM, possibly via ER-PM junctions. How this might be accomplished is mysterious and, although PtdIns4KIIIα binds to a PtdIns transfer protein (Aikawa et al. 1999), such proteins are not known to transfer PtdIns4P. However, there is also evidence that Golgi PtdIns4P can replenish that consumed at the PM during acute signalling, because depletion of the Golgi pool of PtdIns4P using targeted Sac1 PtdIns4P-phosphatase markedly affects the replenishment of PM PtdIns(4,5)P_2 (Szentpetery et al. 2010). PtdIns4P is regarded as the dominant PI species at the Golgi apparatus and the cell's ability to use this pool of precursor during sustained signalling indicates that metabolic compartmentation of PtdIns4P may be leaky under certain circumstances. Fascinating though this phenomenon is, the mechanism by which Golgi PtdIns4P reaches the signalling pool at the PM remains to be determined and, since no protein is known to transfer PtdIns4P, vesicular traffic remains a likely prospect. An earlier study investigating the membrane localisation of PtdIns4P-binding PH domains from FAPP1 and OSBP determined that the domains bound PLC-sensitive lipid pools when transfected cells were challenged with Ca^{2+} ionophores. The PM-localised domains re-associated particularly strongly following Ca^{2+} ionophore treatment and this re-association was inhibited by siRNA downregulation of PtdIns4KIIIα but not PtdIns4KIIIβ or PtdIns4KIIα (Balla et al. 2005).

PtdIns4KIIIα has been found in a complex containing the $P2X_7$ ion channel (Kim et al. 2001). The significance of this interaction is not clear but $P2X_7$ activation leads to rapid cytoskeletal rearrangements, suggesting that the receptor is linked, directly or indirectly, to second messenger signalling. PtdIns4KIIIβ is also stimulated by the tumour-associated elongation factor eEF1A2 (Jeganathan and Lee 2007). This occurs through direct interaction of eEF1A2 with PtdIns4KIIIβ and has been mapped to a region between residues 163–320 (Jeganathan et al. 2008). It is not clear whether this interaction takes place in the Golgi complex but it is responsible for a relatively large increase in PM PtdIns(4,5)P_2 accompanied by actin remodelling.

1.4.2 PtdIns4Ks in Intracellular Traffic

Whilst all PtdIns4Ks have been detected at the Golgi complex (Table 1.2), only PtdIns4KIIα and PtdIns4KIIIβ are known to synthesise PtdIns4P at this location. The PtdIns4P they generate is responsible for the localisation of a number of PtdIns4P-binding proteins involved in anterograde traffic, the biosynthesis of sphingomyelin and glycosphingolipids. In many cases the PtdIns4P-binding proteins in question have a relatively low affinity and specificity for the lipid *in vitro* and rely on additional protein factors or sequences outside the lipid binding domain to bind membranes specifically and with sufficient affinity (Levine and Munro 2002; Carlton and Cullen 2005).

1.4.2.1 Type II PtdIns4Ks and Intracellular Trafficking Pathways

In cultured cells, endogenous PtdIns4KIIα protein has a broad distribution localising primarily to membranes of the TGN (Wang et al. 2003), endosomes (Balla et al. 2002) and, to a lesser extent, the PM (Minogue et al. 2006) and ER (Waugh et al. 2003a)

(Table 1.2). It is likely that the distribution of the enzyme between these organelles is cell-type specific because PtdIns4KIIα is a component of synaptic vesicles (SVs) (Guo et al. 2003); PtdIns4KIIα is also enriched in Bergmann glia, astrocytes, Purkinje cell bodies and selected populations of neurons in rat brain (Simons et al. 2009). This and the fact that PtdIns4KIIα is dynamically trafficked may partially explain the discrepancies between subcellular localisations in published work.

The first evidence that PtdIns4KIIα is involved in membrane traffic was provided by the finding that overexpression of kinase-inactive mutant PtdIns4KIIα or PtdIns4KIIβ affected the traffic of transferrin and AngII (Balla et al. 2002). Subsequently, knockdown of PtdIns4KIIα was shown to decrease levels of Golgi PtdIns4P and block recruitment of the major TGN adaptor AP-1 to Golgi membranes (Wang et al. 2003). AP-1 was shown to be a PtdIns4P binding protein *in vitro* and exogenous synthetic PtdIns4P, but not other lipids, were sufficient for rescue of AP-1 recruitment to membranes. AP-1 is a clathrin adaptor that mediates TGN-to-endosome traffic but the Golgi is also a key organelle mediating protein processing and late secretory traffic to the PM. Knockdown of PtdIns4KIIα did not affect intra-Golgi traffic of haemaglutinin cargo but did affect the non-AP-1 dependent transport of GFP-VSVG to the PM. Interestingly, this pathway could only be rescued with synthetic PtdIns$(4,5)P_2$ indicating that PtdIns4KIIα can also contribute PtdIns4P for generation of PtdIns$(4,5)P_2$ in the late secretory pathway (Wang et al. 2003). PtdIns4KIIα knockdown affects the localisation of the Golgi-localised, γ-ear-containing, Arf-binding protein (GGA) family of adaptors which are also involved in TGN-to-endosomal traffic. GGA1, GGA2 and GGA3 binding to the TGN is dependent on a PtdIns4P and Arf1 coincidence mechanism (Wang et al. 2007).

1.4.2.2 PtdIns4KIIα and Endosomal Traffic

PtdIns4KIIα localises to endosomal membranes staining positive for the early endosomal autoantigen EEA1, transferrin, AngII (Balla et al. 2002) and particularly well to late endosomal compartments staining for lamp-1, CD63, syntaxin 8 and internalised EGF (Minogue et al. 2006). In live cells, PtdIns4KIIα traffics on highly dynamic endosomal carriers and inhibition of PtdIns4KIIα with an inhibitory antibody or by knockdown with siRNA prevents the correct degradation of the EGFR, presumably by impairing traffic to the lysosome (Minogue et al. 2006). PtdIns4KIIα knockdown also affects the steady-state distribution of lamp-1 and CD63, two other markers of lysosomal traffic (Minogue et al. 2006) and this suggests a general trafficking defect is responsible for the effects on receptor degradation.

PtdIns4KIIα has been identified as a component of AP-3-containing vesicles from PC12 cells and the two proteins show substantial colocalisation together, and with, AP-3 cargoes such as the Zn^{2+} transporter ZnT3 and lamp-1 (Salazar et al. 2005). PtdIns4KIIα also co-precipitates with AP-3 and ZnT3 in *in vivo* crosslinked complexes (Craige et al. 2008). Forms of AP-3 are expressed in both neuronal and non-neuronal cells where they control the selection of lysosomal cargo, often defined by the presence of a dileucine sequence for lysosomal traffic. The normal steady-state localisation of PtdIns4KIIα is AP-3 dependent and conversely, the subcellular localisation of AP-3 is PtdIns4KIIα dependent, thus the two proteins appear to control their localisations in a reciprocal manner. This relationship is supported by the discovery of a dileucine AP-3 sorting motif in the N-terminus of PtdIns4KIIα required for endosomal localisation; therefore, the enzymatic product of PtdIns4KIIα may regulate AP-3 function through recruitment in a manner analogous to AP-1 at the TGN. However, despite the observation that kinase activity is required for the traffic of lysosomal cargoes (Craige et al. 2008), no evidence for direct interaction of AP-3 with PtdIns4P or other D4 PIs exists. Sections of brain tissue from the AP-3-deficient mocha mouse lack PtdIns4KIIα staining in specific areas including nerve termini of mossy fibre neurons of the hilus and hippocampus (Salazar et al. 2005). In this respect it is interesting that PtdIns4KIIα localises to SVs (Guo et al. 2003) because the correct generation of SVs and synaptic-like micro vesicles in neuroendocrine cells is dependent on an AP-3 pathway, suggesting that PtdIns4KIIα functions along with the BLOC-1 and AP-3 trafficking machinery in the biogenesis of SVs (Salazar et al. 2009).

1.4.2.3 Type III PtdIns4Ks and Membrane Traffic

PtdIns4KIIIα

Although PtdIns4KIIIα is the isoform most strongly implicated in agonist-dependent signalling (see Sect. 4.1.2), the protein localises to the ER and Golgi apparatus (Wong et al. 1997). Nevertheless, PtdIns4KIIIα may participate in certain trafficking functions in COS-7 cells because the protein colocalises with the Golgi brefeldin A resistance factor (GBF1) in the Golgi region. siRNA of PtdIns4KIIIα ablates Golgi localisation of GBF1 (and also the adaptor GGA3) (Dumaresq-Doiron et al. 2010), suggesting that the kinase drives recruitment of GBF1. GBF1 is not known to bind PtdIns4P but it is an exchange factor for Arf1 and overexpression of constitutively active Arf1 leads to a small increase in Golgi targeting of a FAPP1 PH domain (Dumaresq-Doiron et al. 2010). Another possible role for PtdIns4KIIIα comes from a siRNA screen of membrane trafficking host factors required for hepatitis C virus (HCV) replication. Knockdown of PtdIns4KIIIα resulted in a large reduction in HCV replication and virus production but did not affect virus entry as might be expected for an enzyme responsible for PtdIns$(4,5)P_2$ production at the PM. GFP-PtdIns4KIIIα localised to intracellular membranes containing viral dsRNA and the HCV protein NS5A in HCV replicon-transfected hepatoma cells. Thus given its known ER localisation, PtdIns4KIIIα may be involved in the formation of an ER-derived membrane domain in which HCV replicates (Berger et al. 2009). Supporting this is the earlier finding that PtdIns4KIIIα was found in a yeast two-hybrid screen for binding partners of the HCV NS5A protein (Ahn et al. 2004).

PtdIns4KIIIβ

PtdIns4KIIIβ is a peripheral enzyme recruited to the Golgi of COS-7 cells by GTP-bound Arf1. This is accompanied by activation of the kinase generating both PtdIns4P and PtdIns$(4,5)P_2$ (Godi et al. 1999). Although PtdIns4KIIIβ localises to other subcellular compartments (Table 1.2), it functions principally at the Golgi complex where it maintains structural integrity of the organelle and mediates key transport processes such as TGN-to-PM of nascent proteins and sphingolipids. These functions are directly dependent on either PtdIns4P or its metabolic derivatives and there is some functional overlap with PtdIns4KIIα, the other major TGN-localised PtdIns4K (Wang et al. 2003). In MDCK cells, PtdIns4KIIIβ regulates multiple trafficking steps (Bruns et al. 2002) including the intra-Golgi traffic of influenza haemagglutinin and the basolateral to apical delivery of vesicular stomatitis virus glycoprotein cargo (VSVG) (Weixel et al. 2005). Further evidence of pleiotropic roles of PtdIns4KIIIβ comes from the observation that it functions in regulated exocytosis from mast cells (Kapp-Barnea et al. 2006) and neuroendocrine cells (de Barry et al. 2006; Gromada et al. 2005) implying that the enzyme can perform PtdIns(4,5)P_2-dependent roles outside of the Golgi in specialised cell types.

The yeast orthologue of PtdIns4KIIIβ, Pik1, accounts for ∼45% of total PtdIns4P and ∼40% of PtdIns(4,5)P_2 production (Strahl and Thorner 2007) and the gene is essential for cell survival. Like PtdIns4KIIIβ, Pik1p is a TGN enzyme which also localises to the nucleus (Strahl et al. 2005). Pik1p is a downstream effector of a small N-myristoylated protein frequenin (Frq1p), which not only determines Pik1p localisation to the TGN but also tightly binds and stimulates PtdIns4K activity (Hendricks et al. 1999). Some aspects of this interaction are evolutionarily conserved because the mammalian orthologue of Frq1p, the neuronal calcium sensor NCS-1, interacts with and stimulates PtdIns4KIIIβ (Weisz et al. 2000; Haynes et al. 2005). NCS-1 and Frq1p belong to a family of myristoylated proteins that bind multiple targets, often in a Ca^{2+}-dependent manner. PtdIns4KIIIβ is actually under the dual control of Arf1 and NCS-1 because the enzyme binds both in a Ca^{2+}-dependent manner and both stimulate kinase activity, however, Arf1 and NCS-1 do not act synergistically to increase activity: Arf1 antagonises NCS-1 stimulation of regulated secretion of recombinant human growth hormone from PC12 cells whilst, conversely, Arf1 overexpression antagonises the constitutive trafficking of VSVG from the TGN (Haynes et al. 2005). The implications of these data are that Arf1 or NCS-1 and Ca^{2+} but not both, can activate PtdIns4KIIIβ. It is interesting that the Golgi complex can act as a Ca^{2+} store and localised Ca^{2+} release may affect numerous trafficking steps (Pinton et al. 1998); PtdIns4KIIIβ function may therefore also be under the control of local Ca^{2+} signals. In this regard it is worth mentioning that the other major Golgi activity, PtdIns4KIIα, is extremely sensitive to Ca^{2+} *in vitro,* suggesting that these two activities may be differentially regulated in order to control PtdIns4P synthesis in the Golgi complex.

Many neuronal and secretory functions have been described for NCS-1 and the protein is highly expressed in brain tissues, localising to numerous intracellular membranes including synaptic vesicles in rat axon terminals (Taverna et al. 2002). NCS-1 plays a role in regulated secretion in neuroendocrine cells (McFerran et al. 1998) but is also present in non-neuronal cell types. In different cell systems NCS-1 appears to promote both secretion and endocytic recycling (Kapp-Barnea et al. 2006). It localises to the TGN of rat hippocampal neurons (Martone et al. 1999) and the TGN and PM of COS-7 (Bourne et al. 2001). NCS-1 and PtdIns4KIIIβ increase glucose-induced insulin secretion by increasing the pool of available vesicles in pancreatic β-cells (Gromada et al. 2005), indicating that the enzyme has roles in regulated exocytosis as well as constitutive secretion.

PtdIns4KIIIβ can be phosphorylated by protein kinase D (PKD) isoforms PKD1 and PKD2, which are known to regulate the fission of TGN-derived transport intermediates (Liljedahl et al. 2001). PKD phosphorylation at PtdIns4KIIIβ S294 stimulates PtdIns4K activity and enhances PM traffic of VSVG compared to a mutant PtdIns4KIIIβ in which the S294 is mutated to an alanine residue. Interestingly, activation of PtdIns4KIIIβ is stabilised by interaction with 14-3-3 proteins, which bind phosphorylated PtdIns4KIIIβ. However, the extensive Golgi tubulation characteristic of PKD mutants is not phenocopied by expression of a PtdIns4KIIIβ(S294A) mutant suggesting that other Golgi-localised PtdIns4Ks are sufficient to supply the PtdIns4P required for this trafficking process.

PtdIns4K activities are implicated in the regulation of sphingolipid synthesis via a class of lipid transfer proteins, which contain an N-terminal PtdIns4P-binding PH domain along with a C-terminal lipid transfer domain. These unrelated proteins have been dubbed the 'COF' family after the original members, ceramide transfer protein (CERT), oxysterol binding protein (OSBP) and the 'four phosphate adaptor proteins' (FAPPs) (De Matteis et al. 2007). CERT is important in the non-vesicular transport of ceramide from the ER to the Golgi (Hanada et al. 2003) where it is used in the synthesis of sphingomyelin (SM). CERT is targeted to the Golgi via its PH domain but also associates with the ER-bound membrane protein VAP-A. These interactions may define a transport junction between the ER and Golgi membranes or allow CERT to shuttle between the two compartments and PtdIns4KIIIβ appears to be the main activity responsible for generating the pool of PtdIns4P required for CERT recruitment (D'Angelo et al. 2007; Toth et al. 2006). Like PtdIns4KIIIβ, CERT is a substrate for PKD phosphorylation, which negatively regulates both its ceramide transfer activity and affinity for PtdIns4P. Meanwhile, the synthesis of SM is intimately linked to the generation of diacylglycerol (DAG) at the Golgi that recruits and activates PKD, and together this leads to the inactivation of CERT and the phosphorylation and activation of Golgi-localised PtdIns4KIIIβ (Hausser 2005, p. 69). It is notable that PtdIns4KII, the other Golgi-resident PtdIns4P-synthetic activity, interacts with PKD (Nishikawa et al. 1998) therefore the synthesis of SM and PtdIns4P appear to be co-regulated in the TGN.

OSBP1 binds and transfers sterols and associates with the Golgi via its PH domain (Levine and Munro 1998), translocating there from cytosolic and vesicular membranes in response to treatment with the cholesterol precursor 25-hydroxysterol (Ridgway et al. 1992). OSBP also interacts with VAP-A and VAP-B in the ER and therefore may act in a similar way to CERT. As well as PtdIns4KIIIβ, evidence exists that PtdIns4KIIα is responsible for the Golgi recruitment of CERT and therefore SM synthesis (Banerji et al. 2010). At steady-state, PtdIns4KIIα localises to cholesterol-poor membranes and the increased sterol content of these, as the result of transfer by OSBP, may stimulate PtdIns4P synthesis by PtdIns4KIIα at the TGN where both lipids play important roles.

FAPP1 and FAPP2 are ubiquitously expressed metazoan proteins that are part of a conserved gene family; the proteins play a role in constitutive TGN to PM traffic (Godi et al. 2004) and to the apical surface in MDCK cells (Vieira et al.

2005). The TGN localisation of FAPPs is dependent on an intact PH domain, which binds both PtdIns4*P* and Arf1 and inhibition of PtdIns4K activity using the relatively non-specific compound PAO or expression of dominant negative kinase dead PtdIns4KIIIβ prevents the correct TGN targeting of FAPPs. FAPP2 contains a glycolipid transfer homology domain and the protein is required for non-vesicular transport of glucosylceramide to from its site of synthesis in the Golgi complex to later Golgi compartments where it is converted into complex glycosphingolipids. In this process, the localisation of FAPP2 requires Golgi PtdIns4*P* generated by PtdIns4KIIα and/or PtdIns4KIIIβ (D'Angelo et al. 2007). Together the COF proteins represent PtdIns4*P* effector proteins that are recruited to the Golgi complex and, by virtue of their lipid transfer activities, dictate the lipid composition of membranes that function in important trafficking steps.

1.5 PtdIns4K Biology in the Whole Organism

Genetic models of PtdIns4K function have now been reported in *S. cerevisiae*, *Drosophila*, zebrafish and mice and whilst these have led to a number of very interesting phenotypes, the molecular mechanisms underlying knockout or knockdown phenotypes are not yet fully understood in most cases. Nevertheless, these models provide important clues to the more complex functions of PtdIns4K in whole organisms and there can be no doubt that, in the future, these phenotypes will be explained mechanistically at the molecular level.

In zebrafish, both the PI3K inhibitor LY294002 and morpholinos targeting expression of the zebrafish PtdIns4KIIIα homolog (pi4ka) cause multiple developmental abnormalities including pectoral fin phenotypes reminiscent of defects in fibroblast growth factor (FGF) signalling (Ma et al. 2009). LY294002 was used at a concentration not expected to affect PtdIns4KIIIs, but this treatment effectively phenocopied the pi4ka morpholinos. FGF morphogens are known to control limb development through the balance of two parallel pathways: the mitogen-activated protein kinase (MAPK) pathway and the PI3K-Akt survival pathway. Thus the investigators concluded that pi4ka functions in a PI3K-Akt pathway that is regulated by FGF, probably by supplying PtdIns4*P* to PtdIns4*P* 5K and subsequently PI3Ks. Although not definitively linked to PI3K signalling, developmental defects have also been identified in *Drosophila* where the PtdIns4*P*-phosphatase Sac1 and the *Drosophila* PtdIns4KIIIα orthologue are components of the Hedgehog signalling pathway (Yavari et al. 2010). As with the *S. cerevisiae* mutation (Foti et al. 2001), inactivation of Sac1 protein leads to an accumulation of PtdIns4*P*. Loss of Sac1 in *Drosophila* causes the normally vesicular Smoothened protein to translocate to the PM, resulting in the activation of Hedgehog signalling in multiple tissues during development (Yavari et al. 2010). Conversely, loss of the *Drosophila* PtdIns4KIIIα orthologue leads to a phenotype similar to Hedgehog and Smoothened loss-of-function mutants (Yavari et al. 2010). When the PM receptor for Hedgehog, and target for derepression by Smoothened, is mutated, there is a marked increase in PtdIns4*P* synthesised by PtdIns4KIIIα. This genetic model indicates that PI metabolism plays an important role in the in the positive regulation of Hedgehog signalling at the level of PtdIns4*P* biosynthesis (Yavari et al. 2010). These findings are of great interest because the Patched/Smoothened interaction is conserved and commonly disrupted in human cancers (Rohatgi and Scott 2007).

Knockout of PtdIns4KIIα in mice by Gene Trap does not cause developmental defects but instead leads to a late-onset neurodegenerative phenotype accompanied by some loss of cerebellar Purkinje cells and degeneration of spinal cord axons (Simons et al. 2009). Although mutations in the PtdIns4KIIα gene have yet to be linked to human disease, the Gene Trap mouse phenotype has marked similarities with symptoms of hereditary spastic paraplegia (HSP).

The *Drosophila* PtdIns4KIIIβ homolog four-wheel drive (Fwd) is required for spermatocyte cytokinesis (Brill et al. 2000). Male *fwd* null flies are viable but sterile with multinucleate spermatocytes, a phenotype that is the likely result of failure to form a proper cleavage furrow during meiosis. Fwd localises to the Golgi in spermatocytes where it is required for the formation of PtdIns4*P* and membrane organelles localising to the midzone. A catalytically inactive PtdIns4KIIIβ partially rescues the *fwd* mutant, suggesting that the protein probably functions by recruiting DmRab11 that binds Fwd and colocalises with PtdIns4*P* at the midzone (Brill et al. 2000).

1.6 Conclusions

The products of PtdIns4Ks have assumed great importance in intracellular membrane traffic, indeed this role currently eclipses the traditional second messenger functions of D4 PIs in PLC and PI3K signalling pathways. Several clear themes have emerged from the large body of work that has been enabled by the molecular identification of the PtdIns4Ks: (i) PtdIns4*P* plays fundamental roles in membrane traffic by acting as a specific recruitment and localisation signal, principally in the Golgi complex and often in concert with small GTPases such as Arf1. This dual localisation code, which is based on the coincidence of a PI and a GTPase, is able to determine the recruitment of trafficking machinery to a specific region or domain within an organelle and promote the formation of membrane carriers from that locus, thereby effecting vectorial traffic between membranes. In many cases this traffic is constitutive, maintaining membrane composition and therefore organelle structure and function. (ii) Whilst many proteins recruited to the Golgi by PtdIns4*P* are adaptors that recruit cargo, a class of lipid transfer proteins, the COFs, share a similar mechanism to effect the non-vesicular traffic of ceramide and complex glycosphingolipids, lipids which play important roles in the organisation of intra-membrane domains. (iii) Some PtdIns4Ks may be able to compensate for the loss of each other: although PtdIns4Ks are required for the acute supply of PIs in processes such as regulated secretion, PLC and PI3K signalling, the requirement for individual isoforms is less than clear and it seems that some redundancy exists amongst the PtdIns4Ks; however, this probably occurs at the level of PtdIns4*P* and may be dependent on non-vesicular traffic to replenish distal membrane pools.

Despite the great advances made in recent years, there are still many unanswered questions: foremost is the role of PtdIns4Ks in agonist-dependent PLC and PI3K signalling. The paradox of PtdIns4KIIIα localisation and the apparent redundancy amongst the PtdIns4K isoforms observed in some experiments (Balla et al. 2008a), suggest that there is much to be learned about these pathways. The finding that PtdIns4*P* is an important determinant of Golgi localisation critical for several fundamental trafficking processes has become a central paradigm in cell biology. However, many as yet uncharacterised proteins contain potential D4 PI-binding domains and the PtdIns4K isoforms controlling them are unknown.

It has been a long journey for the PtdIns4Ks and it is somewhat ironic that the first PtdInsK activity to be described in tissue extracts should be the last PtdInsK cloned. This delay can be largely explained by the technical difficulties experienced when purifying PtdIns4KII activities and the striking lack of sequence similarity to other kinases, which made identification difficult. Also, whilst *S. cerevisiae* was the first eukaryotic genome to be fully characterised, the sequence data contained nothing resembling a kinase that could account for the type II activity previously identified by the Carman lab (Nickels et al. 1992). *S. cerevisiae*, the system that taught us so much, therefore held on to its secret till the last (Han et al. 2002). So don't call it a comeback: the PtdIns4Ks have been around for years.

Acknowlegements We are grateful to Prof. Justin Hsuan, Dr. Emma Clayton and Dr. Emily Chu for comments on the manuscript. SM and MGW acknowledge the support of the BBSRC (award BB/G021163/1).

References

Ahn J, Chung KS, Kim DU, Won M, Kim L, Kim KS, Nam M, Choi SJ, Kim HC, Yoon M, Chae SK, Hoe KL (2004) Systematic identification of hepatocellular proteins interacting with NS5A of the hepatitis C virus. J Biochem Mol Biol 37:741–748

Aikawa Y, Kuraoka A, Kondo H, Kawabuchi M, Watanabe T (1999) Involvement of PITPnm, a mammalian homologue of Drosophila rdgB, in phosphoinositide synthesis on Golgi membranes. J Biol Chem 274:20569–20577

Audhya A, Foti M, Emr SD (2000) Distinct roles for the yeast phosphatidylinositol 4-kinases, Stt4p and Pik1p, in secretion, cell growth, and organelle membrane dynamics. Mol Biol Cell 11:2673–2689

Balla A, Balla T (2006) Phosphatidylinositol 4-kinases: old enzymes with emerging functions. Trends Cell Biol 16:351–361

Balla A, Kim YJ, Varnal P, Szentpetery Z, Knight Z, Shokat KM, Balla T (2008a) Maintenance of hormone-sensitive phosphoinositide pools in the plasma membrane requires phosphatidylinositol 4-kinase IIIalpha. Mol Biol Cell 19:711–721

Balla A, Tuymetova G, Barshishat M, Geiszt M, Balla T (2002) Characterization of type II phosphatidylinositol 4-kinase isoforms reveals association of the enzymes with endosomal vesicular compartments. J Biol Chem 277:20041–20050

Balla A, Tuymetova G, Toth B, Szentpetery Z, Zhao X, Knight ZA, Shokat K, Steinbach PJ, Balla T (2008b) Design of drug-resistant alleles of type-III phosphatidylinositol 4-kinases using mutagenesis and molecular modeling. Biochemistry 47:1599–1607

Balla A, Tuymetova G, Tsiomenko A, Varnai P, Balla T (2005) A plasma membrane pool of phosphatidylinositol 4-phosphate is generated by phosphatidylinositol 4-kinase type-III alpha: studies with the PH domains of the oxysterol binding protein and FAPP1. Mol Biol Cell 16:1282–1295

Balla A, Vereb G, Gulkan H, Gehrmann T, Gergely P, Heilmeyer LM, Jr., Antal M (2000a) Immunohistochemical localisation of two phosphatidylinositol 4-kinase isoforms, PI4K230 and PI4K92, in the central nervous system of rats. Exp Brain Res 134:279–288

Balla T, Bondeva T, Varnai P (2000b) How accurately can we image inositol lipids in living cells? Trends Pharmacol Sci 21:238–241

Balla T, Downing GJ, Jaffe H, Kim S, Zolyomi A, Catt KJ (1997) Isolation and molecular cloning of wortmannin-sensitive bovine type III phosphatidylinositol 4-kinases. J Biol Chem 272:18358–18366

Banerji S, Ngo M, Lane CF, Robinson CA, Minogue S, Ridgway ND (2010) Oxysterol binding protein (OSBP)-dependent activation of sphingomyelin synthesis in the golgi apparatus requires PtdIns 4-kinase IIα. Mol Biol Cell 21:4141–4150

Barylko B, Gerber SH, Binns DD, Grichine N, Khvotchev M, Sudhof TC, Albanesi JP (2001) A novel family of phosphatidylinositol 4-kinases conserved from yeast to humans. J Biol Chem 276:7705–7708

Barylko B, Mao YS, Wlodarski P, Jung G, Binns DD, Sun HQ, Yin HL, Albanesi JP (2009) Palmitoylation controls the catalytic activity and subcellular distribution of phosphatidylinositol 4-kinase IIα. J Biol Chem 284:9994–10003

Barylko B, Wlodarski P, Binns DD, Gerber SH, Earnest S, Sudhof TC, Grichine N, Albanesi JP (2002) Analysis of the catalytic domain of phosphatidylinositol 4-kinase type II. J Biol Chem 277:44366–44375

Berger KL, Cooper JD, Heaton NS, Yoon R, Oakland TE, Jordan TX, Mateu G, Grakoui A, Randall G (2009) Roles for endocytic trafficking and phosphatidylinositol 4-kinase III alpha in hepatitis C virus replication. Proc Natl Acad Sci U S A 106:7577–7582

Bourne Y, Dannenberg J, Pollmann V, Marchot P, Pongs O (2001) Immunocytochemical localization and crystal structure of human frequenin (neuronal calcium sensor 1). J Biol Chem 276:11949–11955

Brill JA, Hime GR, Scharer-Schuksz M, Fuller MT (2000) A phospholipid kinase regulates actin organization and intercellular bridge formation during germline cytokinesis. Development 127:3855–3864

Bruns JR, Ellis MA, Jeromin A, Weisz OA (2002) Multiple roles for phosphatidylinositol 4-kinase in biosynthetic transport in polarized Madin-Darby canine kidney cells. J Biol Chem 277:2012–2018

Carlton JG, Cullen PJ (2005) Coincidence detection in phosphoinositide signaling. Trends Cell Biol 15:540–547

Chahdi A, Choi WS, Kim YM, Beaven MA (2003) Mastoparan selectively activates phospholipase D2 in cell membranes. J Biol Chem 278:12039–12045

Cochet C, Filhol O, Payrastre B, Hunter T, Gill GN (1991) Interaction between the epidermal growth factor receptor and phosphoinositide kinases. J Biol Chem 266:637–644

Craige B, Salazar G, Faundez V (2008) Phosphatidylinositol-4-kinase type II alpha contains an AP-3-sorting motif and a kinase domain that are both required for endosome traffic. Mol Biol Cell 19:1415–1426

D'angelo G, Polishchuk E, Di Tullio G, Santoro M, Di Campli A, Godi A, West G, Bielawski J, Chuang CC, Van Der Spoel AC, Platt FM, Hannun YA, Polishchuk R, Mattjus P, de Matteis MA (2007) Glycosphingolipid synthesis requires FAPP2 transfer of glucosylceramide. Nature 449:62–67

De Barry J, Janoshazi A, Dupont JL, Procksch O, Chasserot-Golaz S, Jeromin A, Vitale N (2006) Functional implication of neuronal calcium sensor-1 and phosphoinositol 4-kinase-beta interaction in regulated exocytosis of PC12 cells. J Biol Chem 281:18098–18111

De Graaf P, Klapisz EE, Schulz TK, Cremers AF, Verkleij AJ, van Bergen en Henegouwen PM (2002) Nuclear localization of phosphatidylinositol 4-kinase beta. J Cell Sci 115:1769–1775

De Graaf P, Zwart WT, van Dijken RA, Deneka M, Schulz TK, Geijsen N, Coffer PJ, Gadella BM, Verkleij AJ, Van Der Sluijs P, van Bergen En Henegouwen PM (2004) Phosphatidylinositol 4-kinasebeta is critical for functional association of rab11 with the Golgi complex. Mol Biol Cell 15:2038–2047

De Matteis MA, Di Campli A, D'angelo G (2007) Lipid-transfer proteins in membrane trafficking at the Golgi complex. Biochim Biophys Acta 1771:761–768

Dumaresq-Doiron K, Savard MF, Akam S, Costantino S, Lefrancois S (2010) The phosphatidylinositol 4-kinase PI4KIIIalpha is required for the recruitment of GBF1 to Golgi membranes. J Cell Sci 123:2273–2280

Endemann G, Dunn SN, Cantley LC (1987) Bovine brain contains two types of phosphatidylinositol kinase. Biochemistry 26:6845–6852

Flanagan CA, Schnieders EA, Emerick AW, Kunisawa R, Admon A, Thorner J (1993) Phosphatidylinositol 4-kinase: gene structure and requirement for yeast cell viability. Science 262:1444–1448

Foti M, Audhya A, Emr SD (2001) Sac1 lipid phosphatase and Stt4 phosphatidylinositol 4-kinase regulate a pool of phosphatidylinositol 4-phosphate that functions in the control of the actin cytoskeleton and vacuole morphology. Mol Biol Cell 12:2396–2411

Garcia-Bustos JF, Marini F, Stevenson I, Frei C, Hall MN (1994) PIK1, an essential phosphatidylinositol 4-kinase associated with the yeast nucleus. EMBO J 13:2352–2361

Gehrmann T, Gulkan H, Suer S, Herberg FW, Balla A, Vereb G, Mayr GW, Heilmeyer LM Jr (1999) Functional expression and characterisation of a new human phosphatidylinositol 4-kinase PI4K230. Biochim Biophys Acta 1437:341–356

Godi A, Di Campli A, Konstantakopoulos A, Di Tullio G, Alessi DR, Kular GS, Daniele T, Marra P, Lucocq JM, De Matteis MA (2004) FAPPs control Golgi-to-cell-surface membrane traffic by binding to ARF and PtdIns(4)P. Nat Cell Biol 6:393–404

Godi A, Pertile P, Meyers R, Marra P, Di Tullio G, Iurisci C, Luini A, Corda D, de Matteis MA (1999) ARF mediates recruitment of PtdIns-4-OH kinase-beta and stimulates synthesis of PtdIns(4,5)P2 on the Golgi complex. Nat Cell Biol 1:280–287

Gromada J, Bark C, Smidt K, Efanov AM, Janson J, Mandic SA, Webb DL, Zhang W, Meister B, Jeromin A, Berggren PO (2005) Neuronal calcium sensor-1 potentiates glucose-dependent exocytosis in pancreatic beta cells through activation of phosphatidylinositol 4-kinase beta. Proc Natl Acad Sci USA 102:10303–10308

Guo J, Wenk MR, Pellegrini L, Onofri F, Benfenati F, De Camilli P (2003) Phosphatidylinositol 4-kinase type IIalpha is responsible for the phosphatidylinositol 4-kinase activity associated with synaptic vesicles. Proc Natl Acad Sci USA 100:3995–4000

Hammond GR, Dove SK, Nicol A, Pinxteren JA, Zicha D, Schiavo G (2006) Elimination of plasma membrane phosphatidylinositol (4,5)-bisphosphate is required for exocytosis from mast cells. J Cell Sci 119:2084–2094

Hammond GR, Schiavo G, Irvine RF (2009) Immunocytochemical techniques reveal multiple, distinct cellular pools of PtdIns4P and PtdIns(4,5)P(2). Biochem J 422:23–35

Han GS, Audhya A, Markley DJ, Emr SD, Carman GM (2002) The Saccharomyces cerevisiae LSB6 gene encodes phosphatidylinositol 4-kinase activity. J Biol Chem 277:47709–47718

Hanada K, Kumagai K, Yasuda S, Miura Y, Kawano M, Fukasawa M, Nishijima M (2003) Molecular machinery for non-vesicular trafficking of ceramide. Nature 426:803–809

Harwood JL, Hawthorne JN (1969) The properties and subcellular distribution of phosphatidylinositol kinase in mammalian tissues. Biochim Biophys Acta 171:75–88

Haynes LP, Thomas GM, Burgoyne RD (2005) Interaction of neuronal calcium sensor-1 and ADP-ribosylation factor 1 allows bidirectional control of phosphatidylinositol 4-kinase beta and trans-Golgi network-plasma membrane traffic. J Biol Chem 280:6047–6054

Heasman SJ, Ridley AJ (2008) Mammalian Rho GTPases: new insights into their functions from in vivo studies. Nat Rev Mol Cell Biol 9:690–701

Hendricks KB, Wang BQ, Schnieders EA, Thorner J (1999) Yeast homologue of neuronal frequenin is a regulator of phosphatidylinositol-4-OH kinase. Nat Cell Biol 1:234–241

Hokin MR, Hokin LE (1953) Enzyme secretion and the incorporation of P32 into phospholipides of pancreas slices. J Biol Chem 203:967–977

Hunyady L, Balla T, Spat A (1983) Angiotensin II stimulates phosphatidylinositol turnover in adrenal glomerulosa cells by a calcium-independent mechanism. Biochim Biophys Acta 753:133–135

Jeganathan S, Lee JM (2007) Binding of elongation factor eEF1A2 to phosphatidylinositol 4-kinase beta stimulates lipid kinase activity and phosphatidylinositol 4-phosphate generation. J Biol Chem 282:372–380

Jeganathan S, Morrow A, Amiri A, Lee JM (2008) Eukaryotic elongation factor 1A2 cooperates with phosphatidylinositol-4 kinase III beta to stimulate production of filopodia through increased phosphatidylinositol-4,5 bisphosphate generation. Mol Cell Biol 28:4549–4561

Jung G, Wang J, Wlodarski P, Barylko B, Binns DD, Shu H, Yin HL, Albanesi JP (2008) Molecular determinants of activation and membrane targeting of phosphoinositol 4-kinase IIbeta. Biochem J 409:501–509

Kakuk A, Friedlander E, Vereb G Jr, Kasa A, Balla A, Balla T, Heilmeyer LM Jr, Gergely P, Vereb G (2006) Nucleolar localization of phosphatidylinositol 4-kinase PI4K230 in various mammalian cells. Cytometry A 69:1174–1183

Kapp-Barnea Y, Ninio-Many L, Hirschberg K, Fukuda M, Jeromin A, Sagi-Eisenberg R (2006) Neuronal calcium sensor-1 and phosphatidylinositol 4-kinase beta stimulate extracellular signal-regulated kinase 1/2 signaling by accelerating recycling through the endocytic recycling compartment. Mol Biol Cell 17:4130–4141

Kauffmann-Zeh A, Klinger R, Endemann G, Waterfield MD, Wetzker R, Hsuan JJ (1994) Regulation of human type II phosphatidylinositol kinase activity by epidermal growth factor-dependent phosphorylation and receptor association. J Biol Chem 269:31243–31251

Kikuchi A, Yamamoto H, Sato A (2009) Selective activation mechanisms of Wnt signaling pathways. Trends Cell Biol 19:119–129

Kim M, Jiang LH, Wilson HL, North RA, Surprenant A (2001) Proteomic and functional evidence for a P2X7 receptor signalling complex. Embo J 20:6347–6358

Knight ZA, Gonzalez B, Feldman ME, Zunder ER, Goldenberg DD, Williams O, Loewith R, Stokoe D, Balla A, Toth B, Balla T, Weiss WA, Williams RL, Shokat KM (2006) A pharmacological map of the PI3-K family defines a role for p110alpha in insulin signaling. Cell 125:733–747

Levine T, Rabouille C (2005) Endoplasmic reticulum: one continuous network compartmentalized by extrinsic cues. Curr Opin Cell Biol 17:362–368

Levine TP, Munro S (1998) The pleckstrin homology domain of oxysterol-binding protein recognises a determinant specific to Golgi membranes. Curr Biol 8:729–739

Levine TP, Munro S (2002) Targeting of Golgi-specific pleckstrin homology domains involves both PtdIns 4-kinase-dependent and -independent components. Curr Biol 12:695–704

Liljedahl M, Maeda Y, Colanzi A, Ayala I, Van Lint J, Malhotra V (2001) Protein kinase D regulates the fission of cell surface destined transport carriers from the trans-Golgi network. Cell 104:409–420

Ma H, Blake T, Chitnis A, Liu P, Balla T (2009) Crucial role of phosphatidylinositol 4-kinase IIIalpha in development of zebrafish pectoral fin is linked to phosphoinositide 3-kinase and FGF signaling. J Cell Sci 122:4303–4310

Martone ME, Edelmann VM, Ellisman MH, Nef P (1999) Cellular and subcellular distribution of the calcium-binding protein NCS-1 in the central nervous system of the rat. Cell Tissue Res 295:395–407

Mazzocca A, Liotta F, Carloni V (2008) Tetraspanin CD81-regulated cell motility plays a critical role in intrahepatic metastasis of hepatocellular carcinoma. Gastroenterology 135:244–256 e1

Mcferran BW, Graham ME, Burgoyne RD (1998) Neuronal Ca^{2+} sensor 1, the mammalian homologue of frequenin, is expressed in chromaffin and PC12 cells and regulates neurosecretion from dense-core granules. J Biol Chem 273:22768–22772

Meyers R, Cantley LC (1997) Cloning and characterization of a wortmannin-sensitive human phosphatidylinositol 4-kinase. J Biol Chem 272:4384–4390

Michell RH, Harwood JL, Coleman R, Hawthorne JN (1967) Characteristics of rat liver phosphatidylinositol kinase and its presence in the plasma membrane. Biochim Biophys Acta 144:649–658

Minogue S, Anderson JS, Waugh MG, Dos Santos M, Corless S, Cramer R, Hsuan JJ (2001) Cloning of a human type II phosphatidylinositol 4-kinase reveals a novel lipid kinase family. J Biol Chem 276:16635–16640

Minogue S, Chu KM, Westover EJ, Covey DF, Hsuan JJ, Waugh MG (2010) Relationship between phosphatidylinositol 4-phosphate synthesis, membrane organization, and lateral diffusion of PI4KIIalpha at the trans-Golgi network. J Lipid Res 51:2314–2324

Minogue S, Waugh MG, de Matteis MA, Stephens DJ, Berditchevski F, Hsuan JJ (2006) Phosphatidylinositol 4-kinase is required for endosomal trafficking and degradation of the EGF receptor. J Cell Sci 119:571–581

Nakagawa T, Goto K, Kondo H (1996) Cloning, expression, and localization of 230-kDa phosphatidylinositol 4-kinase. J Biol Chem 271:12088–12094

Nakanishi S, Catt KJ, Balla T (1995) A wortmannin-sensitive phosphatidylinositol 4-kinase that regulates hormone-sensitive pools of inositolphospholipids. Proc Natl Acad Sci U S A 92:5317–5321

Nickels JT Jr, Buxeda RJ, Carman GM (1992) Purification, characterization, and kinetic analysis of a 55-kDa form of phosphatidylinositol 4-kinase from Saccharomyces cerevisiae. J Biol Chem 267:16297–16304

Nishikawa K, Toker A, Wong K, Marignani PA, Johannes FJ, Cantley LC (1998) Association of protein kinase Cmu with type II phosphatidylinositol 4-kinase and type I phosphatidylinositol-4-phosphate 5-kinase. J Biol Chem 273:23126–23133

Olsen JV, Blagoev B, Gnad F, Macek B, Kumar C, Mortensen P, Mann M (2006) Global, in vivo, and site-specific phosphorylation dynamics in signaling networks. Cell 127:635–648

Olsson H, Martinez-Arias W, Jergil B (1993) Phosphatidylcholine enhances the activity of rat liver type II phosphatidylinositol-kinase. FEBS Lett 327:332–336

Pan W, Choi SC, Wang H, Qin Y, Volpicelli-Daley L, Swan L, Lucast L, Khoo C, Zhang X, Li L, Abrams CS, Sokol SY, Wu D (2008) Wnt3a-mediated formation of phosphatidylinositol 4,5-bisphosphate regulates LRP6 phosphorylation. Science 321:1350–1353

Pike LJ (1992) Phosphatidylinositol 4-kinases and the role of polyphosphoinositides in cellular regulation. Endocr Rev 13:692–706

Pinton P, Pozzan T, Rizzuto R (1998) The Golgi apparatus is an inositol 1,4,5-trisphosphate-sensitive Ca^{2+} store, with functional properties distinct from those of the endoplasmic reticulum. EMBO J 17:5298–5308

Qin Y, Li L, Pan W, Wu D (2009) Regulation of phosphatidylinositol kinases and metabolism by Wnt3a and Dvl. J Biol Chem 284:22544–22548

Ridgway ND, Dawson PA, Ho YK, Brown MS, Goldstein JL (1992) Translocation of oxysterol binding protein to Golgi apparatus triggered by ligand binding. J Cell Biol 116:307–319

Rohatgi R, Scott MP (2007) Patching the gaps in Hedgehog signalling. Nat Cell Biol 9:1005–1009

Roth AF, Wan J, Green WN, Yates JR, Davis NG (2006) Proteomic identification of palmitoylated proteins. Methods 40:135–142

Salazar G, Craige B, Wainer BH, Guo J, De Camilli P, Faundez V (2005) Phosphatidylinositol-4-kinase type II alpha is a component of adaptor protein-3-derived vesicles. Mol Biol Cell 16:3692–3704

Salazar G, Zlatic S, Craige B, Peden AA, Pohl J, Faundez V (2009) Hermansky-Pudlak syndrome protein complexes associate with phosphatidylinositol 4-kinase type II alpha in neuronal and non-neuronal cells. J Biol Chem 284:1790–1802

Simons JP, Al-Shawi R, Minogue S, Waugh MG, Wiedemann C, Evangelou S, Loesch A, Sihra TS, King R, Warner TT, Hsuan JJ (2009) Loss of phosphatidylinositol 4-kinase 2alpha activity causes late onset degeneration of spinal cord axons. Proc Natl Acad Sci U S A 106:11535–11539

Stevenson-Paulik J, Love J, Boss WF (2003) Differential regulation of two Arabidopsis type III phosphatidylinositol 4-kinase isoforms. A regulatory role for the pleckstrin homology domain. Plant Physiol 132:1053–1064

Stevenson JM, Perera IY, Boss WF (1998) A phosphatidylinositol 4-kinase pleckstrin homology domain that binds phosphatidylinositol 4-monophosphate. J Biol Chem 273:22761–22767

Strahl T, Hama H, Dewald DB, Thorner J (2005) Yeast phosphatidylinositol 4-kinase, Pik1, has essential roles at the Golgi and in the nucleus. J Cell Biol 171:967–979

Strahl T, Thorner J (2007) Synthesis and function of membrane phosphoinositides in budding yeast, Saccharomyces cerevisiae. Biochim Biophys Acta 1771:353–404

Szentpetery Z, Balla A, Kim YJ, Lemmon MA, Balla T (2009) Live cell imaging with protein domains capable of recognizing phosphatidylinositol 4,5-bisphosphate; a comparative study. BMC Cell Biol 10:67

Szentpetery Z, Varnai P, Balla T (2010) Acute manipulation of Golgi phosphoinositides to assess their importance in cellular trafficking and signaling. Proc Natl Acad Sci USA 107:8225–8230

Taverna E, Francolini M, Jeromin A, Hilfiker S, Roder J, Rosa P (2002) Neuronal calcium sensor 1 and phosphatidylinositol 4-OH kinase beta interact in neuronal cells and are translocated to membranes during nucleotide-evoked exocytosis. J Cell Sci 115:3909–3922

Toth B, Balla A, Ma H, Knight ZA, Shokat KM, Balla T (2006) Phosphatidylinositol 4-kinase IIIbeta regulates the transport of ceramide between the endoplasmic reticulum and Golgi. J Biol Chem 281:36369–36377

Varnai P, Balla T (2006) Live cell imaging of phosphoinositide dynamics with fluorescent protein domains. Biochim Biophys Acta 1761:957–967

Vieira OV, Verkade P, Manninen A, Simons K (2005) FAPP2 is involved in the transport of apical cargo in polarized MDCK cells. J Cell Biol 170:521–526

VIllen J, Beausoleil SA, Gerber SA, Gygi SP (2007) Large-scale phosphorylation analysis of mouse liver. Proc Natl Acad Sci USA 104:1488–1493

Wang J, Sun HQ, Macia E, Kirchhausen T, Watson H, Bonifacino JS, Yin HL (2007) PI4P promotes the recruitment of the GGA adaptor proteins to the trans-Golgi network and regulates their recognition of the ubiquitin sorting signal. Mol Biol Cell 18:2646–2655

Wang YJ, Wang J, Sun HQ, Martinez M, Sun YX, Macia E, Kirchhausen T, Albanesi JP, Roth MG, Yin HL (2003) Phosphatidylinositol 4 phosphate regulates targeting of clathrin adaptor AP-1 complexes to the Golgi. Cell 114:299–310

Watt SA, Kular G, Fleming IN, Downes CP, Lucocq JM (2002) Subcellular localization of phosphatidylinositol 4,5-bisphosphate using the pleckstrin homology domain of phospholipase C delta1. Biochem J 363:657–666

Waugh MG, Minogue S, Anderson JS, Balinger A, Blumenkrantz D, Calnan DP, Cramer R, Hsuan JJ (2003a) Localization of a highly active pool of type II phosphatidylinositol 4-kinase in a p97/valosin-containing-protein-rich fraction of the endoplasmic reticulum. Biochem J 373: 57–63

Waugh MG, Minogue S, Blumenkrantz D, Anderson JS, Hsuan JJ (2003b) Identification and characterization of differentially active pools of type IIalpha phosphatidylinositol 4-kinase activity in unstimulated A431 cells. Biochem J 376:497–503

Waugh MG, Minogue S, Chotai D, Berditchevski F, Hsuan JJ (2006) Lipid and peptide control of phosphatidylinositol 4-kinase IIalpha activity on Golgi-endosomal Rafts. J Biol Chem 281:3757–3763

Wei YJ, Sun HQ, Yamamoto M, Wlodarski P, Kunii K, Martinez M, Barylko B, Albanesi JP, Yin HL (2002) Type II phosphatidylinositol 4-kinase beta is a cytosolic and peripheral membrane protein that is recruited to the plasma membrane and activated by Rac-GTP. J Biol Chem 277:46586–46593

Weisz OA, Gibson GA, Leung SM, Roder J, Jeromin A (2000) Overexpression of frequenin, a modulator of phosphatidylinositol 4-kinase, inhibits biosynthetic delivery of an apical protein in polarized madin-darby canine kidney cells. J Biol Chem 275:24341–24347

Weixel KM, Blumental-Perry A, Watkins SC, Aridor M, Weisz OA (2005) Distinct Golgi populations of phosphatidylinositol 4-phosphate regulated by phosphatidylinositol 4-kinases. J Biol Chem 280:10501–10508

Whitman M, Kaplan D, Roberts T, Cantley L (1987) Evidence for two distinct phosphatidylinositol kinases in fibroblasts. Implications for cellular regulation. Biochem J 247:165–174

Wong K, Cantley LC (1994) Cloning and characterization of a human phosphatidylinositol 4-kinase. J Biol Chem 269:28878–28884

Wong K, Meyers R, Cantley LC (1997) Subcellular locations of phosphatidylinositol 4-kinase isoforms. J Biol Chem 272:13236–13241

Yamakawa A, Nishizawa M, Fujiwara KT, Kawai S, Kawasaki H, Suzuki K, Takenawa T (1991) Molecular cloning and sequencing of cDNA encoding the phosphatidylinositol kinase from rat brain. J Biol Chem 266:17580–17583

Yavari A, Nagaraj R, Owusu-Ansah E, Folick A, Ngo K, Hillman T, Call G, Rohatgi R, Scott MP, Banerjee U (2010) Role of lipid metabolism in smoothened depression in hedgehog signaling. Dev Cell 19:54–65

Yoshida S, Ohya Y, Goebl M, Nakano A, Anraku Y (1994) A novel gene, STT4, encodes a phosphatidylinositol 4-kinase in the PKC1 protein kinase pathway of Saccharomyces cerevisiae. J Biol Chem 269:1166–1172

Zolyomi A, Zhao X, Downing GJ, Balla T (2000) Localization of two distinct type III phosphatidylinositol 4-kinase enzyme mRNAs in the rat. Am J Physiol Cell Physiol 278:C914–C920

Chapter 2
PIP Kinases from the Cell Membrane to the Nucleus

Mark Schramp, Andrew Hedman, Weimin Li, Xiaojun Tan and Richard Anderson

Abstract Phosphatidylinositol 4,5-bisphosphate (PIP_2) is a membrane bound lipid molecule with capabilities to affect a wide array of signaling pathways to regulate very different cellular processes. PIP_2 is used as a precursor to generate the second messengers PIP_3, DAG and IP_3, indispensable molecules for signaling events generated by membrane receptors. However, PIP_2 can also directly regulate a vast array of proteins and is emerging as a crucial messenger with the potential to distinctly modulate biological processes critical for both normal and pathogenic cell physiology. PIP_2 directly associates with effector proteins via unique phosphoinositide binding domains, altering their localization and/or enzymatic activity. The spatial and temporal generation of PIP_2 synthesized by the phosphatidylinositol phosphate kinases (PIPKs) tightly regulates the activation of receptor signaling pathways, endocytosis and vesicle trafficking, cell polarity, focal adhesion dynamics, actin assembly and 3′ mRNA processing. Here we discuss our current understanding of PIPKs in the regulation of cellular processes from the plasma membrane to the nucleus.

Keywords Cell Migration · mRNA processing · Phosphatidylinositol phosphate kinase (PIPK) · Phosphatidylinositol 4,5-bisphosphate (PIP_2) · Vesicle trafficking

2.1 Introduction

Studies by Hokin and Hokin on exocrine tissue in the 1950s brought to light how changes in phospholipids could regulate cellular processes (Hokin and Hokin 1953). Later, discoveries in the 1980s advanced our understanding of how phosphatidyli-

---

R. Anderson (✉) · M. Schramp · W. Li
Department of Pharmacology, School of Medicine and Public Health,
3710 Medical Sciences Center, University of Wisconsin Medical School,
1300 University Ave., Madison, WI 53706
e-mail: raanders@wisc.edu

A. Hedman · X. Tan
Program in Molecular and Cellular Pharmacology
University of Wisconsin-Madison
1300 University Ave., Madison, WI 53706

nositol 4,5-bisphosphate (PIP$_2$) can be utilized by phospholipase C to generate IP$_3$ (Streb et al. 1983), a molecule that mobilizes Ca^{2+} stores from the endoplasmic reticulum, and by PI3K to generate PIP$_3$, a signaling molecule initially discovered downstream of the Src oncoprotein critical for cell proliferation and transformation (Whitman et al. 1988). In 1985, Lassing and Lindberg discovered that PIP$_2$ could directly associate with profilactin and profilin (Lassing and Lindberg 1985). These studies provided the backbone of PIP$_2$ biology and illuminate the importance of PIP$_2$ as a signaling nexus that not only is modified to generate PIP$_3$, DAG and IP$_3$, but can also directly modulate the activities of an ever growing array of proteins to regulate virtually every cellular process. The past few years have seen a resurgent interest in the regulation of PIP$_2$ synthesis. Unlike PIP$_3$, PIP$_2$ levels are relatively high and undergo only small changes in total cellular content upon stimulation (Heck et al. 2007). However, the spatial and temporal targeting of phosphatidyl-inositol phosphate kinases (PIPKs), the molecules that generate the vast majority of cellular PIP$_2$, via the actions of its functional protein associates can result in a dramatic and localized surge in PIP$_2$ levels to coordinate the activation of its regulated signaling pathways (Doughman et al. 2003b; Honda et al. 1999; Kisseleva et al. 2005; Sasaki et al. 2005). The findings that PIPK activity is enhanced by Rho and Arf family GTPases provided regulatory mechanisms and clues to the processes that PIP$_2$ generation might affect (Krauss et al. 2003; Weernink et al. 2004). Perhaps even more surprising, roles for PIPKs and PIP$_2$ within the nucleus are defining how these molecules can regulate mRNA processing and gene expression (Mellman et al. 2008). Thus how PIPKs are regulated has emerged as crucial yet relatively unexplored frontier in molecular biology that could dramatically impact how we view normal and pathogenic behavior. This chapter will focus on the role that PIPKs and PIP$_2$ generation play in modulating signaling pathways and cellular process from the plasma membrane to the nucleus.

2.2 The Enzymes. Sequence, Structure and Enzymology

2.2.1 Sequence and Structure of Phosphatidylinositol Phosphate Kinases

There are three known families of PIP kinases, type I, II and III, which catalyze the formation of phosphatidylinositol bisphosphates. Type I PIP kinases contain three separate genes, α, β and γ, and primarily catalyze phosphorylation of PI4P to PI4,5P$_2$ (Anderson et al. 1999; Doughman et al. 2003a; Heck et al. 2007; Ishihara et al. 1996; Jenkins et al. 1991; Loijens et al. 1996; Schill and Anderson 2009b). Type II PIP kinases, also have three separate genes, α, β and γ, and generate PI4,5P$_2$ by phosphorylating the 4' OH position of PI5P (Boronenkov and Anderson 1995; Rameh et al. 1997). First identified in yeast for their function in vacuole morphology, the most recently discovered type III PIP kinases generate PI3,5P$_2$ by phosphorylating PI3P (Cooke et al. 1998; Gary et al. 1998; McEwen et al. 1999; Yamamoto et al. 1995). Within this family of kinases are the yeast Fab1p and the human PikFYVE

(Cabezas et al. 2006; Sbrissa et al. 1999). At ∼2200 amino acids the type III PIP Kinases are much larger in size than type I and II kinases and contain an N-terminal FYVE domain that allows for association with endosomal compartments (Gary et al. 1998; Odorizzi et al. 1998; Sbrissa et al. 1999; Yamamoto et al. 1995). The conserved feature of all PIP kinases is ∼280 amino acid kinase domain (Ishihara et al. 1996).

2.2.1.1 Sequences of PIP Kinases

There is little conserved sequence among PIPK subtypes, with 60% identity among PIPKI, 60–77% identity among PIPKII, while PikFYVE and Fab1p only share 20% identity. Compared to the yeast MSS4p, human type I PIPKs share only 20% identity even though they catalyze the same reaction and can complement Mss4p deficient cells (Homma et al. 1998). Little identity is conserved between PIPK subtypes, PIPKI and PIPKII share less than 30% identity, while type PIPKIII shares less than 20% identity with human PIPKI and PIPKII. In the C-terminal region of type I, II and III PIPKs is the kinase homology domain, which is responsible for catalytic activity and is the only region that shares significant homology between family members (Anderson et al. 1999; Boronenkov and Anderson 1995; Ishihara et al. 1998; Loijens et al. 1996). PIPKIα, β and γ are approximately 75–80% identical, PIPKIIα, β and γ are 66–78% identical, and 41% identity is shared between the PikFYVE and Fab1p kinase domains. There is ∼30% shared identity in the kinase domain between family members.

While there is little overall sequence conservation among PIPKs, there are highly conserved regions within the kinase domain that are important for catalytic activity and substrate recognition. These essential components include the G-loop, catalytic residues and the activation loop. Deletion of regions N-terminal and C-terminal to the 280 amino acid kinase domain abrogate activity, indicating the importance of these regions in maintaining structural integrity (Ishihara et al. 1998). Starting from the N-terminus, PIPK kinase domains (highlighted in yellow) (Fig. 2.1) contain the consensus sequence of GxSGS in PIPKI and a conserved I**K**, corresponding to a region found in protein kinases known as the G-loop, which consists of a glycine patch followed by a downstream lysine that mediates nucleotide binding (Anderson et al. 1999; Hanks et al. 1988; Heck et al. 2007; Ishihara et al. 1998; Rao et al. 1998; Saraste et al. 1990). Mutations in this region can inhibit kinase activity. Mutation of the murine PIPKIα Glycine 124 in the GxSGS consensus to Valine reduced kinase activity by one third, while mutating the downstream Lysine, K138, to Alanine completely abolished kinase activity (Ishihara et al. 1998). This Lysine in the I**K** sequence corresponds with Lysine 72 in Protein Kinase A, which together with Mg^{2+} coordinates the α and β phosphates in ATP. Mutation of K72 also abolishes kinase activity of PKA and mutation of the corresponding residue abolishes activity of all PIP kinases (Anderson et al. 1999; Heck et al. 2007; Iyer et al. 2005; Knighton et al. 1991a, 1991b; Rao et al. 1998; Taylor et al. 1992, 1993).

C-terminal to the G-loop are two highly conserved regions responsible for PIPK catalytic activity (highlighted in magenta) (Fig. 2.1). These are the **DLKGS** and

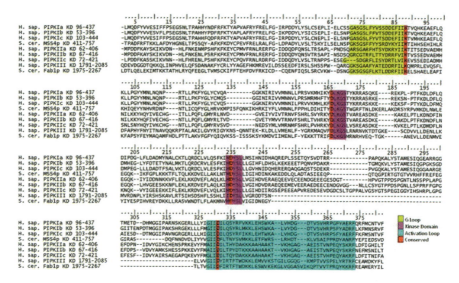

Fig. 2.1 ClustalW multiple sequence alignment for human and yeast PIPK kinase domains. Conserved regions are highlighted—G-loop (*yellow*) for ATP binding, catalytic kinase domain sequences (*Magenta*), activation loop (*cyan*). Conserved residues (*Red*) PIPKIIβ K150, D216, D278 and D369 are required for PIP kinase activity

MDYSL motifs. The **DLKGS** motif shows similarity to the **HRDLK** catalytic sequence in PKA, but is not equivalent based on structure (Anderson et al. 1999; Rao et al. 1998; Yamamoto et al. 1995). However the conserved Aspartate in the **DLKGS** motif has been shown to be required for kinase activity in PIPKIβ and PIPKIγ (Anderson et al. 1999; Narkis et al. 2007). Furthermore, human D253 N mutations in PIPKIγ that result in an **NLKGS** sequence lead to lethal contractural syndrome type 3 disease (Narkis et al. 2007). The PKA Aspartate 166 in the **HRDLK** motif functions as a catalytic base, corresponding to Aspartate 278 within the **MDYSL** motif in PIPKIIβ that superimposes on its structure (Krupa et al. 2004; Rao et al. 1998). Additionally, the catalytic Aspartate 369 in PIPKIIβ, within the (G/A)IIDIL motif, corresponds to Aspartate 184 in PKA found within a conserved **DFG** motif. This region coordinates Mg^{2+} ions and is required for kinase activity (Heck et al. 2007; Rao et al. 1998; Zheng et al. 1991). These four conserved residues (K138, D216, D278 and D369 in PIPKIIβ) in PIPKs are essential for kinase activity.

Finally, at the C-terminus of the kinase homology domain is a region of 20–30 amino acids (highlighted in cyan) (Fig. 2.1), known as the activation loop, responsible for substrate specificity and PIPK intracellular targeting (Anderson et al. 1999; Burden et al. 1999; Kunz et al. 2000, 2002; Rao et al. 1998). The activation loop is common amongst protein kinases (Cox and Taylor 1994; Hanks and Hunter 1995; Johnson et al. 1996; Rao et al. 1998) and its modification is one means to regulate these proteins. The activation loop of PIP kinases begins with the consensus GIIDIL with the aspartate conserved among other protein kinases, and ending with an RF that is conserved in PIP kinases. In experiments where the activation loop of PIPKIβ and PIPKIIβ were swapped, the specific activity of these chimeric PIP kinases matched its activation loop, for example, PIPKIβ containing the IIβ activation loop phosphorylated PI5P, while chimeric PIPKIIβ containing the Iβ activation loop phosphorylated PI4P (Kunz et al. 2000). Additionally, these domain swaps altered PIPKI targeting (Kunz et al. 2000). Furthermore, specific residues conserved within PIPKI and PIPKII are essential for specificity. PIPKIIβ A381 to E point mutations alter substrate specificity from PI5P to PI4P, while the reverse mutation, E362A, in PIPKIβ allowed for PIP_2 synthesis from PI4P and PI5P, demonstrating the importance of this Glutamate in PI4P specificity (Kunz et al. 2002). The role of the activation loop was further shown through complementation studies in yeast deficient in Mss4p. Yeast cells lacking Mss4p can be rescued by the PIPKIIβ chimera containing the Iβ specificity loop, thereby restoring type I activity, but not the PIPKIβ chimera with the IIβ loop (Kunz et al. 2000). There is also a conserved KK motif in the activation loop found among PIPKs. Mutation of this KK to NN reduced substrate affinity, while a conservative KK to RR mutation did not alter function (Kunz et al. 2000). This suggests that this KK motif is important for lipid substrate interaction, but not for recognition of specific PIP isomers.

2.2.1.2 PIP Kinase Structure

Structures have been solved for human PIPKIIβ (PDB ID: 1BO1) and PIPKIIγ (PDB ID: 2GK9), with publications focusing on PIPKIIβ (Burden et al. 1999; Kunz et al. 2000; Rao et al. 1998). The PIPKIIβ structure reveals a homodimer, with two PIPKIIβ monomers flush against one another. A set of antiparallel beta sheets and a single alpha helix form a flat face that allows for membrane association (Burden et al. 1999). Ten basic residues result in a net charge of +14 on this face; mutation of three lysines on this face to glutamates (K72/76/78) inhibited lipid binding and partially inhibited phosphoinositide binding, suggesting the basic face plays a role in membrane association but is not the only requirement for substrate binding (Burden et al. 1999). Figure 2.2a, b show that the planar face that interacts with the plasma membrane consists primarily of a beta sheet and a single helix. Individual monomers are shown in blue and green, with the basic residues found on the planar face, shown in red. Figure 2.2c shows the molecules from the side, with the planar face that interacts with the membrane facing downwards.

The conserved regions of the kinase domain described above are highlighted in Fig. 2.3. The G-loop (yellow) links two beta strands. The kinase catalytic DLKGS and MDYSL motifs (shown in magenta) are also in this pocket and the activation loop (shown in cyan) faces the membrane. The conserved catalytic residues within PIPKIIβ (K150, D216, D278 and D369) are highlighted in red. The structures of ATP and the PIPKIIβ substrate, PI5P, were modeled onto the structure (Fig. 2.3b) (Rao et al. 1998). This reveals a pocket, where the γ phosphate of ATP is oriented such that it faces the membrane, while PI5P fits such that the four hydroxyl of the myo-inositol ring faces this phosphate, consistent with PIPKIIβ function (Heck et al.

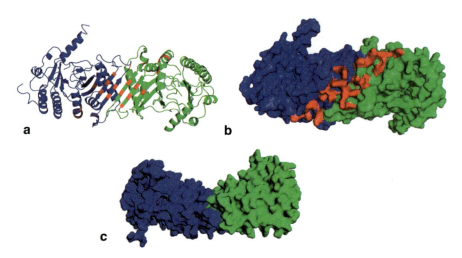

Fig. 2.2 PIPKIIβ structure. **a**, **b** and **c** Pymol rendering of the PIPKIIβ structure (PDB: 1BO1). PIPKIIβ forms a dimer (individual monomers in *blue* and *green*). a and b show the planar face that binds to the membrane, with key basic residues highlighted in *red*. c shows the side view of PIPKIIβ dimer, with membrane binding plane shown facing down

2007; Rao et al. 1998). Proposed residues necessary for ATP (133G, 136G, 139F, 148V, 150K, 201T, 203N, 204V, 205F, 278D, 282L, 368I, 369D) (Blue) and PI5P binding (134R, 218K, 224R, 239K, 278D, 372T) (Red) in the model are shown in Fig. 2.3b (Rao et al. 1998). The planar face of the dimer allows for the pocket to interact with the phosphoinositide head group and allow for catalytic activity without interfering with membrane structure.

2.2.2 Enzymology of PIPKs

The basic reaction catalyzed by PIP kinases is the ATP-dependent phosphorylation of phosphatidylinositol monophosphates to produce phosphatidylinositol bisphosphates (Heck et al. 2007). The substrate specificity of PIPKI, II and III was discovered through kinase activity on purified phosphatidylinositol phosphates from biological sources and synthetic phosphoinositides. These studies identified the primary substrate activity of PI4P for PIPKI, PI5P for PIPKII and PI3P for PIPKIII (Bazenet et al. 1990; Cabezas et al. 2006; Rameh et al. 1997; Sbrissa et al. 1999; Zhang et al. 1997). However, the PIP kinases can also utilize other PIP substrates with lower activity. For example, PIPKIs can phosphorylate PI3P to produce $PI3,4P_2$, and subsequently $PI3,4,5P_3$, or $PI3,5P_2$ (Tolias et al. 1998; Zhang et al. 1997). While the PIPKI affinity for PI3P is similar to PI4P, the reaction rate is higher for PI4P (Zhang et al. 1997). Similarly, PIPKII can also catalyze the phosphorylation of PI3P to $PI3,4P_2$ but is not the preferred substrate (Rameh et al. 1997). Finally, PIPKIII has been reported to generate PI5P from PI (Sbrissa et al. 1999; Shisheva et al. 1999). Though the enzymes can make multiple products, the *in vivo* importance of enzyme specificity can be observed through yeast complementation studies. Yeast deficient

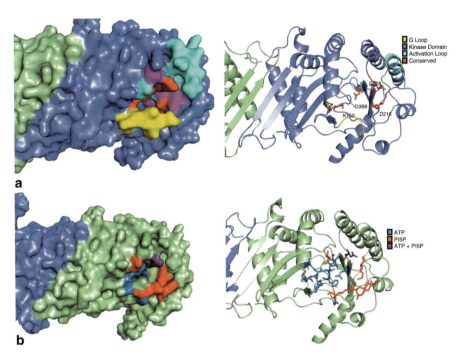

Fig. 2.3 PIPKIIβ structure. **a** and **b** Pymol rendering of PIPKIIβ highlighting conserved regions—G loop (*yellow*), catalytic kinase domain sequences (*magenta*) and activation loop (*cyan*). Conserved residues required for PIP kinase activity are shown in *red* (PIPKIIβ K150, D216, D278 and D369). Highlighting residues believed to mediate ATP (*blue*) and PI5P (*red*) binding on PIPKIIβ structure

in MSS4p can be rescued using human PIPKI but not PIPKII (Homma et al. 1998), while yeast deficient in Fab1p can only be rescued by human PIPKIII (McEwen et al. 1999). However, the function of substrate diversity for each kinase *in vivo* has not been examined closely.

The activity of PIP kinases is regulated by a variety of stimuli, including lipids, proteins and post-translational modifications. Following their initial purification, phosphatidic acid was shown to enhance PIPKI kinase activity (Jenkins et al. 1994; Moritz et al. 1992a, b). More recently, phosphorylation has been shown to regulate PIPKs. Activation of Protein Kinase A (PKA) resulted in phosphorylation of PIPKIα on Serine 214 in the kinase domain, reducing activity, and activation of PKC via LPA treatment reduced PIPKIα phosphorylation and enhanced kinase activity (Park et al. 2001). Also, PKD can phosphorylate the activation loop of PIPKIIα at Thr 376, and mutation of this site reduced kinase activity (Hinchliffe and Irvine 2006). Phosphorylation of PIPKs does not always affect kinase activity as Casein Kinase (CK) II phosphorylation of PIPKIIα at Ser304 affects localization, but not kinase activity (Hinchliffe et al. 1999a, b). PIPKI and PIPKIII also have protein kinase activity, as these enzymes can auto-phosphorylate, reducing enzyme activity (Itoh et al. 2000; Sbrissa et al. 2000). Protein-protein interactions are also critical regulators

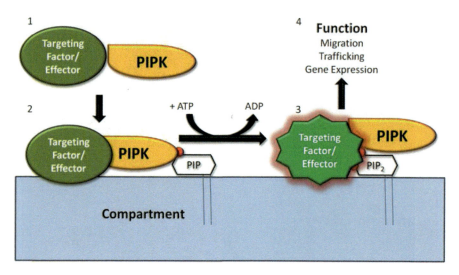

Fig. 2.4 PIPKs are recruited to intracellular compartments by PIP_2-effector proteins. *1* PIPKs interact with proteins that mediate their localization within the cell. *2* Once at a specific site, PIPKs generate PIP_2. *3* The interacting proteins that targeted PIPKs to specific sites are then modulated by the newly synthesized PIP_2. *4* The activated PIPK effectors regulate various cellular processes

of PIP kinase activity. Interactions with GTPases and specific proteins such as Talin and adaptor protein (AP) complexes have been shown to enhance PIPK activity (Di Paolo et al. 2002; Krauss et al. 2006; Weernink et al. 2004). The roles of phosphorylation and protein-protein interactors in regulating PIPK activity will be described in detail later.

The unique sequences in PIPKs allow for specific interactions with proteins that serve to target PIPKs to specific compartments. Often, these same interacting proteins are PIP_2 effectors (Heck et al. 2007). These protein-protein interactions target PIP kinases to specific compartments to generate a local pool of PIP_2, which can then regulate the activity of its effectors. This hypothesis is illustrated in Fig. 2.4. Multiple examples of regulation of PIP kinase targeted PIP_2 production have been demonstrated with diverse roles in cell-adhesion, migration, protein trafficking and nuclear signaling.

2.3 Membrane Associated PIPKs Drive Cell Migration and Vesicle Trafficking

2.3.1 PIPKs Help Regulate Directional Migration

Cell migration is a critical process for normal development, wound healing, cell survival and immunological responses, but can also have deleterious effects as evidenced by its role seen in metastatic tumor formation. Cell migration requires a tight coordination of many molecular biological processes, including the establishment of cell polarity, the organized formation and turnover of adhesive structures and a regulated control of dynamic cytoskeletal rearrangements (Fig. 2.5a). Regulation of these processes produces the well-characterized migration pattern defined by a continuous cycling between membrane protrusions at the leading cell front, anchoring of the newly formed front, the rolling of the cell body forward, and the release of posterior adhesions to retract the trailing cell membrane. Many signal transduction pathways can influence these processes in large part due to their ability to regulate the spatio-temporal generation of the key lipid messenger, PI4,5P$_2$, herein referred to as PIP$_2$.

2.3.1.1 PIPKs Regulate PIP$_2$ Synthesis at the Leading Edge to Drive Membrane Protrusion

In 1985 Lassing and Lindberg discovered that PIP$_2$ specifically interacts with profilin, thus starting a long and multifaceted look at the role of phosphoinositides in migration and invasion (Lassing and Lindberg 1985). Since then, PIP$_2$ has been shown to directly bind many actin-associated proteins that both directly and indirectly regulate the cellular cytoskeletal machinery (Yin and Janmey 2003). Increased PIP$_2$ levels promote actin polymerization, whereas a decrease results in actin disassembly. PIP$_2$ controls actin polymerization at many levels, including inactivating proteins involved in actin severing or depolymerization as well as enhancing actin polymerization, branching and bundling (Sechi and Wehland 2000; Takenawa and Itoh 2001). During cell migration, rampant F-actin synthesis at the migrating front pushes the cell membrane forward (Fig. 2.5) resulting in the formation of subcellular structures including membrane ruffles, lamellipodia, microvilli, motile actin comets, filopodia, microspikes and dorsal ruffles (Janmey and Lindberg 2004; Nicholson-Dykstra et al. 2005). All of these membranous protrusions have been observed in different cell types overexpressing type I PIPKs, a family of kinases responsible for the vast majority of PIP$_2$ produced in a cell (Honda et al. 1999; Matsui et al. 1999; Rozelle et al. 2000; Shibasaki et al. 1997; Yamamoto et al. 2001). Using GFP-tagged PH domains or antibodies that specifically recognize PIP$_2$, this lipid messenger was found to concentrate in dynamic, actin-rich regions of the cell (Tall et al. 2000).

PIP$_2$ plays a multifaceted role in regulating F-actin dynamics at these protrusive cell fronts. PIP$_2$ directly binds gelsolin, a capping protein found at the barbed end of F-actin filaments, blocking its association with F-actin thereby inducing a rapid and local actin polymerization (Niggli 2005). The association of PIP$_2$ with profilin frees up actin monomers so that they can be incorporated into the growing filaments (Lambrechts et al. 2002; Skare and Karlsson 2002). Using *in vitro* actin co-sedimentation assays, PIP$_2$ was shown to attenuate the association between F-actin and cofilin, an actin-severing and depolymerizing factor (Gorbatyuk et al. 2006). However, cofilin activity enhances migration (Sidani et al. 2007). The hydrolysis of PIP$_2$ by Phospholipase C (PLC) releases and activates a distinct membrane-bound pool of cofilin, triggering the formation of membrane protrusions and the establishment of a leading edge (van Rheenen et al. 2007). A localized activation of cofilin enhances the

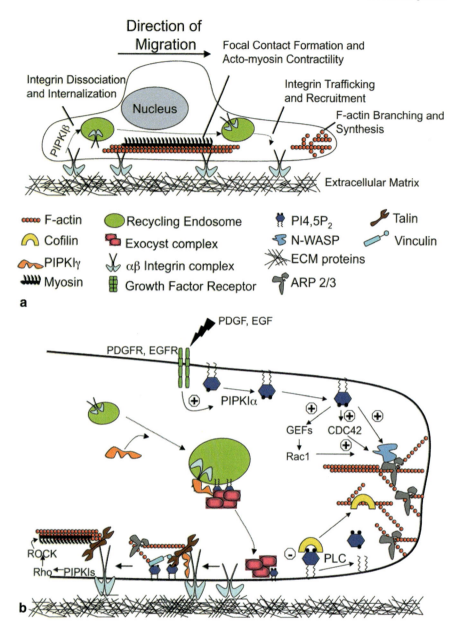

Fig. 2.5 PIPKs are involved in multiple processes associated with cell migration. **a** PIPKs are integral components at each step in the migration process. Cell migration involves a continuous cycle of membrane protusion, adhesion formation and maturation at the leading edge, adhesion disassembly (mediated by PIPKIβ) at the trailing edge and contractile forces to pull the cell body forward. **b** At the leading edge, PIPKIα regulates Rac and Cdc42 to initiate F-actin synthesis. PIPKIγ isoforms are involved in the trafficking of proteins to the leading edge, the assembly of proteins that mediate adhesions and the maturation of these adhesions to induce acto-myosin contractions

number of F-actin barbed ends, setting a site for membrane protrusion, however, curtailing that activity is also required to promote enhanced filament formation and actin bundling required to push the membrane forward. The dynamic interplay between actin severing and F-actin synthesis is further observed in the association of PIP_2 with villin. PIP_2 binding to villin differentially regulates the protein's activities, inhibiting its actin severing and capping activity while enhancing its ability to bundle actin (Kumar and Khurana 2004; Kumar et al. 2004).

While uncovering roles for PIP_2 in regulating actin-associated proteins has been widely successful, defining the specific proteins involved in its synthesis during cell migration has been less concrete. PIPKIα is recruited to membrane ruffles and increases PIP_2 levels in response to EGF and PDGF (Fig. 2.5b) (Doughman et al. 2003b; Honda et al. 1999). Depletion of PIPKIα in mouse embryo fibroblasts blocks cell migration, and PIPKIα-deficient mast cells display atypical actin organization (Kisseleva et al. 2005; Sasaki et al. 2005). Both PIPKIα and PIPKIβ have been shown to directly bind Rac1, a member of the Rho family of small GTPases that help regulate actin-cytoskeletal dynamics (Hall and Nobes 2000), and are thus thought to be Rac1 effectors (Weernink et al. 2004). PIPKIα is targeted to lamellipodia and its presence within signaling cascades that include Rho-GTPases is widely known (Doughman et al. 2003b). In addition, the uncapping of actin filaments in cells expressing a constitutively active mutant of Rac1 could be blocked by the overexpression of a dominant negative mutant of PIPKIβ but not a kinase-dead mutant of PIPKIα (Tolias et al. 1998, 2000). However, conflicting evidence fails to define whether PIPKIα is upstream or downstream of these proteins. Mutations in the PIPKIα binding region on Rac1 prevents translocation to the plasma membrane and fails to activate downstream signaling responses to integrin-mediated adhesion (Del Pozo et al. 2002). A more recent report shows that PIPKIα regulates Rac1 activation and downstream F-actin assembly, focal adhesion formation and directed migration (Chao et al.). Other reports identify PIPKIα as an effector of Rho-GTPases downstream of G protein-coupled and growth factor receptor activation (Chatah and Abrams 2001). GTPase activation is facilitated by guanine-nucleotide exchange factors (GEFs) that are often activated by phosphoinositides including PIP_2 (Di Paolo and De Camilli 2006). However, effectors of GTPases are often the PIPKs that generate phosphoinositides (Di Paolo and De Camilli 2006). This positive feedback loop results in the generation of membrane domains enriched for specific phosphoinosites. Oude Weernink et al. have shown that all type I PIPK isoforms can associate with RhoA and Rac1 but not Cdc42, another Rho family GTPase (Weernink et al. 2004). However, Cdc42 can stimulate PIP5K activity (Weernink et al. 2004). The concomitant stimulation of neuronal Wiskott-Aldrich syndrome protein (N-WASP) by Cdc42 and PIP_2 increases actin filament nucleation and F-actin synthesis via activation of the actin-related protein 2/3 complex (Arp2/3) (Prehoda et al. 2000).

2.3.1.2 PIPKIγ Regulates the Formation and Maturation of Integrin-mediated Contacts

Following membrane protrusion at the migrating front, the cell anchors this region to the underlying matrix via the formation of specialized adhesive structures, including focal adhesions, focal complexes, podosomes, invadopodia and hemidesmosomes (Fig. 2.5) (Broussard et al. 2008; Webb et al. 2002). Cell migration requires a tight coordination of the formation/turnover of these structures (Webb et al. 2002). Focal adhesions are large, multi-protein complexes that physically link the cellular cytoskeleton to the extra-cellular matrix (ECM). Their formation generally involves integrin molecules and is dependent on PIP_2 (Ling et al. 2002). Many of the proteins required for focal contact formation, such as α-actinin, vinculin, talin and ezrin, are positively regulated by PIP_2 (Ling et al. 2006). PIPKIγi2 is required for EGF-stimulated directional migration and is phosphorylated by the EGFR and Src within its C-terminus (Sun et al. 2007). PIPKIγi2 phosphorylation modulates an association with talin, facilitating its assembly into dynamic adhesive complexes at the leading edge (Ling et al. 2002, 2003). Talin is a scaffolding protein that can bind both β integrin, resulting in its activation and engagement to the ECM, and F-actin or vinculin at nascent adhesions (Nayal et al. 2004). Furthermore, talin binds PIP_2 and this interaction is required for targeting talin to nascent contacts and enhancing its association with β integrin (Ling et al. 2002; Martel et al. 2001). Expression of a PIP_2-specific PH domain blocked the membrane localization of talin essential for focal adhesion maintenance (Martel et al. 2001). In addition, a local enrichment of PIP_2 enhances vinculin assembly into newly formed focal contacts (Chandrasekar et al. 2005). PIP_2 binding to vinculin disrupts its auto-inhibition, exposing protein binding domains that mediate actin and talin association (Bakolitsa et al. 2004). Additionally, PIP_2 binding to vinculin exposes a proline rich domain resulting in a vinculin/VASP association that recruits profilin/G-actin or Arp2/3, facilitating their assembly into nascent contacts and enhancing F-actin nucleation and synthesis (DeMali et al. 2002; Millard et al. 2004). This suggests that PIPKIγi2 is recruited to newly forming contacts and the generation of PIP_2 aides their maturation. The result is a rigid adhesive complex that strengthens the cell infrastructure, a critical step in migration as contractile forces use these anchors to pull a cell in the direction of migration.

2.3.1.3 Trafficking of Integrin-containing Vesicles to the Leading Edge Is Mediated by PIPKIγ and PIP_2

Coordinated cell migration requires the trafficking of newly synthesized and endocytosed proteins to the leading edge (Caswell and Norman 2006; Prigozhina and Waterman-Storer 2004; Schmoranzer et al. 2003). Impairment of the endocytosis, trafficking and exocytosis of integrins drastically impairs the establishment of polarity and directionality of cell migration (Tayeb et al. 2005; Zovein et al.). Furthermore, PIP_2 generation is required for vesicular trafficking (Wenk and De Camilli 2004).

PIPKIγ knockout studies have revealed vesicle trafficking defects both at the plasma membrane and in endosome-like structures in neuronal cells (Di Paolo et al. 2004; Wenk et al. 2001). Furthermore, several studies define a pivotal role for the exocyst protein complex in the polarized trafficking of transmembrane proteins during cell polarization and migration (Letinic et al. 2009). The exocyst complex consists of eight subunits and facilitates the exocytosis of post-Golgi and endocytic recycling endosomes (including integrin-containing vesicles) to regions of rapid membrane expansion, such as occurs in migrating cells (Yeaman et al. 2001). Depletion of exocyst complex components inhibits wound healing and migration (Rosse et al. 2006). Rab11 and Arf6, two small GTPases that regulate PIPKs, also regulate the exocyst complex and integrin trafficking (Oztan et al. 2007). Furthermore, at least two exocyst subunits, Sec3 and Exo70, bind PIP_2 via conserved basic residues in their C-termini (Liu et al. 2007). PIPKIγi2 is able to associate with the exocyst complex and regulate the trafficking of integrin molecules in migrating cells (unpublished data). PIPKIγ could play a role in the assembly of the exocyst complex onto vesicles (Fig. 2.5). In epithelial cells, PIPKIγi2 is required for the basolateral sorting and endocytosis of E-cadherin via the AP1B and AP2 adaptor protein (AP) complexes respectively (Ling et al. 2007). This occurs by a novel mechanism where PIPKIγ interacts directly with E-cadherin at regions within the PIP kinase domain and then recruits specific AP complex subunits through interactions within its unique C-terminus. This may be a mechanism for regulation of both cell adhesion and migration. In addition, a localized generation of PIP_2 may be important for vesicle fusion at the plasma membrane (Fig. 2.5). PIP_2 and type PIPKI activity have been shown to be critical during Ca^{2+}-activated secretion of large dense-core vesicles (Hay et al. 1995). A critical player in this process, synaptotagmin, binds PIP_2 via C2 domains and mediates membrane fusion (Bai et al. 2004). Furthermore, Gong et al. showed that chromaffin cells taken from the adrenal gland of PIPKIγ –/– mice displayed a defect in the readily releasable pool of vesicles containing catecholamine and a delay in fusion pore expansion (Gong et al. 2005). Both Sec3 and Exo70 were shown to bind PIP_2 at the plasma membrane in yeast, a critical interaction for normal exocytosis and maintenance of cell morphology (He et al. 2007; Zhang et al. 2008).

2.3.1.4 Rear Retraction of the Cell Requires the Dissociation and Internalization of Integrin and Acto-myosin Contractility

During cell migration, new adhesions are formed along the protruding edge. Their maturation develops a focal point used to generate of forces required to push the membrane in the direction of migration. In addition, the maturation of focal adhesions stabilizes forces used to pull the rear of the cell forward. Overexpression of type I PIPKs has been shown to induce the formation of F-actin stress fibers, though the formation of actin-based protrusions is inhibited at the cell rear (Shibasaki et al. 1997; Yamamoto et al. 2001). Instead, rear retraction occurs at least in part through RhoA-regulated acto-myosin contractility (Yoshinaga-Ohara et al. 2002). In fibroblasts, integrin-mediated adhesion to fibronectin results in a rapid increase in PIP_2 synthesis and previous work has shown that PIPKIγ interacts with RhoA resulting in enhanced PIP_2 production (Fig. 2.5) (Weernink et al. 2004). Lokuta et al. show that PIPKIγi2 is enriched in the uropod of directionally migrating neutrophils (Lokuta et al. 2007). Overexpression of a kinase-inactive mutant of PIPKIγi2 compromised uropod formation and retraction of the cell rear (Lokuta et al. 2007). When bound to GTP, RhoA can signal through ROCK to initiate acto-myosin contractility via its ability to enhance the phosphorylation of the myosin light chain both directly and indirectly through MLCK (Schramp et al. 2008). ROCK also plays a critical role in the RhoA-mediated activation of PIP5 K to synthesize PIP_2, a signaling pathway utilized during neurite remodeling (van Horck et al. 2002; Yamazaki et al. 2002).

In addition, focal contacts at the rear of the cell need to be disassembled. A prevailing theory of cell migration adheres to the idea that disassembled adhesions are internalized via endocytosis and recycled to the protruding edge where they once again are utilized in the formation of nascent adhesions (Pierini et al. 2000). The protease calpain, FAK, phosphatases and kinases that regulate FAK, microtubules and dynamin 2 have all been shown to regulate focal adhesion disassembly (Broussard et al. 2008; Burridge et al. 2006; Ezratty et al. 2005; Franco et al. 2004). All three isoforms of type I PIPKs have been implicated in clathrin-mediated endocytosis (Franco and Huttenlocher 2005). Recently, Chao et al. showed that PIPKIβ is required for b1 integrin uptake and adhesion turnover during cell migration (Chao et al.). Localized PIPK recruitment and PIP_2 formation can recruit molecules that directly regulate endocytosis to the plasma membrane and modulate proteins involved in the inactivation of focal adhesions (Fig. 2.5). PIP_2 can directly bind many endocytic clathrin adaptors, including AP-2, AP180 and epsin (Qualmann et al. 2000). Clathrin adaptor components and dynamin become enriched at focal adhesion sites prior to internalization (Chao and Kunz 2009; Ezratty et al. 2005). Loss of PIPKIβ blocked the assembly of clathrin adaptors at adhesion plaques and prevented an association between dynamin and FAK, an interaction required for endocytosis (Chao et al.). Interestingly, in addition to its role in promoting F-actin synthesis at nascent contacts, a role for vinculin in promoting focal adhesion disassembly is also emerging. The introduction of a vinculin mutant unable to bind PIP_2 into mouse melanoma cells repressed its translocation from the membrane to the cytosol and blocked the disassembly of focal adhesions, cell spreading and migration (Chandrasekar et al. 2005). Calpain is another PIP_2 regulated protein that directly mediates focal adhesion disassembly (Franco and Huttenlocher 2005; Franco et al. 2004). Calpain is a protease that can cleave adhesion proteins including talin, vinculin, paxillin and β integrin and is required for focal adhesion turnover and cell migration (Franco and Huttenlocher 2005; Franco et al. 2004). However, the activation cues to signal these processes remain to be fully uncovered.

PIPKs are proving to be integral components in helping regulate the many processes required for cell migration. The existence of multiple PIPK isoforms may enable cells to coordinate PIP_2 synthesis at specific sites to regulate the forces that induce membrane protrusion, the trafficking events that localize proteins at the leading edge, the assembly, maturation and endocytosis of adhesions and the forces generated to retract the trailing rear of the cell.

2.3.2 *PIP Kinases in Adaptor Protein Complex Assembly and Protein Trafficking*

In the past decade, considerable evidence has demonstrated that PIPKs and their lipid biproducts, PIP_2, are actively involved in various protein trafficking processes.

Table 2.1 Adaptor complexes consist of different subunits

AP complex	Subunits
AP-1A	β1, γ, σ1, μ1A
AP-1B*	β1, γ, σ1, μ1B
AP-2	β2, α, σ2, μ2
AP-3A	β3A, δ, σ3, μ3A
AP-3B*	β3B, δ, σ3, μ3B
AP-4	β4, ε, σ4, μ4

*Tissue specific AP complexes

Among these, the most widely characterized is the role of PIPKs in clathrin-mediated endocytosis and endosomal recycling, both of which are dependent on AP complexes. PIPKs directly bind AP complexes and their ability to synthesize a localized pool of PIP_2 results in AP complex assembly and activation. Here we will summarize how PIPKs regulate different AP complexes to modulate distinct protein trafficking processes.

2.3.2.1 AP Complexes

AP complexes are indispensible components of clathrin-coated vesicles in the endocytic and post-Golgi trafficking pathways. AP complexes not only link clathrin to the membrane, but also determine the specificity of cargo selection at various membrane compartments (Nakatsu and Ohno 2003; Ohno 2006). To date, four ubiquitously expressed AP complexes (AP-1A, AP-2, AP-3A, and AP-4) have been identified (Dell'Angelica et al. 1999; Nakatsu and Ohno 2003), with two additional complexes, AP-1B and AP-3B, found specifically in epithelial and neuronal cells respectively (Fölsch et al. 1999; Nakatsu and Ohno 2003; Ohno 2006). Each AP complex consists of two large adaptin subunits, one derived from the β-class (β1-4) and one of γ, α, δ or ε, together with one medium subunit (μ1-4) and one small subunit (σ1-4) (Table 2.1) (Jackson 1998; Ohno 2006). In each AP complex, the large adaptin subunits are responsible for membrane association through PIP_2 (Gaidarov et al. 1999; Ohno 2006; Rohde et al. 2002). The μ subunit determines specific cargo selection, while the β subunit is required for clathrin recruitment (Ohno 2006). Therefore, different AP complexes direct distinct cargo proteins at the membrane surface of vesicles (Fig. 2.6).

2.3.2.2 PIP Kinases in Regulation of AP Complex Assembly

Recent studies have demonstrated that PIP kinases are involved in clathrin-mediated vesicle trafficking at the plasma membrane. PIPKs have been found to directly bind AP complexes, regulating AP complex assembly and the formation of clathrin-coated pits. Here we will discuss how PIPKs regulate protein trafficking at the plasma membrane through specific interactions with AP complexes.

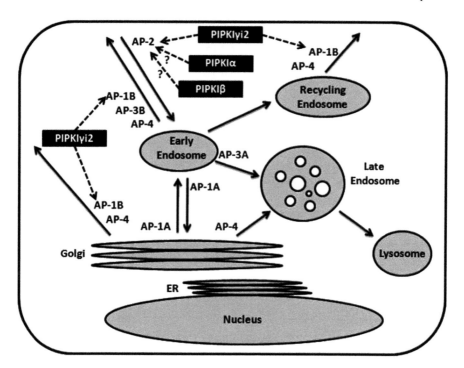

Fig. 2.6 AP complexes mediate endocytic and post-Golgi trafficking. Different AP complexes are involved in clathrin-mediated protein trafficking between different compartments. PIPKIγi2 interacts with AP-1B and AP-2 complexes, modulating basolateral transport and endocytosis

E-cadherin maintains epithelial cell morphology by forming adherens junctions (AJs) with adjacent cells. The amount of E-cadherin on the plasma membrane directly determines the strength of AJs (Yap et al. 2007). In addition to E-cadherin gene expression, post-translational regulations such as exocytosis, endocytosis, recycling and lysosomal degradation have also been implicated as important factors affecting the stability of AJs (Schill and Anderson 2009a). Furthermore, PIP_2 has been shown to be an essential regulator of different E-cadherin trafficking processes (Schill and Anderson 2009a). PIPKIγ directly binds all type I classical cadherins and colocalizes with E-cadherin in epithelial cells (Akiyama et al. 2005; Ling et al. 2007). Specifically, PIPKIγi2 has been found to regulate AJ assembly by modulating E-cadherin trafficking between endosomes and the plasma membrane (Ling et al. 2007). Expression of wild-type PIPKIγi2 promotes both internalization and recycling of E-cadherin, while a kinase dead mutant of PIPKIγi2 inhibits both of these processes (Ling et al. 2007), indicating that phosphoinositide generation is required for E-cadherin trafficking.

PIPKIγi2 can directly associate with both E-cadherin and the μ1B subunit of AP-1B (Ling et al. 2007), suggesting that it could act as a scaffold to link these proteins (Fig. 2.7). Within the unique PIPKIγi2 C-terminus is the sequence YSPL, Yxxϕ tyrosine-sorting motif (x represents any and ϕ represents a bulky hydrophobic amino

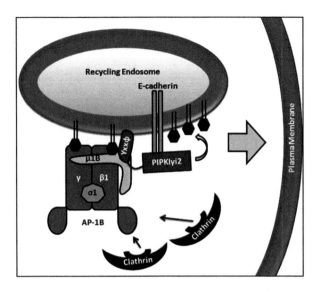

Fig. 2.7 PIPKIγi2/AP-1B interaction regulates E-cadherin trafficking. PIPKIγi2 interacts with the μ1B subunit of AP-1B through the Yxxφ motif in the C-terminus and recruits E-cadherin via the kinase domain. Additionally, PIPKIγi2-generated PIP_2 assembles PIP_2-interacting proteins of the AP-1B and clathrin vesicle complexes at specific sites

acid) recognized by the μ1B subunit of AP-1B (Bairstow et al. 2006; Bonifacino and Traub 2003; Ling et al. 2007; Sugimoto et al. 2002). It has been demonstrated that the PIPKIγi2 YSPL sequence is responsible for the μ1B interaction, as a mutation of tyrosine to a phenylalanine disrupted their interaction and subsequently decreased the association between E-cadherin and μ1B (Ling et al. 2007). Interestingly, PIPKIγi2 also binds to the β2 subunit of AP-2 (Nakano-Kobayashi et al. 2007). Whether the PIPKIγi2/AP-2 interaction is also involved in E-cadherin trafficking is not clear.

Current evidence supports a model where PIPKIγi2 acts as both a scaffold and direct regulator of complex assembly through the generation of PIP_2 during E-cadherin trafficking to the plasma membrane (Fig. 2.7). The μ subunit of the AP complex is the predominate molecule mediating selection through a direct interaction with cargo proteins (Ohno 2006; Sugimoto et al. 2002). However, PIPKIγi2 could function as a scaffold protein that links E-cadherin to the μ1B subunit of the AP-1B complex. At the same time, PIPKIγi2 also functions by temporally and specifically producing PIP_2, which in turn recruits and regulates AP complex proteins and other components of the clathrin-coated vesicle. Certain patients suffering from hereditary diffuse gastric cancer have been found to contain a mutation within the PIPKIγi2-binding site in the C terminus of E-cadherin (Yabuta et al. 2002; Ling et al. 2007). The mutated E-cadherin shows inhibited interaction with PIPKIγi2, and localizes mostly to cytoplasmic compartments rather than the plasma membrane, losing the ability to form AJs (Ling et al. 2007; Suriano et al. 2003; Yabuta et al. 2002). These results underscore the role of PIPKIγi2 in E-cadherin trafficking to the plasma membrane.

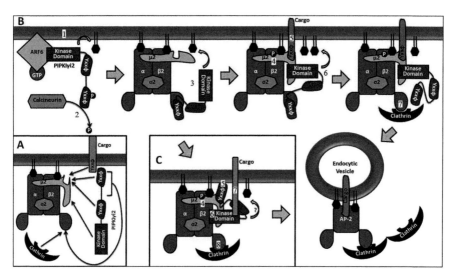

Fig. 2.8 PIPKIγi2/AP2 interaction modulates endocytosis. **a** PIPKIγi2 directly interacts with the AP-2 complex through multiple sites. **b** PIPKIγi2 facilitates clathrin-AP-2-mediated endocytosis of Yxxϕ-motif-containing cargos. Activated Arf6 recruits and activates PIPKIγi2, leading to PI4,5P$_2$ generation, which in turn recruits AP-2 (*1*). Dephosphorylation of PIPKIγi2 by calcineurin or another upstream molecule (*2*) induces PIPKIγi2-binding to the Ear domain of the β2 subunit of AP-2, further activating PIPKIγi2 (*3*). Conformational change in μ2 caused by Thr156 phosphorylation (*4*) promotes cargo binding (*5*) to μ2 through Yxxϕ motif with the help of PI4,5P$_2$. This results in μ2 binding to and activating the PIPKIγi2 kinase domain (*6*). Clathrin-engagement to AP-2 disrupts PIPKIγi2 association with the β2 subunit (*7*). PI4,5P$_2$ accumulation in step 1, 3, and 6 recruits proteins involved in clathrin vesicle formation. PIPKIγi2 is removed from AP-2 by phosphorylation at Ser645 or other unidentified mechanisms before maturation of the clathrin-coated vesicle (*8*). **c** PIPKIγi2 may enhance clathrin-AP-2-mediated endocytosis of Yxxϕ-independent cargos. Conformational change in μ2 caused by Thr156 phosphorylation (*4*) promotes PIPKIγi2 association with μ2 through one of the two Yxxϕ motifs in the C-terminus of PIPKIγi2 (*5*), which may in turn enhance the interaction between PIPKIγi2 kinase domain and the μ2 subunit (*6*). Yxxϕ engagement to μ2 may allow cargo selection through Yxxϕ-independent mechanisms (*7*). Clathrin-binding to β2 (*8*) might facilitate these processes

During clathrin-mediated endocytosis, PIPKIγi2 regulates AP-2 complex assembly by directly binding multiple subunits in the AP-2 complex (Fig. 2.8a). The kinase domain of PIPKIγi2 binds to the μ2 subunit, an association that does not block cargo engagement to μ2 (Krauss et al. 2006). In this situation, cargo proteins bind to μ2 through tyrosine or dileucine sorting motifs. This interaction can then potentially activate PIPKIγi2 activity (Kahlfeldt et al. 2010; Krauss et al. 2006). In addition to this, the tyrosine-based sorting motif (^{641}SWVYSPL^{647}) within the PIPKIγi2 specific C-terminal tail has also been found to directly bind to the μ2 subunit of AP-2 (Bairstow et al. 2006). A recent study discovered another tyrosine-based sorting motif (^{495}RSYPTLED^{502}) within the C terminus of PIPKIγi2 that can bind to the μ2 subunit (Kahlfeldt et al. 2010). Moreover, the C-terminal tail of PIPKIγi2 can also interact with the β2 subunit of AP-2, enhancing PIPKIγi2 activity (Kahlfeldt et al. 2010;

Nakano-Kobayashi et al. 2007; Thieman et al. 2009). Where PIPKIγi2 interacts with the AP-2 complex is believed to mediate the formation of different clathrin-coated vesicles that regulate the trafficking of various cargo proteins (Kahlfeldt et al. 2010; Kwiatkowska 2010).

The PIPKIγi2/AP-2 interaction is subject to regulation by phosphorylation of the PIPKIγi2 C-terminus. Src-mediated phosphorylation of PIPKIγi2 at Tyr644 (Tyr 649 in the human isoform) inhibits its interaction with the μ2 subunit (Bairstow et al. 2006; Ling et al. 2003) and based on crystallographic data could impair PIPKIγi2 binding to the β2 subunit (Kahlfeldt et al. 2010). In addition, phosphorylation of PIPKIγi2 at Ser645 (Ser650 in the human isoform) by cyclin-dependent kinase-5 (Cdk5) diminishes PIPKIγi2 interaction with the β2 subunit (Nakano-Kobayashi et al. 2007; Thieman et al. 2009), while dephosphorylation of PIPKIγi2, possibly by calcineurin upon plasma membrane depolarization, enhances this binding affinity in clathrin-mediated endocytosis at the presynapse (Nakano-Kobayashi et al. 2007). Interestingly, it has been shown that phosphorylation of either Tyr644 or Ser645 inhibits the phosphorylation of the other (Lee et al. 2005). Therefore, dephosphorylation of both Tyr644 and Ser645 promotes PIPKIγi2 interaction with both β2 and μ2, while Tyr644 phosphorylation alone inhibits this interaction and Ser645 phosphorylation alone might increase PIPKIγi2 binding to the μ2 subunit. As clathrin competes with PIPKIγi2 for the same binding site in the β2 subunit (Thieman et al. 2009), a PIPKIγi2 binding switch from β2 to μ2 via phosphorylation changes of Tyr644 and Ser645 might occur during AP-2 complex assembly (Kwiatkowska 2010).

PIPKIγi2-mediated PIP_2 generation recruits AP-2 complex components (Fig. 2.8b) (Höning et al. 2005). Dephosphorylation of PIPKIγi2 by upstream signals triggers its binding to the Ear domain of the β2 subunit, which activates PIPKIγi2 to produce PIP_2. In the presence of PIP_2, phosphorylation of μ2 subunit at Thr156 causes a conformational change that favors cargo engagement through tyrosine sorting motifs (Höning et al. 2005; Olusanya et al. 2001). Although cargo association with μ2 subunit via acidic dileucine motifs is not enhanced by the Thr156-phosphorylation-mediated conformational change, the μ2 subunit has the highest affinity towards both cargo signals when PIP_2 is present (Höning et al. 2005). Cargo engagement induces μ2 to interact with the kinase core domain of PIPKIγi2 enhancing its activity. Dynamic PIP_2 production recruits other proteins, such as AP180, epsin and dynamin-2, to facilitate the formation of the clathrin-coated vesicles during endocytosis. It is noteworthy that although PIPKIγi2 plays important roles in clathrin-AP-2 complex assembly, it is not enriched in clathrin-coated vesicles (Thieman et al. 2009; Wenk et al. 2001), suggesting that PIPKIγi2 functions in the dynamic assembly of AP-2 complex and clathrin-coated pits, but it is not a component of mature vesicles. Clathrin binding to the β2 subunit has been found to eliminate PIPKIγi2 association with β2 through the C-terminus (Thieman et al. 2009), yet whether this removes PIPKIγi2 from the AP complex or only causes a binding switch of PIPKIγi2 from β2 to μ2 still needs to be determined.

An alternative hypothesis is that PIPKIγi2 Yxxφ motifs associates with μ2 like a cargo protein (Kahlfeldt et al. 2010; Kwiatkowska 2010) (Fig. 2.8c). This alternative Yxxφ/μ2 interaction might provide a mechanism for Yxxφ sorting motif-independent cargo selections. Yxxφ/μ2 interactions have been suggested to facilitate μ2 interactions with the kinase core domain of PIPKIγi2 and activate PIPKIγi2 (Kahlfeldt et al.

2010; Kwiatkowska 2010). Kahlfeldt et al. showed that PIPKIγi2 activity could be increased when purified μ2 and the C-tail of PIPKIγi2 were incubated with cell lysates containing overexpressed PIPKIγi2. However, it is possible that the PIPKIγi2 C-terminus mimicks the effects of a cargo protein like EGFR, as they did not see any increase in PIPKIγi2 activity when the lysates were incubated with μ2 alone (Kahlfeldt et al. 2010). Although evidence shows that PIPKIγi2 can interact with μ2 through either the kinase core domain or through one of the two Yxxϕ motifs (Bairstow et al. 2006; Kahlfeldt et al. 2010), whether the μ2-PIPKIγi2 complex alone can lead to PIP kinase activation *in vivo* or whether inhibiting the μ2 association with the PIPKIγi2 C-terminus impairs PIPKIγi2 functions in AP-2 complex assembly is not clear. Future studies are needed to understand how the interaction between μ2 and the Yxxϕ motifs in the C-terminus of PIPKIγi2 correlates with Yxxϕ sorting motif-independent cargo selections.

2.4 Nuclear Localized PIPKs Regulate Gene Expression and mRNA Processing

2.4.1 Nuclear PIP Kinases and Phosphoinositides

Much as how the discovery of the phosphatidylinositol (PI) cycle became fundamental to our understanding of signal transduction and overall cell biology (Gurr et al. 1963; Hokin and Hokin 1953; Kleinig 1970), the discovery of nuclear phospholipid signaling has revolutionized our view of the processes regulated by phospholipids (Cocco et al. 1987, 1988). Still, nuclear phospholipid signaling is fulfilled by lipid second messengers and requires an array of coordinated activities from lipid kinases, phosphatases, and effectors. However, the lipid kinases within the nucleus appear to be more selective, in that not all identical lipid kinases exist in the nucleus and the phosphoinositides they generate differ both spatially and temporally. Interestingly, a substantial pool of nuclear phosphoinositides does not appear to be associated with the nuclear membrane (Payrastre et al. 1992). These observations are helping to unravel another facet of phospholipid signaling found in cells.

2.4.1.1 PIPKs and PIP$_2$ at the Nuclear Envelope

Continuous with the ER, the nuclear envelope is a bilayered membranous structure. Pioneering studies on isolated rat liver nuclei revealed the existence of PI and PIP pools in the nuclear envelope (Smith and Wells 1983a). A recent study in sea urchin sperm identified an atypical composition of polyphosphoinositides in the nuclear membrane, as much as 51% of the total phospholipids, an enrichment conserved in mammalian sperm (Garnier-Lhomme et al. 2009). Incubation of isolated nuclear envelopes with [γ–^{32}P]ATP resulted in rapid labeling of phopholipid products, later identified as phosphatidylinositol 4-phosphate (PI4P), phosphotidylinositol 4, 5-bisphosphate (PI4,5P$_2$), and phosphatidic acid (PA), suggesting nuclear envelopes contain PIK, PIPK, and diacylglycerol kinase (DGK) activity (Smith and Wells 1983b). Additionally, substantial evidence demonstrates the requirement of PIP$_2$, diacylglycerol (DAG), and phospholipase C (PLCγ) in the membrane fusion events leading to nuclear envelope formation during mitosis (Dumas et al. 2010). In conjunction with this, PKC βII was identified as a mitotic lamin kinase and its phosphorylation of lamin B is critical for nuclear envelope disassembly, providing evidence that phosphoinositde signaling extended from the nuclear membrane into the nucleus (Goss et al. 1994; Thompson and Fields 1996). The hydrolysis of PIP by phosphomonoesterases and a PIP-dependant ATPase are also associated with the nuclear envelope (Smith and Wells 1984a, b). Besides the canonical PIPKs, phosphatidylinositol 3-kinase (PI3K) may also produce PI3,4P$_2$ at the nuclear surface and therefore act as a "PIPK" (Yokogawa et al. 2000). These studies suggest that a crucial phosphotidylinositide metabolic cycle is present at the nuclear membrane and is required for normal cellular processes. Interestingly, PIPKs and PIP$_2$ were not found to be associated with the invaginations of the nuclear envelope (Boronenkov et al. 1998), indicating that PIPK activity and PIP$_2$ production are temporally and spatially regulated. At present, it is unclear which phosphoinositides (other than PI4P, PI3,4P$_2$, and PI4,5P$_2$) and which PIPKs are involved in the synthesis of phosphoinositides at the nuclear envelope and even less is known about how nuclear PIPKs are regulated.

2.4.1.2 The Intra-nuclear PIPKs and PIP$_2$

Studies by Cocco et al. in the late 1980s revealed that not all PIPKs in the nucleus appeared to be associated with the nuclear membrane. Using highly purified nuclei from mouse erythroleukemia (MEL) and Swiss 3T3 cells, nuclear membranes were stripped away using detergents and the remaining matrix was able to produce both PI4P and PIP$_2$ (Cocco et al. 1987, 1988). The existence of phosphoinositides in non-membranous structures within the nucleus raised the question of where these second messengers and the enzymes that generate them localize. Plausible hypotheses include the nuclear matrix/nucleoskeleton, chromatin, and protein complexes that are associated with these nuclear microenvironments (Fig. 2.9). Payrastre and colleagues showed nuclear matrix-associated PI4 K, PIPKIs, DGK, and PLC in mouse NIH 3T3-fibroblasts and rat liver cells (Payrastre et al. 1992). Using laser scanning confocal microscopy and immunofluorescence staining, PIP, PIPKIα (PIPKIβ in mouse), PIPKIIα (PIP4Kα), PIPKIIβ (PIP4Kβ), PI(4,5)P$_2$, and PI(3,4)P$_2$ were identified at nuclear speckles (interchromatin granule clusters) (Boronenkov et al. 1998). However, the amount of PI4P is about 20-fold higher than that of PI5P in nucleus, and [γ–^{32}P]ATP incubation with isolated rat liver nuclei determined the relative labeling ratio for PI4,5P$_2$ generation at the 5 vs. 4 OH position was ~1.8, suggesting that PIPKIs are the major kinase for PIP$_2$ synthesis within the nucleus (Keune et al. 2010; Vann et al. 1997).

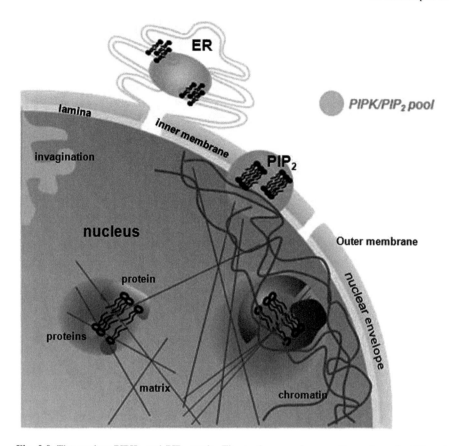

Fig. 2.9 The nuclear PIPKs and PIP$_2$ pools. The nuclear envelope continues from the ER and contains PIP$_2$ as well as associated PIPK activities in both outer and inner bilayers. The interchromatin granule clusters (nuclear speckles) nest another PIPK and PIP$_2$ pool. Some protein complexes docked on the nuclear matrix may also serve to recruit PIPK and PIP$_2$ or that the PIP$_2$/PIPK can regulate the assembly of such matrices

So far, nuclear PIPKIα has been identified in complex with a *s*peckle *t*argeted PIPKI*a*lpha *r*egulated-*p*oly(A) *p*olymerase (Star-PAP), a PIP$_2$-sensitive protein kinase, CKIα, and mRNA processing factors (Mellman et al. 2008). The yeast Mss4p (mammalian PIPK homologue) contains a functional nuclear localization signal (NLS) and can shuttle between the cytoplasm and the nucleus (Audhya and Emr 2003). However, it remains to be determined whether PIPKIα can shuttle between the cytoplasm and nucleus or is capable of generating the majority of the nuclear PIP$_2$. Recently, another PIPKI family member, PIPKIγi4, was identified in the nucleus (Schill and Anderson 2009b). PIPKIγi4 is the only PIPKIγ subfamily member to be detected in the nucleus though its function remains to be characterized. PIPKIIβ predominantly localizes in the nucleus in certain cell lines owing to a specific 17-amino-acid α-helix insertion (Ciruela et al. 2000). PIPKIIα also partially localizes

Table 2.2 The identified nuclear PIP kinases

PIPK	Substrate	Product	Localization	Reference
PIPKI α	PI4P	PIP_2	Nuclear speckle	Boronenkov et al. 1998; Mellman et al. 2008
PIPKIγi4	PI4P	PIP_2	Nucleus	Schill and Anderson 2009
PIPKII β/α	PI5P	PIP_2	Nucleus	Boronenkov et al. 1998; Ciruella et al. 2000; Bultsma et al. 2010

to the nucleus via PIPKIIβ an interaction that is indispensible for protecting PIPKIIβ from degradation via the nuclear ubiquitin ligase, Cul3-SPOP (cullin 3-speckle-type POZ domain protein) (Bultsma et al. 2010; Bunce et al. 2008). It appears that the principle role of nuclear PIPKIIβ is to regulate PI5P levels rather than generate PIP_2 (Bultsma et al. 2010; Keune et al. 2010). The different PIPK isoforms in the nucleus and potentially different localizations of these kinases (Table 2.2) suggest the production of PIP_2 is spatially and temporally regulated, which could result in differential control of the cellular signaling and biological functions.

2.4.2 Regulation and Functions of the Nuclear PIPKs

Compared to the vast amount of data detailing the functions of PIPK and PIP_2 in the cytoplasm, the nuclear functions of these molecules are less understood. Emerging data indicate that nuclear PIP_2 regulates diverse processes including stress response, cell cycle control and mitosis, transcription, mRNA processing and export, DNA repair, chromatin remodeling, and gene expression (Barlow et al. 2010; Bunce et al. 2006b; D'Santos et al. 1998; Gonzales and Anderson 2006; Irvine 2003; Keune et al. 2010). It is becoming clearer that specific nuclear PIPKs produce localized PIP_2 in response to different external and internal signals. Nuclear PIPKs could be associated with scaffolds containing other proteins including effector proteins or even phospholipids, which are organized into the nuclear matrix. The resources of PIPs can be supplied locally or delivered by phosphoinositide-carrier proteins. Upon PIPK stimulation and PIP_2 production, conformational/stoichiometric changes in effector proteins could lead to their activation and the induction of nuclear biological processes (Bunce et al. 2006a).

2.4.2.1 Regulation of the Nuclear PIPKIα

PIPKIα has been shown to be targeted to nuclear speckles where it complexes with the non-canonical poly(A) polymerase Star-PAP, CKIα, RNA polymerase II, and splicing factors (Gonzales et al. 2008; Mellman et al. 2008). The PIPKIα C-terminus directly interacts with Star-PAP. It regulates the expression of select genes induced by oxidative stress through the generation of PIP_2 and modulation of Star-PAP activities at the 3'-end of pre-mRNAs (Mellman et al. 2008). However, the upstream regulator of this PIPKIα intra-nuclear activity has not yet been identified. CKIα is one of the potential site-specific regulators of PIPKIα, but it needs another protein kinase for priming phosphorylation of PIPKIα. Another possibility could be PIP_2 downstream signaling molecules, which can mediate PIPKIα activities by a feedback mechanism. PIPKIα also associates with pre-mRNA splicing factors at nuclear speckles (Boronenkov et al. 1998; Mellman et al. 2008). Speckles are storage sites for these factors whose structure dynamically changes during the cell cycle concomitantly with changes in nuclear PIP_2 levels (Clarke et al. 2001; Lamond and Spector 2003). Interestingly, PIP_2 and PI3K all assemble in nuclear speckles (Boronenkov et al. 1998; Didichenko and Thelen 2001; Osborne et al. 2001). These observations suggest multiple roles for nuclear PIPKIα and PIP_2 in addition to mediating Star-PAP activity.

The tumor suppressor retinoblastoma protein RB (pRB) is another PIPKIα regulator. pRb interacts with and highly activates PIPKIα in a large T antigen-regulated manner (Divecha et al. 2002). In line with this, PIP_2 is sufficient to target the chromatin remodeling complex BAF, which can then be recruited to gene transcription sites by pRB (Zhao et al. 1998), and may further promote an interaction between chromatin and the nuclear matrix via the BAF complex subunit BRG-1 (Rando et al. 2002). The results implicate a role of PIPKIα in transcriptional control and in the regulation of proteins via their ability to bind PIP_2. PIP_2 can also bind histones H1 and H3 alleviating the suppression of transcription (Yu et al. 1998). These studies suggest PIPKIα and its product PIP_2 are involved in the regulation of different phases of gene transcription.

2.4.2.2 Regulation of the Nuclear PIPKIIβ/IIα

PIPKIIβ in combination with PIPKIIα (as addressed formerly) potentially contribute to the generation of a second nuclear PIP_2 pool. Studies in MEL cells demonstrate that the function of PIPKII in the nucleus is mainly to regulate nuclear PI5P levels, because depletion of PIPKIIβ by RNAi increased PI5P levels while overexpression of PIPKIIβ decreased the nuclear PI5P levels (Jones et al. 2006). Like nuclear PIP_2, nuclear PI5P levels also change during the cell cycle (Clarke et al. 2001), implying a role of PIPKIIβ and PI5P in cell cycle regulation. PIPKIIβ can be activated by phosphorylation at Ser326 by p38-MAPK in response to UV irradiation resulting in nuclear accumulation of PI5P (Jones et al. 2006). The Cul3-SPOP ubiquitin ligase complex also regulates PIPKIIβ levels and PI5P production (Bunce et al. 2008). In concert with PIPKIIβ, the type I $PI4,5P_2$ 4-phosphatase converts nuclear PIP_2 to PI5P and therefore modulates PI5P levels (Barlow et al. 2010; Bunce et al. 2008; Zou et al. 2007). PI5P can induce the activation of the p38-MAPK pathway and stimulate Cul3-SPOP activity for PIPKIIβ ubiquitylation (Bunce et al. 2008). As a PIPKIIβ interacting partner, PIPKIIα nuclear targeting seems to be required for suppressing the Cul3-SPOP-dependent PIPKIIβ ubiquitylation, because expression of the kinase-dead PIPKIIα and a nuclear targeting-defective PIPKIIα were not

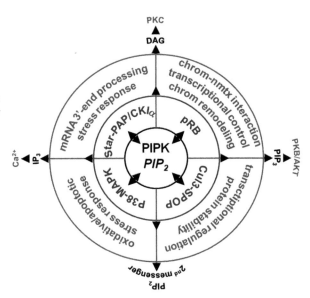

Fig. 2.10 PIPKs generate PIP$_2$ in the nucleus to control many signaling events. Nuclear PIPKs and PIP$_2$ regulate different yet specific cellular molecules and processes through interactions with select nuclear proteins or by producing other second messengers. chrom=chromatin; nmtx=nuclear matrix

able to suppress ubiquitylation of PIPKIIβ, whereas overexpression of the wild-type PIPKIIα did (Bultsma et al. 2010). An extension of PIPKIIβ regulated signaling lies in that PI5P can interact with inhibitor of growth protein 2 (ING2) and regulate the localization of ING2 to chromatin. ING2, in turn, modulates the acetylation of the tumor suppressor protein p53, linking PIPKIIβ activities to transcriptional regulation and gene expression (Bultsma et al. 2010; Jones et al. 2006).

2.4.3 Downstream Signaling of the Nuclear PIPK and PIP$_2$

As discussed above, PIP$_2$ can directly regulate interacting proteins or it can be processed to generate other second messengers including PIP$_3$ (Fig. 2.10). Class I and class II PI3Ks have been found in nucleus with the class IA PI3K being responsible for generating the majority of nuclear PIP$_3$ [reviewed in (D'Santos et al. 1998; Gonzales and Anderson 2006)]. Another bifurcated signaling pathway from nuclear PIP$_2$ is mediated by the nuclear phosphoinositide-specific phospholipase C (PI-PLC). PI-PLCs break down PIP$_2$ into DAG and inositol-1,4,5-triphosphate (IP$_3$). DAG is a direct activator of nuclear PKCs, which are known regulators of gene transcription, DNA synthesis, stress response, mitosis and cell cycle progression (D'Santos et al. 1998). IP$_3$ receptors have been identified on both the outer and inner nuclear membrane (Humbert et al. 1996). The nuclear IP$_3$ is recognized to be involved in Ca^{2+} import into nucleus and nuclear Ca^{2+} homeostasis, processes important for gene expression, DNA synthesis and repair, chromatin condensation, protein import, and apoptosis (Bading et al. 1997; D'Santos et al. 1998; Hardingham et al. 1997; Malviya and Rogue 1998) (Fig. 2.10).

The PIPK and PIP$_2$ downstream signaling cascades are far more intricate than we have thought so far. Identification of more PIPK regulators and effectors as well as other roles for PIPKs, perhaps even independent of their kinase activity, will enhance our understanding of phosphoinositide signaling in the nucleus.

Acknowledgments We apologize to those whose work could not be cited due to space limitations. We thank Suyong Choi, Yue Sun and Rakesh Singh for their scientific discussions and comments on the manuscript prior to submission. Structures rendered in The PyMOL Molecular Graphics System, Version 1.3, Schrödinger, LLC. Research in the authors' lab is supported by NIH grants GM057549, GM051968 and CA104708.

References

Akiyama C, Shinozaki-Narikawa N, Kitazawa T, Hamakubo T, Kodama T, Shibasaki Y (2005) Phosphatidylinositol-4-phosphate 5-kinase gamma is associated with cell-cell junction in A431 epithelial cells. Cell Biol Int 29:514–520

Anderson RA, Boronenkov IV, Doughman SD, Kunz J, Loijens JC (1999) Phosphatidylinositol phosphate kinases, a multifaceted family of signaling enzymes. J Biol Chem 274:9907–9910

Audhya A, Emr SD (2003) Regulation of PI4,5P2 synthesis by nuclear-cytoplasmic shuttling of the Mss4 lipid kinase. EMBO J 22:4223–4236

Bading H, Hardingham GE, Johnson CM, Chawla S (1997) Gene regulation by nuclear and cytoplasmic calcium signals. Biochem Biophys Res Commun 236:541–543

Bai J, Tucker WC, Chapman ER (2004) PIP2 increases the speed of response of synaptotagmin and steers its membrane-penetration activity toward the plasma membrane. Nat Struct Mol Biol 11:36–44

Bairstow SF, Ling K, Su X, Firestone AJ, Carbonara C, Anderson RA (2006) Type Igamma661 phosphatidylinositol phosphate kinase directly interacts with AP2 and regulates endocytosis. J Biol Chem 281:20632–20642

Bakolitsa C, Cohen DM, Bankston LA, Bobkov AA, Cadwell GW, Jennings L, Critchley DR, Craig SW, Liddington RC (2004) Structural basis for vinculin activation at sites of cell adhesion. Nature 430:583–586

Barlow CA, Laishram RS, Anderson RA (2010) Nuclear phosphoinositides: a signaling enigma wrapped in a compartmental conundrum. Trends Cell Biol 20:25–35

Bazenet CE, Ruano AR, Brockman JL, Anderson RA (1990) The human erythrocyte contains two forms of phosphatidylinositol-4-phosphate 5-kinase which are differentially active toward membranes. J Biol Chem 265:18012–18022

Bonifacino J, Traub L (2003) Signals for sorting of transmembrane proteins to endosomes and lysosomes. Annu Rev Biochem 72:395–447

Boronenkov IV, Anderson RA (1995) The sequence of phosphatidylinositol-4-phosphate 5-kinase defines a novel family of lipid kinases. J Biol Chem 270:2881–2884

Boronenkov IV, Loijens JC, Umeda M, Anderson RA (1998) Phosphoinositide signaling pathways in nuclei are associated with nuclear speckles containing pre-mRNA processing factors. Mol Biol Cell 9:3547–3560

Broussard JA, Webb DJ, Kaverina I (2008) Asymmetric focal adhesion disassembly in motile cells. Curr Opin Cell Biol 20:85–90

Bultsma Y, Keune WJ, Divecha N (2010) PIP4Kbeta interacts with and modulates nuclear localization of the high-activity PtdIns5P-4-kinase isoform PIP4Kalpha. Biochem J 430:223–235

Bunce MW, Bergendahl K, Anderson RA (2006a) Nuclear PI(4,5)P(2): a new place for an old signal. Biochim Biophys Acta 1761:560–569

Bunce MW, Gonzales ML, Anderson RA (2006b) Stress-ING out: phosphoinositides mediate the cellular stress response. Sci STKE 2006:pe46

Bunce MW, Boronenkov IV, Anderson RA (2008) Coordinated activation of the nuclear ubiquitin ligase Cul3-SPOP by the generation of phosphatidylinositol 5-phosphate. J Biol Chem 283:8678–8686

Burden LM, Rao VD, Murray D, Ghirlando R, Doughman SD, Anderson RA, Hurley JH (1999) The flattened face of type II beta phosphatidylinositol phosphate kinase binds acidic phospholipid membranes. Biochemistry 38:15141–15149

Burridge K, Sastry SK, Sallee JL (2006) Regulation of cell adhesion by protein-tyrosine phosphatases. I. Cell-matrix adhesion. J Biol Chem 281:15593–15596

Cabezas A, Pattni K, Stenmark H (2006) Cloning and subcellular localization of a human phosphatidylinositol 3-phosphate 5-kinase, PIKfyve/Fab1. Gene 371:34–41

Caswell PT, Norman JC (2006) Integrin trafficking and the control of cell migration. Traffic 7:14–21

Chandrasekar I, Stradal TE, Holt MR, Entschladen F, Jockusch BM, Ziegler WH (2005) Vinculin acts as a sensor in lipid regulation of adhesion-site turnover. J Cell Sci 118:1461–1472

Chao WT, Kunz J (2009) Focal adhesion disassembly requires clathrin-dependent endocytosis of integrins. FEBS Lett 583:1337–1343

Chao WT, Ashcroft F, Daquinag AC, Vadakkan T, Wei Z, Zhang P, Dickinson ME, Kunz, J (2010) Type I phosphatidylinositol phosphate kinase beta regulates focal adhesion disassembly by promoting beta1 integrin endocytosis. Mol Cell Biol 30:4463–4479

Chao WT, Daquinag AC, Ashcroft F, Kunz, J (2010) Type I PIPK-alpha regulates directed cell migration by modulating Rac1 plasma membrane targeting and activation. J Cell Biol 190:247–262

Chatah NE, Abrams CS (2001) G-protein-coupled receptor activation induces the membrane translocation and activation of phosphatidylinositol-4-phosphate 5-kinase I alpha by a Rac- and Rho-dependent pathway. J Biol Chem 276:34059–34065

Ciruela A, Hinchliffe KA, Divecha N, Irvine RF (2000) Nuclear targeting of the beta isoform of type II phosphatidylinositol phosphate kinase (phosphatidylinositol 5-phosphate 4-kinase) by its alpha-helix 7. Biochem J 346(Pt 3):587–591

Clarke JH, Letcher AJ, D'Santos C S, Halstead JR, Irvine RF, Divecha N (2001) Inositol lipids are regulated during cell cycle progression in the nuclei of murine erythroleukaemia cells. Biochem J 357:905–910

Cocco L, Gilmour RS, Ognibene A, Letcher AJ, Manzoli FA, Irvine RF (1987) Synthesis of polyphosphoinositides in nuclei of Friend cells. Evidence for polyphosphoinositide metabolism inside the nucleus which changes with cell differentiation. Biochem J 248:765–770

Cocco L, Martelli AM, Gilmour RS, Ognibene A, Manzoli FA, Irvine RF (1988) Rapid changes in phospholipid metabolism in the nuclei of Swiss 3T3 cells induced by treatment of the cells with insulin-like growth factor I. Biochem Biophys Res Commun 154:1266–1272

Cooke FT, Dove SK, McEwen RK, Painter G, Holmes AB, Hall MN, Michell RH, Parker PJ (1998) The stress-activated phosphatidylinositol 3-phosphate 5-kinase Fab1p is essential for vacuole function in S. cerevisiae. Curr Biol 8:1219–1222

Cox S, Taylor SS (1994) Holoenzyme interaction sites in the cAMP-dependent protein kinase. Histidine 87 in the catalytic subunit complements serine 99 in the type I regulatory subunit. J Biol Chem 269:22614–22622

D'Santos CS, Clarke JH, Divecha N (1998) Phospholipid signalling in the nucleus. Een DAG uit het leven van de inositide signalering in de nucleus. Biochim Biophys Acta 1436:201–232

Del Pozo MA, Kiosses WB, Alderson NB, Meller N, Hahn KM, Schwartz MA (2002) Integrins regulate GTP-Rac localized effector interactions through dissociation of Rho-GDI. Nat Cell Biol 4:232–239

Dell'Angelica E, Mullins C, Bonifacino J (1999) AP-4, a novel protein complex related to clathrin adaptors. J Biol Chem 274:7278–7285

DeMali KA, Barlow CA, Burridge K (2002) Recruitment of the Arp2/3 complex to vinculin: coupling membrane protrusion to matrix adhesion. J Cell Biol 159:881–891

Di Paolo G, De Camilli P (2006) Phosphoinositides in cell regulation and membrane dynamics. Nature 443:651–657

Di Paolo G, Pellegrini L, Letinic K, Cestra G, Zoncu R, Voronov S, Chang S, Guo J, Wenk MR, De Camilli P (2002) Recruitment and regulation of phosphatidylinositol phosphate kinase type 1 gamma by the FERM domain of talin. Nature 420:85–89

Di Paolo G, Moskowitz HS, Gipson K, Wenk MR, Voronov S, Obayashi M, Flavell R, Fitzsimonds RM, Ryan TA, De Camilli P (2004) Impaired PtdIns(4,5)P2 synthesis in nerve terminals produces defects in synaptic vesicle trafficking. Nature 431:415–422

Didichenko SA, Thelen M (2001) Phosphatidylinositol 3-kinase c2alpha contains a nuclear localization sequence and associates with nuclear speckles. J Biol Chem 276:48135–48142

Divecha N, Roefs M, Los A, Halstead J, Bannister A, D'Santos C (2002) Type I PIPkinases interact with and are regulated by the retinoblastoma susceptibility gene product-pRB. Curr Biol 12:582–587

Doughman RL, Firestone AJ, Anderson RA (2003a) Phosphatidylinositol phosphate kinases put PI4,5P(2) in its place. J Membr Biol 194:77–89

Doughman RL, Firestone AJ, Wojtasiak ML, Bunce MW, Anderson RA (2003b) Membrane ruffling requires coordination between type Ialpha phosphatidylinositol phosphate kinase and Rac signaling. J Biol Chem 278:23036–23045

Dumas F, Byrne RD, Vincent B, Hobday TM, Poccia DL, Larijani B (2010) Spatial regulation of membrane fusion controlled by modification of phosphoinositides. PLoS One 5:e12208

Ezratty EJ, Partridge MA, Gundersen GG (2005) Microtubule-induced focal adhesion disassembly is mediated by dynamin and focal adhesion kinase. Nat Cell Biol 7:581–590

Fölsch H, Ohno H, Bonifacino J, Mellman I (1999) A novel clathrin adaptor complex mediates basolateral targeting in polarized epithelial cells. Cell 99:189–198

Franco SJ, Huttenlocher A (2005) Regulating cell migration: calpains make the cut. J Cell Sci 118:3829–3838

Franco SJ, Rodgers MA, Perrin BJ, Han J, Bennin DA, Critchley DR, Huttenlocher A (2004) Calpain-mediated proteolysis of talin regulates adhesion dynamics. Nat Cell Biol 6:977–983

Gaidarov I, Krupnick JG, Falck JR, Benovic JL, Keen JH (1999) Arrestin function in G protein-coupled receptor endocytosis requires phosphoinositide binding. Embo J 18:871–881

Garnier-Lhomme M, Byrne RD, Hobday TM, Gschmeissner S, Woscholski R, Poccia DL, Dufourc EJ, Larijani B (2009) Nuclear envelope remnants: fluid membranes enriched in sterols and polyphosphoinositides. PLoS One 4:e4255

Gary JD, Wurmser AE, Bonangelino CJ, Weisman LS, Emr SD (1998) Fab1p is essential for PtdIns(3)P 5-kinase activity and the maintenance of vacuolar size and membrane homeostasis. J Cell Biol 143:65–79

Gong LW, Di Paolo G, Diaz E, Cestra G, Diaz ME, Lindau M, De Camilli P, Toomre D (2005) Phosphatidylinositol phosphate kinase type I gamma regulates dynamics of large dense-core vesicle fusion. Proc Natl Acad Sci U S A 102:5204–5209

Gonzales ML, Anderson RA (2006) Nuclear phosphoinositide kinases and inositol phospholipids. J Cell Biochem 97:252–260

Gonzales ML, Mellman DL, Anderson RA (2008) CKIalpha is associated with and phosphorylates star-PAP and is also required for expression of select star-PAP target messenger RNAs. J Biol Chem 283:12665–12673

Gorbatyuk VY, Nosworthy NJ, Robson SA, Bains NP, Maciejewski MW, Dos Remedios CG, King GF (2006) Mapping the phosphoinositide-binding site on chick cofilin explains how PIP2 regulates the cofilin-actin interaction. Mol Cell 24:511–522

Goss VL, Hocevar BA, Thompson LJ, Stratton CA, Burns DJ, Fields AP (1994) Identification of nuclear beta II protein kinase C as a mitotic lamin kinase. J Biol Chem 269:19074–19080

Gurr MI, Finean JB, Hawthorne JN (1963) The phospholipids of liver-cell fractions. I. The phospholipid composition of the liver-cell nucleus. Biochim Biophys Acta 70:406–416

Hall A, Nobes CD (2000) Rho GTPases: molecular switches that control the organization and dynamics of the actin cytoskeleton. Philos Trans R Soc Lond B Biol Sci 355:965–970

Hanks SK, Hunter T (1995) Protein kinases 6. The eukaryotic protein kinase superfamily: kinase (catalytic) domain structure and classification. FASEB J 9:576–596

Hanks SK, Quinn AM, Hunter T (1988) The protein kinase family: conserved features and deduced phylogeny of the catalytic domains. Science 241:42–52

Hardingham GE, Chawla S, Johnson CM, Bading H (1997) Distinct functions of nuclear and cytoplasmic calcium in the control of gene expression. Nature 385:260–265

Hay JC, Fisette PL, Jenkins GH, Fukami K, Takenawa T, Anderson RA, Martin TF (1995) ATP-dependent inositide phosphorylation required for $Ca^{(2+)}$-activated secretion. Nature 374:173–177

He B, Xi F, Zhang X, Zhang J, Guo W (2007) Exo70 interacts with phospholipids and mediates the targeting of the exocyst to the plasma membrane. EMBO J 26:4053–4065

Heck JN, Mellman DL, Ling K, Sun Y, Wagoner MP, Schill NJ, Anderson RA (2007) A conspicuous connection: structure defines function for the phosphatidylinositol-phosphate kinase family. Crit Rev Biochem Mol Biol 42:15–39

Hinchliffe KA, Irvine RF (2006) Regulation of type II PIP kinase by PKD phosphorylation. Cell Signal 18:1906–1913

Hinchliffe KA, Ciruela A, Letcher AJ, Divecha N, Irvine RF (1999a) Regulation of type IIalpha phosphatidylinositol phosphate kinase localisation by the protein kinase CK2. Curr Biol 9:983–986

Hinchliffe KA, Ciruela A, Morris JA, Divecha N, Irvine RF (1999b) The type II PIPkins (PtdIns5P 4-kinases): enzymes in search of a function? Biochem Soc Trans 27:657–661

Hokin MR, Hokin LE (1953) Enzyme secretion and the incorporation of P32 into phospholipides of pancreas slices. J Biol Chem 203:967–977

Homma K, Terui S, Minemura M, Qadota H, Anraku Y, Kanaho Y, Ohya Y (1998) Phosphatidylinositol-4-phosphate 5-kinase localized on the plasma membrane is essential for yeast cell morphogenesis. J Biol Chem 273:15779–15786

Honda A, Nogami M, Yokozeki T, Yamazaki M, Nakamura H, Watanabe H, Kawamoto K, Nakayama K, Morris AJ, Frohman MA, Kanaho Y (1999) Phosphatidylinositol 4-phosphate 5-kinase alpha is a downstream effector of the small G protein ARF6 in membrane ruffle formation. Cell 99:521–532

Höning S, Ricotta D, Krauss M, Späte K, Spolaore B, Motley A, Robinson M, Robinson C, Haucke V, Owen D (2005) Phosphatidylinositol-(4,5)-bisphosphate regulates sorting signal recognition by the clathrin-associated adaptor complex AP2. Mol Cell 18:519–531

Humbert JP, Matter N, Artault JC, Koppler P, Malviya AN (1996) Inositol 1,4,5-trisphosphate receptor is located to the inner nuclear membrane vindicating regulation of nuclear calcium signaling by inositol 1,4,5-trisphosphate. Discrete distribution of inositol phosphate receptors to inner and outer nuclear membranes. J Biol Chem 271:478–485

Irvine RF (2003) Nuclear lipid signalling. Nat Rev Mol Cell Biol 4:349–360

Ishihara H, Shibasaki Y, Kizuki N, Katagiri H, Yazaki Y, Asano T, Oka Y (1996) Cloning of cDNAs encoding two isoforms of 68-kDa type I phosphatidylinositol-4-phosphate 5-kinase. J Biol Chem 271:23611–23614

Ishihara H, Shibasaki Y, Kizuki N, Wada T, Yazaki Y, Asano T, Oka Y (1998) Type I phosphatidylinositol-4-phosphate 5-kinases. Cloning of the third isoform and deletion/substitution analysis of members of this novel lipid kinase family. J Biol Chem 273:8741–8748

Itoh T, Ishihara H, Shibasaki Y, Oka Y, Takenawa T (2000) Autophosphorylation of type I phosphatidylinositol phosphate kinase regulates its lipid kinase activity. J Biol Chem 275:19389–19394

Iyer GH, Moore MJ, Taylor SS (2005) Consequences of lysine 72 mutation on the phosphorylation and activation state of cAMP-dependent kinase. J Biol Chem 280:8800–8807

Jackson T (1998) Transport vesicles: coats of many colours. Curr Biol 8:R609–R612

Janmey PA, Lindberg U (2004) Cytoskeletal regulation: rich in lipids. Nat Rev Mol Cell Biol 5:658–666

Jenkins GH, Subrahmanyam G, Anderson RA (1991) Purification and reconstitution of phosphatidylinositol 4-kinase from human erythrocytes. Biochim Biophys Acta 1080:11–18

Jenkins GH, Fisette PL, Anderson RA (1994) Type I phosphatidylinositol 4-phosphate 5-kinase isoforms are specifically stimulated by phosphatidic acid. J Biol Chem 269:11547–11554

Johnson LN, Noble ME, Owen DJ (1996) Active and inactive protein kinases: structural basis for regulation. Cell 85:149–158

Jones DR, Bultsma Y, Keune WJ, Halstead JR, Elouarrat D, Mohammed S, Heck AJ, D'Santos CS, Divecha N (2006) Nuclear PtdIns5P as a transducer of stress signaling: an in vivo role for PIP4Kbeta. Mol Cell 23:685–695

Kahlfeldt N, Vahedi-Faridi A, Koo S, Schäfer J, Krainer G, Keller S, Saenger W, Krauss M, Haucke V (2010) Molecular basis for association of PIPKI gamma-p90 with clathrin adaptor AP-2. J Biol Chem 285:2734–2749

Keune W, Bultsma Y, Sommer L, Jones D, Divecha N (2010) Phosphoinositide signalling in the nucleus. Adv Enzyme Regul 51(1):91–99

Kisseleva M, Feng Y, Ward M, Song C, Anderson RA, Longmore GD (2005) The LIM protein Ajuba regulates phosphatidylinositol 4,5-bisphosphate levels in migrating cells through an interaction with and activation of PIPKI alpha. Mol Cell Biol 25:3956–3966

Kleinig H (1970) Nuclear membranes from mammalian liver. II. Lipid composition. J Cell Biol 46:396–402

Knighton DR, Zheng JH, Ten Eyck LF, Ashford VA, Xuong NH, Taylor SS, Sowadski JM (1991a) Crystal structure of the catalytic subunit of cyclic adenosine monophosphate-dependent protein kinase. Science 253:407–414

Knighton DR, Zheng JH, Ten Eyck LF, Xuong NH, Taylor SS, Sowadski JM (1991b) Structure of a peptide inhibitor bound to the catalytic subunit of cyclic adenosine monophosphate-dependent protein kinase. Science 253:414–420

Krauss M, Kinuta M, Wenk MR, De Camilli P, Takei K, Haucke V (2003) ARF6 stimulates clathrin/AP-2 recruitment to synaptic membranes by activating phosphatidylinositol phosphate kinase type Igamma. J Cell Biol 162:113–124

Krauss M, Kukhtina V, Pechstein A, Haucke V (2006) Stimulation of phosphatidylinositol kinase type I-mediated phosphatidylinositol (4,5)-bisphosphate synthesis by AP-2mu-cargo complexes. Proc Natl Acad Sci U S A 103:11934–11939

Krupa A, Preethi G, Srinivasan N (2004) Structural modes of stabilization of permissive phosphorylation sites in protein kinases: distinct strategies in Ser/Thr and Tyr kinases. J Mol Biol 339:1025–1039

Kumar N, Khurana S (2004) Identification of a functional switch for actin severing by cytoskeletal proteins. J Biol Chem 279:24915–24918

Kumar N, Zhao P, Tomar A, Galea CA, Khurana S (2004) Association of villin with phosphatidylinositol 4,5-bisphosphate regulates the actin cytoskeleton. J Biol Chem 279:3096–3110

Kunz J, Wilson MP, Kisseleva M, Hurley JH, Majerus PW, Anderson RA (2000) The activation loop of phosphatidylinositol phosphate kinases determines signaling specificity. Mol Cell 5:1–11

Kunz J, Fuelling A, Kolbe L, Anderson RA (2002) Stereo-specific substrate recognition by phosphatidylinositol phosphate kinases is swapped by changing a single amino acid residue. J Biol Chem 277:5611–5619

Kwiatkowska K (2010) One lipid, multiple functions: how various pools of PI(4,5)P(2) are created in the plasma membrane. Cell Mol Life Sci 67:3927–3946

Lambrechts A, Jonckheere V, Dewitte D, Vandekerckhove J, Ampe C (2002) Mutational analysis of human profilin I reveals a second PI(4,5)-P2 binding site neighbouring the poly(L-proline) binding site. BMC Biochem 3:12

Lamond AI, Spector DL (2003) Nuclear speckles: a model for nuclear organelles. Nat Rev Mol Cell Biol 4:605–612

Lassing I, Lindberg U (1985) Specific interaction between phosphatidylinositol 4,5-bisphosphate and profilactin. Nature 314:472–474

Lee S, Voronov S, Letinic K, Nairn A, Di Paolo G, De Camilli P (2005) Regulation of the interaction between PIPKI gamma and talin by proline-directed protein kinases. J Cell Biol 168:789–799

Letinic K, Sebastian R, Toomre D, Rakic P (2009) Exocyst is involved in polarized cell migration and cerebral cortical development. Proc Natl Acad Sci U S A 106:11342–11347

Ling K, Doughman RL, Firestone AJ, Bunce MW, Anderson RA (2002) Type I gamma phosphatidylinositol phosphate kinase targets and regulates focal adhesions. Nature 420:89–93

Ling K, Doughman RL, Iyer VV, Firestone AJ, Bairstow SF, Mosher DF, Schaller MD, Anderson RA (2003) Tyrosine phosphorylation of type Igamma phosphatidylinositol phosphate kinase by Src regulates an integrin-talin switch. J Cell Biol 163:1339–1349

Ling K, Schill NJ, Wagoner MP, Sun Y, Anderson RA (2006) Movin' on up: the role of PtdIns(4,5)P(2) in cell migration. Trends Cell Biol 16:276–284

Ling K, Bairstow S, Carbonara C, Turbin D, Huntsman D, Anderson R (2007) Type I gamma phosphatidylinositol phosphate kinase modulates adherens junction and E-cadherin trafficking via a direct interaction with mu 1B adaptin. J Cell Biol 176:343–353

Liu J, Zuo X, Yue P, Guo W (2007) Phosphatidylinositol 4,5-bisphosphate mediates the targeting of the exocyst to the plasma membrane for exocytosis in mammalian cells. Mol Biol Cell 18:4483–4492

Loijens JC, Boronenkov IV, Parker GJ, Anderson RA (1996) The phosphatidylinositol 4-phosphate 5-kinase family. Adv Enzyme Regul 36:115–140

Lokuta MA, Senetar MA, Bennin DA, Nuzzi PA, Chan KT, Ott VL, Huttenlocher A (2007) Type Igamma PIP kinase is a novel uropod component that regulates rear retraction during neutrophil chemotaxis. Mol Biol Cell 18:5069–5080

Malviya AN, Rogue PJ (1998) "Tell me where is calcium bred": clarifying the roles of nuclear calcium. Cell 92:17–23

Martel V, Racaud-Sultan C, Dupe S, Marie C, Paulhe F, Galmiche A, Block MR, Albiges-Rizo C (2001) Conformation, localization, and integrin binding of talin depend on its interaction with phosphoinositides. J Biol Chem 276:21217–21227

Matsui T, Yonemura S, Tsukita S, Tsukita S (1999) Activation of ERM proteins in vivo by Rho involves phosphatidyl-inositol 4-phosphate 5-kinase and not ROCK kinases. Curr Biol 9:1259–1262

McEwen RK, Dove SK, Cooke FT, Painter GF, Holmes AB, Shisheva A, Ohya Y, Parker PJ, Michell RH (1999) Complementation analysis in PtdInsP kinase-deficient yeast mutants demonstrates that Schizosaccharomyces pombe and murine Fab1p homologues are phosphatidylinositol 3-phosphate 5-kinases. J Biol Chem 274:33905–33912

Mellman DL, Gonzales ML, Song C, Barlow CA, Wang P, Kendziorski C, Anderson RA (2008) A PtdIns4,5P2-regulated nuclear poly(A) polymerase controls expression of select mRNAs. Nature 451:1013–1017

Millard TH, Sharp SJ, Machesky LM (2004) Signalling to actin assembly via the WASP (Wiskott-Aldrich syndrome protein)-family proteins and the Arp2/3 complex. Biochem J 380:1–17

Moritz A, De Graan PN, Gispen WH, Wirtz KW (1992a) Phosphatidic acid is a specific activator of phosphatidylinositol-4-phosphate kinase. J Biol Chem 267:7207–7210

Moritz A, Westerman J, De Graan PN, Wirtz KW (1992b) Phosphatidylinositol 4-kinase and phosphatidylinositol-4-phosphate 5-kinase from bovine brain membranes. Methods Enzymol 209:202–211

Nakano-Kobayashi A, Yamazaki M, Unoki T, Hongu T, Murata C, Taguchi R, Katada T, Frohman M, Yokozeki T, Kanaho Y (2007) Role of activation of PIP5Kgamma661 by AP-2 complex in synaptic vesicle endocytosis. EMBO J 26:1105–1116

Nakatsu F, Ohno H (2003) Adaptor protein complexes as the key regulators of protein sorting in the post-Golgi network. Cell Struct Funct 28:419–429

Narkis G, Ofir R, Manor E, Landau D, Elbedour K, Birk OS (2007) Lethal congenital contractural syndrome type 2 (LCCS2) is caused by a mutation in ERBB3 (Her3), a modulator of the phosphatidylinositol-3-kinase/Akt pathway. Am J Hum Genet 81:589–595

Nayal A, Webb DJ, Horwitz AF (2004) Talin: an emerging focal point of adhesion dynamics. Curr Opin Cell Biol 16:94–98

Nicholson-Dykstra S, Higgs HN, Harris ES (2005) Actin dynamics: growth from dendritic branches. Curr Biol 15:R346–R357

Niggli V (2005) Regulation of protein activities by phosphoinositide phosphates. Annu Rev Cell Dev Biol 21:57–79

Odorizzi G, Babst M, Emr SD (1998) Fab1p PtdIns(3)P 5-kinase function essential for protein sorting in the multivesicular body. Cell 95:847–858

Ohno H (2006) Physiological roles of clathrin adaptor AP complexes: lessons from mutant animals. J Biochem 139:943–948

Olusanya O, Andrews P, Swedlow J, Smythe E (2001) Phosphorylation of threonine 156 of the mu2 subunit of the AP2 complex is essential for endocytosis in vitro and in vivo. Curr Biol 11:896–900

Osborne SL, Thomas CL, Gschmeissner S, Schiavo G (2001) Nuclear PtdIns(4,5)P2 assembles in a mitotically regulated particle involved in pre-mRNA splicing. J Cell Sci 114:2501–2511

Oztan A, Silvis M, Weisz OA, Bradbury NA, Hsu SC, Goldenring JR, Yeaman C, Apodaca G (2007) Exocyst requirement for endocytic traffic directed toward the apical and basolateral poles of polarized MDCK cells. Mol Biol Cell 18:3978–3992

Park SJ, Itoh T, Takenawa T (2001) Phosphatidylinositol 4-phosphate 5-kinase type I is regulated through phosphorylation response by extracellular stimuli. J Biol Chem 276:4781–4787

Payrastre B, Nievers M, Boonstra J, Breton M, Verkleij AJ, Van Bergen en Henegouwen PM (1992) A differential location of phosphoinositide kinases, diacylglycerol kinase, and phospholipase C in the nuclear matrix. J Biol Chem 267:5078–5084

Pierini LM, Lawson MA, Eddy RJ, Hendey B, Maxfield FR (2000) Oriented endocytic recycling of alpha5beta1 in motile neutrophils. Blood 95:2471–2480

Prehoda KE, Scott JA, Mullins RD, Lim WA (2000) Integration of multiple signals through cooperative regulation of the N-WASP-Arp2/3 complex. Science 290:801–806

Prigozhina NL, Waterman-Storer CM (2004) Protein kinase D-mediated anterograde membrane trafficking is required for fibroblast motility. Curr Biol 14:88–98

Qualmann B, Kessels MM, Kelly RB (2000) Molecular links between endocytosis and the actin cytoskeleton. J Cell Biol 150:F111–F116

Rameh LE, Tolias KF, Duckworth BC, Cantley LC (1997) A new pathway for synthesis of phosphatidylinositol-4,5-bisphosphate. Nature 390:192–196

Rando OJ, Zhao K, Janmey P, Crabtree GR (2002) Phosphatidylinositol-dependent actin filament binding by the SWI/SNF-like BAF chromatin remodeling complex. Proc Natl Acad Sci U S A 99:2824–2829

Rao VD, Misra S, Boronenkov IV, Anderson RA, Hurley JH (1998) Structure of type IIbeta phosphatidylinositol phosphate kinase: a protein kinase fold flattened for interfacial phosphorylation. Cell 94:829–839

Rohde G, Wenzel D, Haucke V (2002) A phosphatidylinositol (4,5)-bisphosphate binding site within mu2-adaptin regulates clathrin-mediated endocytosis. J Cell Biol 158:209–214

Rosse C, Hatzoglou A, Parrini MC, White MA, Chavrier P, Camonis J (2006) RalB mobilizes the exocyst to drive cell migration. Mol Cell Biol 26:727–734

Rozelle AL, Machesky LM, Yamamoto M, Driessens MH, Insall RH, Roth MG, Luby-Phelps K, Marriott G, Hall A, Yin HL (2000) Phosphatidylinositol 4,5-bisphosphate induces actin-based movement of raft-enriched vesicles through WASP-Arp2/3. Curr Biol 10:311–320

Saraste M, Sibbald PR, Wittinghofer A (1990) The P-loop—a common motif in ATP- and GTP-binding proteins. Trends Biochem Sci 15:430–434

Sasaki J, Sasaki T, Yamazaki M, Matsuoka K, Taya C, Shitara H, Takasuga S, Nishio M, Mizuno K, Wada T et al (2005) Regulation of anaphylactic responses by phosphatidylinositol phosphate kinase type I {alpha}. J Exp Med 201:859–870

Sbrissa D, Ikonomov OC, Shisheva A (1999) PIKfyve, a mammalian ortholog of yeast Fab1p lipid kinase, synthesizes 5-phosphoinositides. Effect of insulin. J Biol Chem 274:21589–21597

Sbrissa D, Ikonomov OC, Shisheva A (2000) PIKfyve lipid kinase is a protein kinase: down-regulation of 5'-phosphoinositide product formation by autophosphorylation. Biochemistry 39:15980–15989

Schill N, Anderson R (2009a) Out, in and back again: PtdIns(4,5)P(2) regulates cadherin trafficking in epithelial morphogenesis. Biochem J 418:247–260

Schill NJ, Anderson RA (2009b) Two novel phosphatidylinositol-4-phosphate 5-kinase type Igamma splice variants expressed in human cells display distinctive cellular targeting. Biochem J 422(3):473–482

Schmoranzer J, Kreitzer G, Simon SM (2003) Migrating fibroblasts perform polarized, microtubule-dependent exocytosis towards the leading edge. J Cell Sci 116:4513–4519

Schramp M, Ying O, Kim TY, Martin GS (2008) ERK5 promotes Src-induced podosome formation by limiting Rho activation. J Cell Biol 181:1195–1210

Sechi AS, Wehland J (2000) The actin cytoskeleton and plasma membrane connection: PtdIns(4,5)P(2) influences cytoskeletal protein activity at the plasma membrane. J Cell Sci 113(Pt 21):3685–3695

Shibasaki Y, Ishihara H, Kizuki N, Asano T, Oka Y, Yazaki Y (1997) Massive actin polymerization induced by phosphatidylinositol-4-phosphate 5-kinase in vivo. J Biol Chem 272:7578–7581

Shisheva A, Sbrissa D, Ikonomov O (1999) Cloning, characterization, and expression of a novel Zn^{2+}-binding FYVE finger-containing phosphoinositide kinase in insulin-sensitive cells. Mol Cell Biol 19:623–634

Sidani M, Wessels D, Mouneimne G, Ghosh M, Goswami S, Sarmiento C, Wang W, Kuhl S, El-Sibai M, Backer JM et al (2007) Cofilin determines the migration behavior and turning frequency of metastatic cancer cells. J Cell Biol 179:777–791

Skare P, Karlsson R (2002) Evidence for two interaction regions for phosphatidylinositol(4,5)-bisphosphate on mammalian profilin I. FEBS Lett 522:119–124

Smith CD, Wells WW (1983a) Phosphorylation of rat liver nuclear envelopes. I. Characterization of in vitro protein phosphorylation. J Biol Chem 258:9360–9367

Smith CD, Wells WW (1983b) Phosphorylation of rat liver nuclear envelopes. II. Characterization of in vitro lipid phosphorylation. J Biol Chem 258:9368–9373

Smith CD, Wells WW (1984a) Characterization of a phosphatidylinositol 4-phosphate-specific phosphomonoesterase in rat liver nuclear envelopes. Arch Biochem Biophys 235:529–537

Smith CD, Wells WW (1984b) Solubilization and reconstitution of a nuclear envelope-associated ATPase. Synergistic activation by RNA and polyphosphoinositides. J Biol Chem 259:11890–11894

Streb H, Irvine RF, Berridge MJ, Schulz I (1983) Release of Ca^{2+} from a nonmitochondrial intracellular store in pancreatic acinar cells by inositol-1,4,5-trisphosphate. Nature 306:67–69

Sugimoto H, Sugahara M, Fölsch H, Koide Y, Nakatsu F, Tanaka N, Nishimura T, Furukawa M, Mullins C, Nakamura N et al (2002) Differential recognition of tyrosine-based basolateral signals by AP-1B subunit mu1B in polarized epithelial cells. Mol Biol Cell 13:2374–2382

Sun Y, Ling K, Wagoner MP, Anderson RA (2007) Type I gamma phosphatidylinositol phosphate kinase is required for EGF-stimulated directional cell migration. J Cell Biol 178:297–308

Suriano G, Mulholland D, De Wever O, Ferreira P, Mateus A, Bruyneel E, Nelson C, Mareel M, Yokota J, Huntsman D, Seruca R (2003) The intracellular E-cadherin germline mutation V832 M lacks the ability to mediate cell-cell adhesion and to suppress invasion. Oncogene 22:5716–5719

Takenawa T, Itoh T (2001) Phosphoinositides, key molecules for regulation of actin cytoskeletal organization and membrane traffic from the plasma membrane. Biochim Biophys Acta 1533:190–206

Tall EG, Spector I, Pentyala SN, Bitter I, Rebecchi MJ (2000) Dynamics of phosphatidylinositol 4,5-bisphosphate in actin-rich structures. Curr Biol 10:743–746

Tayeb MA, Skalski M, Cha MC, Kean MJ, Scaife M, Coppolino MG (2005) Inhibition of SNARE-mediated membrane traffic impairs cell migration. Exp Cell Res 305:63–73

Taylor SS, Knighton DR, Zheng J, Ten Eyck LF, Sowadski JM (1992) Structural framework for the protein kinase family. Annu Rev Cell Biol 8:429–462

Taylor SS, Radzio-Andzelm E, Knighton DR, Ten Eyck LF, Sowadski JM, Herberg FW, Yonemoto W, Zheng J (1993) Crystal structures of the catalytic subunit of cAMP-dependent protein kinase reveal general features of the protein kinase family. Receptor 3:165–172

Thieman JR, Mishra SK, Ling K, Doray B, Anderson RA, Traub LM (2009) Clathrin regulates the association of PIPKIgamma661 with the AP-2 adaptor beta2 appendage. J Biol Chem 284:13924–13939

Thompson LJ, Fields AP (1996) betaII protein kinase C is required for the G2/M phase transition of cell cycle. J Biol Chem 271:15045–15053

Tolias KF, Couvillon AD, Cantley LC, Carpenter CL (1998) Characterization of a Rac1- and RhoGDI-associated lipid kinase signaling complex. Mol Cell Biol 18:762–770

Tolias KF, Hartwig JH, Ishihara H, Shibasaki Y, Cantley LC, Carpenter CL (2000) Type Ialpha phosphatidylinositol-4-phosphate 5-kinase mediates Rac-dependent actin assembly. Curr Biol 10:153–156

Van Horck FP, Lavazais E, Eickholt BJ, Moolenaar WH, Divecha N (2002) Essential role of type I(alpha) phosphatidylinositol 4-phosphate 5-kinase in neurite remodeling. Curr Biol 12:241–245

Van Rheenen J, Song X, Van Roosmalen W, Cammer M, Chen X, Desmarais V, Yip SC, Backer JM, Eddy RJ, Condeelis JS (2007) EGF-induced PIP2 hydrolysis releases and activates cofilin locally in carcinoma cells. J Cell Biol 179:1247–1259

Vann LR, Wooding FB, Irvine RF, Divecha N (1997) Metabolism and possible compartmentalization of inositol lipids in isolated rat-liver nuclei. Biochem J 327(Pt 2):569–576

Webb DJ, Parsons JT, Horwitz AF (2002) Adhesion assembly, disassembly and turnover in migrating cells—over and over and over again. Nat Cell Biol 4:E97–E100

Weernink PA, Meletiadis K, Hommeltenberg S, Hinz M, Ishihara H, Schmidt M, Jakobs KH (2004) Activation of type I phosphatidylinositol 4-phosphate 5-kinase isoforms by the Rho GTPases, RhoA, Rac1, and Cdc42. J Biol Chem 279:7840–7849

Wenk MR, De Camilli P (2004) Protein-lipid interactions and phosphoinositide metabolism in membrane traffic: insights from vesicle recycling in nerve terminals. Proc Natl Acad Sci U S A 101:8262–8269

Wenk MR, Pellegrini L, Klenchin VA, Di Paolo G, Chang S, Daniell L, Arioka M, Martin TF, De Camilli P (2001) PIP kinase Igamma is the major PI(4,5)P(2) synthesizing enzyme at the synapse. Neuron 32:79–88

Whitman M, Downes CP, Keeler M, Keller T, Cantley L (1988) Type I phosphatidylinositol kinase makes a novel inositol phospholipid, phosphatidylinositol-3-phosphate. Nature 332:644–646

Yabuta T, Shinmura K, Tani M, Yamaguchi S, Yoshimura K, Katai H, Nakajima T, Mochiki E, Tsujinaka T, Takami M et al (2002) E-cadherin gene variants in gastric cancer families whose probands are diagnosed with diffuse gastric cancer. Int J Cancer 101:434–441

Yamamoto A, DeWald DB, Boronenkov IV, Anderson RA, Emr SD, Koshland D (1995) Novel PI(4)P 5-kinase homologue, Fab1p, essential for normal vacuole function and morphology in yeast. Mol Biol Cell 6:525–539

Yamamoto M, Hilgemann DH, Feng S, Bito H, Ishihara H, Shibasaki Y, Yin HL (2001) Phosphatidylinositol 4,5-bisphosphate induces actin stress-fiber formation and inhibits membrane ruffling in CV1 cells. J Cell Biol 152:867–876

Yamazaki M, Miyazaki H, Watanabe H, Sasaki T, Maehama T, Frohman MA, Kanaho Y (2002) Phosphatidylinositol 4-phosphate 5-kinase is essential for ROCK-mediated neurite remodeling. J Biol Chem 277:17226–17230

Yap A, Crampton M, Hardin J (2007) Making and breaking contacts: the cellular biology of cadherin regulation. Curr Opin Cell Biol 19:508–514

Yeaman C, Grindstaff KK, Wright JR, Nelson WJ (2001) Sec6/8 complexes on trans-Golgi network and plasma membrane regulate late stages of exocytosis in mammalian cells. J Cell Biol 155:593–604

Yin HL, Janmey PA (2003) Phosphoinositide regulation of the actin cytoskeleton. Annu Rev Physiol 65:761–789

Yokogawa T, Nagata S, Nishio Y, Tsutsumi T, Ihara S, Shirai R, Morita K, Umeda M, Shirai Y, Saitoh N, Fukui Y (2000) Evidence that 3′-phosphorylated polyphosphoinositides are generated at the nuclear surface: use of immunostaining technique with monoclonal antibodies specific for PI 3,4-P(2). FEBS Lett 473:222–226

Yoshinaga-Ohara N, Takahashi A, Uchiyama T, Sasada M (2002) Spatiotemporal regulation of moesin phosphorylation and rear release by Rho and serine/threonine phosphatase during neutrophil migration. Exp Cell Res 278:112–122

Yu H, Fukami K, Watanabe Y, Ozaki C, Takenawa T (1998) Phosphatidylinositol 4,5-bisphosphate reverses the inhibition of RNA transcription caused by histone H1. Eur J Biochem 251:281–287

Zhang X, Loijens JC, Boronenkov IV, Parker GJ, Norris FA, Chen J, Thum O, Prestwich GD, Majerus PW, Anderson RA (1997) Phosphatidylinositol-4-phosphate 5-kinase isozymes catalyze the synthesis of 3-phosphate-containing phosphatidylinositol signaling molecules. J Biol Chem 272:17756–17761

Zhang X, Orlando K, He B, Xi F, Zhang J, Zajac A, Guo W (2008) Membrane association and functional regulation of Sec3 by phospholipids and Cdc42. J Cell Biol 180:145–158

Zhao K, Wang W, Rando OJ, Xue Y, Swiderek K, Kuo A, Crabtree GR (1998) Rapid and phosphoinositol-dependent binding of the SWI/SNF-like BAF complex to chromatin after T lymphocyte receptor signaling. Cell 95:625–636

Zheng JH, Knighton DR, Parello J, Taylor SS, Sowadski JM (1991) Crystallization of catalytic subunit of adenosine cyclic monophosphate-dependent protein kinase. Methods Enzymol 200:508–521

Zou J, Marjanovic J, Kisseleva MV, Wilson M, Majerus PW (2007) Type I phosphatidylinositol-4,5-bisphosphate 4-phosphatase regulates stress-induced apoptosis. Proc Natl Acad Sci U S A 104:16834–16839

Zovein AC, Luque A, Turlo KA, Hofmann JJ, Yee KM, Becker MS, Fassler R, Mellman I, Lane TF, Iruela-Arispe ML (2010) Beta1 integrin establishes endothelial cell polarity and arteriolar lumen formation via a Par3-dependent mechanism. Dev Cell 18:39–51

Chapter 3
The Phospholipase C Isozymes and Their Regulation

Aurelie Gresset, John Sondek and T. Kendall Harden

Abstract The physiological effects of many extracellular neurotransmitters, hormones, growth factors, and other stimuli are mediated by receptor-promoted activation of phospholipase C (PLC) and consequential activation of inositol lipid signaling pathways. These signaling responses include the classically described conversion of phosphatidylinositol(4,5)P_2 to the Ca^{2+}-mobilizing second messenger inositol(1,4,5)P_3 and the protein kinase C-activating second messenger diacylglycerol as well as alterations in membrane association or activity of many proteins that harbor phosphoinositide binding domains. The 13 mammalian PLCs elaborate a minimal catalytic core typified by PLC-δ to confer multiple modes of regulation of lipase activity. PLC-β isozymes are activated by Gαq- and Gβγ-subunits of heterotrimeric G proteins, and activation of PLC-γ isozymes occurs through phosphorylation promoted by receptor and non-receptor tyrosine kinases. PLC-ε and certain members of the PLC-β and PLC-γ subclasses of isozymes are activated by direct binding of small G proteins of the Ras, Rho, and Rac subfamilies of GTPases. Recent high resolution three dimensional structures together with biochemical studies have illustrated that the X/Y linker region of the catalytic core mediates autoinhibition of most if not all PLC isozymes. Activation occurs as a consequence of removal of this autoinhibition.

Keywords Phospholipase C · Inositol lipid signaling · Heterotrimeric G protein · Ras GTPase · Tyrosine kinase · X/Y-linker-mediated autoinhibition

3.1 Introduction

The physiological effects of many hormones, neurotransmitters, growth factors, and other extracellular stimuli are initiated through receptor-promoted inositol lipid signaling. The ground-breaking work of Hokin and Hokin in the 1950s/1960s (Hokin

---

T. K. Harden (✉) · A. Gresset · J. Sondek
Department of Pharmacology,
University of North Carolina School of Medicine, Chapel Hill, NC 27599, USA
e-mail: tkh@med.unc.edu

J. Sondek
Departments of Pharmacology and Biochemistry and the Lineberger Cancer Center,
University of North Carolina School of Medicine, Chapel Hill, NC 27599, USA

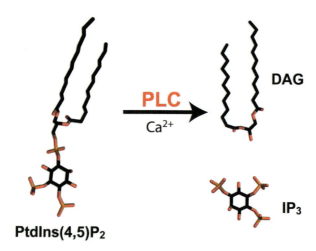

Fig. 3.1 The enzyme activity of phospholipase C. Phospholipase C (*PLC*) isozymes convert membrane phosphatidylinositol (4,5)bisphosphate (PtdIns(4,5)P$_2$) into the Ca^{2+}-mobilizing second messenger inositol(1,4,5) trisphosphate (IP$_3$) and the protein kinase C-activating second messenger diacylglycerol (*DAG*). PtdIns(4,5)P$_2$ also acts as a second messenger that binds to a broad range of membrane, cytoskeletal, and cytosolic proteins to change their activities

and Hokin 1953) and of Berridge, Michell, Nishizuka, and many other investigators in the 1970s/1980s established the importance of membrane inositol lipids in hormone action (Berridge 1987; Michell 1975; Nishizuka 1992). Receptors for extracellular stimuli promote activation of phospholipase C (PLC), which converts (Fig. 3.1) phosphatidylinositol (4,5)bisphosphate (PtdIns(4,5)P$_2$) into the Ca^{2+}-mobilizing second messenger, inositol (1,4,5)trisphosphate (Ins(1,4,5)P$_3$), and the protein kinase-activating second messenger, diacylglycerol (DAG). Although PLC-catalyzed formation of second messengers from PtdIns(4,5)P$_2$ constitutes one of the major mammalian cell signaling responses, inositol lipids themselves also carry out important signaling functions. Indeed, PtdIns(4,5)P$_2$ selectively binds to PH, FYVE, and PX domains (Lemmon 2003), and the activities and/or subcellular localization of a broad range of proteins involved in cell signaling (e.g. PTEN and Ca^{2+}, K$^+$, and Na$^+$ channels), actin assembly and remodeling, and vesicle trafficking are modified (Ling et al. 2006; Suh and Hille 2008; Yin and Janmey 2003).

Receptor-mediated regulation of inositol lipid signaling historically has been considered to occur through two major mechanisms (Exton 1996; Rhee 2001). First, PLC-γ isozymes are activated by a panoply of growth factors and immunological stimuli that signal through receptor and non-receptor tyrosine kinases. Second, the G protein-coupled receptors (GPCR) represent one of the largest classes of proteins in the mammalian genome, and a large proportion of these receptors produce their major cellular responses through activation of PLC-β isozymes. Although these two major components of inositol lipid signaling make enormous contributions to the cell "signalsome", the existence of at least 13 different mammalian PLC isozymes suggests much more extensive modes of regulation (Harden and Sondek 2006). Indeed, multiple Ras superfamily GTPases directly activate PLC-ε, as well as certain of the $G\alpha_q$- (e.g. PLC-β2) and tyrosine kinase- (e.g. PLC-γ2) activated PLCs (Harden et al. 2009), and the inositol lipid signaling field continues to expand in surprising directions. This chapter focuses on the mammalian PLC isozymes, their complex modes of regulation, and their physiological functions.

3.2 Phosphoinositide-specific PLC

PLC enzymes were purified initially several decades ago from a variety of tissues and were shown to exist in at least three major isozyme forms based on size and immunoreactivity (Hofmann and Majerus 1982; Ryu et al. 1986, 1987a, 1987b; Takenawa and Nagai 1981). These included enzymes that are activated by tyrosine phosphorylation (Meisenhelder et al. 1989; Wahl et al. 1989) or by G proteins (Morris et al. 1990). The deduced amino acid sequences obtained from initial cloning of the cDNAs of several of these enzymes revealed the existence of PLC-β, -δ, and -γ isozymes (Suh et al. 1988). Multiple subtypes eventually were shown to exist in each of these isozyme classes, but a protein originally designated as PLC-α proved to not be a PLC. Additional PLC isozymes (-ε, -ζ, -η) were subsequently discovered and cloned (Harden and Sondek 2006), and a total of 13 isozymes in six mammalian PLC family members are currently recognized (Fig. 3.2).

3.3 Catalytic Function and Structure of Conserved Core Domains of PLC

PLC enzymes are calcium-dependent phosphodiesterases that preferentially hydrolyze PtdIns(4,5)P_2 into DAG and Ins(1,4,5)P_3 (Fig. 3.1). The core structure of these isozymes (Fig. 3.2) consists of an N-terminal PH domain, an array of four EF-hands, a catalytic triose phosphate isomerase (TIM) barrel comprised of two halves (X and Y boxes), and a C-terminal C2 domain (Katan and Williams 1997). Additional regulatory domains evolved that engender unique regulatory mechanisms to individual isozymes (Harden and Sondek 2006). The structure of PLC-β2 (Fig. 3.3) highlights the conserved core structure found in all PLC isozymes.

The conserved tertiary structure of PH domains is composed of a sandwich of seven β-strands capped on one end by a C-terminal α-helix and on the other end by three loops, which diverge in both length and amino acid sequence (Rebecchi and Scarlata 1998). Although the structural folds of PH domains are generally conserved, they carry out diverse functions. Thus, the N-terminal PH domains of isozymes of the PLC-δ and PLC-γ families bind PtdIns(4,5)P_2 and PtdIns(3,4,5)P_3, respectively (Essen et al. 1996; Falasca et al. 1998; Singh and Murray 2003), whereas the PH domain of PLC-β2 binds Rac GTPases with high affinity (Illenberger et al. 2003a; Snyder et al. 2003). The PH domain of PLC-δ1 is tethered to the rest of the protein via a flexible linker and is highly mobile (Essen et al. 1996; Ferguson et al. 1995). In contrast, the PH domain of PLC-β2 makes direct contacts with the other domains of

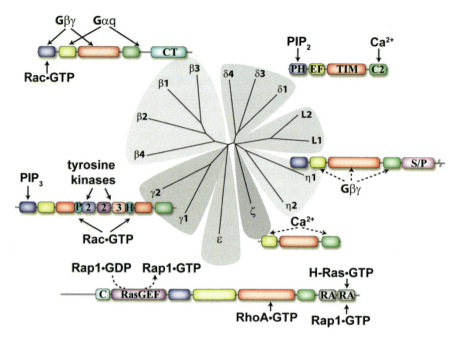

Fig. 3.2 The mammalian PLC isozymes and their modes of regulation. The human PLC isozymes were aligned based upon conservation of protein sequence, and a dendrogram that clusters similar sequences within shared branches is presented. The common core of these isozymes includes a pleckstrin homology (PH) domain (*purple*), a series of four EF-hands (*yellow*), a catalytic TIM barrel (*pink*), and a C2 domain (*green*). The four PLC-β isozymes contain a long C-terminal (CT) domain (*light blue*). The two PLC-γ isozymes contain conserved domains inserted within the TIM barrel that include a split PH domain, two Src-homology 2 (SH2) domains and a single Src-homology 3 (SH3) domain. PLC-ε contains a guanine nucleotide exchange domain (RasGEF) that activates Rap1 and possibly other GTPases and two C-terminal Ras-association (RA) domains that bind activated Ras GTPases. A cysteine-rich (C) domain of unestablished function occurs at the N-terminus. PLC-ζ is the only mammalian PLC that lacks a PH domain. PLC-η isozymes contain a serine/proline (S/P) rich region in the C-terminus. The PLC-like (PLC-L) proteins exhibit the common core of other PLC isozymes but are catalytically inactive due to mutations of critical residues in the active site. Established modes of regulation are indicated for each of the PLC isozyme classes

the catalytic core and remains tightly associated during activation (Hicks et al. 2008; Jezyk et al. 2006).

Typical EF-hands are calcium-binding motifs composed of two helixes (E and F) joined by a loop and divided into pairwise lobes (Kawasaki and Kretsinger 1994). The electron density of the loops connecting the secondary elements of the EF hands is incomplete in the structures of both PLC-δ1 and PLC-β2 indicating a high degree of flexibility in this region (Essen et al. 1996; Hicks et al. 2008; Jezyk et al. 2006). Little evidence exists for Ca^{2+}-promoted regulation of PLC isozymes through the EF-hands, and Ca^{2+} is not bound in the EF-hands in the structures of PLC-δ1, PLC-β2, or PLC-β3. In contrast, the structure of PLC-β3 in an activated complex with $G\alpha_q$

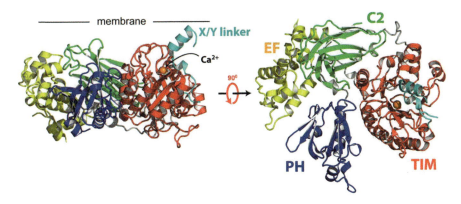

Fig. 3.3 Three-dimensional structure of PLC-β2. *Left panel*, A ribbon diagram is illustrated of the three dimensional structure (PDB 2ZKM) of PLC-β2 solved at 1.6 Å resolution by Hicks and coworkers (Hicks et al. 2008). The PH domain (*purple*), EF hands (*yellow*), TIM barrel (*red*) and C2 domain (*green*) are colored as in Fig. 3.2. The Ca^{2+} co-factor (*orange sphere*) within the active site and the X/Y linker region (*cyan*) that occludes the active site also are shown. The approximate membrane-binding surface is indicated. *Right panel*, The structure is rotated 90° with respect to the *left panel*. This view emphasizes occlusion of the active site within the TIM barrel by the X/Y linker

highlights a novel function for EF hands in G protein-dependent signaling (Waldo et al. 2010). A cassette uniquely present between the third and fourth EF hands of the four mammalian PLC-β isozymes contains the structural determinants necessary for PLC-β-promoted enhancement of GTP hydrolysis by it activator $G\alpha_q$.

The C2 domain folds into an eight β-stranded antiparallel sandwich, with three loops at one end of the sandwich forming calcium binding sites (Nalefski and Falke 1996). The C2 domains of PLC-δ1 and PLC-β2 exhibit very similar structures tightly packed against the TIM barrel, likely to maintain the structural integrity of the catalytic core (Essen et al. 1996; Jezyk et al. 2006). The C2 domain of PLC-δ1 binds Ca^{2+} and promotes translocation of the isozyme to the plasma membrane (Essen et al. 1997). However, the residues coordinating Ca^{2+} in PLC-δ1 are not generally conserved among other PLC isozymes, and it is unclear whether Ca^{2+} regulates any of the other mammalian isozymes in a similar fashion. Although not formally part of the C2 domain, conserved sequences found at its N- and C-terminal ends in PLC-β isozymes provide the major binding surface for activated $G\alpha_q$ (Waldo et al. 2010).

3.4 Mechanism of PtdIns(4,5)P₂ Hydrolysis

The catalytic TIM barrel is the most highly conserved region among PLC isozymes with 60–70% sequence identity. The X and Y boxes fold together in an alternative pattern of α-helices on the outside and β-strands in the inner part of the barrel to constitute the active site of the lipase (Essen et al. 1996; Wierenga 2001). The

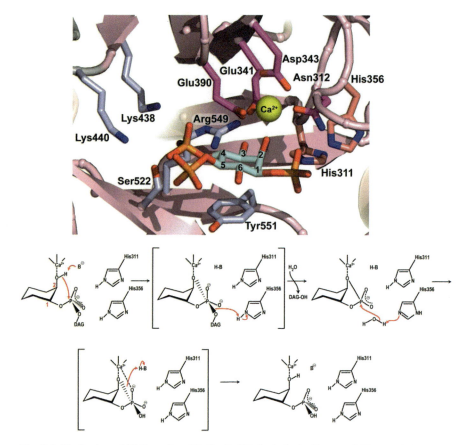

Fig. 3.4 Mechanism of PLC-catalyzed PtdIns(4,5)P_2 hydrolysis. *Top panel*, The catalytic site of PLC-δ1 (Essen et al. 1996) is shown. The residues that ligate the soluble head group (Ins(1,4,5)P_3) of the substrate are colored in *light blue*. The residues that ligate the essential Ca^{2+} cofactor (*yellow sphere*) are colored in *magenta*. The residues that are essential for the acid-base mechanism of catalysis are colored in *salmon*. The oxygen atoms of the side chains are colored in *red*, and the nitrogen atoms of the side chains are colored in *blue*. *Bottom panel*, The mechanism of PtdIns(4,5)P_2 hydrolysis as proposed by Essen et al. (1996) is presented

structure of PLC-δ1 first illustrated the organization of the active site of a PLC and revealed the mechanism for PtdIns(4,5)P_2 hydrolysis (Essen et al. 1996).

The active site is formed as a solvent-accessible depression at the C-terminal ends of the β-strands (Figs. 3.3 and 3.4). Indeed, a ridge of hydrophobic residues, Leu320, Tyr358, Phe360, Leu529, and Trp555 surrounding the active site of PLC-δ1 facilitates insertion of the catalytic domain into the lipid bilayer (Essen et al. 1996). A PLC-δ1 mutant containing alanine substitution of these bulky nonpolar residues exhibited activity similar to wild-type PLC-δ1 when PtdIns(4,5)P_2 hydrolysis was measured in detergent-mixed micelles but was a much less effective enzyme in assays of PtdIns(4,5)P_2 hydrolysis using phospholipid vesicles (Ellis et al. 1998). These residues are conserved across all PLC isozymes and assume the same orientation in the structures of PLC-β2 and PLC-β3 (Hicks et al. 2008; Jezyk et al. 2006; Waldo et al. 2010).

Eukaryotic PLC enzymes preferentially hydrolyze PtdIns(4,5)P$_2$, but also hydrolyze PtdIns(4)P and to a much lesser extent PdtIns (Ryu et al. 1987b). Ins(1,4,5)P$_3$ buried at the bottom of the active site in PLC-δ1 provided the initial structural snapshot of substrate recognition (Essen et al. 1996). An extensive network of H-bonds and salt-bridge interactions formed through the side chains of Lys438, Lys440, Ser522, and Arg549 with the 4′ and 5′ phosphorylated hydroxyl groups of the inositol ring favors interaction with lipids phosphorylated at both positions. The aromatic ring of Tyr551 also is parallel with the inositol ring and forms numerous van der Waals contacts with it (Fig. 3.4).

The essential Ca^{2+} cofactor is ligated by surrounding acidic residues, Asn312, Glu341, Asp343, and Glu390, and PLC activity is completely lost after, for example, mutation of Glu341 to Gly (Cheng et al. 1995). Ligation of Ca^{2+} with the 2′-hydroxyl group of the inositol ring also is essential for PLC activity (Essen et al. 1996). In contrast, prokaryotic PLC enzymes utilize basic amino acids to fulfill the functional requirements of the Ca^{2+} cofactor and thus are Ca^{2+}-independent (Heinz et al. 1998).

PtdIns(4,5)P$_2$ hydrolysis follows a general acid/base catalytic scheme (Fig. 3.4). Ca^{2+} lowers the pK$_a$ of the 2-hydroxyl group of the inositol ring to facilitate its deprotonation, and Glu341 is a putative general base carrying out nucleophilic attack on the 1-phosphate (Ellis et al. 1995, 1998). This initial step in hydrolysis leads to formation of a cyclic intermediate, which is stabilized by His311 and Ca^{2+} through ligation of the 1-phosphate. His356 then utilizes a proton from water to promote nucleophilic attack on the pentavalent cyclic intermediate and DAG and Ins(1,4,5)P$_3$ are formed (Cheng et al. 1995; Ellis et al. 1995; Essen et al. 1996; Heinz et al. 1998). All of the residues participating in substrate specificity, Ca^{2+} coordination, and the catalytic reaction are strictly conserved across the PLC family (Fig. 3.4) and are positioned in similar orientation in PLC-β2 structures (Jezyk et al. 2006; Hicks et al. 2008), indicating that the mechanism of PtdIns(4,5)P$_2$ hydrolysis is conserved throughout eurakyotic PLC enzymes.

3.5 PLC Subfamilies and Their Regulation

3.5.1 *PLC-δ Isozymes*

PLC-δ is found in early eukaryotes, including yeast (*Saccharomyces cerevisiae* and *Schizosaccharomyces pombe*) (Andoh et al. 1995; Payne and Fitzgerald-Hayes 1993; Yoko-o et al. 1993) and slime mold (*Dictyostelium discoideum*) (Drayer and van Haastert 1992). Three PLC-δ isozymes (PLC-δ1, -δ3, and -δ4) exist in mammals (Harden and Sondek 2006) ("PLC-δ2" proved to be a species homologue of PLC-δ4 (Irino et al. 2004)). Most, if not all, cells express at least one of these isoforms, and all three are broadly, if not ubiquitously, expressed (Suh et al. 2008). PLC-δ1 is mainly a cytoplasmic protein, whereas PLC-δ3 is detected in membrane fractions. PLC-δ4 is principally located in the nucleus where its expression is directly linked with the cell cycle.

3.5.1.1 Regulation

Activation of PLC-δ1 occurs through association with membrane surfaces driven by PtdIns(4,5)P$_2$ binding by the PH domain and Ca^{2+} binding by the C2 domain. The structure of the isolated PH domain of PLC-δ1 in complex with Ins(1,4,5)P$_3$ highlighted its capacity to bind phospholipids with high affinity and therefore serve as a plasma membrane anchor for PLC-δ isoforms (Ferguson et al. 1995). PH domains are typically highly polarized with a positively charged surface interacting with the negatively charged inner face of the plasma membrane. Consistent with this idea, the binding site for the soluble head group of PtdIns(4,5)P$_2$, Ins(1,4,5)P$_3$, is located opposite to the C-terminal α-helix and at the center of the positively charged region of the PH domain of PLC-δ1. The residues coordinating Ins(1,4,5)P$_3$ in the β1/β2 and β3/β4 loops are poorly conserved among PH domains. Moreover, these residues, which ligate PtdIns(4,5)P$_2$ with high affinity in PLC-δ1 are not conserved in the PH domains of the isozymes of the other PLC subfamilies, strongly suggesting that the N-terminal PH domain is not functionally redundant (Harden and Sondek 2006).

Several studies have highlighted the functional role of the PH domain in modulation of lipase activity of PLC-δ isoforms. These isozymes carry out a scooting mode of substrate hydrolysis facilitated by the PH domain. Binding of PtdIns(4,5)P$_2$ to the PH domain anchors PLC-δ1 at the plasma membrane and multiple PtdIns(4,5)P$_2$ molecules are processively hydrolyzed by the catalytic site (Lomasney et al. 1996). Thus, PLC-δ isoforms hydrolyze numerous PtdIns(4,5)P$_2$ molecules during a single binding event at the membrane interface. This two-substrate process also allows feedback regulation of enzymatic activity through decreases in local PtdIns(4,5)P$_2$ concentration. Experimental evidence for this model includes observation of a concentration-dependent increase in PLC-δ1 activity with increases in the mole fraction of PtdIns(4,5)P$_2$, but not PtdIns(4)P (Lomasney et al. 1996). Truncation of the PH domain, substitution at residues of the PH domain that ligate PtdIns(4,5)P$_2$, or competitive addition of Ins(1,4,5)P$_3$ all impair PtdIns(4,5)P$_2$-stimulated PLC-δ1 activity, but do not affect function of the catalytic site (Yagisawa et al. 1998).

The PH domain of PLC-δ1 is the prototypical PtdIns(4,5)P$_2$-binding module, and the high affinity and selective interaction between PtdIns(4,5)P$_2$ and this PH domain has been exploited to generate fluorescent probes for monitoring local phospholipid signaling. Thus, the PH domain of PLC-δ1 fused to green fluorescent protein (GFP) reveals the localization and dynamics of PtdIns(4,5)P$_2$ in living cells (Stauffer et al. 1998). More recent studies utilized an enhanced GFP-tag to delimit the cellular localization of PtdIns(4,5)P$_2$ to distinct regions of the plasma membrane, such as membrane ruffles. PtdIns(4,5)P$_2$ associates with the cytoskeleton and binds and modulates many actin-regulatory proteins, including gelsolin, cofilin, profilin, the Arp2/3 complex, and Wiskott-Aldrich syndrome protein (Nebl et al. 2000). Indeed,

PtdIns(4,5)P$_2$ acts as a second messenger regulating adhesion between the plasma membrane and cytoskeletal structure (Raucher et al. 2000).

The C2 domain has been proposed to mediate Ca^{2+}-stimulated membrane association of PLC-δ isoforms, and the crystal structure of PLC-δ1 reveals the existence of three Ca^{2+} binding regions (CBRs) in the loops of the C2 domain (Essen et al. 1997; Grobler et al. 1996). Membrane binding studies with isolated C2 domains from PLC-δ1, -δ3, and -δ4 indicate that Ca^{2+} binding switches the electrostatic potential to favor non-specific electrostatic interactions with the plasma membrane (Ananthanarayanan et al. 2002). The C2 domains of PLC-δ1 and -δ3 also form a protein-Ca^{2+} complex with the anionic lipid phosphatidylserine (PS) and therein target these isozymes to specific regions of the plasma membrane (Lomasney et al. 1999). In contrast, the C2 domain of PLC-δ4 lacks two of the four aspartic acid residues coordinating Ca^{2+} and does not exhibit Ca^{2+}-dependent translocation to PS-enriched membrane regions (Ananthanarayanan et al. 2002).

To date, Ca^{2+} is the only regulator that directly enhances the activity of PLC-δ isoforms. Reconstitution assays using permeabilized cells depleted of PLC isozymes illustrated that the lipase activity of PLC-δ1, but not PLC-β1 or PLC-γ1, is stimulated in a concentration dependent manner by physiological concentrations (10 nM to 10 μM) of Ca^{2+}, indicating that Ca^{2+} alone is sufficient to promote increased lipase activity of PLC-δ isoforms (Allen et al. 1997; Kim et al. 1999). Whether Ca^{2+}-mediated activation is further potentiated through interaction of these isozymes with other regulators has not been clearly established.

Several lines of evidence suggest that PLC-δ isoforms are regulated by G-protein mediated signaling. For example, stimulation of the α1-adrenergic receptor has been proposed to directly regulate PLC-δ1 through activation of the atypical G-protein transglutaminase II, also called Gα$_H$ (Feng et al. 1996). Additional studies indicate that GPCR indirectly activate PLC-δ through signaling pathways selective for other PLC isoforms. For example, inositol phosphate production downstream of the angiotensin II type 1 receptor requires activation of PLC-β isoforms via Gα$_{q/11}$ and is potentiated by interaction between RalA and PLC-δ1 (Godin et al. 2010; Sidhu et al. 2005). Thus, PLC-δ1 may serve as an amplifier of signaling initiated or mediated by other PLC isoforms (Guo et al. 2010).

3.5.1.2 Physiology

PLC-δ1 null mice exhibit a hairless phenotype that is reminiscent of nude mice in which loss of function of Foxn1, a member of the winged helix/forkhead family of transcription factors, leads to hair loss and an inborn dysgenesis of the thymus (Nehls et al. 1994). In fact, PLC-δ1 is a Foxn1-inducible gene that regulates the expression of hair keratins, although the molecular mechanism underlying this effect remains to be identified.

PLC-δ1 null mice also display symptoms of skin inflammation (Ichinohe et al. 2007). Exogenous expression of PLC-δ1 attenuates LPS-induced upregulation of IL-1β, a pro-inflammatory cytokine that typically induces expression of IL-6, which in turn promotes keratinocyte proliferation. It is likely that the lack of PLC-δ1 in keratinocytes results in aberrant production of IL-1β and subsequent upregulation of IL-6 expression, leading to skin inflammation and epidermal hyperplasia. Together, these results suggest that PLC-δ1 regulates homeostasis of the immune system in the skin.

Although the localization of PLC-δ1 is primarily cytoplasmic, PLC-δ1 contains both nuclear export and import sequences that allow it to shuttle between the nucleus and the cytoplasm (Yamaga et al. 1999). PLC-δ1 accumulates in the nucleus at the G_1/S boundary of the cell cycle, and its accumulation is positively correlated with the level of nuclear PtdIns(4,5)P$_2$ (Stallings et al. 2005). Depletion of PLC-δ1 in the nucleus delays the completion of S phase and transition into G_2/M phase leading to decreased cell proliferation and growth rate (Stallings et al. 2008). Levels of cyclin E, a key regulator of the G_2/M transition, are also elevated. Therefore, PLC-δ1 modulates nuclear phospholipid metabolism critical for cell cycle progression. Consistent with this idea, PtdIns(4,5)P$_2$ inhibits histone H1-mediated basal transcription initiated by RNA polymerase II through a direct interaction with its C-terminal tail (Yu et al. 1998).

Several studies also implicate PLC-δ1 in neurodegenerative disorders (Shimohama et al. 1993). Specifically, PLC-δ1 accumulates in the neurofibrillary tangles of Alzheimer patients (Shimohama et al. 1991). Overexpression of PLC-δ1 protein also is associated with high PLC activity in Alzheimer patients, suggesting that PLC-δ1-mediated phospholipid turnover plays an important role in the development of this neurodegenerative disease. A recent study indicated a direct link between activation of the N-methyl-D-aspartic acid receptor under oxidative stress conditions and an increase in PLC-δ1 protein levels (Nagasawa et al. 2004). This result provides a putative molecular link between neurons responding to oxidative stress and the accumulation of neurofibrillary tangles and senile plaques in Alzheimer's disease.

PLC-δ1 was recently identified as a tumor suppressor located at chromosome 3p22, an important tumor suppressor locus. Down-regulation of PLC-δ1 in esophageal squamous cell carcinoma is associated with promoter hypermethylation and frequent allelic loss at the PLC-δ1 locus (Fu et al. 2007). Epigenetic regulation of PLC-δ1 also was linked to cancer progression in other tissues. For example, PLC-δ1 expression is greatly reduced through hypermethylation of its gene in both gastric cancer cell lines and primary tumors. This gene silencing is associated with later stages of gastric cancer (Hu et al. 2009). Moreover, this study showed that PLC-δ1 decreases cell motility, which is consistent with the idea that PLC-δ1 regulates proteins, such as actin-regulated protein, Rho GTPases, or metalloprotease proteins that in turn modulate cytoskeletal rearrangement. Together, these studies indicate that PLC-δ1 is frequently silenced by epigenetic alteration in a tumor-specific manner. Another study reported that expression of mRNA for PLC-δ1 and -δ3 directly correlates with the metastatic state of human breast cell lines and that PLC-δ1 and -δ3 are more highly expressed in transformed cell lines (Rebecchi et al. 2009). This study also supported an important role for PLC-δ1 and -δ3 in cell growth and migration.

Disruption of the PLC-δ3 gene in mice has not resulted to date in a reported abnormality. However, embryonic lethality occurs after disruption of both PLC-δ1 and PLC-δ3 due to developmental failure of the placenta (Nakamura et al. 2005). The labyrinth trophoblast layer of the placenta was poorly vascularized in the PLC-δ1/PLC-δ3 double knock-out mouse and exhibited reduced cell proliferation and abnormal cell death (Nakamura et al. 2005).

Disruption of the PLC-δ4 gene leads to male infertility, whereas female mice remain fertile. *In vitro* fertilization studies indicated that sperm from PLC-δ4-deficient mice were unable to maintain a sustained influx of Ca^{2+}, which is a critical component of the interaction of the sperm with the zona pellucida in the acrosome reaction (Fukami et al. 2003). Therefore, PLC-δ4 is important in the early steps of fertilization.

3.5.2 PLC-β Isozymes

The four PLC-β isozymes differ in expression pattern and regulation. While PLC-β1 is highly expressed in the cerebral cortex and hippocampus (Homma et al. 1989), PLC-β2 expression is largely, but not entirely, limited to hematopoietic cells (Park et al. 1992). PLC-β3 is broadly expressed (Jhon et al. 1993), while PLC-β4 expression is enriched in the cerebellum and the retina (Adamski et al. 1999). Historically, PLC-β isozymes have been structurally characterized by the unique presence of a C-terminal (CT) coiled-coil domain thought to be important for dimerization, membrane association, and activation by Gα-subunits (Ilkaeva et al. 2002; Singer et al. 2002).

3.5.2.1 Regulation

PLC-β isozymes are effectors of heterotrimeric G-proteins downstream of GPCR belonging to the rhodopsin superfamily of seven transmembrane receptors. These isozymes are activated by Gα-subunits of the G_q subfamily (Smrcka et al. 1991; Taylor et al. 1991; Waldo et al. 1991) as well as by Gβγ (Boyer et al. 1992; Camps et al. 1992) and mediate the physiological actions of many extracellular stimuli.

The Gq family consists of four different Gα-subunits ($G\alpha_q$, $G\alpha_{11}$, $G\alpha_{14}$, $G\alpha_{16}$) of closely related sequence (Hepler and Gilman 1992). All four of these G proteins markedly activate PLC-β isozymes in intact cells as well as in assays with purified components using phospholipid vesicles. All four PLC-β isozymes are activated by $G\alpha_q$, although there may be selectivity for PLC-β1 and -β3 over PLC-β2 (Paterson et al. 1995; Smrcka and Sternweis 1993). The interface between $G\alpha_q$ and PLC-β isoforms was originally thought to be contained within the isozyme-specific CT domain (Ilkaeva et al. 2002; Paulssen et al. 1996; Singer et al. 2002), but the recent crystal structure of an activated complex of $G\alpha_q$ with PLC-β3 (Waldo et al. 2010) revealed a novel interface outside the CT domain. Thus, $G\alpha_q$ interacts with a unique extension of the C2 domain which forms a helix-turn-helix. This motif is found as a highly conserved insert in all PLC-β isozymes including the two PLC-βs of *C. elegans*; conversely, it is not found in other PLC isozymes. A similar structure also is present in other Gα_q effectors, including p63RhoGEF (Lutz et al. 2007) and G protein-coupled receptor kinase 2 (Tesmer et al. 2005). In all three effectors, the helical motif interacts with switch 2 and α3 from Gα_q and is necessary and sufficient to confer Gα_q binding.

The CT domain of PLC-β3 is not necessary for binding of Gα_q (Waldo et al. 2010) as was suggested in earlier studies, and the role of this domain in signaling by the PLC-β subgroup of isozymes remains incompletely understood. Clearly, this domain is polybasic and is important for membrane association. A three-dimensional structure of the CT domain of avian PLC-β2 highlights three α-helices packed together to form a coiled-coil likely important for dimerization (Singer et al. 2002). However, the functional significance of this dimerization remains unclear. As is the case with the PH domain in PLC-δ isoforms, the polybasic CT domain of PLC-β isoforms likely provides an anchor point for interaction with the plasma membrane and almost certainly works in coordination with G protein binding to orient the active site for efficient enzymatic activity.

PLC-β1 was the first GTPase-activating protein (GAP) identified for heterotrimeric G proteins (Berstein et al. 1992) and all PLC-β isoforms robustly stimulate the hydrolysis of GTP by Gα_q-subunits of the Gq family (Biddlecome et al. 1996; Ross 2008). Not only does this activity result in rapid turn-off of Gα_q-promoted signaling once agonist-dependent stimulation of GPCR is terminated, it also markedly alters the dynamics of PLC-β-mediated signaling nodes. The rates of activation and deactivation are robustly increased and signaling acuity is sharpened. Indeed, the magnitude of signaling may paradoxically increase as a consequence of a phenomenon known as "kinetic scaffolding" (Ross 2008). Gα_q remains in a signaling complex with the activated GPCR through multiple cycles of activation/deactivation and the steady state amount of Gα_q in the GTP state is increased. The recent crystal structure of the PLC-β3 • Gα_q complex indicates that the structural requirements for GAP activity exist within a unique insertion between the third and fourth EF hands (Waldo et al. 2010). This eight amino acid sequence is conserved in all PLC-β isoforms but not in other PLC isozymes. An asparagine (Asp260 in PLC-β3) in this loop directly interacts with the catalytic Gln209 of Gα_q and stabilizes the pentameric transition state necessary for GTP hydrolysis. The same mechanism independently evolved in the very large family of regulator of G-protein signaling (RGS) proteins (Tesmer et al. 1997).

Overexpression of PLC-β2 with G$\beta\gamma$-subunits resulted in large increases in inositol phosphate accumulation, and reconstitution assays with purified avian and mammalian PLC-β isozymes illustrated that G$\beta\gamma$ stimulates inositol lipid signaling by directly binding to PLC-β isozymes (Boyer et al. 1992; Camps et al. 1992; Smrcka and Sternweis 1993). This effect is more prominent with PLC-β2 and PLC-β3 than with PLC-β1 and PLC-β4. Indeed, physiological signaling through PLC-β1 and PLC-β4 appears to be mostly, if not entirely, mediated through the Gα-subunits of the Gq family. In contrast, inositol lipid signaling downstream of Gi-linked GPCR (e.g. receptors for chemotactic peptides in neutrophils) occurs through G$\beta\gamma$-mediated activation of PLC-β2 and/or PLC-β3. The binding interface between PLC-β isoforms and G$\beta\gamma$ has not been firmly established and may include both the N-terminal PH domain as well as part of the catalytic TIM barrel (Barr et al. 2000; Wang et al. 2000b). Interestingly, PLC-β2 and PLC-β3 can be simultaneously activated by Gα$_q$ and Gβγ, and studies with purified proteins illustrate that the combined presence of activated Gα$_q$ and Gβγ results in supra-additive stimulation of PLC-β3 (Philip et al. 2010). This cooperative activation apparently accounts for synergistic activation of PLC-β often observed in cells during simultaneous activation of Gq- and Gi-activating GPCR (Rebres et al. 2011).

Members of the Rac subfamily of small GTPases also activate PLC-β2 (Illenberger et al. 1998) and potentially PLC-β3, but no detectable binding is observed with PLC-β1 or PLC-β4 (Illenberger et al. 2003a; Snyder et al. 2003). Biochemical approaches illustrated that the PH domain of PLC-β2 is both necessary and sufficient for Rac binding (Illenberger et al. 2003a; Snyder et al. 2003). A structure of a complex of GTP-bound Rac2 with PLC-β2 confirmed that the switch regions of Rac directly engage the PH domain of PLC-β2 (Jezyk et al. 2006). Rac-dependent activation of PLC-β2 occurs through recruitment to the plasma membrane since the active site of the lipase observed in a structure of PLC-β2 alone (Hicks et al. 2008) is superimposable with that of Rac2-bound isozyme (Jezyk et al. 2006). Recent studies quantified fluorescence recovery after photobleaching to illustrate that, whereas Rac recruits PLC-β2 to specific regions of the plasma membrane, recruitment promoted by Gβγ is more diffuse throughout the plasma membrane (Gutman et al. 2010; Illenberger et al. 2003b).

Hydrolysis of PtdIns(4)P and PtdIns(4,5)P$_2$ by nuclear PLC-β1 established the existence of nuclear inositol lipid signaling (Martelli et al. 1992). PLC-β1 apparently is the most abundant PLC-β isozyme in the nucleus, although the presence (in decreasing order of abundance) of PLC-β3, -β2, and -β4 has also been detected (Cocco et al. 1999). Although both the significance and regulation of nuclear signaling of these lipases needs further clarification, several studies suggest a role for PLC-β1 in controlling cell cycle progression, specifically at the G2/M boundary (Faenza et al. 2000; Fiume et al. 2009).

3.5.2.2 Physiology

PLC-β1-null mice experience sudden death preceded by epileptic seizures. This phenotype resembles responses observed with GABA$_A$ receptor antagonists, suggesting that PLC-β1 is essential for the normal function of inhibitory neuronal pathways (Kim et al. 1997). A recent study suggested that PLC-β1 regulates the plasticity of M1-muscarinic receptor expression in the adult neocortex, resulting in an imbalance between the muscarinic and dopaminergic systems, as often seen in schizophrenia (McOmish et al. 2008).

PLC-β2-deficient mice are viable but display a reduction in chemoattractant-stimulated inositol phosphate accumulation, intracellular Ca^{2+} levels, superoxide production, and cell surface MAC-1 expression, suggesting that PLC-β2 is critical for chemoattractant-elicited signals in leukocytes (Jiang et al. 1997). Neutrophils lacking PLC-β2 exhibit increased rates of chemotaxis, which is consistent with the idea the PLC-β2-dependent signaling negatively influences chemotaxis (Li et al. 2000).

PLC-β3 null mice exhibit higher sensitivity to morphine, suggesting that PLC-β3 suppresses μ-opioid receptor signaling possibly downstream of Gβγ (Xie et al. 1999). Conversely, PLC-β3(-/-) mice do not respond to sensory stimuli that induce itch responses (Han et al. 2006). PLC-β3 deficiency leads to premature death in mice and is associated with lymphomas and carcinomas (Xiao et al. 2009). PLC-β3-deficient mice also develop myeloproliferative disease due to the loss of Stat5 regulation by a PLC-β3•SHP-1 complex (Xiao et al. 2009).

PLC-β4 null mice develop ataxia (Jiang et al. 1996), motor defects, and impaired visual processing (Kim et al. 1997). A recent genome-wide profiling study of pancreatic cancer revealed a genetic substitution of Arg254 in PLC-β4 (Jones et al. 2008). This substitution diminishes GAP activity of PLC-β isozymes leading to aberrant $G\alpha_q$-mediated signaling (Waldo et al. 2010).

3.5.3 PLC-γ Isozymes

Two PLC-γ isozymes (PLC-γ1 and PLC-γ2) exist in mammals. Whereas PLC-γ1 is found ubiquitously, PLC-γ2 expression primarily is restricted to cells of the haematopoietic system (Homma et al. 1989). PLC-γ isozymes are structurally characterized by a large insertion between the two halves of the catalytic TIM barrel consisting of a split PH domain, two SH2 domains, and a SH3 domain. They are primarily regulated through phosphorylation by receptor and non-receptor tyrosine kinases in a mechanism that involves the unique domain insert of the linker region.

3.5.3.1 Regulation

Almost all growth factor receptors with intrinsic tyrosine kinase activity (RTKs) have been linked to stimulation of PLC-γ isozymes (Kamat and Carpenter 1997). Agonist binding stimulates dimerization and tyrosine autophosphorylation of these RTKs on the cytoplasmic side of the receptor, which creates docking sites for SH2-containing proteins like PLC-γ (Hubbard and Till 2000). Specifically, autophosphorylation of Tyr766 in fibroblast growth factor receptor 1 (Mohammadi et al. 1991), Tyr992 in epidermal growth factor receptor (Rotin et al. 1992), and Tyr1021 in platelet-derived growth factor receptor (Larose et al. 1993) confer a high-specificity interaction with the N-terminal SH2 (nSH2) domain of PLC-γ1 (Bae et al. 2009; Poulin et al. 2000). This interaction is essential for both membrane recruitment (Matsuda et al. 2001; Todderud et al. 1990) and tyrosine phosphorylation of PLC-γ1 (Larose et al. 1993; Poulin et al. 2000).

Activated RTKs phosphorylate PLC-γ1 at five residues, Tyr472, Tyr771, Tyr775, Tyr783, and Tyr1254 (Bae et al. 2009; Matsuda et al. 2001; Todderud et al. 1990) leading to increased catalytic activity (Gresset et al. 2010). Recent studies highlight differences between requirements for tyrosine phosphorylation *in vitro* versus *in vivo*. Specifically, reconstitution assays with purified protein indicated that only phosphorylation of Tyr783 increases PLC-γ1 activity (Gresset et al. 2010), whereas phosphorylation at both Tyr775 and Tyr783 was necessary to increase lipase activity in intact cells (Matsuda et al. 2001). Although the nature of this discrepancy remains unclear, evolutionary analysis suggests that Tyr783 is the primary regulatory residue in PLC-γ1 since its conservation extends to *C. elegans*, whereas Tyr775 first appears in arthropods (*Aedes aegypti*), then in chordates (*Xenopus laevis*). Phosphorylation of Tyr775 in intact cells likely acts in concert with Tyr783 to sustain PLC-γ1 activity.

The molecular details linking tyrosine phosphorylation and increased lipase activity of PLC-γ isozymes were recently described (Gresset et al. 2010). Phosphorylation of Tyr783 results in high affinity interaction with the C-terminal SH2 (cSH2) domain, which in turn results in a conformational change responsible for removal of autoinhibition. PLC-γ1 and PLC-γ2 share high sequence conservation, and although not fully elucidated, the mechanism of phosphorylation-dependent activation of PLC-γ2 is likely to be analogous to that of PLC-γ1.

PLC-γ1 and PLC-γ2 are also activated downstream of cytosolic tyrosine kinases in large signaling complexes located at the plasma membrane (Marrero et al. 1996; Park et al. 1991; Roifman and Wang 1992; Venema et al. 1998). For example, in haematopoietic cells most receptors that activate PLC-γ isozymes do so through non-receptor tyrosine kinases coupled to proteins containing immunoreceptor tyrosine-based activation motifs (ITAMs) of the consensus sequence $DX_2YXLX_{6-12}YDXL$ (X = any amino acid). PLC-γ1 is the predominant isoform activated in T cells downstream of the T-cell antigen receptor (TCR, CD3) (Secrist et al. 1991; Weiss et al. 1991). The Src-family tyrosine kinases Lck, and to a lesser extent Fyn (Shiroo et al. 1992), phosphorylate the ITAM motifs within the TCR complex. This recruits T-cell-specific ZAP-70 into the signaling complex through its tandem SH2 domains (Chan et al. 1991). ZAP-70 phosphorylates tyrosines in two adaptor proteins, LAT (Zhang et al. 1998) and SLP-76 (Bubeck Wardenburg et al. 1996), which act as scaffolding proteins. PLC-γ1 is recruited to this membrane complex through its N-terminal SH2 domain binding LAT, while SLP-76 interacts with PLC-γ1 through its SH3 domain (Braiman et al. 2006; Stoica et al. 1998; Yablonski et al. 2001). Although the kinases responsible for directly phosphorylating PLC-γ1 remain undefined, phosphopeptide mapping studies indicate the major sites of PLC-γ1 tyrosine phosphorylation in human T cells are the same as those described for cells treated with growth factors (Park et al. 1991).

Cross-linking of the B-cell antigen receptor results in tyrosine phosphorylation and activation of PLC-γ2 rather than PLC-γ1 (Coggeshall et al. 1992). Cytosolic tyrosine kinases, e.g. Lyn, Syk, or Btk, are recruited to a signaling complex and phosphorylation of PLC-γ2 ensues (Kim et al. 2004). Phosphorylation of both Tyr753 and Tyr759 in PLC-γ2 (the equivalent of Tyr775 and Tyr783 in PLC-γ1) appear to be essential for functional B-cell signaling.

PLC-γ isozymes are also activated by mechanisms that do not require tyrosine phosphorylation. Activation of PtdIns 3-kinase generates $PtdIns(3,4,5)P_3$, which functions as a specific ligand for the N-terminal PH domain of PLC-γ isozymes and mediates translocation to the plasma membrane (Falasca et al. 1998). Studies using truncation constructs of PLC-γ1 indicate that PtdIns(3,4,5)P_3 also binds to the cSH2 domain of PLC-γ1 (Bae et al. 1998) providing an additional anchor point to the plasma membrane. However, the mechanistic details of such activation remain unknown.

Rac GTPases directly activate PLC-γ2, but not PLC-γ1 (Piechulek et al. 2005). Rac-dependent activation requires the split PH domain of PLC-γ2 (Walliser et al. 2008), and Rac2 is the most potent GTPase activator (Piechulek et al. 2005). Structural studies revealed that the switch regions of activated Rac2 directly engage the β5-strand and α-helix of the isolated split PH domain of PLC-γ2 through hydrophobic interactions (Bunney et al. 2009). Interestingly, this interface is distinct from the engagement of activated Rac1 with the N-terminal PH domain of PLC-β isozymes (Jezyk et al. 2006). The mechanism leading to increased PLC-γ2 activity after Rac binding also apparently involves translocation to the plasma membrane and removal of auto-inhibition mediated by the X/Y-linker (Everett et al. 2011).

3.5.3.2 Physiology

Animals homozygous for the PLC-γ1 null allele die by embryonic day 9 due to generalized growth failure (Ji et al. 1997). Closer examination of PLC-γ1 embryos indicated that the embryonic lethality might be attributed to the loss of both erythroid progenitors and endothelial cells, necessary for vasculogenesis and erythropoiesis (Liao et al. 2002). Vascular endothelial growth factor (VEGF) activates PLC-γ1 via the RTKs FLT-1 and FLK-1, and VEGF is produced and secreted by myocardiocytes during development to enhance cardiac vascularization (Rottbauer et al. 2005). A zebrafish model also suggests that VEGF signaling through PLC-γ1 modulates cardiac contractility since zebrafish deficient in functional PLC-γ1 lose ventricular contractility and are defective in vasculogenesis (Rottbauer et al. 2005).

Several studies implicate PLC-γ1 as a critical component of cellular transformation downstream of EGFR/erbB2 activation. Increased expression of PLC-γ1 occurs in a number of EGF-dependent breast cancer tissues (Arteaga et al. 1991). Overexpressed PLC-γ1 also was highly phosphorylated in correlation with upregulation of both EGFR and erbB2, suggesting that PLC-γ1 activity drives breast tumor formation.

PLC-γ2 null mice remain viable after birth but exhibit strong deficiencies in signaling responses of B cells to immunoglobulins (Wang et al. 2000a). Collagen-induced platelet aggregation is also compromised in PLC-γ2-deficient mice.

3.5.4 PLC-ε

PLC-ε initially was discovered in *C. elegans* (Shibatohge et al. 1998), and the mammalian homologue was later cloned independently by three research groups (Kelley et al. 2001; Lopez et al. 2001; Song et al. 2001). A single isoform of PLC-ε exists. It is expressed relatively ubiquitously with highest levels found in heart, liver, and lung (Kelley et al. 2001; Lopez et al. 2001; Song et al. 2001). PLC-ε is a complex signaling protein since in addition to the core lipase domains it contains an N-terminal cysteine-rich domain, an N-terminal CDC25 domain, and two C-terminal Ras-associating domains (Wing et al. 2003a). PLC-ε provides a unique signaling node that integrates phospholipid signaling with pathways involving heterotrimeric and Ras family G proteins.

3.5.4.1 Regulation

CDC25 domains typically exhibit guanine nucleotide exchange factor (GEF) activity, resulting in exchange of GTP for GDP on small GTPases of the Ras family (Boguski and McCormick 1993). The GTPase specificity of the CDC25 domain is not well-defined, but several studies suggest that PLC-ε acts as a GEF for Rap1 and/or Ras (Jin et al. 2001; Satoh et al. 2006). Therefore, pathways downstream of these GTPases are activated as a consequence of activation of PLC-ε.

PLC-ε possesses tandem Ras-associating (RA) domains at its C-terminus (Shibatohge et al. 1998). RA domains are known effector sites for members of the Ras subfamily of small GTPases, and both H-Ras and Rap1 were shown in early studies to bind to the RA2 domain of PLC-ε in a GTP-dependent manner (Kelley et al. 2001; Lopez et al. 2001; Shibatohge et al. 1998; Song et al. 2001). Binding of H-Ras to the RA1 domain also was observed albeit with an affinity much lower than measured for the RA2 domain (Kelley et al. 2001). A later study confirmed the capacity of the RA2 domain to bind both H-Ras and Rap1b, whereas the RA1 domain exhibited binding to neither protein (Wohlgemuth et al. 2005). The RA2 domain binds H-Ras with an affinity eightfold higher than for Rap1 (Bunney et al. 2006). The activity of the CDC25 domain of PLC-ε produces GTP-bound Ras GTPases. Therefore, upstream activators position PLC-ε for CDC25-dependent activation of GTPases that in turn bind the second RA domain of the C-terminus of the isozyme to produce long-lasting activation through a feed-forward mechanism.

Cellular studies indicate that EGF increases PLC-ε activity through H-Ras and Rap1-promoted translocation to proximal membranes, which is dependent on the C-terminal RA domain (Kelley et al. 2001; Song et al. 2001). Whereas H-Ras mediated translocation of PLC-ε to the plasma membrane, Rap1 promoted translocation to the perinuclear region (Song et al. 2001). The detailed mechanism for Ras-mediated activation of PLC-ε remains unknown, but it likely involves RA2 domain-dependent translocation of PLC-ε to the plasma membrane and orientation of the active site for substrate hydrolysis.

Early studies illustrated that co-transfection of PLC-ε with a GTPase-deficient mutant of $G\alpha_{12}$ (but not $G\alpha_{13}$) leads to increased lipase activity, suggesting that $G\alpha_{13}$-coupled GPCR, such as lysophosphatidic acid and thrombin receptors, signal to this lipase (Lopez et al. 2001). Subsequent studies revealed that both $G\alpha_{12}$ and $G\alpha_{13}$ mediate activation of PLC-ε (Wing et al. 2001), but this effect is not direct. Rather, $G\alpha_{12/13}$ activates RhoGEFs, e.g., p115RhoGEF or LARG, which activate Rho, and then Rho directly binds to and activates PLC-ε (Seifert et al. 2004; Wing et al. 2003b). Thus, C3 botulinum toxin, which ADP ribosylates and inactivates Rho, blocks PLC-ε-mediated inositol lipid signaling responses in intact cells promoted by activation of GPCR, $G\alpha_{12/13}$, or Rho (Hains et al. 2006). GTP-dependent activation of PLC-ε can be recapitulated with purified PLC-ε and Rho in a phospholipid vesicle reconstitution system, and is independent of the Ras/Rap binding RA-domains since it occurs with truncation mutants of PLC-ε lacking the entire C-terminal region (Seifert et al. 2008). Although activation requires binding of Rho in the catalytic core of PLC-ε and is lost when a unique 62-residue insert within the Y box of PLC-ε is removed, the mechanism of this activation remains undefined.

One complexity of Rho dependent-activation of PLC-ε is that it robustly occurs through activation of two different classes of GPCRs that nonetheless converge on a common mechanism. Thus, Gα subunits of the G_{12} family of G proteins activate Rho, via activation of p115RhoGEF or LARG, whereas Gα subunits of the G_q family of G proteins do so by activating p63RhoGEF (Aittaleb et al. 2010). The physiological significance of GPCR simultaneously signaling through PLC-β- and PLC-ε-dependent pathways is unclear, but obviously provides opportunities for cross-talk and synergy between pathways. Kelley and his colleagues illustrated that inositol lipid signaling downstream of certain GPCR, e.g. thrombin and lysophosphatidic acid receptors, in Rat-1 cells involves both PLC-β- and PLC-ε-dependent responses; PLC-β3 mediates acute phospholipid signaling, whereas PLC-ε mediates sustained signaling (Kelley et al. 2006). Although PLC-ε also is activated by cotransfection with Gβγ, this effect apparently does not occur via a direct interaction and its mechanism and potential importance remain uncertain (Wing et al. 2001).

Members of the Ras and Rho subfamilies of GTPases activate PLC-ε by binding to two distinct regions of the isozyme, and assays in intact cells as well as in reconstitution assays with purified components indicate that at least additive effects on PLC-ε activity occur during simultaneous activation by H-Ras and RhoA (Seifert et al. 2008). The fact that PLC-ε is more sensitive to stimulation by H-Ras following RhoA binding also suggests the potential for signaling synergy between these two GTPases. Indeed, cooperative activation of Rho and Ras/Rap through PLC-ε clearly occurs. For example, thrombin induces astrocyte proliferation by stimulating $G\alpha_{12/13}$-dependent activation of a RhoGEF, which activates Rho, which in turn activates PLC-ε (Citro et al. 2007). The CDC25 domain of PLC-ε promotes activation of Rap1, which in turn leads to activation of ERK and DNA synthesis.

Activation of the $β_2$-adrenergic receptor or forskolin-promoted activation of adenylyl cyclase leads to PLC-ε-dependent increases in inositol lipid hydrolysis (Schmidt et al. 2001). This effect is mediated by cyclic AMP-dependent activation of a RapGEF, which in turn activates Rap1B (Evellin et al. 2002), and consequently, activates PLC-ε through binding to the RA2 domain.

3.5.4.2 Physiology

Targeted disruption of PLC-ε in mice leads to developmental defects of the aortic and pulmonary cardiac valves (Tadano et al. 2005). This phenotype is similar to that of mice deficient in heparin-binding epidermal growth factor, an agonist of the EGF receptor (Jackson et al. 2003), suggesting that growth factor-mediated activation of PLC-ε is important in heart development. PLC-ε(-/-) mice exhibit reduced cardiac contraction in response to activation of β-adrenergic receptors, which renders these animals susceptible to hypertrophy (Wang et al. 2005). The involvement of PLC-ε in the action of catecholamines occurs downstream of the activation of adenylyl cyclase. That is, β1-adrenergic receptor-promoted elevation of cyclic AMP levels activates the RasGEF, Epac, which activates Rap2B and subsequently, PLC-ε (Oestreich et al. 2007, 2009).

Targeted inactivation of PLC-ε in the skin resulted in dampened cell proliferation and markedly reduced incidence of squamous tumors in a chemical carcinogenesis model (Bai et al. 2004). The idea that PLC-ε might be a tumor suppressor gene also is supported by the observation that the PLC-ε gene was significantly down-regulated in patients with sporadic colorectal cancer (Wang et al. 2008).

Positional cloning also identified PLC-ε as a targeted gene in nephrotic syndrome (Hinkes et al. 2006). Patients with severe kidney disease have missense or truncating mutations in the PLC-ε gene that lead to loss of PLC-ε function; normal glomerular development is arrested and early-onset nephrotic syndrome occurs. Knockdown of PLC-ε in zebrafish resulted in lack of development of a functional kidney barrier, suggesting that a role of PLC-ε in podocyte development is conserved evolutionarily (Hinkes et al. 2006).

3.5.5 PLC-ζ Isozymes

PLC-ζ was first isolated and cloned from human and mouse testis and exists as a gamete-specific PLC only expressed in spermatids (Saunders et al. 2002). Fluorescence microscopy studies indicate that PLC-ζ accumulates in the pronucleus (Yoda et al. 2004). It is the smallest PLC isozyme, and is the only one that lacks an N-terminal PH domain. The absence of this domain suggests that PLC-ζ is more closely related to plant PLCs (Mueller-Roeber and Pical 2002; Tasma et al. 2008) than the mammalian PLC-δ isozymes; it is 33% identical with PLC-δ1 (Saunders et al. 2002).

3.5.5.1 Regulation

Enzymatic characterization of PLC-ζ using purified proteins indicates that the EC_{50} of PLC-ζ for Ca^{2+}-dependent PtdIns(4,5)P$_2$ hydrolysis is ∼ 100-fold lower than that of PLC-δ1, suggesting that PLC-ζ exhibits the highest sensitivity to Ca^{2+} of all PLC isoforms (Kouchi et al. 2004). PLC-ζ activity was stimulated with Ca^{2+} concentrations as low as 10 nM and reached a maximum at 1 μM. Deletion of the EF-hands or the C2 domain abrogated this Ca^{2+}-dependent PLC-ζ activation, indicating a potential role for these domains in regulating PLC-ζ activity (Kouchi et al. 2004, 2005; Nomikos et al. 2007). However, structural knowledge from PLC-δ1 (Essen et al. 1996) and PLC-β2 (Hicks et al. 2008; Jezyk et al. 2006) illustrate a supportive role of the EF-hands and the C2 domain in overall structural integrity of the PLC isozymes, and it is likely that their absence in PLC-ζ would result in impaired substrate hydrolysis.

PLC-ζ contains two nuclear localization signals: the first (residues 299–308 (KFKILVKNRK)) is in the C-terminus of the X domain, and the second (residues 374–381 (KKRKRKMK)) is located in the X/Y-linker (Kuroda et al. 2006). Point mutations within the two regions abrogate the nuclear localization of PLC-ζ, but the mutant proteins retain capacity to initiate Ca^{2+} signaling. It is unclear how PLC-ζ targets to the plasma membrane where substrate $PtdIns(4,5)P_2$ resides.

3.5.5.2 Physiology

Injection of RNA encoding PLC-ζ in mouse eggs induces intracellular Ca^{2+} oscillations and egg activation (Saunders et al. 2002). Additional fluorescent studies indicate that PLC-ζ located in the perinucleus disperses to the cytoplasm upon nuclear envelope breakdown and translocates back into the nucleus after cleavage in a cell-cycle dependent manner (Sone et al. 2005). The shuttling of PLC-ζ in and out of the nucleus coincides with Ca^{2+} oscillations. PLC-ζ is associated with the perinucleus in interphase, and no Ca^{2+} is detected; PLC-ζ translocates to the cytoplasm in the mitotic phase, and Ca^{2+} oscillations are initiated.

3.5.6 PLC-η Isozymes

A novel class of PLC isozymes that includes two isoforms, PLC-η1 and PLC-η2, was discovered in 2005 (Hwang et al. 2005; Nakahara et al. 2005; Stewart et al. 2005; Zhou et al. 2005). Both PLC-η isoforms are enriched in neuron-enriched regions of the brain, suggesting a role for these proteins in neuronal development (Nakahara et al. 2005; Zhou et al. 2005). PLC-η isozymes are structurally similar to PLC-δ isozymes with the addition of an extended C-terminus after the C2 domain that includes a putative PDZ domain-interacting sequence at the end (Zhou et al. 2005).

3.5.6.1 Regulation

The extent and identity of upstream regulators of PLC-η isozymes has not been fully established. However, coexpression of PLC-η2 with Gβγ in COS-7 cells resulted in increases in inositol lipid hydrolysis (Zhou et al. 2005). Moreover, purified PLC-η2 is robustly activated by Gβγ in reconstitution assays with model phospholipid vesicles, and therefore, this isozyme signals downstream of GPCR (Zhou et al. 2008). The N-terminal PH domain and the C-terminal extension are dispensable for Gβγ-mediated activation of PLC-η2, and therefore, the interface for Gβγ lies within the catalytic core of PLC-η2. Deletion of the N-terminal PH domain resulted in appearance of PLC-η2 in the cytosol, suggesting that the PH domain of PLC-η2 functions as a localization signal for the plasma membrane, analogous to the PH domain of PLC-δ isozymes (Nakahara et al. 2005). Further studies are necessary to confirm the role of the PH domain of PLC-η isozymes as a membrane anchor and to determine its specificity of interaction with phospholipids.

3.5.6.2 Physiology

PLC-η2 knockout mice are viable and no detectable abnormalities have been identified to date (Kanemaru et al. 2010). Transcriptional reporter assays, *in-situ* hybridization, and immunohistochemistry illustrate that PLC-η2 is highly expressed in the habenula and retina. Further analysis highlighted the previously unappreciated role of PLC-η2 in the development and maturation of the retina (Kanemaru et al. 2010).

3.6 Auto-inhibition of PLC Isozymes by Their X/Y-linker

All PLC isoforms are soluble enzymes while their substrate, PtdIns(4,5)P$_2$, is membrane-bound. Therefore, recruitment to plasma (and other) membranes is a required step in the action of these signaling proteins. Many of the activators discussed above are membrane-associated proteins, but the molecular details that accompany membrane association and activation of PLC isozymes remain unclear. Nonetheless, biochemical and structural studies suggest a central role of the X/Y-linker in autoinhibition of most if not all PLC isozymes. The most parsimonious model for activation centers on mechanisms that remove this autoinhibition (Fig. 3.5).

Multiple biochemical experiments indicate that disruption of the X/Y-linker of PLC isozymes enhances their enzymatic activities. Specifically, limited proteolysis within PLC-δ1 (Ellis et al. 1993), PLC-β2 (Schnabel and Camps 1998), PLC-γ1 (Fernald et al. 1994), and PLC-ζ (Kurokawa et al. 2007) targets the X/Y-linker and resulted in PLC isozymes that exhibit higher enzymatic activity. Similarly, independently expressed polypeptide chains encompassing the N-terminus to the X-box and the Y-box to the C-terminus of PLC-β2 (Zhang and Neer 2001) or PLC-γ1 (Horstman et al. 1996) reassemble in the absence of their respective X/Y-linker regions and reconstitute as functional isozymes that exhibit higher basal activity than the holoenzymes. Moreover, systematic deletion of the X/Y-linker dramatically enhances the lipase activity of PLC-β2, -β3, -δ1, -ε, -γ1, and -γ2 both in intact cells and with purified proteins (Gresset et al. 2010; Hicks et al. 2008; Waldo et al. 2010). The X/Y-linker-deleted versions of several PLC isoforms retain capacity to be activated further by G proteins (Hicks et al. 2008), indicating that deletion of the X/Y-linker does not

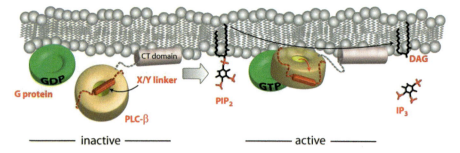

Fig. 3.5 General model of auto-inhibition and activation of PLC isozymes. The model presents the general mechanism whereby a G protein, e.g., Gαq or Rac1, activates a PLC-β isozyme. *Left side*, The G protein (*green toroid*) is shown in an inactive GDP-bound state, and PLC-β is presented as a gold toroid except for its C-terminal (CT) domain (*light pink*) and X/Y linker (*red cylinder* and *dotted lines*). The CT domain of PLC-β basally associates with membranes, and the X/Y linker blocks the active site. *Right side*, GTP binding activates the G protein, the active G protein forms a complex with the main portion of PLC-β, the lipase active site is anchored and oriented at the membrane surface, and the X/Y linker is repulsed by the membrane surface therein freeing the active site to hydrolyze PtdIns(4,5)P$_2$ into diacylglycerol (DAG) and Ins(1,4,5)P$_3$ (IP$_3$)

impair the structural integrity or the signaling capacity of the remainder of the protein. Together, these results indicate that the X/Y-linker mediates auto-inhibition of PLC isoforms.

A regulatory role of the X/Y-linker is supported by several high-resolution structures that highlight interactions between a portion of the X/Y-linker and the active site of PLC-β isozymes (Hicks et al. 2008; Jezyk et al. 2006; Waldo et al. 2010). The absence of conformational rearrangements of the catalytic core of PLC isoforms in the high-resolution structures of activated complexes of PLC-β isozymes with Rac1 or Gαq (Jezyk et al. 2006; Waldo et al. 2010) also is consistent with activation occurring as a direct consequence of the removal of the auto-inhibitory X/Y-linker.

Although biochemical and structural studies highlight a common inhibitory role of the X/Y-linker in all PLC isozymes, how is this function effected by the very divergent primary sequences of the various X/Y-linkers?

PLC-δ and -β isozymes share a high density of negative charge in their X/Y-linker, but they diverge structurally. The entire X/Y-linker of PLC-δ1 is disordered with no electron density visible in the structure of PLC-δ1 (Essen et al. 1996). In contrast, a small portion of the X/Y-linker was visible in the different structures of PLC-β isozymes (Hicks et al. 2008; Jezyk et al. 2006; Waldo et al. 2010). For example, 22 of the 70 residues of the X/Y-linker are ordered in the inactive form of PLC-β2 (Hicks et al. 2008). Fourteen of these residues form an α-helix that runs perpendicular to the TIM barrel and the last eight residues lay on the surface of the catalytic cleft and form a small 3$_{10}$ helix that makes direct hydrogen-bond contacts with active site residues. These interactions are conserved in the structure of an activated complex between PLC-β2 and Rac1 (Jezyk et al. 2006). The PLC-β3 structure in complex with Gα$_q$ also displays residues making direct contacts with the active site (Waldo et al. 2010). The persistence of a small ordered portion directly occluding the active site in both the absence and presence of G protein-binding suggests that binding of activators is not sufficient to remove auto-inhibition. Consistent with this idea, superposition of the three structures of PLC-β isozymes highlights no conformational changes within the catalytic TIM barrel in the absence or presence of activators (Hicks et al. 2008; Jezyk et al. 2006; Waldo et al. 2010).

The most parsimonious model (Fig. 3.5) for activation of these lipases involves an interfacial mechanism driven by electrostatic repulsions between the negatively charged X/Y-linker and the proximal membranes. Thus, activators do not induce general conformational changes, but rather, recruit and optimally orient PLC isoforms at the plasma membrane to facilitate electrostatic repulsion and removal of the X/Y-linker from the active site. Consistent with this model, removal of monovalent acidic phospholipids from the monolayer decreases by threefold the initial rate of PtdIns(4,5)P$_2$ hydrolysis by PLC-β1 and -δ1 (Boguslavsky et al. 1994), indicating that the presence of negative charges in the inner leaflet of the membrane is essential for maximal activities in PLC-β and -δ isozymes. Furthermore, Rac-mediated activation of PLC-β2 was abolished in detergent-mixed micelles, but not in phospholipid vesicles, which is also consistent with the idea that recruitment to the membrane is critical in interfacial activation (Hicks et al. 2008).

Although three-dimensional structures are not yet available for PLC-ε, its mechanism of activation is apparently similar to that of PLC-β and -δ isozymes. Thus, removal of the X/Y-linker robustly activates PLC-ε but does not prevent activation by its upstream regulators, RhoA and H-Ras (Seifert et al. 2008).

PLC-ζ is the only PLC isozyme with a highly basic X/Y-linker, although two splice forms of this isozyme contain an insertion in the X/Y-linker that is highly acidic. Conflicting results suggest that the basic patches of the X/Y-linker of PLC-ζ serve as a membrane targeting signal to facilitate interaction with phospholipids (Nomikos et al. 2007) or as a nuclear localization signal (Kuroda et al. 2006). Thus, interfacial activation as described for PLC-β and -δ isozymes likely applies to PLC-ζ isozymes harboring the acidic insertion. Alternatively, PLC-ζ isozymes lacking the acidic insertion might be solely regulated by proteolytic processing of their X/Y-linker to enhance PLC-ζ catalytic activity (Kurokawa et al. 2007).

PLC-γ isozymes contain a unique X/Y-linker composed of modular domains—a split PH domain, tandem SH2 domains, and an SH3 domain. These isozymes are recruited to the plasma membrane by their activators, either receptor or cytosolic tyrosine kinases, through engagement of the nSH2 domain. Although structural data are not available, the cSH2 domain mediates auto-inhibition of PLC-γ isozymes since its deletion recapitulates the high degree of constitutive activation observed after removal of the entire X/Y-linker (Gresset et al. 2010).

PLC-γ isozymes also harbor a unique mechanism of activation since they link tyrosine phosphorylation with enhanced catalytic activity. Interfacial activation is unlikely to be important for PLC-γ isozymes. Instead, tyrosine kinases phosphorylate PLC-γ isozymes at a specific tyrosine within the X/Y-linker, Tyr783 in PLC-γ1 (Tyr759 in PLC-γ2), and a high affinity interaction with the phosphotyrosine and the cSH2 domain ensues (Gresset et al. 2010). Substitution of the invariant Argβ5 within the cSH2 domain that coordinates the phosphate oxygens of the phosphorylated tyrosine and is critical for high affinity binding of the phosphorylated tyrosine to the SH2 domain (Booker et al. 1992; Waksman et al. 1992) eliminates phosphorylation-mediated increases in PLC-γ activity (Gresset et al. 2010). In addition, increased PLC-γ isozyme activity is associated with a large conformational rearrangement of the X/Y-linker with respect to the rest of the protein (Gresset et al. 2010). Overall, PLC-γ isozymes have elaborated on the common mechanism of auto-inhibition observed in other PLC isoforms and now couple tyrosine phosphorylation and release of a highly complex X/Y-linker from its autoinhibited state.

3.7 Conclusion

Activation of receptors for hundreds of extracellular signaling molecules promotes activation of PLC by mechanisms involving heterotrimeric and Ras superfamily G proteins, tyrosine kinases, Ca^{2+}, and/or other stimuli. Although the importance of the PLC isozymes was first established for Ca^{2+}- and PKC-mediated cell signaling, PLC-mediated changes in membrane phosphoinositide levels alter the activities of many membrane, cytoskeletal, and cytosolic proteins. Thus, PLC isozymes function as major signaling nexuses from which a panoply of downstream signals radiate. How these signaling nodes are organized spatially and functionally is largely unknown but assuredly occurs in a cell-specific manner. The existence of 13 different isozymes that are differentially regulated and co-expressed across tissues adds complexity to understanding of these signaling networks. The core function of PLC isozymes as major cell signaling proteins is well-established, and beginning insight into the physiological roles played by individual PLC isozymes has accrued recently from genetic studies. However, much is yet to be learned about how these proteins function in the larger context of human physiology and pathophysiology. Identification of pharmacological agents that selectively inhibit the function of individual PLC isozymes would provide important new reagents for the study of these signaling proteins.

References

Adamski FM, Timms KM, Shieh BH (1999) A unique isoform of phospholipase C-β4 highly expressed in the cerebellum and eye. Biochim Biophys Acta 1444:55–60
Aittaleb M, Boguth CA, Tesmer JJ (2010) Structure and function of heterotrimeric G protein-regulated Rho guanine nucleotide exchange factors. Mol Pharmacol 77:111–125
Allen V, Swigart P, Cheung R, Cockcroft S, Katan M (1997) Regulation of inositol lipid-specific phospholipase C-δ by changes in Ca^{2+} ion concentrations. Biochem J 327:545–552
Ananthanarayanan B, Das S, Rhee SG, Murray D, Cho W (2002) Membrane targeting of C2 domains of phospholipase C-δ isoforms. J Biol Chem 277:3568–3575
Andoh T, Yoko T, Matsui Y, Toh A (1995) Molecular cloning of the plc1+ gene of Schizosaccharomyces pombe, which encodes a putative phosphoinositide-specific phospholipase C. Yeast 11:179–185

Arteaga CL, Johnson MD, Todderud G, Coffey RJ, Carpenter G, Page DL (1991) Elevated content of the tyrosine kinase substrate phospholipase C-γ1 in primary human breast carcinomas. Proc Natl Acad Sci USA 88:10435–10439

Bae YS, Cantley LG, Chen CS, Kim SR, Kwon KS, Rhee SG (1998) Activation of phospholipase C-γ by phosphatidylinositol 3,4,5-trisphosphate. J Biol Chem 273:4465–4469

Bae JH, Lew ED, Yuzawa S, Tome F, Lax I, Schlessinger J (2009) The selectivity of receptor tyrosine kinase signaling is controlled by a secondary SH2 domain binding site. Cell 138:514–524

Bai Y, Edamatsu H, Maeda S, Saito H, Suzuki N, Satoh T, Kataoka T (2004) Crucial role of phospholipase C-ε in chemical carcinogen-induced skin tumor development. Cancer Res 64:8808–8810

Barr AJ, Ali H, Haribabu B, Snyderman R, Smrcka AV (2000) Identification of a region at the N-terminus of phospholipase C-β3 that interacts with G protein βγ subunits. Biochemistry 39:1800–1806

Berridge MJ (1987) Inositol trisphosphate and diacylglycerol: two interacting second messengers. Annu Rev Biochem 56:159–193

Berstein G, Blank JL, Jhon DY, Exton JH, Rhee SG, Ross EM (1992) Phospholipase C-β1 is a GTPase-activating protein for Gq/11, its physiologic regulator. Cell 70:411–418

Biddlecome GH, Berstein G, Ross EM (1996) Regulation of phospholipase C-β1 by Gq and m1 muscarinic cholinergic receptor. Steady-state balance of receptor-mediated activation and GTPase-activating protein-promoted deactivation. J Biol Chem 271:7999–8007

Boguski MS, McCormick F (1993) Proteins regulating Ras and its relatives. Nature 366:643–654

Boguslavsky V, Rebecchi M, Morris AJ, Jhon DY, Rhee SG, McLaughlin S (1994) Effect of monolayer surface pressure on the activities of phosphoinositide-specific phospholipase C-β1, -γ1, and -δ1. Biochemistry 33:3032–3037

Booker GW, Breeze AL, Downing AK, Panayotou G, Gout I, Waterfield MD, Campbell ID (1992) Structure of an SH2 domain of the p85 alpha subunit of phosphatidylinositol-3-OH kinase. Nature 358:684–687

Boyer JL, Waldo GL, Harden TK (1992) βγ-subunit activation of G-protein-regulated phospholipase C. J Biol Chem 267:25451–25456

Braiman A, Barda-Saad M, Sommers CL, Samelson LE (2006) Recruitment and activation of PLC-γ1 in T cells: a new insight into old domains. EMBO J 25:774–784

Bubeck Wardenburg J, Fu C, Jackman JK, Flotow H, Wilkinson SE, Williams DH, Johnson R, Kong G, Chan AC, Findell PR (1996) Phosphorylation of SLP-76 by the ZAP-70 protein-tyrosine kinase is required for T-cell receptor function. J Biol Chem 271:19641–19644

Bunney TD, Harris R, Gandarillas NL, Josephs MB, Roe SM, Sorli SC, Paterson HF, Rodrigues-Lima F, Esposito D, Ponting CP, Gierschik P, Pearl LH, Driscoll PC, Katan M (2006) Structural and mechanistic insights into Ras association domains of phospholipase C-ε. Mol Cell 21:495–507

Bunney TD, Opaleye O, Roe SM, Vatter P, Baxendale RW, Walliser C, Everett KL, Josephs MB, Christow C, Rodrigues-Lima F, Gierschik P, Pearl LH, Katan M (2009) Structural insights into formation of an active signaling complex between Rac and phospholipase C-γ2. Mol Cell 34:223–233

Camps M, Carozzi A, Schnabel P, Scheer A, Parker PJ, Gierschik P (1992) Isozyme-selective stimulation of phospholipase C-β2 by G protein βγ-subunits. Nature 360:684–686

Chan AC, Irving BA, Fraser JD, Weiss A (1991) The ζ chain is associated with a tyrosine kinase and upon T-cell antigen receptor stimulation associates with ZAP-70, a 70-kDa tyrosine phosphoprotein. Proc Natl Acad Sci USA 88:9166–9170

Cheng HF, Jiang MJ, Chen CL, Liu SM, Wong LP, Lomasney JW, King K (1995) Cloning and identification of amino acid residues of human phospholipase C-δ1 essential for catalysis. J Biol Chem 270:5495–5505

Citro S, Malik S, Oestreich EA, Radeff-Huang J, Kelley GG, Smrcka AV, Brown JH (2007) Phospholipase C-ε is a nexus for Rho and Rap-mediated G protein-coupled receptor-induced astrocyte proliferation. Proc Natl Acad Sci USA 104:15543–15548

Cocco L, Rubbini S, Manzoli L, Billi AM, Faenza I, Peruzzi D, Matteucci A, Artico M, Gilmour RS, Rhee SG (1999) Inositides in the nucleus: presence and characterisation of the isozymes of phospholipase-β family in NIH 3T3 cells. Biochim Biophys Acta 1438:295–299

Coggeshall KM, McHugh JC, Altman A (1992) Predominant expression and activation-induced tyrosine phosphorylation of phospholipase C-γ2 in B lymphocytes. Proc Natl Acad Sci USA 89:5660–5664

Drayer AL, Haastert PJ van (1992) Molecular cloning and expression of a phosphoinositide-specific phospholipase C of Dictyostelium discoideum. J Biol Chem 267:18387–18392

Ellis MV, Carne A, Katan M (1993) Structural requirements of phosphatidylinositol-specific phospholipase C-δ1 for enzyme activity. Eur J Biochem 213:339–347

Ellis MV, Katan SU, Katan M (1995) Mutations within a highly conserved sequence present in the X region of phosphoinositide-specific phospholipase C-δ1. Biochem J 307:69–75

Ellis MV, James SR, Perisic O, Downes CP, Williams RL, Katan M (1998) Catalytic domain of phosphoinositide-specific phospholipase C. Mutational analysis of residues within the active site and hydrophobic ridge of PLC-δ1. J Biol Chem 273:11650–11659

Essen LO, Perisic O, Cheung R, Katan M, Williams RL (1996) Crystal structure of a mammalian phosphoinositide-specific phospholipase C-δ. Nature 380:595–602

Essen LO, Perisic O, Lynch DE, Katan M, Williams RL (1997) A ternary metal binding site in the C2 domain of phosphoinositide-specific phospholipase C-δ1. Biochemistry 36:2753–2762

Evellin S, Nolte J, Tysack K, vom Dorp F, Thiel M, Weernink PA, Jakobs KH, Webb EJ, Lomasney JW, Schmidt M (2002) Stimulation of phospholipase C-ε by the M3 muscarinic acetylcholine receptor mediated by cyclic AMP and the GTPase Rap2B. J Biol Chem 277:16805–16813

Everett KL, Buehler A, Bunney TD, Margineanu A, Baxendale RW, Vatter P, Retlich M, Walliser C, Manning HB, Neil MA, Dunsby C, French PM, Gierschik P, Katan M (2011) Membrane environment exerts an important influence on Rac-mediated activation of phospholipase C-γ2. Mol Cell Biol 31(6):1240–1251

Exton JH (1996) Regulation of phosphoinositide phospholipases by hormones, neurotransmitters, and other agonists linked to G proteins. Annu Rev Pharmacol Toxicol 36:481–509

Faenza I, Matteucci A, Manzoli L, Billi AM, Aluigi M, Peruzzi D, Vitale M, Castorina S, Suh PG, Cocco L (2000) A role for nuclear phospholipase C-β1 in cell cycle control. J Biol Chem 275:30520–30524

Falasca M, Logan SK, Lehto VP, Baccante G, Lemmon MA, Schlessinger J (1998) Activation of phospholipase C-γ by PI 3-kinase-induced PH domain-mediated membrane targeting. EMBO J 17:414–422

Feng JF, Rhee SG, Im MJ (1996) Evidence that phospholipase-δ1 is the effector in the Gh (transglutaminase II)-mediated signaling. J Biol Chem 271:16451–16454

Ferguson KM, Lemmon MA, Schlessinger J, Sigler PB (1995) Structure of the high affinity complex of inositol trisphosphate with a phospholipase C pleckstrin homology domain. Cell 83:1037–1046

Fernald AW, Jones GA, Carpenter G (1994) Limited proteolysis of phospholipase C-γ1 indicates stable association of X and Y domains with enhanced catalytic activity. Biochem J 302:503–509

Fiume R, Ramazzotti G, Teti G, Chiarini F, Faenza I, Mazzotti G, Billi AM, Cocco L (2009) Involvement of nuclear PLC-β1 in lamin B1 phosphorylation and G2/M cell cycle progression. FASEB J 23:957–966

Fu L, Qin YR, Xie D, Hu L, Kwong DL, Srivastava G, Tsao SW, Guan XY (2007) Characterization of a novel tumor-suppressor gene PLC-δ1 at 3p22 in esophageal squamous cell carcinoma. Cancer Res 67:10720–10726

Fukami K, Yoshida M, Inoue T, Kurokawa M, Fissore RA, Yoshida N, Mikoshiba K, Takenawa T (2003) Phospholipase C-δ4 is required for Ca^{2+} mobilization essential for acrosome reaction in sperm. J Cell Biol 161:79–88

Godin CM, Ferreira LT, Dale LB, Gros R, Cregan SP, Ferguson SS (2010) The small GTPase Ral couples the angiotensin II type 1 receptor to the activation of phospholipase C-δ1. Mol Pharmacol 77:388–395

Gresset A, Hicks SN, Harden TK, Sondek J (2010) Mechanism of phosphorylation-induced activation of phospholipase C-γ isozymes. J Biol Chem 285:35836–35847

Grobler JA, Essen LO, Williams RL, Hurley JH (1996) C2 domain conformational changes in phospholipase C-δ1. Nat Struct Biol 3:788–795

Guo Y, Golebiewska U, D'Amico S, Scarlata S (2010) The small G protein Rac1 activates phospholipase C-δ1 through phospholipase C-β2. J Biol Chem 285:24999–25008

Gutman O, Walliser C, Piechulek T, Gierschik P, Henis YI (2010) Differential regulation of phospholipase C-β2 activity and membrane interaction by Gαq, Gβ1γ2, and Rac2. J Biol Chem 285:3905–3915

Hains MD, Wing MR, Maddileti S, Siderovski DP, Harden TK (2006) Gα12/13- and Rho-dependent activation of phospholipase C-ε by lysophosphatidic acid and thrombin receptors. Mol Pharmacol 69:2068–2075

Han SK, Mancino V, Simon MI (2006) Phospholipase C-β3 mediates the scratching response activated by the histamine H1 receptor on C-fiber nociceptive neurons. Neuron 52:691–703

Harden TK, Sondek J (2006) Regulation of phospholipase C isozymes by Ras superfamily GTPases. Annu Rev Pharmacol Toxicol 46:355–379

Harden TK, Hicks SN, Sondek J (2009) Phospholipase C isozymes as effectors of Ras superfamily GTPases. J Lipid Res 50:S243–S248

Heinz DW, Essen LO, Williams RL (1998) Structural and mechanistic comparison of prokaryotic and eukaryotic phosphoinositide-specific phospholipases C. J Mol Biol 275:635–650

Hepler JR, Gilman AG (1992) G proteins. Trends Biochem Sci 17:383–387

Hicks SN, Jezyk MR, Gershburg S, Seifert JP, Harden TK, Sondek J (2008) General and versatile autoinhibition of PLC isozymes. Mol Cell 31:383–394

Hinkes B, Wiggins RC, Gbadegesin R, Vlangos CN, Seelow D, Nurnberg G, Garg P, Verma R, Chaib H, Hoskins BE, Ashraf S, Becker C, Hennies HC, Goyal M, Wharram BL, Schachter AD, Mudumana S, Drummond I, Kerjaschki D, Waldherr R, Dietrich A, Ozaltin F, Bakkaloglu A, Cleper R, Basel-Vanagaite L, Pohl M, Griebel M, Tsygin AN, Soylu A, Muller D, Sorli CS, Bunney TD, Katan M, Liu J, Attanasio M, O'Toole J F, Hasselbacher K, Mucha B, Otto EA, Airik R, Kispert A, Kelley GG, Smrcka AV, Gudermann T, Holzman LB, Nurnberg P, Hildebrandt F (2006) Positional cloning uncovers mutations in PLCE1 responsible for a nephrotic syndrome variant that may be reversible. Nat Genet 38:1397–1405

Hofmann SL, Majerus PW (1982) Identification and properties of two distinct phosphatidylinositol-specific phospholipase C enzymes from sheep seminal vesicular glands. J Biol Chem 257:6461–6469

Hokin MR, Hokin LE (1953) Enzyme secretion and the incorporation of ^{32}P into phospholipides of pancreas slices. J Biol Chem 203:967–977

Homma Y, Takenawa T, Emori Y, Sorimachi H, Suzuki K (1989) Tissue- and cell type-specific expression of mRNAs for four types of inositol phospholipid-specific phospholipase C. Biochem Biophys Res Commun 164:406–412

Horstman DA, DeStefano K, Carpenter G (1996) Enhanced phospholipase C-γ1 activity produced by association of independently expressed X and Y domain polypeptides. Proc Natl Acad Sci USA 93:7518–7521

Hu XT, Zhang FB, Fan YC, Shu XS, Wong AH, Zhou W, Shi QL, Tang HM, Fu L, Guan XY, Rha SY, Tao Q, He C (2009) Phospholipase C-δ1 is a novel 3p22.3 tumor suppressor involved in cytoskeleton organization, with its epigenetic silencing correlated with high-stage gastric cancer. Oncogene 28:2466–2475

Hubbard SR, Till JH (2000) Protein tyrosine kinase structure and function. Annu Rev Biochem 69:373–398

Hwang JI, Oh YS, Shin KJ, Kim H, Ryu SH, Suh PG (2005) Molecular cloning and characterization of a novel phospholipase C, PLC-η. Biochem J 389:181–186

Ichinohe M, Nakamura Y, Sai K, Nakahara M, Yamaguchi H, Fukami K (2007) Lack of phospholipase C-δ1 induces skin inflammation. Biochem Biophys Res Commun 356:912–918

Ilkaeva O, Kinch LN, Paulssen RH, Ross EM (2002) Mutations in the carboxyl-terminal domain of phospholipase C-β1 delineate the dimer interface and a potential Gαq interaction site. J Biol Chem 277:4294–4300

Illenberger D, Schwald F, Pimmer D, Binder W, Maier G, Dietrich A, Gierschik P (1998) Stimulation of phospholipase C-β2 by the Rho GTPases Cdc42Hs and Rac1. EMBO J 17:6241–6249

Illenberger D, Walliser C, Nurnberg B, Diaz Lorente M, Gierschik P (2003a) Specificity and structural requirements of phospholipase C-β stimulation by Rho GTPases versus G protein βγ dimers. J Biol Chem 278:3006–3014

Illenberger D, Walliser C, Strobel J, Gutman O, Niv H, Gaidzik V, Kloog Y, Gierschik P, Henis YI (2003b) Rac2 regulation of phospholipase C-β2 activity and mode of membrane interactions in intact cells. J Biol Chem 278:8645–8652

Irino Y, Cho H, Nakamura Y, Nakahara M, Furutani M, Suh PG, Takenawa T, Fukami K (2004) Phospholipase C δ-type consists of three isozymes: bovine PLC-δ2 is a homologue of human/mouse PLC-δ4. Biochem Biophys Res Commun 320:537–543

Jackson LF, Qiu TH, Sunnarborg SW, Chang A, Zhang C, Patterson C, Lee DC (2003) Defective valvulogenesis in HB-EGF and TACE-null mice is associated with aberrant BMP signaling. EMBO J 22:2704–2716

Jezyk MR, Snyder JT, Gershberg S, Worthylake DK, Harden TK, Sondek J (2006) Crystal structure of Rac1 bound to its effector phospholipase C-β2. Nat Struct Mol Biol 13:1135–1140

Jhon DY, Lee HH, Park D, Lee CW, Lee KH, Yoo OJ, Rhee SG (1993) Cloning, sequencing, purification, and Gq-dependent activation of phospholipase C-β3. J Biol Chem 268:6654–6661

Ji QS, Winnier GE, Niswender KD, Horstman D, Wisdom R, Magnuson MA, Carpenter G (1997) Essential role of the tyrosine kinase substrate phospholipase C-γ1 in mammalian growth and development. Proc Natl Acad Sci USA 94:2999–3003

Jiang H, Lyubarsky A, Dodd R, Vardi N, Pugh E, Baylor D, Simon MI, Wu D (1996) Phospholipase C-β4 is involved in modulating the visual response in mice. Proc Natl Acad Sci USA 93:14598–14601

Jiang H, Kuang Y, Wu Y, Xie W, Simon MI, Wu D (1997) Roles of phospholipase C-β2 in chemoattractant-elicited responses. Proc Natl Acad Sci USA 94:7971–7975

Jin TG, Satoh T, Liao Y, Song C, Gao X, Kariya K, Hu CD, Kataoka T (2001) Role of the CDC25 homology domain of phospholipase C-ε in amplification of Rap1-dependent signaling. J Biol Chem 276:30301–30307

Jones S, Zhang X, Parsons DW, Lin JC, Leary RJ, Angenendt P, Mankoo P, Carter H, Kamiyama H, Jimeno A, Hong SM, Fu B, Lin MT, Calhoun ES, Kamiyama M, Walter K, Nikolskaya T, Nikolsky Y, Hartigan J, Smith DR, Hidalgo M, Leach SD, Klein AP, Jaffee EM, Goggins M, Maitra A, Iacobuzio-Donahue C, Eshleman JR, Kern SE, Hruban RH, Karchin R, Papadopoulos N, Parmigiani G, Vogelstein B, Velculescu VE, Kinzler KW (2008) Core signaling pathways in human pancreatic cancers revealed by global genomic analyses. Science 321:1801–1806

Kamat A, Carpenter G (1997) Phospholipase C-γ1: regulation of enzyme function and role in growth factor-dependent signal transduction. Cytokine Growth Factor Rev 8:109–117

Kanemaru K, Nakahara M, Nakamura Y, Hashiguchi Y, Kouchi Z, Yamaguchi H, Oshima N, Kiyonari H, Fukami K (2010) Phospholipase C-η2 is highly expressed in the habenula and retina. Gene Expr Patterns 10:119–126

Katan M, Williams RL (1997) Phosphoinositide-specific phospholipase C: structural basis for catalysis and regulatory interactions. Semin Cell Dev Biol 8:287–296

Kawasaki H, Kretsinger RH (1994) Calcium-binding proteins. 1: EF-hands. Protein Profile 1:343–517

Kelley GG, Reks SE, Ondrako JM, Smrcka AV (2001) Phospholipase C-ε: a novel Ras effector. EMBO J 20:743–754

Kelley GG, Kaproth-Joslin KA, Reks SE, Smrcka AV, Wojcikiewicz RJ (2006) G-protein-coupled receptor agonists activate endogenous phospholipase C-ε and phospholipase C-β3 in a temporally distinct manner. J Biol Chem 281:2639–2648

Kim D, Jun KS, Lee SB, Kang NG, Min DS, Kim YH, Ryu SH, Suh PG, Shin HS (1997) Phospholipase C isozymes selectively couple to specific neurotransmitter receptors. Nature 389:290–293

Kim YH, Park TJ, Lee YH, Baek KJ, Suh PG, Ryu SH, Kim KT (1999) Phospholipase C-δ1 is activated by capacitative calcium entry that follows phospholipase C-β activation upon bradykinin stimulation. J Biol Chem 274:26127–26134

Kim YJ, Sekiya F, Poulin B, Bae YS, Rhee SG (2004) Mechanism of B-cell receptor-induced phosphorylation and activation of phospholipase C-γ2. Mol Cell Biol 24:9986–9999

Kouchi Z, Fukami K, Shikano T, Oda S, Nakamura Y, Takenawa T, Miyazaki S (2004) Recombinant phospholipase C-ζ has high Ca^{2+} sensitivity and induces Ca^{2+} oscillations in mouse eggs. J Biol Chem 279:10408–10412

Kouchi Z, Shikano T, Nakamura Y, Shirakawa H, Fukami K, Miyazaki S (2005) The role of EF-hand domains and C2 domain in regulation of enzymatic activity of phospholipase C-ζ. J Biol Chem 280:21015–21021

Kuroda K, Ito M, Shikano T, Awaji T, Yoda A, Takeuchi H, Kinoshita K, Miyazaki S (2006) The role of X/Y linker region and N-terminal EF-hand domain in nuclear translocation and Ca^{2+} oscillation-inducing activities of phospholipase C-ζ, a mammalian egg-activating factor. J Biol Chem 281:27794–27805

Kurokawa M, Yoon SY, Alfandari D, Fukami K, Sato K, Fissore RA (2007) Proteolytic processing of phospholipase C-ζ and $[Ca^{2+}]i$ oscillations during mammalian fertilization. Dev Biol 312:407–418

Larose L, Gish G, Shoelson S, Pawson T (1993) Identification of residues in the beta platelet-derived growth factor receptor that confer specificity for binding to phospholipase C-γ1. Oncogene 8:2493–2499

Lemmon MA (2003) Phosphoinositide recognition domains. Traffic 4:201–213

Li Z, Jiang H, Xie W, Zhang Z, Smrcka AV, Wu D (2000) Roles of PLC-β2 and -β3 and PI3K-γ in chemoattractant-mediated signal transduction. Science 287:1046–1049

Liao HJ, Kume T, McKay C, Xu MJ, Ihle JN, Carpenter G (2002) Absence of erythrogenesis and vasculogenesis in PLC-γ1-deficient mice. J Biol Chem 277:9335–9341

Ling K, Schill NJ, Wagoner MP, Sun Y, Anderson RA (2006) Movin' on up: the role of $PtdIns(4,5)P_2$ in cell migration. Trends Cell Biol 16:276–284

Lomasney JW, Cheng HF, Wang LP, Kuan Y, Liu S, Fesik SW, King K (1996) Phosphatidylinositol 4,5-bisphosphate binding to the pleckstrin homology domain of phospholipase C-δ1 enhances enzyme activity. J Biol Chem 271:25316–25326

Lomasney JW, Cheng HF, Roffler SR, King K (1999) Activation of phospholipase C-δ1 through C2 domain by a Ca^{2+}-enzyme-phosphatidylserine ternary complex. J Biol Chem 274:21995–22001

Lopez I, Mak EC, Ding J, Hamm HE, Lomasney JW (2001) A novel bifunctional phospholipase C that is regulated by Gα12 and stimulates the Ras/mitogen-activated protein kinase pathway. J Biol Chem 276:2758–2765

Lutz S, Shankaranarayanan A, Coco C, Ridilla M, Nance MR, Vettel C, Baltus D, Evelyn CR, Neubig RR, Wieland T, Tesmer JJ (2007) Structure of Gαq-p63RhoGEF-RhoA complex reveals a pathway for the activation of RhoA by GPCRs. Science 318:1923–1927

Marrero MB, Schieffer B, Ma H, Bernstein KE, Ling BN (1996) ANG II-induced tyrosine phosphorylation stimulates phospholipase C-γ1 and Cl-channels in mesangial cells. Am J Physiol 270:C1834–C1842

Martelli AM, Gilmour RS, Bertagnolo V, Neri LM, Manzoli L, Cocco L (1992) Nuclear localization and signalling activity of phosphoinositidase C-β in Swiss 3T3 cells. Nature 358:242–245

Matsuda M, Paterson HF, Rodriguez R, Fensome AC, Ellis MV, Swann K, Katan M (2001) Real time fluorescence imaging of PLC-γ translocation and its interaction with the epidermal growth factor receptor. J Cell Biol 153:599–612

McOmish CE, Burrows E, Howard M, Scarr E, Kim D, Shin HS, Dean B, Buuse M van den, Hannan AJ (2008) Phospholipase C-β1 knockout mice exhibit endophenotypes modeling schizophrenia which are rescued by environmental enrichment and clozapine administration. Mol Psychiatry 13:661–672

Meisenhelder J, Suh PG, Rhee SG, Hunter T (1989) Phospholipase C-γ is a substrate for the PDGF and EGF receptor protein-tyrosine kinases in vivo and in vitro. Cell 57:1109–1122

Michell RH (1975) Inositol phospholipids and cell surface receptor function. Biochim Biophys Acta 415:81–47

Mohammadi M, Honegger AM, Rotin D, Fischer R, Bellot F, Li W, Dionne CA, Jaye M, Rubinstein M, Schlessinger J (1991) A tyrosine-phosphorylated carboxy-terminal peptide of the fibroblast growth factor receptor is a binding site for the SH2 domain of phospholipase C-γ1. Mol Cell Biol 11:5068–5078

Morris AJ, Waldo GL, Downes CP, Harden TK (1990) A receptor and G-protein-regulated polyphosphoinositide-specific phospholipase C from turkey erythrocytes. II. P2Y-purinergic receptor and G-protein-mediated regulation of the purified enzyme reconstituted with turkey erythrocyte ghosts. J Biol Chem 265:13508–13514

Mueller-Roeber B, Pical C (2002) Inositol phospholipid metabolism in Arabidopsis. Characterized and putative isoforms of inositol phospholipid kinase and phosphoinositide-specific phospholipase C. Plant Physiol 130:22–46

Nagasawa K, Nishida K, Nagai K, Shimohama S, Fujimoto S (2004) Differential expression profiles of PLC-β1 and -δ1 in primary cultured rat cortical neurons treated with N-methyl-D-aspartate and peroxynitrite. Neurosci Lett 367:246–249

Nakahara M, Shimozawa M, Nakamura Y, Irino Y, Morita M, Kudo Y, Fukami K (2005) A novel phospholipase C, PLC-η2, is a neuron-specific isozyme. J Biol Chem 280:29128–29134

Nakamura Y, Hamada Y, Fujiwara T, Enomoto H, Hiroe T, Tanaka S, Nose M, Nakahara M, Yoshida N, Takenawa T, Fukami K (2005) Phospholipase C-δ1 and -δ3 are essential in the trophoblast for placental development. Mol Cell Biol 25:10979–10988

Nalefski EA, Falke JJ (1996) The C2 domain calcium-binding motif: structural and functional diversity. Protein Sci 5:2375–2390

Nebl T, Oh SW, Luna EJ (2000) Membrane cytoskeleton: PIP_2 pulls the strings. Curr Biol 10:R351–R354

Nehls M, Pfeifer D, Schorpp M, Hedrich H, Boehm T (1994) New member of the winged-helix protein family disrupted in mouse and rat nude mutations. Nature 372:103–107

Nishizuka Y (1992) Intracellular signaling by hydrolysis of phospholipids and activation of protein kinase C. Science 258:607–614

Nomikos M, Mulgrew-Nesbitt A, Pallavi P, Mihalyne G, Zaitseva I, Swann K, Lai FA, Murray D, McLaughlin S (2007) Binding of phosphoinositide-specific phospholipase C-ζ to phospholipid membranes: potential role of an unstructured cluster of basic residues. J Biol Chem 282:16644–16653

Oestreich EA, Wang H, Malik S, Kaproth-Joslin KA, Blaxall BC, Kelley GG, Dirksen RT, Smrcka AV (2007) Epac-mediated activation of phospholipase C-ε plays a critical role in β-adrenergic receptor-dependent enhancement of Ca^{2+} mobilization in cardiac myocytes. J Biol Chem 282:5488–5495

Oestreich EA, Malik S, Goonasekera SA, Blaxall BC, Kelley GG, Dirksen RT, Smrcka AV (2009) Epac and phospholipase C-ε regulate Ca^{2+} release in the heart by activation of protein kinase C-ε and calcium-calmodulin kinase II. J Biol Chem 284:1514–1522

Park DJ, Rho HW, Rhee SG (1991) CD3 stimulation causes phosphorylation of phospholipase C-γ1 on serine and tyrosine residues in a human T-cell line. Proc Natl Acad Sci USA 88:5453–5456

Park D, Jhon DY, Kriz R, Knopf J, Rhee SG (1992) Cloning, sequencing, expression, and Gq-independent activation of phospholipase C-β2. J Biol Chem 267:16048–16055

Paterson A, Boyer JL, Watts VJ, Morris AJ, Price EM, Harden TK (1995) Concentration of enzyme-dependent activation of PLC-β1 and PLC-β2 by Gα11 and βγ-subunits. Cell Signal 7:709–720

Paulssen RH, Woodson J, Liu Z, Ross EM (1996) Carboxyl-terminal fragments of phospholipase C-β1 with intrinsic Gq GTPase-activating protein (GAP) activity. J Biol Chem 271:26622–26629

Payne WE, Fitzgerald-Hayes M (1993) A mutation in PLC1, a candidate phosphoinositide-specific phospholipase C gene from Saccharomyces cerevisiae, causes aberrant mitotic chromosome segregation. Mol Cell Biol 13:4351–4364

Philip F, Kadamur G, Silos RG, Woodson J, Ross EM (2010) Synergistic activation of phospholipase C-β3 by Gαq and Gβγ describes a simple two-state coincidence detector. Curr Biol 20:1327–1335

Piechulek T, Rehlen T, Walliser C, Vatter P, Moepps B, Gierschik P (2005) Isozyme-specific stimulation of phospholipase C-γ2 by Rac GTPases. J Biol Chem 280:38923–38931

Poulin B, Sekiya F, Rhee SG (2000) Differential roles of the Src homology 2 domains of phospholipase C-γ in platelet-derived growth factor-induced activation of PLC-γ1 in intact cells. J Biol Chem 275:6411–6416

Raucher D, Stauffer T, Chen W, Shen K, Guo S, York JD, Sheetz MP, Meyer T (2000) Phosphatidylinositol 4,5-bisphosphate functions as a second messenger that regulates cytoskeleton-plasma membrane adhesion. Cell 100:221–228

Rebecchi MJ, Scarlata S (1998) Pleckstrin homology domains: a common fold with diverse functions. Annu Rev Biophys Biomol Struct 27:503–528

Rebecchi MJ, Raghubir A, Scarlata S, Hartenstine MJ, Brown T, Stallings JD (2009) Expression and function of phospholipase C in breast carcinoma. Adv Enzyme Regul 49:59–73

Rebres RA, Roach TI, Fraser ID, Philip F, Moon C, Lin KM, Liu J, Santat L, Cheadle L, Ross EM, Simon MI, Seaman WE (2011) Synergistic Ca^{2+} responses by Gαi- and Gαq-coupled G-protein-coupled receptors require a single PLC-β isoform that is sensitive to both Gβγ and Gαq. J Biol Chem 286:942–951

Rhee SG (2001) Regulation of phosphoinositide-specific phospholipase C. Annu Rev Biochem 70:281–312

Roifman CM, Wang G (1992) Phospholipase C-γ1 and phospholipase C-γ2 are substrates of the B cell antigen receptor associated protein tyrosine kinase. Biochem Biophys Res Commun 183:411–416

Ross EM (2008) Coordinating speed and amplitude in G-protein signaling. Curr Biol 18:R777–R783

Rotin D, Margolis B, Mohammadi M, Daly RJ, Daum G, Li N, Fischer EH, Burgess WH, Ullrich A, Schlessinger J (1992) SH2 domains prevent tyrosine dephosphorylation of the EGF receptor: identification of Tyr992 as the high-affinity binding site for SH2 domains of phospholipase C-γ. EMBO J 11:559–567

Rottbauer W, Just S, Wessels G, Trano N, Most P, Katus HA, Fishman MC (2005) VEGF-PLCγ1 pathway controls cardiac contractility in the embryonic heart. Genes Dev 19:1624–1634

Ryu SH, Cho KS, Lee KY, Suh PG, Rhee SG (1986) Two forms of phosphatidylinositol-specific phospholipase C from bovine brain. Biochem Biophys Res Commun 141:137–144

Ryu SH, Cho KS, Lee KY, Suh PG, Rhee SG (1987a) Purification and characterization of two immunologically distinct phosphoinositide-specific phospholipases C from bovine brain. J Biol Chem 262:12511–12518

Ryu SH, Suh PG, Cho KS, Lee KY, Rhee SG (1987b) Bovine brain cytosol contains three immunologically distinct forms of inositolphospholipid-specific phospholipase C. Proc Natl Acad Sci U S A 84:6649–6653

Satoh T, Edamatsu H, Kataoka T (2006) Phospholipase C-ε guanine nucleotide exchange factor activity and activation of Rap1. Methods Enzymol 407:281–290

Saunders CM, Larman MG, Parrington J, Cox LJ, Royse J, Blayney LM, Swann K, Lai FA (2002) PLC-ζ: a sperm-specific trigger of Ca^{2+} oscillations in eggs and embryo development. Development 129:3533–3544

Schmidt M, Evellin S, Weernink PA, von Dorp F, Rehmann H, Lomasney JW, Jakobs KH (2001) A new phospholipase-C-calcium signalling pathway mediated by cyclic AMP and a Rap GTPase. Nat Cell Biol 3:1020–1024

Schnabel P, Camps M (1998) Activation of a phospholipase C-β2 deletion mutant by limited proteolysis. Biochem J 330(Pt 1):461–468

Secrist JP, Karnitz L, Abraham RT (1991) T-cell antigen receptor ligation induces tyrosine phosphorylation of phospholipase C-γ1. J Biol Chem 266:12135–12139

Seifert JP, Wing MR, Snyder JT, Gershburg S, Sondek J, Harden TK (2004) RhoA activates purified phospholipase C-ε by a guanine nucleotide-dependent mechanism. J Biol Chem 279:47992–47997

Seifert JP, Zhou Y, Hicks SN, Sondek J, Harden TK (2008) Dual activation of phospholipase C-ε by Rho and Ras GTPases. J Biol Chem 283:29690–29698

Shibatohge M, Kariya K, Liao Y, Hu CD, Watari Y, Goshima M, Shima F, Kataoka T (1998) Identification of PLC210, a Caenorhabditis elegans phospholipase C, as a putative effector of Ras. J Biol Chem 273:6218–6222

Shimohama S, Homma Y, Suenaga T, Fujimoto S, Taniguchi T, Araki W, Yamaoka Y, Takenawa T, Kimura J (1991) Aberrant accumulation of phospholipase C-δ in Alzheimer brains. Am J Pathol 139:737–742

Shimohama S, Perry G, Richey P, Takenawa T, Whitehouse PJ, Miyoshi K, Suenaga T, Matsumoto S, Nishimura M, Kimura J (1993) Abnormal accumulation of phospholipase C-δ in filamentous inclusions of human neurodegenerative diseases. Neurosci Lett 162:183–186

Shiroo M, Goff L, Biffen M, Shivnan E, Alexander D (1992) CD45 tyrosine phosphatase-activated p59fyn couples the T cell antigen receptor to pathways of diacylglycerol production, protein kinase C activation and calcium influx. EMBO J 11:4887–4897

Sidhu RS, Clough RR, Bhullar RP (2005) Regulation of phospholipase C-δ1 through direct interactions with the small GTPase Ral and calmodulin. J Biol Chem 280:21933–21941

Singer AU, Waldo GL, Harden TK, Sondek J (2002) A unique fold of phospholipase C-β mediates dimerization and interaction with Gαq. Nat Struct Biol 9:32–36

Singh SM, Murray D (2003) Molecular modeling of the membrane targeting of phospholipase C pleckstrin homology domains. Protein Sci 12:1934–1953

Smrcka AV, Hepler JR, Brown KO, Sternweis PC (1991) Regulation of polyphosphoinositide-specific phospholipase C activity by purified Gq. Science 251:804–807

Smrcka AV, Sternweis PC (1993) Regulation of purified subtypes of phosphatidylinositol-specific phospholipase C-β by G protein α and βγ subunits. J Biol Chem 268:9667–9674

Snyder JT, Singer AU, Wing MR, Harden TK, Sondek J (2003) The pleckstrin homology domain of phospholipase C-β2 as an effector site for Rac. J Biol Chem 278:21099–21104

Sone Y, Ito M, Shirakawa H, Shikano T, Takeuchi H, Kinoshita K, Miyazaki S (2005) Nuclear translocation of phospholipase C-ζ, an egg-activating factor, during early embryonic development. Biochem Biophys Res Commun 330:690–694

Song C, Hu CD, Masago M, Kariyai K, Yamawaki-Kataoka Y, Shibatohge M, Wu D, Satoh T, Kataoka T (2001) Regulation of a novel human phospholipase C, PLC-ε, through membrane targeting by Ras. J Biol Chem 276:2752–2757

Stallings JD, Tall EG, Pentyala S, Rebecchi MJ (2005) Nuclear translocation of phospholipase C-δ1 is linked to the cell cycle and nuclear phosphatidylinositol 4,5-bisphosphate. J Biol Chem 280:22060–22069

Stallings JD, Zeng YX, Narvaez F, Rebecchi MJ (2008) Phospholipase C-δ1 expression is linked to proliferation, DNA synthesis, and cyclin E levels. J Biol Chem 283:13992–14001

Stauffer TP, Ahn S, Meyer T (1998) Receptor-induced transient reduction in plasma membrane PtdIns(4,5)P_2 concentration monitored in living cells. Curr Biol 8:343–346

Stewart AJ, Mukherjee J, Roberts SJ, Lester D, Farquharson C (2005) Identification of a novel class of mammalian phosphoinositol-specific phospholipase C enzymes. Int J Mol Med 15:117–121

Stoica B, DeBell KE, Graham L, Rellahan BL, Alava MA, Laborda J, Bonvini E (1998) The amino-terminal Src homology 2 domain of phospholipase C-γ1 is essential for TCR-induced tyrosine phosphorylation of phospholipase C-γ1. J Immunol 160:1059–1066

Suh BC, Hille B (2008) PIP_2 is a necessary cofactor for ion channel function: how and why? Annu Rev Biophys 37:175–195

Suh PG, Ryu SH, Moon KH, Suh HW, Rhee SG (1988) Cloning and sequence of multiple forms of phospholipase C. Cell 54:161–169

Suh PG, Park JI, Manzoli L, Cocco L, Peak JC, Katan M, Fukami K, Kataoka T, Yun S, Ryu SH (2008) Multiple roles of phosphoinositide-specific phospholipase C isozymes. BMB Rep 41:415–434

Tadano M, Edamatsu H, Minamisawa S, Yokoyama U, Ishikawa Y, Suzuki N, Saito H, Wu D, Masago-Toda M, Yamawaki-Kataoka Y, Setsu T, Terashima T, Maeda S, Satoh T, Kataoka T (2005) Congenital semilunar valvulogenesis defect in mice deficient in phospholipase C-ε. Mol Cell Biol 25:2191–2199

Takenawa T, Nagai Y (1981) Purification of phosphatidylinositol-specific phospholipase C from rat liver. J Biol Chem 256:6769–6775

Tasma IM, Brendel V, Whitham SA, Bhattacharyya MK (2008) Expression and evolution of the phosphoinositide-specific phospholipase C gene family in Arabidopsis thaliana. Plant Physiol Biochem 46:627–637

Taylor SJ, Chae HZ, Rhee SG, Exton JH (1991) Activation of the β1 isozyme of phospholipase C by α-subunits of the Gq class of G proteins. Nature 350:516–518

Tesmer JJ, Berman DM, Gilman AG, Sprang SR (1997) Structure of RGS4 bound to AlF4-activated Gαi1: stabilization of the transition state for GTP hydrolysis. Cell 89:251–261

Tesmer VM, Kawano T, Shankaranarayanan A, Kozasa T, Tesmer JJ (2005) Snapshot of activated G proteins at the membrane: the Gαq-GRK2-Gβγ complex. Science 310:1686–1690

Todderud G, Wahl MI, Rhee SG, Carpenter G (1990) Stimulation of phospholipase C-γ1 membrane association by epidermal growth factor. Science 249:296–298

Venema VJ, Ju H, Sun J, Eaton DC, Marrero MB, Venema RC (1998) Bradykinin stimulates the tyrosine phosphorylation and bradykinin B2 receptor association of phospholipase C-γ1 in vascular endothelial cells. Biochem Biophys Res Commun 246:70–75

Wahl MI, Olashaw NE, Nishibe S, Rhee SG, Pledger WJ, Carpenter G (1989) Platelet-derived growth factor induces rapid and sustained tyrosine phosphorylation of phospholipase C-γ in quiescent BALB/c 3T3 cells. Mol Cell Biol 9:2934–2943

Waksman G, Kominos D, Robertson SC, Pant N, Baltimore D, Birge RB, Cowburn D, Hanafusa H, Mayer BJ, Overduin M et al (1992) Crystal structure of the phosphotyrosine recognition domain SH2 of v-src complexed with tyrosine-phosphorylated peptides. Nature 358:646–653

Waldo GL, Boyer JL, Morris AJ, Harden TK (1991) Purification of an AlF$_4$- and G-protein βγ-subunit-regulated phospholipase C-activating protein. J Biol Chem 266:14217–14225

Waldo GL, Ricks TK, Hicks SN, Cheever ML, Kawano T, Tsuboi K, Wang X, Montell C, Kozasa T, Sondek J, Harden TK (2010) Kinetic scaffolding mediated by a phospholipase C-β and Gq signaling complex. Science 330:974–980

Walliser C, Retlich M, Harris R, Everett KL, Josephs MB, Vatter P, Esposito D, Driscoll PC, Katan M, Gierschik P, Bunney TD (2008) Rac regulates its effector phospholipase C-γ2 through interaction with a split pleckstrin homology domain. J Biol Chem 283:30351–30362

Wang D, Feng J, Wen R, Marine JC, Sangster MY, Parganas E, Hoffmeyer A, Jackson CW, Cleveland JL, Murray PJ, Ihle JN (2000a) Phospholipase C γ2 is essential in the functions of B cell and several Fc receptors. Immunity 13:25–35

Wang T, Dowal L, El-Maghrabi MR, Rebecchi M, Scarlata S (2000b) The pleckstrin homology domain of phospholipase C-β2 links the binding of Gβγ to activation of the catalytic core. J Biol Chem 275:7466–7469

Wang H, Oestreich EA, Maekawa N, Bullard TA, Vikstrom KL, Dirksen RT, Kelley GG, Blaxall BC, Smrcka AV (2005) Phospholipase C-ε modulates β-adrenergic receptor-dependent cardiac contraction and inhibits cardiac hypertrophy. Circ Res 97:1305–1313

Wang X, Zbou C, Qiu G, Fan J, Tang H, Peng Z (2008) Screening of new tumor suppressor genes in sporadic colorectal cancer patients. Hepatogastroenterology 55:2039–2044

Weiss A, Koretzky G, Schatzman RC, Kadlecek T (1991) Functional activation of the T-cell antigen receptor induces tyrosine phosphorylation of phospholipase C-γ1. Proc Natl Acad Sci USA 88:5484–5488

Wierenga RK (2001) The TIM-barrel fold: a versatile framework for efficient enzymes. FEBS Lett 492:193–198

Wing MR, Houston D, Kelley GG, Der CJ, Siderovski DP, Harden TK (2001) Activation of phospholipase C-ε by heterotrimeric G protein βγ subunits. J Biol Chem 276:48257–48261

Wing MR, Bourdon DM, Harden TK (2003a) PLC-ε: a shared effector protein in Ras-, Rho-, and Gαβγ-mediated signaling. Mol Interv 3:273–280

Wing MR, Snyder JT, Sondek J, Harden TK (2003b) Direct activation of phospholipase C-ε by Rho. J Biol Chem 278:41253–41258

Wohlgemuth S, Kiel C, Kramer A, Serrano L, Wittinghofer F, Herrmann C (2005) Recognizing and defining true Ras binding domains I: biochemical analysis. J Mol Biol 348:741–758

Xiao W, Hong H, Kawakami Y, Kato Y, Wu D, Yasudo H, Kimura A, Kubagawa H, Bertoli LF, Davis RS, Chau LA, Madrenas J, Hsia CC, Xenocostas A, Kipps TJ, Hennighausen L, Iwama A, Nakauchi H, Kawakami T (2009) Tumor suppression by phospholipase C-β3 via SHP-1-mediated dephosphorylation of Stat5. Cancer Cell 16:161–171

Xie W, Samoriski GM, McLaughlin JP, Romoser VA, Smrcka A, Hinkle PM, Bidlack JM, Gross RA, Jiang H, Wu D (1999) Genetic alteration of phospholipase C-β3 expression modulates behavioral and cellular responses to mu opioids. Proc Natl Acad Sci USA 96:10385–10390

Yablonski D, Kadlecek T, Weiss A (2001) Identification of a phospholipase C-γ1 SH3 domain-binding site in SLP-76 required for T-cell receptor-mediated activation of PLC-γ1 and NFAT. Mol Cell Biol 21:4208–4218

Yagisawa H, Sakuma K, Paterson HF, Cheung R, Allen V, Hirata H, Watanabe Y, Hirata M, Williams RL, Katan M (1998) Replacements of single basic amino acids in the pleckstrin homology domain of phospholipase C-δ1 alter the ligand binding, phospholipase activity, and interaction with the plasma membrane. J Biol Chem 273:417–424

Yamaga M, Fujii M, Kamata H, Hirata H, Yagisawa H (1999) Phospholipase C-δ1 contains a functional nuclear export signal sequence. J Biol Chem 274:28537–28541

Yin HL, Janmey PA (2003) Phosphoinositide regulation of the actin cytoskeleton. Annu Rev Physiol 65:761–789

Yoda A, Oda S, Shikano T, Kouchi Z, Awaji T, Shirakawa H, Kinoshita K, Miyazaki S (2004) Ca^{2+} oscillation-inducing phospholipase C-ζ expressed in mouse eggs is accumulated to the pronucleus during egg activation. Dev Biol 268:245–257

Yoko-o T, Matsui Y, Yagisawa H, Nojima H, Uno I, Toh-e A (1993) The putative phosphoinositide-specific phospholipase C gene, PLC1, of the yeast Saccharomyces cerevisiae is important for cell growth. Proc Natl Acad Sci USA 90:1804–1808

Yu H, Fukami K, Watanabe Y, Ozaki C, Takenawa T (1998) Phosphatidylinositol 4,5-bisphosphate reverses the inhibition of RNA transcription caused by histone H1. Eur J Biochem 251:281–287

Zhang W, Neer EJ (2001) Reassembly of phospholipase C-β2 from separated domains: analysis of basal and G protein-stimulated activities. J Biol Chem 276:2503–2508

Zhang W, Sloan-Lancaster J, Kitchen J, Trible RP, Samelson LE (1998) LAT: the ZAP-70 tyrosine kinase substrate that links T cell receptor to cellular activation. Cell 92:83–92

Zhou Y, Wing MR, Sondek J, Harden TK (2005) Molecular cloning and characterization of PLC-η2. Biochem J 391:667–676

Zhou Y, Sondek J, Harden TK (2008) Activation of human phospholipase C-η2 by Gβγ. Biochemistry 47:4410–4417

Chapter 4
Phosphoinositide 3-Kinases—A Historical Perspective

Alex Toker

Abstract The phosphoinositide 3-kinase (PI 3-K) signal relay pathway represents arguably one of the most intensely studied mechanisms by which extracellular signals elicit cellular responses through the generation of second messengers that are associated with cell growth and transformation. This chapter reviews the many landmark discoveries in the PI 3-K signaling pathway in biology and disease, from the identification of a novel phosphoinositide kinase activity associated with transforming oncogenes in the 1980s, to the identification of oncogenic mutations in the catalytic subunit of PI 3-K in the mid 2000s. Two and a half decades of intense research have provided clear evidence that the PI 3-K pathway controls virtually all aspects of normal cellular physiology, and that deregulation of one or more proteins that regulate or transduce the PI 3-K signal ultimately leads to human pathology. The most recent efforts have focused on the development of specific PI 3-K inhibitors that are currently being evaluated in clinical trials for a range of disease states.

This chapter is devoted to a historical review of the landmark findings in the PI 3-K from its relatively humble beginnings in the early to mid 1980s up until the present day. When considering the key findings in the history of PI 3-K, it is essential to recognize the landmark studies by Lowell and Mabel Hokin in the 1950s who were the first to describe that extracellular agonists such as acetylcholine could stimulate the incorporation of radiolabeled phosphate into phospholipids (Hokin and Hokin 1953). Their work initiated an entirely new field of lipid signaling, and subsequent studies in the 1970s by Michell and Lapetina who linked phosphoinositide turnover to membrane-associated receptors that initiate intracellular calcium mobilization (Lapetina and Michell 1973). Later studies revealed that the phospholipase-mediated breakdown of the same minor membrane phospholipids such as PtdIns-4,5-P_2 (phosphatidylinositol-4,5-bisphosphate) is responsible for the release of two additional key second messengers, diacylglycerol (DG) and IP_3 (inositol-1,4,5-trisphosphate) (Kirk et al. 1981; Berridge 1983; Berridge et al. 1983).

A. Toker (✉)
Department of Pathology, Beth Israel Deaconess Medical Center, Harvard Medical School,
330 Brookline Avenue, EC/CLS-633A, 02130 Boston, MA, USA
e-mail: atoker@bidmc.harvard.edu

Berridge, Irvine and Schulz then revealed that one of the byproducts of this lipid signal relay pathway is the release of calcium from intracellular stores such as the endoplasmic reticulum (Streb et al. 1983). Finally, pioneering studies by Nishizuka in the late 1970s identified PKC (protein kinase C) as a phospholipid and diacylglycerol-activated serine/threonine protein kinase (Inoue et al. 1977; Takai et al. 1977). At this point, it probably seemed to most at the time that the story was complete, such that hydrolysis of phosphoinositides such as PtdIns-4,5-P_2 and PtdIns-4-P would account for the major mechanisms of agonist-stimulated lipid signaling leading to physiological responses. On the contrary, the story was far from complete and was about to become a lot more complex.

Keywords Phosphorylation · Phosphoinositide 3-kinase · Pleckstrin homology · Akt · PTEN

4.1 The Discovery of PtdIns 3-OH Phosphorylation and PI 3-Kinases

The discovery of an enzymatic activity that could phosphorylate the 3-OH position of the inositol head ring of phosphoinositides has its roots in initial work by Macara and Balduzzi who showed that a phosphatidylinositol kinase activity was found associated with an avian sarcoma virus when expressed in cells (Macara et al. 1984). Simultaneously, Erickson and Cantley, working on the recently-discovered oncogenic tyrosine kinase Src, showed that it could phosphorylate phosphatidylinositol and phosphatidylglycerol in immunoprecipitates (Sugimoto et al. 1984). This suggested to them that Src itself has lipid kinase activity, yet as it turned out later it was the Src-associated PI 3-K activity that was responsible for this reaction. This result prompted Cantley to extend these findings, and in collaboration with Roberts, Schaffhausen and Kaplan they showed that the phosphatidylinositol kinase activity was due to an 85 kDa phosphoprotein bound to the middle T antigen of the polyoma tumor DNA virus (Kaplan et al. 1987). Despite these exciting new findings demonstrating an interaction between lipid-derived signals and transforming oncoproteins, PI 3-K had not yet formally been discovered. It was Whitman, a graduate student in the laboratory of Cantley who realized that the product of the phosphatidylinositol activity in Src immunoprecipitates migrated slightly differently on thin layer chromatography plates from PtdIns-4-P, which was the presumed product of the reaction. A collaboration with Downes, an expert in inositol lipids, proved beyond doubt that the lipid product of the reaction was in fact PtdIns-3-P, the first new phosphoinositide identified in some 30 years (Whitman et al. 1988). Downes, Hawkins and Stephens working independently also provided evidence of the existence of PtdIns-3-P in astrocytoma cells (Stephens et al. 1989). As became evident later, that PtdIns-3-P transduces the PI 3-K signal in response to growth factor was somewhat of a red herring, since the primary activity of class I PI 3-Ks that are activated by both receptor tyrosine kinases (RTK) and G protein-coupled receptors (GPCR) is to interconvert PtdIns-4,5-P_2 into PtdIns-3,4,5-P_3. PtdIns-3-P, as it turned out later, is generated by class III PI 3-Ks, the prototype being the yeast ortholog *vps34p*. In this context, it is often overlooked that PtdIns-3,4,5-P_3 was actually discovered by Traynor-Kaplan and Sklar, who identified an inositol tetrakisphosphate (IP_4)-containing phospholipid in stimulated neutrophils (Traynor-Kaplan et al. 1988). They showed that PtdIns-3,4,5-P_3 is a new phosphoinositide that contains four phosphates, making it the most negatively charged lipid in the cell. Both the Cantley, Stephens and Sklar papers were published within a few months of each other in a period between 1988 and 1989, representing the birth of the PI 3-K field.

There then followed a flurry of activity by the laboratories highlighted above, though arguably the rest of the signaling world did not take much notice at first. At the time most were heavily invested in understanding the molecular mechanisms by which oncogenes such as Src and Ras transduce mitogenic signals, and delineating the mechanisms of activation of ERK (extracellular-regulated kinase). In the PI 3-K world, it was shown that exposure of a variety of cells with mitogenic growth factors such as PDGF (platelet-derived growth factor) (Auger et al. 1989) and GPCR (G protein coupled receptor) agonists (such as fMLP, formylmethionyl leucyl phenylalanine) (Stephens et al. 1991) stimulated the rapid accumulation of PtdIns-3,4,5-P_3, and also PtdIns-3,4-P_2, the latter primarily due to the activity of class II PI 3-Ks and/or breakdown of PtdIns-3,4,5-P_3 by lipid phosphatases. In contrast, the levels of PtdIns-3-P remained relatively constant in most cells studied. The conclusion was therefore that PtdIns-3,4-P_2, and primarily PtdIns-3,4,5-P_3, are true second messengers that transduce the PI 3-K signal. The initial identification of the 85 kDa phosphoprotein associated with Src and polyoma middle T led to its biochemical purification and subsequent cloning by several laboratories (Escobedo et al. 1991; Otsu et al. 1991; Skolnik et al. 1991). During these efforts, Carpenter and Cantley recognized that p85, that we now know represents one of the regulatory subunits of class I PI 3-Ks, was found tightly associated with a p110 protein (Carpenter et al. 1990). Protein microsequencing of purified p110 allowed for the design of degenerate oligonucleotide probes, and using this information Waterfield's laboratory was the first to clone and characterize the catalytic subunit of PI 3-K, in their case p110α (Hiles et al. 1992). Analysis of the p110 sequence revealed significant homology with the yeast gene *vps34p*, which in 1990 Herman and Emr showed is required for vacuolar protein sorting and segregation in budding yeast (Herman and Emr 1990). Subsequent pioneering studies by Emr's laboratory showed that yeast *vps34p* and its mammalian ortholog are class III PI 3-Ks with a unique substrate specificity, incapable of generating PtdIns-3,4-P_2 or PtdIns-3,4,5-P_3, and exclusively responsible for PtdIns-3-P synthesis (Schu et al. 1993; Brown et al. 1995; De Camilli et al. 1996).

4.2 Expansion of the Family of PI 3-Kinases and Their Regulators

In the years that followed the cloning of p85α and p110α, several laboratories described additional PI 3-K isoforms in both mammals, flies, nematodes, yeasts and even plants, and this required reclassification of the PI 3-K family into distinct classes.

We now know there are eight mammalian PI 3-kinases: class Ia enzymes (p110α, p110β, p110δ) and class Ib (p110γ); class II enzymes (PI 3-K-C2α, PI 3-K-C2β, PI 3-K-Cγ); and class III enzymes (*vps34p*). Note that the nomenclature refers to the catalytic subunits, yet distinct regulatory subunits for each class exist: (class Ia, p85α, p55α, p50α, p85β, p55γ); class Ib, p101 and p84; a p150 regulatory subunit for *vpv34p* has also been identified. For more detail, see the following reviews (Vanhaesebroeck et al. 2010; Engelman et al. 2006; Hawkins et al. 2006) and Vol. I, Chap. 5 by Wymann.

The protein sequencing and cloning of the p85α regulatory subunit coincided with the discovery of SH2 (Src homology 2) domains in pioneering studies by the Hanafusa and Pawson laboratories (Moran et al. 1990; DeClue et al. 1987; O'Brien et al. 1990; Matsuda et al. 1990). Recognizing that p85 subunits have two copies of SH2 domains provided the mechanism by which PI 3-K is relocalized to sites of activated, phosphorylated receptor tyrosine kinases. Zhou and Cantley then used this information to decode the specificity of SH2 domains, including that of p85α, towards phosphotyrosine motifs in RTKs (Zhou et al. 1993). At the same time, Kazlauskas and Cooper developed an approach in which individual tyrosines in the PDGF-receptor are individually mutated to prevent binding to SH2 domains in proteins including PI 3-K (Kazlauskas and Cooper 1989). This showed that both PI 3-K and PLCγ-1 (phospholipase C-γ) are required to transduce the PDGF signal to DNA synthesis and cell cycle progression (Valius and Kazlauskas 1993). These PDGF-R mutants were then subsequently used by many other laboratories to probe the mechanisms by which PI 3-K engages the cell cycle, and also activates secondary effectors such as S6K1 (p70 ribosomal protein S6 kinase-1) (Chung et al. 1994). Importantly, Kazlauskas later went to show that receptors such as PDGF-R stimulate two waves of PI 3-K activity, one immediately after ligand stimulation lasting about an hour, and a second much larger wave typically 4–8 h post stimulation (Jones et al. 1999). Using PDGF-R mutants as well as PI 3-K inhibitors, his laboratory showed that the first wave is actually dispensable for cell cycle progression to S phase, instead it is the second wave that is responsible.

The next major discovery in the field was wortmannin. In 1993, two laboratories, Wymann and Arcaro as well as Matsuda's group showed that the furanosteroid metabolite isolated from the fungus *Penicillium funiculosum*, known as wortmannin, potently inhibits fMLP-stimulated PtdIns-3,4,5-P_3 synthesis in neutrophils at low nanomolar doses (Arcaro and Wymann 1993; Yano et al. 1993). Other laboratories subsequently confirmed the inhibitory activity of wortmannin on PI 3-K. A year later, Vlahos at Ely Lilly published a structurally distinct compound termed LY294002, that also directly inhibits PI 3-K (Vlahos et al. 1994). Remarkably, both compounds are still widely used in the field, considering that they have since been shown to have off-target effects on other kinases. Indeed much more specific PI 3-K inhibitors have since been described, to the extent that isoform-specific compounds are now available, many of them in phase I and II clinical trials. Regardless, the availability of wortmannin and LY294002 stimulated much progress in the still nascent PI 3-K signaling field.

The discovery that PI 3-K functions a direct effector of GTP-loaded Ras by Rodriguez-Viciana and Downward added further credence to the model that the PI 3-K pathway contributes to cancer etiology, considering the numerous human solid tumors that harbor oncogenic activating mutations in the Ras GTPase (Rodriguez et al. 1994). Although largely overlooked, it was initially Lapetina who first showed that p21 Ras can associate with PI 3-K in cells (Sjolander et al. 1991). This was later confirmed in the seminal paper by Downward's group who showed that GTP-Ras is required for efficient PI 3-K activation by growth factors, and they further identified a direct PI 3-K binding site on Ras (Rodriguez et al. 1994). Thus PI 3-K was added to the list of Ras effectors which now encompass the Ser/Thr kinase Raf, the nucleotide dissociation factor RalGDS and the phospholipase PLCε. It is also important to note that as first recognized by Hawkins and Stephens (Wennstrom et al. 1994; Welch et al. 2003), PI 3-K and PtdIns-3,4,5-P_3 also contribute to the regulation of other GTPases, most notably Rac, and this in turn has profound effects on the regulation of the actin cytoskeleton and cell motility in virtually all cells. This is best illustrated in both neutrophils and *Dictyostelium* amoeba where a gradient of PtdIns-3,4,5-P_3 and Rac activation is rate-limiting for actin assembly and directed cell migration (Funamoto et al. 2001) (also see Vol. II, Chap. 6 by Yin and Vol. II, Chap. 7 by Parent).

4.3 How Do PI 3-Kinase Lipid Products Affect Cell Function?

In the mid 1990s, the field of PI 3-K was firmly established, but was missing a critical piece of the puzzle. How do the PI 3-K lipid products, particularly PtdIns-3,4,5-P_3, transduce the signal to immediate effectors and then onto cellular responses? The first clue came from Fesik and Harlan, who showed that PH (pleckstrin homology) domains bind with high affinity to phosphoinositides, in their case, PtdIns-4,5-P_2 (Harlan et al. 1994). One year later, Tsichlis working with Kaplan and Franke published a seminal finding identifying the first *bona-fide* effector of PI 3-K, the Ser/Thr kinase Akt (Franke et al. 1995). Akt had been originally identified by Tsichlis as the cellular homolog of the transforming retroviral oncogene *v-Akt* (Bellacosa et al. 1991). It was independently identified by Woodgett and termed PKB (protein kinase B), as it was sequence-related to PKA (protein kinase A) and PKC (protein kinase C) (Coffer and Woodgett 1991), and also by Hemmings as Rac (related to A and C kinases (Jones et al. 1991)). Akt contains a PH domain, and this prompted Tsichlis and colleagues to investigate whether Akt is PI 3-K target, and indeed they showed that PDGF stimulates Akt activation in a PI 3-K-dependent manner (Franke et al. 1995). Burgering and Coffer came to the same conclusion independently (Burgering and Coffer 1995). Initially suggested to be mediated by PtdIns-3-P, several groups subsequently showed that the PH domain of Akt has high affinity for PtdIns-3,4-P_2 and PtdIns-3,4,5-P_3 (Frech et al. 1997; Andjelkovic et al. 1997; Franke et al. 1997). Although several groups showed that PI 3-K lipids, particularly PtdIns-3,4-P_2, could weakly activate Akt *in vitro*, it was not until two groups independently discovered PDK-1 (phosphoinositide-dependent kinase-1), the enzyme that is responsible for full Akt activation. Working with Hemmings and Cohen, Alessi had originally shown that Akt is phosphorylated at two key residues, Thr308 in the T loop segment of the kinase domain, and Ser473 in a hydrophobic region at the carboxyl-terminus (Alessi et al. 1996). Later, working in the lab of Cohen, Alessi set about to purify the enzymatic activity capable of phosphorylating and activating Akt. This led to the identification of PDK-1 as the Thr308 upstream kinase (Alessi et al. 1997). Once again, Hawkins and Stephens working with Stokoe independently identified the same activity, and they further showed that PDK-1 phosphorylates Akt at Thr308 in a manner that depends on PtdIns-3,4,5-P_3 binding to the Akt PH domain (Stokoe et al. 1997). Subsequent studies by Alessi and other laboratories demonstrated that PDK-1 is the master upstream T loop kinase for many AGC kinases, such as S6K, PKCs, SGKs and several others (Alessi 2001; Mora et al. 2004). As for Ser473, whose phosphorylation contributes to full Akt activation, several upstream kinases were proposed over the years, and while some of these may function in specific settings, it wasn't until 2005 when Sabatini's group showed that the mTORC2 (mammalian target of rapamycin complex 2) complex represents the physiologically-relevant Ser473 kinase, at least in most cells and tissues (Sarbassov et al. 2005). It is also important to note that PH domains with very discrete phosphoinositide-binding specifies are now known to exist (Lemmon 2008). Moreover, since the discovery of PH domains, other domains in signaling proteins have been described that bind other 3 phosphoinositides, such as the PX and FYVE domains, providing the mechanistic basis for the regulation of specific cellular responses to individual PI 3-K lipid products.

Once Akt was identified as a direct effector of PI 3-K and PtdIns-3,4,5-P_3, the field moved quickly to identify the immediate substrates of Akt that would transduce the signal to a specific response. Cohen identified the first Akt substrate, GSK-3β, a rate-limiting enzyme in the regulation of glycogen synthase in insulin-responsive cells and tissues (Cross et al. 1995). Because Cooper initially showed that PI 3-K comprises a key cellular survival pathway (Yao and Cooper 1995), this prompted Greenberg and colleagues to subsequently show that indeed Akt is one of the PI 3-K effectors in cellular survival mechanisms (Dudek et al. 1997). His laboratory also showed that the pro-apoptotic protein BAD is an Akt substrate whereby phosphorylation attenuates its apoptotic function (Datta et al. 1997). Secondly, Brunet also working in Greenberg's lab showed that the pro-apoptotic FOXO3a transcription factor is phosphorylated by Akt, and in turn this blocks its apoptotic function (Brunet et al. 1999). Burgering's lab published the same finding for AFX (Kops et al. 1999; Morris et al. 1996). It is important to note that much of the work on the importance of PI 3-K and Akt signaling in cellular survival was pioneered by Ruvkun and colleagues, who provided irrefutable genetic evidence for a linear pathway in *Caenorhabditis elegans* comprising PI 3-K (*age-1*) (Morris et al. 1996), PDK-1 (*pdk-1*) (Paradis et al. 1999), Akt (*akt-1* and *akt-2*) (Paradis and Ruvkun 1998), and FOXO (*daf-16*) (Ogg et al. 1997) that controls survival and lifespan in the nematode worm. Similar findings using *Drosophila* genetics also corroborated the importance of PI 3-K signaling to cellular growth and survival (Leevers et al. 1996).

At the start of the new millennium, the PI 3-K field was firmly established as one of the central mechanisms by which extracellular stimuli initiate a lipid signaling pathway that culminates in the synthesis of PtdIns-3,4,5-P$_3$, activation of downstream effectors such as Akt, and in turn alterations in cell growth, proliferation, survival, migration and gene transcription. Much of this had been achieved using biochemistry and cell-based assays with overexpression, and inhibitors such as wortmannin. What was missing was a genetic approach to begin to probe the contributions of individual regulatory p85 and catalytic p110 subunits in normal physiology and disease. In 1999, Kadowaki's group published the first p85α (*PIK3R1*) knockout mouse which displayed increased insulin sensitivity and hypoglycemia (Terauchi et al. 1999). Shortly thereafter, Fruman and Cantley reported that ablation of all three splice variants encoded by p85α (p85α, p55α and p50α) results in perinatal lethality due to hepatocyte necrosis and chylous ascites (Fruman et al. 2000). Similar to Kadawaki's finding, they also reported that loss of p85α alone results in hypoglycemia and increased insulin sensitivity. Knockout of p85β has also been reported, with phenotypes similar to p85α null mice (Ueki et al. 2002). At about the same time reports began to emerge on the knockouts of the p110 catalytic subunits. Knockout of p110α and p110β results in early embryonic lethality, and this has precluded a detailed analysis of the role of these isoforms in adult tissues (Bi et al. 1999, 2002). On the other hand, considerable information on p110 isoform-specific signaling has been gained by analysis of the p110γ and p110δ isoforms, which are preferentially, although by no means exclusively, expressed in the immune system. Hirsch and Wymann showed that p110γ knockout mice are viable, yet their neutrophils show an impaired respiratory burst and migration phenotype (Hirsch et al. 2000). Macrophages from these mice reveal impaired motility, and defective accumulation in a sepsis model. Interestingly, both Hirsch, Wymann and Penninger also reported effects on cardiomyocyte contractility that at least in one case was attributed to a non-kinase scaffolding function of p110γ that functions through a cyclic AMP (cAMP) and phosphodiesterase (PDE3B) pathway (Crackower et al. 2002; Patrucco et al. 2004). Two groups also independently deleted p110δ, and also generated 'knock-in' mice, and although the mice are viable, they reveal impaired immune cell development and function, much like the p110γ null mice (Clayton et al. 2002; Jou et al. 2002). In combination, these studies have provided compelling evidence that p110γ and p110δ are viable targets for immunity and cardiac dysfunctions. To this end, specific p110γ and p110δ inhibitors are currently undergoing clinical trials for a range of immunological pathologies (see below and Vol. I, Chap. 6 by Hirsch for more details).

4.4 Linking PtdIns-3,4,5-P$_3$ to Cancer

A major leap forward in our understanding of the importance of PI 3-K signaling in pathophysiology was made by the discovery of the PtdIns-3,4,5-P$_3$ phosphatase PTEN. Parsons group had been studying a region on chromosome 10q23 known to be frequently subjected to loss of heterozygosity (LOH), particularly in prostate cancer. This led to speculation that a yet-to-be-discovered tumor suppressor gene resides in this area whose loss would predispose to malignancy in several cancer types. Li, Parsons and colleagues used positional cloning to identify an open reading frame that encoded a protein with homologies to protein tyrosine phosphatases, chicken tensin and bovine auxilin (Li et al. 1997). They therefore named the gene *PTEN*, for phosphatase and tensin homolog deleted on chromosome 10. The same year Steck and colleagues also independently cloned the same gene and named it *MMAC1*, for mutated in multiple advanced cancers (Steck et al. 1997). Both groups recognized the striking homology to the catalytic domain of protein tyrosine phosphatases. However, the discovery that PTEN is a PtdIns-3,4,5-P_3 phosphatase is attributed to Dixon and Maehama (Maehama and Dixon 1998). Dixon, an expert in protein tyrosine phosphatases, realized that *in vitro* PTEN has a much higher specific activity towards acidic and thus negatively-charged phosphopeptides compared to uncharged peptides. He then recognized that PTEN might utilize lipid rather than proteinaceous substrates, and recalled that PtdIns-3,4,5-P_3 is the most negatively charged lipid in the cell. Sure enough, when PtdIns-3,4,5-P_3 was used as substrate, PTEN efficiently dephosphorylated it back to PtdIns-4,5-P_2, providing evidence that PTEN is the first identified 3 phosphoinositide phosphatase (Maehama and Dixon 1998). Given that *PTEN* LOH is a frequent event in various human solid tumors, this once again bolstered the link between PI 3-K, PtdIns-3,4,5-P_3 and tumorigenesis. We now know that inactivating mutations, deletions and LOH are frequent events in the *PTEN* locus, to the extent that over 30% of glioblastoma patients have *PTEN* loss leading to hyperactivation of PI 3-K and Akt signaling (Keniry and Parsons 2008).

It took almost 20 years since the discovery of PI 3-K for researchers to identify oncogenic activating mutations in *PIK3CA*, the gene that encodes p110α. In spite of the findings that PI 3-K associated with transforming oncogenes, that it mediates growth factor signaling in response to mitogens, and the fact that the PtdIns-3,4,5-P_3 phosphatase *PTEN* is a tumor suppressor, the discovery in 2004 that *PIK3CA* is an oncogene frequently mutated in human solid tumors was the first real direct evidence linking this pathway to cancer (Samuels et al. 2004; Campbell et al. 2004). In certain tumors, for example estrogen receptor positive breast cancer, *PIK3CA* is the most frequently mutated oncogene, leading to hyperactivation of Akt and other PI 3-K effectors to drive cellular proliferation, survival and metastasis (reviewed by (Engelman 2009)). Surprisingly, no oncogenic activating mutations have been detected in other p110 isoforms, particularly *PIK3CB*, yet studies have clearly demonstrated that p110β can promote tumorigenesis, particularly in settings on *PTEN* loss (Zhao et al. 2005; Jiang et al. 2010). On the other hand, mutations in PI 3-K regulatory subunits, for example *PIK3R1* are found in human tumors, albeit at relatively low frequencies, and thus may activate p110β and p110δ indirectly (Jaiswal et al. 2009). Similarly, activating oncogenic mutations in Akt isoforms have also been described in various human tumors, again at low frequency (Carpten et al. 2007). The frequency and spectrum of *PTEN* LOH and mutations, *PIK3CA* amplifications and mutations, *PIK3R1* and Akt oncogenic mutations render PI 3-K one of the most frequently mutated and druggable pathways in human cancer. It is not surprising therefore that numerous clinical trials for novel drugs are targeting one or more of these enzymes for therapeutic benefit. Presently the compound in the most advanced trials is the dual mTOR and PI 3-K inhibitor NVP-BEZ235 from Novartis (Maira et al. 2008). Other dual TOR/PI 3-K inhibitors are also being evaluated for targeted therapy, and this is because the TORC1 pathway effects a negative feedback loop on PI 3-K, such that inhibition of TORC1 with rapamycin can actually enhance PI 3-K/Akt activity *in vivo* (O'Reilly et al. 2006). As already discussed, specific p110β inhibitors are also in trials for treatment of thrombosis, and a p110δ inhibitor is being evaluated for hematological malignancies, in addition to a range of next generation p110α and p110β inhibitors (reviewed by (Vanhaesebroeck et al. 2010)). While the outlook for the use of these inhibitors is promising, it is likely that they will be used in dual targeted therapies with drugs that block activated or amplified receptor tyrosine kinases (such as EGF-R or Met) or drugs that interfere with ERK activity. Moreover, while some or possibly even many of the PI 3-K inhibitors currently in trials will fail, this will be of significant benefit to the research community who will use these much more specific inhibitors for basic research. It is also reasonable to argue that the use of these inhibitors, coupled with global RNAi screening approaches in tumor cells as well as model organisms such as flies will result in the identification of previously unappreciated mechanisms by which PI 3-K promotes disease states.

In this context, much of the activity and function of PI 3-K in both normal physiology and human disease has been attributed to Akt, most likely because of the availability of numerous reagents for studying Akt biology. Yet it is clear that a significant proportion of PI 3-K and PtdIns-3,4,5-P_3 and PtdIns-3,4-P_2 signaling is mediated by non-Akt effectors. Indeed a recent RNAi screening experiment revealed that in certain breast cancer cells harboring oncogenic *PIK3CA*, Akt is actually dispensable for growth and proliferation, yet both PDK-1 and SGK3 (serum and glucocorticoid-regulated kinase-3) are required (Vasudevan et al. 2009). Because SGK3 has a PX domain that interacts primarily with PtdIns-3-P in endosomes, the suggestion is that *PIK3CA* may mediate oncogenic signaling in certain human tumors through PtdIns-3-P at the endosome. Moreover, other targets of PI 3-K likely function independently of Akt, for example the tyrosine kinase Btk which binds with high affinity to PtdIns-3,4,5-P_3 and PtdIns-3,4-P_2 (Isakoff et al. 1998; Fruman et al. 2002). A large family of PtdIns-3,4,5-P_3 binding proteins include guanine nucleotide exchange factors (GEFs) and GTPase activating proteins (GAPs) for members of the Rho superfamily of small GTPases. Examples are the proteins pREX2a and P-REX1 which function as GEFs for Rac and integrate PtdIns-3,4,5-P_3 and heterotrimeric G protein signaling in cancer (Fine et al. 2009; Sosa et al. 2010).

4.5 Challenges Ahead

Due to the enormous advances in PI 3-K signaling and biology during the past two decades, we are presently at the stage where this pathway has entered the clinical arena and there is the realistic expectation that current drugs, or their derivatives, that target enzymes in the PI 3-K pathway will ultimately prove to have therapeutic value.

This should not be taken as meaning that we now have learned the major mechanisms by which the PI 3-K pathway controls human biology and disease. There are numerous gaps in our understanding as to how PI 3-K and its lipid products mediate signal relay, and how deregulation of these mechanisms results in pathology, and yet new discoveries are constantly being made. Even though we knew back in the 1990s that PtdIns-3,4-P_2 has a signaling role *per se*, for example in the binding and regulation of Akt, it wasn't until the polyphosphates 4 phosphatase *INPP4B* was shown to function as a tumor suppressor that the exclusive role of this phosphoinositide was directly demonstrated (Gewinner et al. 2009). The implication is that in addition to Akt there likely exist other direct PtdIns-3,4-P_2 effectors. Another major unexplored area of signaling in the field is the specific roles played by p110 isoforms. Much information has been garnered from studies of individual p110 knockout mice, and also cell-based studies using RNAi. What is lacking is a more compete understanding how individual p110 isoforms, perhaps in specific cellular compartments, elicit pools of PtdIns-3,4-P_2 and PtdIns-3,4,5-P_3 at discrete locations, and in turn how this affects downstream signaling. Compartmentalized PI 3-K signaling is likely to have a profound impact on certain cellular responses, for example remodeling of the actin cytoskeleton which requires specific temporal as well as kinetic regulation during directed cell migration. Genetically-encoded FRET-based sensors to monitor both PI 3-K effectors, as well as PtdIns-3,4,5-P_3/PtdIns-3,4-P_2 themselves will likely aid in this pursuit. This issue also applies to the nuclear phosphoinositide cycle, first recognized and pioneered by Irvine, Divecha and Cocco (Cocco et al. 1987; Divecha et al. 1991). Even before the discovery of PI 3-K we knew there exists a nuclear phosphoinositide cycle, and nuclear phosphoinositides had been detected (Cocco et al. 1987). Since then, both PI 3-K and its lipid products have been detected inside the nucleus, consistent with the model that certain targets of Akt, for example FOXO3a, are actually phosphorylated in the nucleus, and subsequently exported to the cytoplasm thus terminating transcriptional activity (Brunet et al. 2002). Yet there remain many gaps in our knowledge of the regulation and function of nuclear PI 3-K signaling. Many of the effectors of PI 3-K also exist as multi-gene families, a good example is Akt (Akt1, Akt2 and Akt3 in mammals). While it was originally assumed that they would function redundantly, we now know that there are specific functions and even specific substrates of Akt isoforms. For example, in the context of cancer progression, Akt2 promotes tumor invasion and metastasis, whereas Akt1 either does not or even attenuates this phenotype (Yoeli-Lerner et al. 2005; Irie et al. 2005; Maroulakou et al. 2007). These findings of opposing or non-redundant function for PI 3-K effectors have profound implications for drugs that are being developed for therapeutic intervention in the PI 3-K/Akt pathway. Whether similar opposing phenotypes exist for other PI 3-K effectors remains to be determined, but is likely to be the case. It is also very likely that additional feedback loops exist in the pathway, analogous to the mTOR/S6K1 negative feedback loop. If this is the case, this will make the use of PI 3-K and Akt inhibitors somewhat of a challenge. Moreover, although powerful genetic models exist, such as p110 isoform knockout as well as transgenes, the next generation models will likely be considerably more complex, and recapitulate the multiple genetic lesions in the PI 3-K pathway that we now know exist in human cancer patients. These can then be used in combination with the latest PI 3-K inhibitors to more accurately model human cancer therapy in preclinical models.

In closing, this chapter has highlighted the major discoveries in the PI 3-K field in some two and a half decades. From the identification of the first novel phosphoinositide PtdIns-3-P in the mid 1980s, to the discovery of oncogenic *PIK3CA* mutations in the mid 2000s, to the present day with clinical trials for p110 inhibitors, the field has matured at a dizzying pace. Yet we have clearly much more to learn as we await the results from the PI 3-K inhibitor clinical trials. We are now at a point where biochemical studies, genome-wide screening approaches and clinical trials are all informing the basic and clinical arena, and the field is poised to capture this information and much more efficiently translate it into therapeutic benefit.

References

Alessi DR (2001) Discovery of PDK1, one of the missing links in insulin signal transduction. Colworth Medal Lecture. Biochem Soc Trans 29:1–14

Alessi DR, Andjelkovic M, Caudwell B, Cron P, Morrice N, Cohen P, Hemmings BA (1996) Mechanism of activation of protein kinase B by insulin and IGF-1. EMBO J 15:6541–6551

Alessi DR, James SR, Downes CP, Holmes AB, Gaffney PR, Reese CB, Cohen P (1997) Characterization of a 3-phosphoinositide-dependent protein kinase which phosphorylates and activates protein kinase Balpha. Curr Biol 7:261–269

Andjelkovic M, Alessi DR, Meier R, Fernandez A, Lamb NJ, Frech M, Cron P, Cohen P, Lucocq JM, Hemmings BA (1997) Role of translocation in the activation and function of protein kinase B. J Biol Chem 272:31515–31524

Arcaro A, Wymann MP (1993) Wortmannin is a potent phosphatidylinositol 3-kinase inhibitor: the role of phosphatidylinositol 3,4,5-trisphosphate in neutrophil responses. Biochem J 296(Pt 2):297–301

Auger KR, Serunian LA, Soltoff SP, Libby P, Cantley LC (1989) PDGF-dependent tyrosine phosphorylation stimulates production of novel polyphosphoinositides in intact cells. Cell 57:167–175

Bellacosa A, Testa JR, Staal SP, Tsichlis PN (1991) A retroviral oncogene, akt, encoding a serine-threonine kinase containing an SH2-like region. Science 254:274–277

Berridge MJ (1983) Rapid accumulation of inositol trisphosphate reveals that agonists hydrolyse polyphosphoinositides instead of phosphatidylinositol. Biochem J 212:849–858

Berridge MJ, Dawson RM, Downes CP, Heslop JP, Irvine RF (1983) Changes in the levels of inositol phosphates after agonist-dependent hydrolysis of membrane phosphoinositides. Biochem J 212:473–482

Bi L, Okabe I, Bernard DJ, Wynshaw-BORIS A, Nussbaum RL (1999) Proliferative defect and embryonic lethality in mice homozygous for a deletion in the p110alpha subunit of phosphoinositide 3-kinase. J Biol Chem 274:10963–10968

Bi L, Okabe I, Bernard DJ, Nussbaum RL (2002) Early embryonic lethality in mice deficient in the p110beta catalytic subunit of PI 3-kinase. Mamm Genome 13:169–172

Brown WJ, Dewald DB, Emr SD, Plutner H, Balch WE (1995) Role for phosphatidylinositol 3-kinase in the sorting and transport of newly synthesized lysosomal enzymes in mammalian cells. J Cell Biol 130:781–796

Brunet A, Bonni A, Zigmond MJ, Lin MZ, Juo P, Hu LS, Anderson MJ, Arden KC, Blenis J, Greenberg ME (1999) Akt promotes cell survival by phosphorylating and inhibiting a Forkhead transcription factor. Cell 96:857–868

Brunet A, Kanai F, Stehn J, Xu J, Sarbassova D, Frangioni JV, Dalal SN, Decaprio JA, Greenberg ME, Yaffe MB (2002) 14-3-3 transits to the nucleus and participates in dynamic nucleocytoplasmic transport. J Cell Biol 156:817–828

Burgering BM, Coffer PJ (1995) Protein kinase B (c-Akt) in phosphatidylinositol-3-OH kinase signal transduction. Nature 376:599–602

Campbell IG, Russell SE, Choong DY, Montgomery KG, Ciavarella ML, Hooi CS, Cristiano BE, Pearson RB, Phillips WA (2004) Mutation of the PIK3CA gene in ovarian and breast cancer. Cancer Res 64:7678–7681

Carpenter CL, Duckworth BC, Auger KR, Cohen B, Schaffhausen BS, Cantley LC (1990) Purification and characterization of phosphoinositide 3-kinase from rat liver. J Biol Chem 265:19704–19711

Carpten JD, Faber AL, Horn C, Donoho GP, Briggs SL, Robbins CM, Hostetter G, Boguslawski S, Moses TY, Savage S, Uhlik M, Lin A, Du J, Qian YW, Zeckner DJ, Tucker-KELLOGG G, Touchman J, Patel K, Mousses S, Bittner M, Schevitz R, Lai MH, Blanchard KL, Thomas JE (2007) A transforming mutation in the pleckstrin homology domain of AKT1 in cancer. Nature 448:439–444

Chung J, Grammer TC, Lemon KP, Kazlauskas A, Blenis J (1994) PDGF- and insulin-dependent pp70S6k activation mediated by phosphatidylinositol-3-OH kinase. Nature 370:71–75

Clayton E, Bardi G, Bell SE, Chantry D, Downes CP, Gray A, Humphries LA, Rawlings D, Reynolds H, Vigorito E, Turner M (2002) A crucial role for the p110delta subunit of phosphatidylinositol 3-kinase in B cell development and activation. J Exp Med 196:753–763

Cocco L, Gilmour RS, Ognibene A, Letcher AJ, Manzoli FA, Irvine RF (1987) Synthesis of polyphosphoinositides in nuclei of Friend cells. Evidence for polyphosphoinositide metabolism inside the nucleus which changes with cell differentiation. Biochem J 248:765–770

Coffer PJ, Woodgett JR (1991) Molecular cloning and characterisation of a novel putative protein-serine kinase related to the cAMP-dependent and protein kinase C families [published erratum appears in Eur J Biochem 1992 May 1;205(3):1217]. Eur J Biochem 201:475–481

Crackower MA, Oudit GY, Kozieradzki I, Sarao R, Sun H, Sasaki T, Hirsch E, Suzuki A, Shioi T, Irie-SASAKI J, Sah R, Cheng HY, Rybin VO, Lembo G, Fratta L, Oliveira-DOS-SANTOS AJ, Benovic JL, Kahn CR, Izumo S, Steinberg SF, Wymann MP, Backx PH, Penninger JM (2002) Regulation of myocardial contractility and cell size by distinct PI3K-PTEN signaling pathways. Cell 110:737–749

Cross DA, Alessi DR, Cohen P, Andjelkovich M, Hemmings BA (1995) Inhibition of glycogen synthase kinase-3 by insulin mediated by protein kinase B. Nature 378:785–789

Datta SR, Dudek H, Tao X, Masters S, Fu H, Gotoh Y, Greenberg ME (1997) Akt phosphorylation of BAD couples survival signals to the cell-intrinsic death machinery. Cell 91:231–241

De Camilli P, Emr SD, McPherson PS, Novick P (1996) Phosphoinositides as regulators in membrane traffic. Science 271:1533–1539

Declue JE, Sadowski I, Martin GS, Pawson T (1987) A conserved domain regulates interactions of the v-fps protein-tyrosine kinase with the host cell. Proc Natl Acad Sci U S A 84:9064–9068

Divecha N, Banfic H, Irvine RF (1991) The polyphosphoinositide cycle exists in the nuclei of Swiss 3T3 cells under the control of a receptor (for IGF-I) in the plasma membrane, and stimulation of the cycle increases nuclear diacylglycerol and apparently induces translocation of protein kinase C to the nucleus. EMBO J 10:3207–3214

Dudek H, Datta SR, Franke TF, Birnbaum MJ, Yao R, Cooper GM, Segal RA, Kaplan DR, Greenberg ME (1997) Regulation of neuronal survival by the serine-threonine protein kinase Akt. Science 275:661–665

Engelman JA (2009) Targeting PI3K signalling in cancer: opportunities, challenges and limitations. Nat Rev Cancer 9:550–562

Engelman JA, Luo J, Cantley LC (2006) The evolution of phosphatidylinositol 3-kinases as regulators of growth and metabolism. Nat Rev Genet 7:606–619

Escobedo JA, Kaplan DR, Kavanaugh WM, Turck CW, Williams LT (1991) A phosphatidylinositol-3 kinase binds to platelet-derived growth factor receptors through a specific receptor sequence containing phosphotyrosine. Mol Cell Biol 11:1125–1132

Fine B, Hodakoski C, Koujak S, Su T, Saal LH, Maurer M, Hopkins B, Keniry M, Sulis ML, Mense S, Hibshoosh H, Parsons R (2009) Activation of the PI3K pathway in cancer through inhibition of PTEN by exchange factor P-REX2a. Science 325:1261–1265

Franke TF, Yang SI, Chan TO, Datta K, Kazlauskas A, Morrison DK, Kaplan DR, Tsichlis PN (1995) The protein kinase encoded by the Akt proto-oncogene is a target of the PDGF-activated phosphatidylinositol 3-kinase. Cell 81:727–736

Franke TF, Kaplan DR, Cantley LC, Toker A (1997) Direct regulation of the Akt proto-oncogene product by phosphatidylinositol-3,4-bisphosphate. Science 275:665–668

Frech M, Andjelkovic M, Ingley E, Reddy KK, Falck JR, Hemmings BA (1997) High affinity binding of inositol phosphates and phosphoinositides to the pleckstrin homology domain of RAC/protein kinase B and their influence on kinase activity. J Biol Chem 272:8474–8481

Fruman DA, Mauvais-Jarvis F, Pollard DA, Yballe CM, Brazil D, Bronson RT, Kahn CR, Cantley LC (2000) Hypoglycaemia, liver necrosis and perinatal death in mice lacking all isoforms of phosphoinositide 3-kinase p85 alpha. Nat Genet 26:379–382

Fruman DA, Ferl GZ, An SS, Donahue AC, Satterthwaite AB, Witte ON (2002) Phosphoinositide 3-kinase and Bruton's tyrosine kinase regulate overlapping sets of genes in B lymphocytes. Proc Natl Acad Sci U S A 99:359–364

Funamoto S, Milan K, Meili R, Firtel RA (2001) Role of phosphatidylinositol 3′ kinase and a downstream pleckstrin homology domain-containing protein in controlling chemotaxis in dictyostelium. J Cell Biol 153:795–810

Gewinner C, Wang ZC, Richardson A, Teruya-FELDSTEIN J, Etemadmoghadam D, Bowtell D, Barretina J, Lin WM, Rameh L, Salmena L, Pandolfi PP, Cantley LC (2009) Evidence that inositol polyphosphate 4-phosphatase type II is a tumor suppressor that inhibits PI3K signaling. Cancer Cell 16:115–125

Harlan JE, Hajduk PJ, Yoon HS, Fesik SW (1994) Pleckstrin homology domains bind to phosphatidylinositol-4,5-bisphosphate. Nature 371:168–170

Hawkins PT, Anderson KE, Davidson K, Stephens LR (2006) Signalling through Class I PI3Ks in mammalian cells. Biochem Soc Trans 34:647–662

Herman PK, Emr SD (1990) Characterization of VPS34, a gene required for vacuolar protein sorting and vacuole segregation in Saccharomyces cerevisiae. Mol Cell Biol 10:6742–6754

Hiles ID, Otsu M, Volinia S, Fry MJ, Gout I, Dhand R, Panayotou G, Ruiz LF, Thompson A, Totty NF et al (1992) Phosphatidylinositol 3-kinase: structure and expression of the 110 kd catalytic subunit. Cell 70:419–429

Hirsch E, Katanaev VL, Garlanda C, Azzolino O, Pirola L, Silengo L, Sozzani S, Mantovani A, Altruda F, Wymann MP (2000) Central role for G protein-coupled phosphoinositide 3-kinase gamma in inflammation. Science 287:1049–1053

Hokin LE, Hokin MR (1953) The incorporation of 32P into the nucleotides of ribonucleic acid in pigeon pancreas slices. Biochim Biophys Acta 11:591–592

Inoue M, Kishimoto A, Takai Y, Nishizuka Y (1977) Studies on a cyclic nucleotide-independent protein kinase and its proenzyme in mammalian tissues. II. Proenzyme and its activation by calcium-dependent protease from rat brain. J Biol Chem 252:7610–7616

Irie HY, Pearline RV, Grueneberg D, Hsia M, Ravichandran P, Kothari N, Natesan S, Brugge JS (2005) Distinct roles of Akt1 and Akt2 in regulating cell migration and epithelial-mesenchymal transition. J Cell Biol 171:1023–1034

Isakoff SJ, Cardozo T, Andreev J, Li Z, Ferguson KM, Abagyan R, Lemmon MA, Aronheim A, Skolnik EY (1998) Identification and analysis of PH domain-containing targets of phosphatidylinositol 3-kinase using a novel in vivo assay in yeast. EMBO J 17:5374–5387

Jaiswal BS, Janakiraman V, Kljavin NM, Chaudhuri S, Stern HM, Wang W, Kan Z, Dbouk HA, Peters BA, Waring P, Dela VEGA T, Kenski DM, Bowman KK, Lorenzo M, Li H, Wu J, Modrusan Z, Stinson J, Eby M, Yue P, Kaminker JS, De Sauvage FJ, Backer JM, Seshagiri S (2009) Somatic mutations in p85alpha promote tumorigenesis through class IA PI3K activation. Cancer Cell 16:463–474

Jiang X, Chen S, Asara JM, Balk SP (2010) Phosphoinositide 3-kinase pathway activation in phosphate and tensin homolog (PTEN)-deficient prostate cancer cells is independent of receptor tyrosine kinases and mediated by the p110beta and p110delta catalytic subunits. J Biol Chem 285:14980–14989

Jones PF, Jakubowicz T, Pitossi FJ, Maurer F, Hemmings BA (1991) Molecular cloning and identification of a serine/threonine protein kinase of the second-messenger subfamily. Proc Natl Acad Sci U S A 88:4171–4175

Jones SM, Klinghoffer R, Prestwich GD, Toker A, Kazlauskas A (1999) PDGF induces an early and a late wave of PI 3-kinase activity, and only the late wave is required for progression through G1. Curr Biol 9:512–521

Jou ST, Carpino N, Takahashi Y, Piekorz R, Chao JR, Wang D, Ihle JN (2002) Essential, nonredundant role for the phosphoinositide 3-kinase p110delta in signaling by the B-cell receptor complex. Mol Cell Biol 22:8580–8591

Kaplan DR, Whitman M, Schaffhausen B, Pallas DC, White M, Cantley L, Roberts TM (1987) Common elements in growth factor stimulation and oncogenic transformation: 85 kd phosphoprotein and phosphatidylinositol kinase activity. Cell 50:1021–1029

Kazlauskas A, Cooper JA (1989) Autophosphorylation of the PDGF receptor in the kinase insert region regulates interactions with cell proteins. Cell 58:1121–1133

Keniry M, Parsons R (2008) The role of PTEN signaling perturbations in cancer and in targeted therapy. Oncogene 27:5477–5485

Kirk CJ, Creba JA, Downes CP, Michell RH (1981) Hormone-stimulated metabolism of inositol lipids and its relationship to hepatic receptor function. Biochem Soc Trans 9:377–379

Kops GJ, De Ruiter ND, De Vries-Smits AM, Powell DR, Bos JL, Burgering BM (1999) Direct control of the Forkhead transcription factor AFX by protein kinase B. Nature 398:630–634

Lapetina EG, Michell RH (1973) A membrane-bound activity catalysing phosphatidylinositol breakdown to 1,2-diacylglycerol, D-myoinositol 1:2-cyclic phosphate an D-myoinositol 1-phosphate. Properties and subcellular distribution in rat cerebral cortex. Biochem J 131:433–442

Leevers SJ, Weinkove D, Macdougall LK, Hafen E, Waterfield MD (1996) The Drosophila phosphoinositide 3-kinase Dp110 promotes cell growth. EMBO J 15:6584–6594

Lemmon MA (2008) Membrane recognition by phospholipid-binding domains. Nat Rev Mol Cell Biol 9:99–111

Li J, Yen C, Liaw D, Podsypanina K, Bose S, Wang SI, Puc J, Miliaresis C, Rodgers L, McCombie R, Bigner SH, Giovanella BC, Ittmann M, Tycko B, Hibshoosh H, Wigler MH, Parsons R (1997) PTEN, a putative protein tyrosine phosphatase gene mutated in human brain, breast, and prostate cancer [see comments]. Science 275:1943–1947

Macara IG, Marinetti GV, Balduzzi PC (1984) Transforming protein of avian sarcoma virus UR2 is associated with phosphatidylinositol kinase activity: possible role in tumorigenesis. Proc Natl Acad Sci U S A 81:2728–2732

Maehama T, Dixon JE (1998) The tumor suppressor, PTEN/MMAC1, dephosphorylates the lipid second messenger, phosphatidylinositol 3,4,5-trisphosphate. J Biol Chem 273:13375–13378

Maira SM, Stauffer F, Brueggen J, Furet P, Schnell C, Fritsch C, Brachmann S, Chene P, De Pover A, Schoemaker K, Fabbro D, Gabriel D, Simonen M, Murphy L, Finan P, Sellers W, Garcia-Echeverria C (2008) Identification and characterization of NVP-BEZ235, a new orally available dual phosphatidylinositol 3-kinase/mammalian target of rapamycin inhibitor with potent in vivo antitumor activity. Mol Cancer Ther 7:1851–1863

Maroulakou IG, Oemler W, Naber SP, Tsichlis PN (2007) Akt1 ablation inhibits, whereas Akt2 ablation accelerates, the development of mammary adenocarcinomas in mouse mammary tumor virus (MMTV)-ErbB2/neu and MMTV-polyoma middle T transgenic mice. Cancer Res 67:167–177

Matsuda M, Mayer B, Fukui Y, Hanafusa H (1990) Binding of transforming protein p47gag-crk to a broad range of phosphotyrosine-containing proteins. Science 248:1537–1539

Mora A, Komander D, Van Aalten DM, Alessi DR (2004) PDK1, the master regulator of AGC kinase signal transduction. Semin Cell Dev Biol 15:161–170

Moran MF, Koch CA, Anderson D, Ellis C, England L, Martin GS, Pawson T (1990) Src homology region 2 domains direct protein-protein interactions in signal transduction. Proc Natl Acad Sci U S A 87:8622–8626

Morris JZ, Tissenbaum HA, Ruvkun G (1996) A phosphatidylinositol-3-OH kinase family member regulating longevity and diapause in Caenorhabditis elegans. Nature 382:536–539

O'Brien MC, Fukui Y, Hanafusa H (1990) Activation of the proto-oncogene p60c-src by point mutations in the SH2 domain. Mol Cell Biol 10:2855–2862

Ogg S, Paradis S, Gottlieb S, Patterson GI, Lee L, Tissenbaum HA, Ruvkun G (1997) The Fork head transcription factor DAF-16 transduces insulin-like metabolic and longevity signals in C. elegans. Nature 389:994–999

O'Reilly KE, Rojo F, She QB, Solit D, Mills GB, Smith D, Lane H, Hofmann F, Hicklin DJ, Ludwig DL, Baselga J, Rosen N (2006) mTOR inhibition induces upstream receptor tyrosine kinase signaling and activates Akt. Cancer Res 66:1500–1508

Otsu M, Hiles I, Gout I, Fry MJ, Ruiz LF, Panayotou G, Thompson A, Dhand R, Hsuan J, Totty N, et al (1991) Characterization of two 85 kd proteins that associate with receptor tyrosine kinases, middle-T/pp60c-src complexes, and PI3-kinase. Cell 65:91–104

Paradis S, Ailion M, Toker A, Thomas JH, Ruvkun G (1999) A PDK1 homolog is necessary and sufficient to transduce AGE-1 PI3 kinase signals that regulate diapause in Caenorhabditis elegans. Genes Dev 13:1438–1452

Paradis S, Ruvkun G (1998) Caenorhabditis elegans Akt/PKB transduces insulin receptor-like signals from AGE-1 PI3 kinase to the DAF-16 transcription factor. Genes Dev 12:2488–2498

Patrucco E, Notte A, Barberis L, Selvetella G, Maffei A, Brancaccio M, Marengo S, Russo G, Azzolino O, Rybalkin SD, Silengo L, Altruda F, Wetzker R, Wymann MP, Lembo G, Hirsch E (2004) PI3Kgamma modulates the cardiac response to chronic pressure overload by distinct kinase-dependent and -independent effects. Cell 118:375–387

Rodriguez VP, Warne PH, Dhand R, Vanhaesebroeck B, Gout I, Fry MJ, Waterfield MD, Downward J (1994) Phosphatidylinositol-3-OH kinase as a direct target of Ras. Nature 370:527–532

Samuels Y, Wang Z, Bardelli A, Silliman N, Ptak J, Szabo S, Yan H, Gazdar A, Powell SM, Riggins GJ, Willson JK, Markowitz S, Kinzler KW, Vogelstein B, Velculescu VE (2004) High frequency of mutations of the PIK3CA gene in human cancers. Science 304:554

Sarbassov DD, Guertin DA, Ali SM, Sabatini DM (2005) Phosphorylation and regulation of Akt/PKB by the rictor-mTOR complex. Science 307:1098–1101

Schu PV, Takegawa K, Fry MJ, Stack JH, Waterfield MD, Emr SD (1993) Phosphatidylinositol 3-kinase encoded by yeast VPS34 gene essential for protein sorting. Science 260:88–91

Sjolander A, Yamamoto K, Huber BE, Lapetina EG (1991) Association of p21ras with phosphatidylinositol 3-kinase. Proc Natl Acad Sci U S A 88:7908–7912

Skolnik EY, Margolis B, Mohammadi M, Lowenstein E, Fischer R, Drepps A, Ullrich A, Schlessinger J (1991) Cloning of PI3 kinase-associated p85 utilizing a novel method for expression/cloning of target proteins for receptor tyrosine kinases. Cell 65:83–90

Sosa MS, Lopez-Haber C, Yang C, Wang H, Lemmon MA, Busillo JM, Luo J, Benovic JL, Klein-Szanto A, Yagi H, Gutkind JS, Parsons RE, Kazanietz MG (2010) Identification of the Rac-GEF P-Rex1 as an essential mediator of ErbB signaling in breast cancer. Mol Cell 40:877–892

Steck PA, Pershouse MA, Jasser SA, Yung WK, Lin H, Ligon AH, Langford LA, Baumgard ML, Hattier T, Davis T, Frye C, Hu R, Swedlund B, Teng DH, Tavtigian SV (1997) Identification of a candidate tumour suppressor gene, MMAC1, at chromosome 10q23.3 that is mutated in multiple advanced cancers. Nat Genet 15:356–362

Stephens L, Hawkins PT, Downes CP (1989) Metabolic and structural evidence for the existence of a third species of polyphosphoinositide in cells: D-phosphatidyl-myo-inositol 3-phosphate. Biochem J 259:267–276

Stephens LR, Hughes KT, Irvine RF (1991) Pathway of phosphatidylinositol(3,4,5)-trisphosphate synthesis in activated neutrophils. Nature 351:33–39

Stokoe D, Stephens LR, Copeland T, Gaffney PR, Reese CB, Painter GF, Holmes AB, McCormick F, Hawkins PT (1997) Dual role of phosphatidylinositol-3,4,5-trisphosphate in the activation of protein kinase B. Science 277:567–570

Streb H, Irvine RF, Berridge MJ, Schulz I (1983) Release of Ca^{2+} from a nonmitochondrial intracellular store in pancreatic acinar cells by inositol-1,4,5-trisphosphate. Nature 306:67–69

Sugimoto Y, Whitman M, Cantley LC, Erikson RL (1984) Evidence that the Rous sarcoma transforming gene product phosphorylates phosphatidylinositol and diacylglycerol. Proc Natl Acad Sci U S A 81:2117–2121

Takai Y, Yamamoto M, Inoue M, Kishimoto A, Nishizuka Y (1977) A proenzyme of cyclic nucleotide-independent protein kinase and its activation by calcium-dependent neutral protease from rat liver. Biochem Biophys Res Commun 77:542–550

Terauchi Y, Tsuji Y, Satoh S, Minoura H, Murakami K, Okuno A, Inukai K, Asano T, Kaburagi Y, Ueki K, Nakajima H, Hanafusa T, Matsuzawa Y, Sekihara H, Yin Y, Barrett JC, Oda H, Ishikawa T, Akanuma Y, Komuro I, Suzuki M, Yamamura K, Kodama T, Suzuki H, Koyasu S, Aizawa S, Tobe K, Fukui Y, Yazaki Y, Kadowaki T (1999) Increased insulin sensitivity and hypoglycaemia in mice lacking the p85 alpha subunit of phosphoinositide 3-kinase. Nat Genet 21:230–235

Traynor-Kaplan AE, Harris AL, Thompson BL, Taylor P, Sklar LA (1988) An inositol tetrakisphosphate-containing phospholipid in activated neutrophils. Nature 334:353–356

Ueki K, Yballe CM, Brachmann SM, Vicent D, Watt JM, Kahn CR, Cantley LC (2002) Increased insulin sensitivity in mice lacking p85beta subunit of phosphoinositide 3-kinase. Proc Natl Acad Sci U S A 99:419–424

Valius M, Kazlauskas A (1993) Phospholipase C-gamma 1 and phosphatidylinositol 3 kinase are the downstream mediators of the PDGF receptor's mitogenic signal. Cell 73:321–334

Vanhaesebroeck B, Guillermet-Guibert J, Graupera M, Bilanges B (2010) The emerging mechanisms of isoform-specific PI3K signalling. Nat Rev Mol Cell Biol 11:329–341

Vasudevan KM, Barbie DA, Davies MA, Rabinovsky R, McNear CJ, Kim JJ, Hennessy BT, Tseng H, Pochanard P, Kim SY, Dunn IF, Schinzel AC, Sandy P, Hoersch S, Sheng Q, Gupta PB, Boehm JS, Reiling JH, Silver S, Lu Y, Stemke-HALE K, Dutta B, Joy C, Sahin AA, Gonzalez-Angulo AM, Lluch A, Rameh LE, Jacks T, Root DE, Lander ES, Mills GB, Hahn WC, Sellers WR, Garraway LA (2009) AKT-independent signaling downstream of oncogenic PIK3CA mutations in human cancer. Cancer Cell 16:21–32

Vlahos CJ, Matter WF, Hui KY, Brown RF (1994) A specific inhibitor of phosphatidylinositol 3-kinase, 2-(4- morpholinyl)-8-phenyl-4H-1-benzopyran-4-one (LY294002). J Biol Chem 269:5241–5248

Welch HC, Coadwell WJ, Stephens LR, Hawkins PT (2003) Phosphoinositide 3-kinase-dependent activation of Rac. FEBS Lett 546:93–97

Wennstrom S, Hawkins P, Cooke F, Hara K, Yonezawa K, Kasuga M, Jackson T, Claesson-WELSH L, Stephens L (1994) Activation of phosphoinositide 3-kinase is required for PDGF-stimulated membrane ruffling. Curr Biol 4:385–393

Whitman M, Downes CP, Keeler M, Keller T, Cantley L (1988) Type I phosphatidylinositol kinase makes a novel inositol phospholipid, phosphatidylinositol-3-phosphate. Nature 332:644–646

Yano H, Nakanishi S, Kimura K, Hanai N, Saitoh Y, Fukui Y, Nonomura Y, Matsuda Y (1993) Inhibition of histamine secretion by wortmannin through the blockade of phosphatidylinositol 3-kinase in RBL-2H3 cells. J Biol Chem 268:25846–25856

Yao R, Cooper GM (1995) Requirement for phosphatidylinositol-3 kinase in the prevention of apoptosis by nerve growth factor. Science 267:2003–2006

Yoeli-Lerner M, Yiu GK, Rabinovitz I, Erhardt P, Jauliac S, Toker A (2005) Akt blocks breast cancer cell motility and invasion through the transcription factor NFAT. Mol Cell 20:539–550

Zhao JJ, Liu Z, Wang L, Shin E, Loda MF, Roberts TM (2005) The oncogenic properties of mutant p110alpha and p110beta phosphatidylinositol 3-kinases in human mammary epithelial cells. Proc Natl Acad Sci U S A 102:18443–18448

Zhou S, Shoelson SE, Chaudhuri M, Gish G, Pawson T, Haser WG, King F, Roberts T, Ratnofsky S, Lechleider RJ et al (1993) SH2 domains recognize specific phosphopeptide sequences. Cell 72:767–778

Chapter 5
PI3Ks—Drug Targets in Inflammation and Cancer

Matthias Wymann

Abstract Phosphoinositide 3-kinases (PI3Ks) control cell growth, proliferation, cell survival, metabolic activity, vesicular trafficking, degranulation, and migration. Through these processes, PI3Ks modulate vital physiology. When over-activated in disease, PI3K promotes tumor growth, angiogenesis, metastasis or excessive immune cell activation in inflammation, allergy and autoimmunity. This chapter will introduce molecular activation and signaling of PI3Ks, and connections to target of rapamycin (TOR) and PI3K-related protein kinases (PIKKs). The focus will be on class I PI3Ks, and extend into current developments to exploit mechanistic knowledge for therapy.

Keywords Cancer inflammation allergy metabolism phosphatidylinositol phosphoinositide phosphoinositide 3-kinase · PI3K target of rapamycin · TOR · mTOR protein kinase B · Akt · PKB 3-phosphoinositide phosphatase and tensin homolog deleted in chromosome ten · Also PTEN wortmannin LY294002 rapamycin pharmacology signal transduction

5.1 PI3Ks—Molecular Mechanisms

5.1.1 Introduction to PI3Ks

The deregulation of phosphoinositide 3-kinase (PI3K) pathways interferes with cellular hemostasis and contributes to the over-activation of many cell types. In this respect, PI3Ks have been shown to play a central role in the control of cellular metabolism, growth, proliferation, survival and migration, intracellular membrane transport, secretion and more.

M. Wymann (✉)
Institute Biochemistry & Genetics, Department Biomedicine,
University of Basel, Mattenstrasse 28, 4058 Basel, Switzerland
e-mail: Matthias.Wymann@UniBas.CH

T. Balla et al. (eds.), *Phosphoinositides I: Enzymes of Synthesis and Degradation*,
Subcellular Biochemistry 58, DOI 10.1007/978-94-007-3012-0_5,
© Springer Science+Business Media B.V. 2012

Although cancer and inflammatory disease include a wide variety of disorders with a broad degree of severity and clinical outcome, both emerge from preexisting, but derailed physiologic repair and defense mechanisms. To enter a tissue repair or host defense mode, cells have to switch from a quiescent to an activated state. Cell surface receptor-controlled PI3Ks produce PtdIns(3,4,5)P_3 at the inner leaflet of the plasma membrane, where the lipid provides a docking site for signaling molecules with lipid receptor domains (Balla and Varnai 2002; Lemmon 2008). The prototype for these is the Ser/Thr protein kinase B (PKB, also called Akt)(Bellacosa et al. 1991), which indirectly relays the activation of PI3K to the target of rapamycin (TOR or mammalian TOR, mTOR), which in turn regulates protein synthesis and growth. Via the activation of guanine nucleotide exchange factors (GEFs), PtdIns(3,4,5)P_3 levels modulate Rho GTPase activities and thus cytoskeletal rearrangements, cell polarity and migration. The importance of PtdIns(3,4,5)P_3 levels in these processes has been validated by the loss of one of the counter-players of PI3Ks: when the expression of the lipid 3'-phosphatase PTEN (Phosphatase and Tensin homolog deleted on chromosome Ten) is lowered, PtdIns(3,4,5)P_3 rises, and develops its oncogenic potential (Stambolic et al. 1998; Di Cristofano et al. 1998; Leslie et al. 2008; Zhang and Yu 2010). In fact, most cases of Cowden Syndrome are due mutations in PTEN (Eng 1998), and result in the formation of hyperplasia and adenoma formation in various tissues, which constitute early forms of cancer (Liaw et al. 1997; Sansal and Sellers 2004; Hobert and Eng 2009; Farooq et al. 2010). But also the genetic attenuation of lipid 3'-phosphatases degrading PtdIns(3,5)P_2 or PtdIns-3-P gives rise to progressive disease, as mutations in the gene coding for myotubularin 1 (MTM1) cause X-linked myotubular myopathy (XLMTM), and loss of phosphatase activity in myotubularin-related protein 2 (Mtmr2) correlates with Charcot-Marie-Tooth disease type 4B1 (CMT4B1, Berger et al. 2002; Cao et al. 2008). This demonstrates that a delicate balance between lipid kinase and lipid phosphatase activities controls the flux through the phosphoinositide pathway, and that phosphoinositide levels play an important part in cellular homeostasis (Fig. 5.1).

Initially, PI3Ks were discovered as lipid kinases associated with viral oncogens (Whitman et al. 1985; Sugimoto et al. 1984; Macara et al. 1984), and for the last two decades the link between cancer and elevated PtdIns(3,4,5)P_3 levels has been corroborated (Vivanco and Sawyers 2002; Wymann and Marone 2005; Cully et al. 2006; Engelman et al. 2006; refer to Vol. 1, Chap. 4). While there is a clear-cut connection between elevated PtdIns(3,4,5)P_3 levels and progression of cancer, chronic inflammation, allergy, metabolic disease, diabetes and cardiovascular problems, it has been challenging to associate specific PI3K isoforms with defined disease states. The combination of genetic and pharmacological approaches has somewhat elucidated the integration of distinct PI3K isoforms in disease-associated signaling cascades during the past decade, which has helped to validate PI3Ks as drug targets in cancer and chronic inflammation.

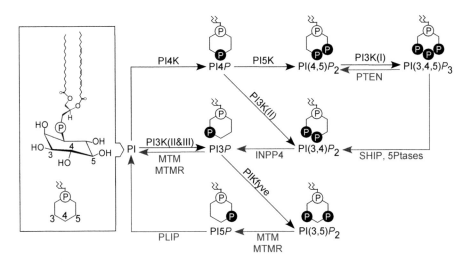

Fig. 5.1 Lipid kinases (indicated in *black*) and lipid phosphatases (*grey*) mediate the flow of phosphoinositides from phosphoinositol (PtdIns, PI) to PtdIns(3,4,5)P_3 [PI(3,4,5)P_3]. The main synthesis pathways relevant in physiology and disease are indicated: PI is phosphorylated by phosphatidylinositol 4-kinase (PI4K) to PI4P (Balla and Balla 2006; Graham and Burd 2011), which is turned over into PI(4,5)P_2 by PI5K (type I phosphatidylinositol phosphate kinase; PIPKI (Anderson et al. 1999; Ling et al. 2006)). PI(4,5)P_2 is then converted to PI(3,4,5)P_3 by class I PI3Ks [PI3K(I)]. The 3′-lipid phosphatase PTEN (Stambolic et al. 1998; Di Cristofano et al. 1998; Leslie et al. 2008; Zhang and Yu 2009) reverses the action of PI3K(I), while the 5′-lipid phosphatase SHIP (Krystal et al. 1999) produces PI(3,4)P_2. PI(3,4)P_2 could eventually also be produced by class II PI3Ks [PI3K(II)] from PI4P. PI(3,4)P_2 is degraded mostly by inositol polyphosphate 4-phosphatases, e.g. INPP4, to form PI3P. The latter is formed directly from PI by class III PI3K (PI3K(III)/Vps34) or class II PI3Ks from PI. PI3P can be converted to PI(3,5)P_2 by PIKfyve (yeast Fab1) as response to cellular stress (Dove and Johnson 2007; Dove et al. 2009). Myotubularins (MTM) and myotubularin-related (MTMR) constitute a family of lipid phosphatases that dephosphorylate PI3P and PI(3,5)P_2 (Laporte et al. 2001; Mruk and Cheng 2010). Simplified, PI(4,5)P_2 and PI(3,4,5)P_3 are localized in the plasma membrane, golgi and ER are rich in PI4P (Graham and Burd 2011); PI3P is a marker for early endosomes; and late endosomes and multi-vesicular bodies contain PI(3,5)P_2. Excellent reviews elucidation phosphoinositide localization and conversions further are (Di Paolo and De Camilli 2006; Sasaki et al. 2009; Bunney and Katan 2010). Other useful source: http://www.genome.jp > KEGG Pathway > map04070

5.1.2 The PI3K Family and PI3K-related Protein Kinases

The core PI3K family consists of three PI3K classes (Wymann and Pirola 1998; Vanhaesebroeck et al. 2001; Wymann et al. 2003b; Vanhaesebroeck et al. 2010), which have been defined according to structural characteristics and their *in vitro* substrate specificity (Fig. 5.2). The complete set of all PI3K family members has first been identified in the fruit fly *Drosophila melanogaster* (MacDougall et al. 1995; Leevers et al. 1996), and work in model organisms like *Caenorhabditiselegans* (Morris et al. 1996; Roggo et al. 2002), *Dictyosteliumdiscoideum* (refs see (Chen et al. 2007)) and yeast (Schu et al. 1993), has much contributed to the elucidation of the function of members of the PI3K family (Engelman et al. 2006).

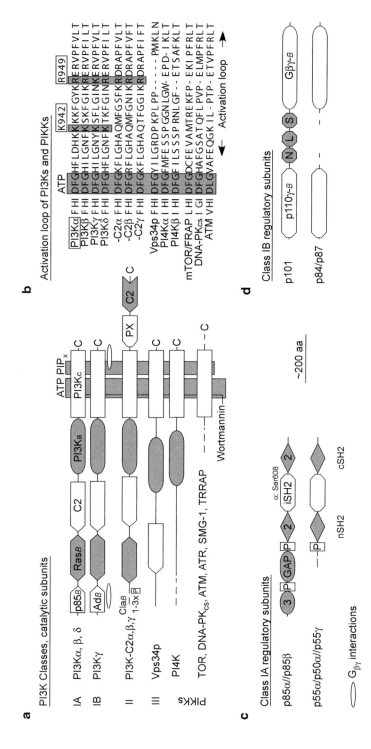

Fig. 5.2 Schematic representation of the domain structure of the phosphoinositide 3-kinase (PI3K) family, and PI3K-related protein kinases (PIKKs). **a** the catalytic subunits of class IA enzymes (PI3Kα, β, δ) are composed of a *N*-terminal p85-binding (p85*B*) region, followed by a Ras-binding (Ras*B*) domain, a C2 domain, a PI3K accessory (or helical, PI3K*a*) domain and a catalytic core (PI3K*c*) domain. The latter constitutes the ATP-binding site, and also accommodates the head group of the class I substrate PtdIns(4,5)P_2 (labeledPIP$_x$-binding site here). Binding of the PI3K inhibitor wortmannin occurs within the ATP-binding

Fig. 5.2 (Continued) site, and wortmannin forms a covalent link with a conserved lysine in the PI3K and PIKK enzymes' catalytic core pocket (see text). All PI3Ks and PIKKs share a core catalytic domain (PI3Kc), which displays lipid or protein kinase activity, except for the TRRAP protein, where two mutations in conserved residues abrogate the catalytic activity completely. Class IB enzyme PI3Kγ (p110γ) does not bind p85-like regulatory subunits, but interacts instead with the p101 or p84(also called p87^{PIKUP}) adapter proteins and Gβγ subunits of trimeric G proteins. PI3Kγ was proposed to bind adaptor proteins mainly, but not exclusively, via its N-terminus (Ad*B*). Class II (PI3K-C2α, β, γ) enzymes lack an adapter-binding site, but were reported to interact with clathrin (Cla*B*) or SH3 domains via N-terminal proline-rich sequences (P, only present in PI3K-C2β). Class II enzymes are characterized by a C-terminal extension containing a phox (PX) and a C2 domain (for a concise review on class II enzymes see (Falasca and Maffucci 2007)). Class III PI3Ks and PI4Ks display PI3K*a* and PI3K*c* domains, and both are only capable to accept PtdIns as a substrate. The PI4K members shown here belong to the type III PI4K enzymes. The type II PI4K enzymes are structurally and biochemically distinct from the PI3K family, are insensitive to wortmannin, and show even little resemblance to other phosphotransferase enzymes (Minogue et al. 2001; Barylko et al. 2001; Balla and Balla 2006; refer to chapter on PI4K, Vol. 1. Chap. 1). Target of rapamycin (TOR or FRAP), the catalytic subunit of DNA-dependent protein kinase (DNA-PK$_{cs}$), the gene mutated in the human genetic disorder ataxia telangiectasia (ATM), and ATM and Rad3-related (ATR) contain a PI3Kc domain, and possess activities as serine/threonine protein kinases. Numerous protein/protein interaction domains present in the N-terminal part of PIKKs are not depicted here (for a reviews on PIKKs see (Zoncu et al. 2011; Shiloh 2003; Cimprich and Cortez 2008)). **b** The activation loop in PI3Ks and PIKKs is located C-terminally to the prominent, ATP-interacting DFG motif (*boxed*). Two positive charges (K942 and R949 in p110α) have to be positioned correctly to allow a productive accommodation of the PtdIns(4,5)P_2 head group (Bondeva et al. 1998; Pirola et al. 2001). Class II PI3K and PI4K provide no charge compensation at the corresponding position to K942, which explains why they make only PtdIns(3,4)P_2 *in vitro*. Class III PI3K and PI4K seem to have a more rigid activation loop by the presence of multiple prolines, and accept only PtdIns as a substrate. Class III PI3K interact with a regulatory kinase, yeast Vps15p or mammalian p150 not depicted here (Vanhaesebroeck et al. 2010; Stack et al. 1995). **c** Class IA PI3K catalytic subunits are tightly associated with the inter-SH2 (iSH2) domain of p85-like regulatory subunits. The long forms of these regulatory subunits present N-terminally an SH3 and a RhoGAP (GAP) domain flanked by proline-rich motifs (P). Although interactions of these domains with Shc, Cbl, dynamin, Grb2 and Src-like protein kinases have been demonstrated (for references see (Wymann and Pirola 1998)), their importance in physiologic processes is ill defined. **d** Class IB PI3K, PI3Kγ, binds adapter subunits p101 and p84 (also named p87^{PIKUP}), which C-terminally interact with Gβγ subunits. Regulatory functions of these adapters are localized N-terminally, but binding to the PI3Kγ catalytic subunits was reported to involve various regions in the adapter subunits (Bohnacker et al. 2009; Krugmann et al. 1999; Voigt et al. 2005). Abbreviations (alphabetical): 2/3 src-homology 2/3 domains, ATM ataxia telangiectasia mutated, ATR ataxia telangiectasia related, C2 protein kinase C homology domain 2, Cla*B* clathrin-binding sites, DNA-PK$_{cs}$ catalytic subunit of DNA-dependent protein kinase, Gβγ-*B* Gβγ-binding domains, iSH2 inter-SH2 domain, NLS nuclear localization signals, P proline-rich region, p85*B* p85-binding domain, PIK3a PI3K accessory domain, also called helical domain, PI3Kc PI3K catalytic domain, PX phox domain or phagocyte oxidase domain, GAP: RhoGAP domain, also referred to as (BH) BCR homology domain, SMG1 suppressor of morphogenesis in genitalia-1, TOR/mTOR mammalian target of rapamycin, TRRAP transformation/transcription domain-associated protein

Ligand-activated receptors located in the plasma membrane relay their signals to class I PI3Ks, which are the only PI3K members able to convert PtdIns(4,5)P_2 to PtdIns(3,4,5)P_3. Although class I PI3K are capable to phosphorylate PtdIns to PtdIns-3-P, and PtdIns-4-P to PtdIns(3,4)P_2 *in vitro*, they have a preference for PtdIns(4,5)P_2 as a substrate *in vivo* (Stephens et al. 1993; Cantley 2002). All members of the class I PI3Ks are heterodimers, contain a catalytic subunit of 100–120 kD (referred to as p110 proteins and genes named *PIK3c*).

Although class II PI3Ks were reported to be direct downstream targets of growth factor receptors like the epidermal growth factor receptor (EGFR) or platelet-derived growth factor receptor (PDGFR) and stem cell factor receptor (SCFR, c-kit; Arcaro et al. 2000, 2002), their main connection to signaling at the plasma membrane is likely to be mediated via clathrin-mediated endocytosis (Domin et al. 2000; Gaidarov et al. 2001; Falasca and Maffucci 2007). Class II PIKs were proposed to produce PtdIns-3-P and PtdIns(3,4)P_2. Class III PI3Ks are represented by Vps34p (vacuolar protein sorting mutant 34; Schu et al. 1993), which generates PtdIns-3-P only. Vps34p is involved in vesicular transport to yeast vacuolar and mammalian early endosomal compartments, protein sorting (Simonsen et al. 2001), autophagy (Backer 2008; Simonsen and Tooze 2009; Funderburk et al. 2010) and has recently been proposed to promote cytokinesis (Sagona et al. 2010; Nezis et al. 2010).

The members of the PI3K family have close relatives, which are the type III phosphoinositide 4-kinases (PI4Ks; Minogue et al. 2001; Barylko et al. 2001; Balla and Balla 2006) refer to Vol. 1, Chap. 1 and protein kinases referred to as class IV PI3Ks or phosphoinositide 3-kinase-related kinases (PIKKs). These include the target of rapamycin (TOR; also dubbed FRAP or mTOR), DNA-dependent protein kinase (DNA-PK$_{cs}$), ATM (ataxia telangiectasia mutated), ataxia telangiectasia-related (ATR), suppressor of morphogenesis in genitalia-1 (SMG-1) and transformation/transcription domain-associated protein (TRRAP). DNA-PK$_{cs}$, ATM, ATR and SMG-1 take part in DNA-damage repair responses. DNA-PK$_{cs}$ and ATM respond mainly to double strand break (DSB), while ATR and SMG-1 are activated due to ultraviolet-induced stress, DNA-damage, and DSB (Durocher and Jackson 2001; Hiom 2005). SMG-1 is involved also in mRNA surveillance mechanisms, genotoxic stress responses and non-sense-mediated mRNA decay (Yamashita et al. 2005; Oliveira et al. 2008). TOR, on the other hand, is regulated by growth factors receptors and the availability of nutrients. Two distinct TOR complexes coordinate protein and lipid synthesis, cell growth and proliferation (Wullschleger et al. 2006; Laplante and Sabatini 2009; Sancak et al. 2010; Kapahi et al. 2010).

5.1.3 The PI3K Catalytic Core, Enzymatic Activities

PI3Ks and PIKKs share a similar catalytic core (PI3Kc), where the ATP- and lipid and protein substrate-binding sites are localized (Fig. 5.2). Interestingly, early PI3K inhibitors like wortmannin (Arcaro and Wymann 1993; Yano et al. 1993; Wymann and Arcaro 1994) and LY294002 (Vlahos et al. 1994) inhibit multiple enzymes of the PI3K and PIKK family at elevated concentrations (Wymann and Arcaro 1994; Marone et al. 2008). For class I PI3Ks wortmannin's IC_{50} is 1–5 nM, for class II PI3K the IC_{50} was reported to be isoform-dependent [PI3K-C2α: $IC_{50} \cong 500$ nM; PI3K-C2β: $IC_{50} \cong 2$–5 nM; for human Vps34p class III PI3K the IC_{50} is ca. 3 nM, while yeast Vps34p was reported to be rather resistant to wortmannin (Panaretou et al. 1997; Woscholski et al. 1994); and for type III PI4Ks $IC_{50} \cong 100$–300 nM (Falasca and Maffucci 2007; Balla and Balla 2006). PIKKs were inhibited at somewhat higher concentrations: IC_{50}s for wortmannin have been reported for SMG-1 around 60 nM, for DNA-PKcsfrom 20–120 nM, for mTOR and ATM from 100–200 nM, while ATR requires ca. 1.8 µM (IC_{50}) of wortmannin (Yamashita et al. 2005; Brunn et al. 1996; Izzard et al. 1999; Chan et al. 2000; Sarkaria et al. 1998).

Wortmannin binds covalently to a conserved lysine residue in the catalytic pocket of PI3Ks (Lys 802 in p110α (Wymann et al. 1996); Lys833 in p110γ (Stoyanova et al. 1997; Walker et al. 2000)) and PIKKs (e.g. Lys2187 in TOR (Brunn et al. 1996)). This active site Lys residue is conserved in protein and lipid kinases, and is critical for the transfer of the γ-phosphate group of ATP to kinase substrates. The ε-amino group of this Lys residue is especially nucleophilic and is capable to form a Schiff-base with the carbon-20 of the furan ring of wortmannin. This adduct is stable at a physiologic pH and can be detected by immunoblotting with anti-wortmannin antibodies (Balla and Balla 2006; Marone et al. 2008; Wymann et al. 1996). Data obtained using wortmannin as a PI3K inhibitor above 100 nM must therefore be interpreted with great caution, last but not least because it was reported that also polo-like protein kinases involved in the control of mitosis are inhibited by wortmannin in the nM range ($IC_{50} \cong 30$–50 nM; Liu et al. 2005). Although wortmannin and LY294002 remain useful tools to explore the importance of PI3K signaling in a cellular context, panels of more selective inhibitors are now available, and should be used complementary (see Table 5.1 and sections below).

A productive transfer of the γ-phosphate group of ATP to the D3-OH position of the phosphoinositide substrate depends also on a tight interaction of the phosphoinositide head group with the PI3K catalytic pocket. For the higher phosphorylated phosphoinositides this poses the problem that the negatively charged phosphate groups must be accommodated. In class I PI3Ks, the so-called activation loop contains two essential positively charged amino acid side chains (for p110α these are Lys942 and Arg949), which interact with the 4- and the 5-phosphate groups of PtdIns(4,5)P_2. If these charges are removed, or if the class I PI3K activation loop is replaced with a sequence derived from other PI3K family members, the resulting mutant enzyme looses its ability to turn over PtdIns(4,5)P_2, but can still phosphorylate phosphatidylinositol (Fig. 5.2b, Bondeva et al. 1998; Pirola et al. 2001). As the experimental transfer of an activation loop from class II PI3Ks (containing one correctly positioned positive charged amino acid side chain corresponding to the Arg949 in p110α) to class I enzymes retains the ability to phosphorylate PtdIns-4-P and to generate PtdIns(3,4)P_2, while the insertion of a class III activation loop only conserves the reactivity towards phosphatidylinositol, it is possible to classify PI3Ks solely on the basis of their activation loop sequences. Based on the analysis of the activation loop, class I PI3K are the only PI3Ks to produce PtdIns(3,4,5)P_3, and class III PI3K can only generate PtdIns-3-P. The *in vitro* products of class II PI3K were reported to be PtdIns-3-P

Table 5.1 Inhibitors and drugs targeting the PI3K/PKB/TOR pathway—stages of developments to therapy

Drug name	References	Target(s)	Stage, phase	Company	Application
PI3K isoform-specific inhibitors, targeted inhibitors					
AS605240 and AS252424	(Fougerat et al. 2008; Camps et al. 2005; Barber et al. 2005; Pomel et al. 2006; Edling et al. 2010)	p110γ	Preclinical		Chronic inflammatory, allergic, cardio-vascular disease, atherosclerosis, pancreatic cancer?
BYL719		p110α	Phase I	Novartis	Advanced solid tumors with PIK3CA mutations
CAL-101 (clinical follow-up of IC87114)	(Lannutti et al. 2011; Ikeda et al. 2010; Herman et al. 2010)	p110δ	Phase I	Calistoga	B cell malignancies, CLL, allergies
CAL-120	(Lannutti et al. 2009)	Dual p110β/p110δ	Preclinical	Calistoga	Solid tumors
CAL-263		p110δ	Phase I	Calistoga	Allergic rhinitis
SF1126	(Garlich et al. 2008; Ozbay et al. 2010)	Targeted pan-PI3K inhibitor	Phase I	Semafore Pharmaceuticals	Solid tumors
TG100-115	(Doukas et al. 2006, 2007)	Dual p110γ/p110δ	Phase I, completed	TargeGen/(Sanofi-Aventis?)	Myocardial ischemia/reperfusion injury
TGX-221	(Jackson et al. 2005)	P110β inhibitor	Preclinical		Platelet aggregation
pan-PI3K inhibitors					
BKM-120	(Aziz et al. 2010; Buonamici et al. 2010)	pan-PI3K	Phase I/II	Novartis	Breast, colon, ovarian, endometrium cancer, solid tumors
GDC-0941	(Raynaud et al. 2009; Folkes et al. 2008)	pan-PI3K (p110α/p110δ)	Phase I/II	Roche/Genentech	Advanced and metastatic solid tumors, NSCLC, Non-Hodgkin's lymphoma
GSK1059615		pan-PI3K	I/II, terminated	GlaxoSmithKline	Advanced solid tumors, lymphoma

Table 5.1 (continued)

Drug name	References	Target(s)	Stage, phase	Company	Application
XL-147	(Shapiro et al. 2009)	pan-PI3K	I/II	Exelixis/Sanofi-Aventis	Solid tumors, advanced breast, endometrial cancer, glioblastoma
ZSTK474	(Marone et al. 2009; Kong et al. 2010; Kong and Yamori 2007, 2009; Yaguchi et al. 2006)	pan-PI3K	Phase I	Zenyaku	Neoplasms
Dual PI3K/mTOR kinase inhibitors					
BEZ235	(Maira et al. 2008; Marone et al. 2009; Engelman et al. 2008; Serra et al. 2008; Brachmann et al. 2009)	PI3K/mTOR	Phase I/II, completed	Novartis	Solid tumors
BGT226		PI3K/mTOR	Phase I/II completed	Novartis	Solid tumors, Cowden disease
GDC-0980	(Wagner et al. 2009)	PI3K/mTOR	Phase I	Roche/Genentech	Refractory solid tumors, non-Hodgkin's lymphoma
GSK2126458	(Knight et al. 2009)	PI3K/mTOR	Phase I	GlaxoSmithKline	Solid malignancies
PF-04691502/	(Cheng et al. 2010)	PI3K/mTOR	Phase I	Pfizer	Advanced solid tumors
PKI-402	(Mallon et al. 2010; Dehnhardt et al. 2010)	PI3K/mTOR	Preclinical	Pfizer/Wyeth	
PKI-587, PF-05212384	(Venkatesan et al. 2010)	PI3K/mTOR	Phase I	Pfizer/Wyeth	Advanced solid tumors
PX-866	(Ihle et al. 2004)	PI3K/mTOR	Phase I/II	Oncothyreon	Solid tumors, glioblastoma
XL-765	(Laird et al. 2008)	PI3K/mTOR	Phase II	Exelixis/Sanofi-Aventis	Solid tumors, glioblastoma
mTORC1/mTORC2 kinase inhibitors					
AR-mTOR-1	(Wallace et al.)	mTOR kinase	Preclinical	Array Biopharma	
AZD2014		mTOR kinase	Phase I	AstraZeneca	Advanced solid tumors

Table 5.1 (continued)

Drug name	References	Target(s)	Stage, phase	Company	Application
AZD8055	(Chresta et al. 2010)	Selective mTOR kinase	Phase I/II	AstraZeneca	Liver cancer; advanced tumors
CC-223		mTOR kinase	Phase I/II	Celgene	Advanced Solid Tumors, Non-Hodgkin Lymphoma, Multiple Myeloma
INK128	(Jessen et al. 2009)	mTOR kinase	Phase I	Intellikine	Advanced tumors, relapsed/refractory multiple myeloma, Waldenström's macroglobulinemia
OSI-027	(Carayol et al. 2010; Vakana et al. 2010)	mTOR kinase	Phase I	OSI Pharmaceuticals	Advanced solid tumors, lymphoma
Palomid 529, P529	(Diaz et al. 2009; Xue et al. 2008)	Affects mTORC1 and mTOR2 activity. Mechanism?	Phase I	Paloma Pharmaceuticals	Advanced Neovascular Age-Related Macular Degeneration (AMD)
PP242/PP30	(Feldman et al. 2009; Hoang et al. 2010)	Selective mTOR kinase inhibitor	Preclinical		
XL388	(Miller 2009)	mTOR kinase inhibitor	Preclinical	Exelixis	
mTORC1 allosteric inhibitors (rapamycin derivatives or rapalogs)					
Everolimus, Afinitor, Certican, RAD001	(Motzer et al. 2008; Ryan et al. 2011; Yao et al. 2010; Ellard et al. 2009; Baselga et al. 2009; Amato et al. 2009; Gridelli et al. 2007)	mTORC1	Phase II/III	Novartis	Pancreatic neuroendocrine tumors/2nd line elderly NSCLC; Approved for organ rejection; was approved for renal cell carcinoma 3/2009 (Motzer et al. 2008)

Table 5.1 (continued)

Drug name	References	Target(s)	Stage, phase	Company	Application
Temsirolimus, Torisel, CCI-779	(Hess et al. 2009; Hudes et al. 2007; Galanis et al. 2005; Witzig et al. 2011, 2005; Atkins et al. 2004; Sarkaria et al. 2010; Johnston et al. 2010a; Ghobrial et al. 2010)	mTORC1	Phase II/III	Wyeth	Approved for renal carcinoma, relapsed mantle cell lymphoma /newly diagnosed GBM/ non-Hodgkin's lymphoma (NHL), Waldenström's macroglobulinemia
Ridaforolimus, Deforolimus, AP23573, MK-8669	(Rizzieri et al. 2008; Sessa et al. 2010)	mTORC1	Phase III, Ib	Merck/ARIAD Pharmaceuticals	Sarcoma, relapsed/refractory hematologic malignancies
PKB/Akt inhibitors					
A-443654	(de Frias et al. 2009; Han et al. 2007; Liu et al. 2008; Zhuang et al. 2010)	pan-PKB/Akt	Preclinical	Abbott Labs	
AT-13148	(Lyons et al. 2007)	PKB/Akt	Phase I	AstraZeneca/Astex Therapeutics	
CCT129254, AT11854	(Davies et al. 2009)	PKB/Akt	Preclinical	AstraZeneca	
GSK690693	(Altomare et al. 2010; Carol et al. 2010; Heerding et al. 2008; Levy et al. 2009; Rhodes et al. 2008)	pan-PKB/Akt	Phase I (withdrawn)	GlaxoSmithKline	
GSK2141795 and GSK21110183 (follow-ups of GSK690693)		pan-PKB/Akt	Phase I	GlaxoSmithKline	Hematologic cancer

Table 5.1 (continued)

Drug name	References	Target(s)	Stage, phase	Company	Application
Perifosine, D21226	(Argiris et al. 2006; Bailey et al. 2006; Elrod et al. 2007; Ernst et al. 2005; Knowling et al. 2006; Posadas et al. 2005; Van Ummersen et al. 2004; Vink et al. 2006; Chee et al. 2007)	PKB/Akt	Phase I/II	Keryx	Advanced solid tumors, ovarian cancer, metastatic melanoma, multiple myeloma
MK-2206	(Tolcher et al. 2009)	PKB/Akt & mTOR	Phase I	Merck	Advanced solid tumors
XL-418		PKB/Akt	I, suspended	Exelixis	Solid tumors

For updates of study progress and status consult http://www.clinicaltrials.gov; http://clinicaltrialsfeeds.org; http://www.cancer.gov; compounds in clinical trials can be retrieved by entering keywords [e.g. Akt, PKB, PI3K, mTOR] at http://nci.nih.gov/drugdictionary. Recent, excellent reviews covering PI3K pathway drug development include (Liu et al. 2009; Albert et al. 2010). Compound structures were reviewed in depth in (Marone et al. 2008).

and PtdIns(3,4)P_2, but it was shown that PI3K-C2α bound to clathrin was able to generate PtdIns(3,4,5)P_3 *in vitro* (Gaidarov et al. 2001). Activation loop sequences would predict that this does not happen in a PI3K-C2α monomer, and suggests that clathrin takes part in the presentation of PtdIns(4,5)P_2 as a substrate for this lipid kinase. Presently, the physiologic role, lipid product(s) and selective downstream targets of class II PI3Ks remain largely undefined.

5.1.4 Structural Features and Activation of Class IA PI3Ks

Class IA PI3K catalytic subunits p110α (encoded by the gene named *PIK3ca*), p110β (*PIK3cb*), and p110δ (*PIK3cd*) bind tightly to a regulatory subunit harboring two src-homology 2 (SH2) domains. The latter docks specifically to phosphorylated tyrosines in pYxxM (phospho-Tyr-x-x-Met) motifs on growth factor receptors or protein tyrosine kinase substrates. Mammalia have three genes encoding five major p85-like regulatory subunits subunits (*PIK3r1* encodes p85α and splice variants p55α, and p50α; *PIK3r2* yields p85β; and *PIK3r3* gives rise to p55γ). Each regulatory subunit contains a coiled-coil region located between the N- and C-terminal SH2 domains (dubbed interSH, or iSH domain), which tightly binds to the N-terminus (designated as p85-binding region, p85B) of the catalytic subunits p110α, p110β, and p110δ (Fig. 5.2). The class IA regulatory p85-like subunits exert an inhibitory action onto the catalytic subunit, which keeps the potentially oncogenic class I enzymes silent (Zhao et al. 2005; Kang et al. 2006). Inhibition is released by the translocation of p85 regulatory subunits to growth factor receptors and binding of the SH2 domains to pYxxM motifs (for reviews see (Wymann and Marone 2005; Vanhaesebroeck et al. 2001; Wymann et al. 2003b; Cantley 2002; Backer 2010)). The isolated domain structures of the regulatory p85 subunit have been determined early on (for the SH3 domain of p85α see (Batra-Safferling et al. 2010); the N-terminal SH2 domain of p85α (nSH2) see (Nolte et al. 1996); for the iSH2 of p85β see (Schauder et al. 2010); for the C-terminal SH2-domain (cSH2) of p85α see (Hoedemaeker et al. 1999)), but the mechanism of this important inhibitory action has only been elucidated recently: the structural determination of p85-fragments bound to p110α (Miled et al. 2007) and the resolution of a p85α iSH-cSH2 fragment in a complex with p110β suggest that both SH2 domains interact with perpendicularly oriented C-terminal α-helices dubbed "the regulatory square" (Zhang et al. 2011). For class I PI3K it is thus clear, that phosphorylated YxxM motifs on growth factor receptors do not only translocate PI3Ks to the plasma membrane to secure access of the catalytic subunit to PtdIns(4,5)P_2, but that they also relieve two to three safety latches from class IA PI3K complexes.

Some growth factor receptors directly recruit class IA PI3Ks, but for other input signals, adaptor molecules are key to the activation and localization of class IA PI3Ks (Fig. 5.3). As such, Grb2-associated binders (Gab1-3) and insulin receptor substrates (IRS1-4) display pYxxM sites to class IA PI3Ks. They belong to aYxxM-multisite adaptor protein family including daughter of sevenless (Dos). The insulin and IGF-1

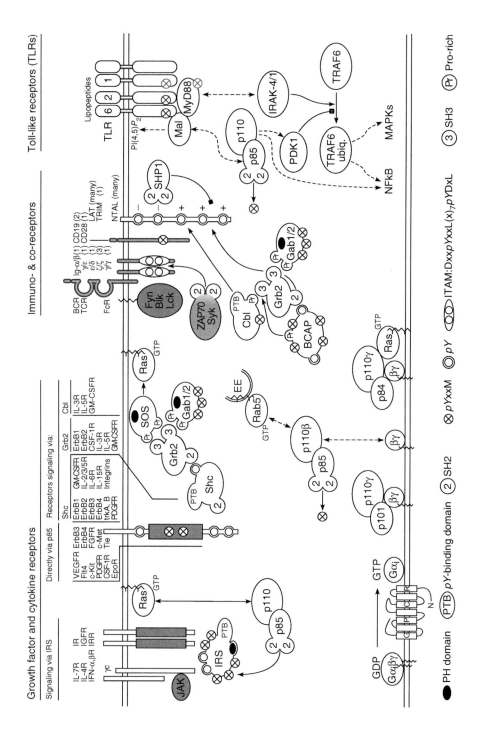

Fig. 5.3 Class I PI3Ks—multiple ways of activation. A plethora of growth factor receptors, cytokine and immune receptors, Toll-like and G protein-coupled receptors trigger PI3K activities. The simplest mode of PI3K activation is mediated by growth factor receptors with intrinsic protein tyrosine kinase activity, of which some present phosphorylated Tyr-x-x-Met motifs (pYxxM; see explanation of symbols at the bottom of the figure) and thus directly interact with the two SH2 domains of p85-like regulatory subunits. Other receptors operate via adapter proteins like Shc, Grb2, and Cbl that are interlinked by SH3/proline-rich regions, SH2/pY and phosphotyrosine-binding/pY interaction pairs. Here, the output to PI3K is often relayed via Gab1/2 (Nishida and Hirano 2003), which also displays a PtdIns(3,4,5)P_3-selective PH domain and docking sites for p85 (pYxxM motifs). Cytokine receptors signal through Janus kinase (JAK) either via Shc/Grb2/Cbl as indicated, or via the phosphorylation of insulin receptor substrate (IRS), which is the main target of insulin receptor (IR), IGFR and IRR. Ras, e.g. activated by SOS (son of sevenless), interacts with the Ras-binding domain of class I PI3Ks. In hematopoietic cells, constitutively membrane-bound src-like protein kinases (Fyn, Blk, Lck) phosphorylate cross-linked immunoreceptors on immunoreceptor tyrosine-based activation motifs (ITAMs; Rolli et al. 2002; Hamerman et al. 2009). The immunoglobulin or antigen-binding portion of the immunoreceptors is shown schematically for the B-cell receptor (BCRs), T-cell receptors (TCRs), and immunoglobulin receptors (FcR). Accessory chains and co-receptor molecules are listed on the right of the respective immune receptors. The phosphorylated ITAM motifs of the accessory chains recruit SH2 containing tyrosine kinases ZAP70 or Syk, supporting the protein tyrosine cascade and extending phosphorylation to CD19, CD28, TRIM, LAT, and NTAL. Inhibitory ITAMs recruit the protein phosphatase SHP1 and the lipid phosphatase SHIP (not shown; Blank et al. 2009). In the BCR, complement receptor 2 (CR2/CD21, not shown) drags along CD19 with two pYxxM motifs. The B-cell adaptor for PI3K (BCAP) and Gab2 mediate translocation of PI3K (mainly p110δ) to the BCR. In T cells, tyrosine-phosphorylated CD28 and TRIM display docking sites for p85. When phosphorylated, LAT recruits phospholipase Cγ1 (PLC-γ1), Gads, Grap, 3BP2, and Shb (not shown) and Grb2. With these components SOS, Vav, SLP-76, Itk (not depicted), Gab and c-Cbl, can be recruited. For reasons of simplicity, other PH domain-containing adapters where omitted (such as SKAPs, Bam32/DAPP1 [binding PtdIns(3,4,5)P_3 and PtdIns(3,4)P_2], and TAPP1/2 [specific for PtdIns(3,4)P_2]; more on their function can be found in (Zhang et al. 2009)). In myeloid cells, ITAMs of immunoglobulin receptors are phosphorylated by Lyn and Syk. The direct activation of PI3K in mast cells by the high affinity IgE receptor requires Gab2 (Gu et al. 2001) and LAT (Wilson et al. 2001). Toll-like receptors (TLRs) play an important role in the recognition of pathogens and "non-self", and prime cell of the innate immune system for activation: TLR 1,2, and 6 have been shown to contain a YxxM sequence, which can potentially recruit p85 (Li et al. 2010). Lipopeptide-dependent activation of PI3K via heterodimeric TLR2/6 has been shown to depend on a direct interaction of p85 and Mal (Santos-Sierra et al. 2009). Mal has a N-terminal PtdIns(4,5)P_2-binding domain that is crucial for membrane translocation. Abbreviations: CSF-1R colony stimulating factor 1 receptor, CD cluster of differentiation, ErbB epidermal growth factor receptor [family], EpoR erythropoietin receptor, FGFR fibroblast growth factor receptor, Gab Grb2-associated binder, GM-CSF granulocyte–macrophage colony-stimulating factor, Grb2 growth factor receptor-bound protein 2, IGFR insulin-like growth factor 1 receptor, IL interleukin, IRR insulin receptor-related receptor, LAT linker for activation of T cells, Mal/TIRAP MyD88-adapter-like/TIR-domain-containing adaptor protein, MyD88 myeloid differentiation primary response gene 88, PDGFR platelet-derived growth factor receptor, PH domain pleckstrin homology domain, PTB domain phosphotyrosine-binding domain, Pr proline-rich, pY phosphotyrosine, SOS son of sevenless, TRIM TCR-interacting molecule, ZAP70 [TCR] zeta chain-associated protein kinase receptors directly phosphorylate IRSs, while members of the IL-4 receptor family (γc-chain-containing cytokine receptors, see Fig. 5.2) require Janus kinases (JAK) to complete the phosphorylation of YxxM motifs. IRS1 and Gab proteins display a pleckstrin homology (PH) domain that is selective for PtdIns(3,4,5)P_3, and can potentially provide amplification of the PI3K response. The importance of Gabs for PI3K signaling has been demonstrated in various systems, e.g. acquired immunity (Pratt et al. 2000), allergy (Gu et al. 2001) and transformation of myeloid cells by the p190$^{BCR/Abl}$ protein tyrosine kinase driving chronic myelogenousleukemia (CML; Sattler et al. 2002). The E3 ligase Cbl, has also been shown to recruit PI3K to growth factor and cytokine receptors. *In vivo*, however, c-Cbl and Cbl-b significantly reduces PI3K activity downstream of the T cell receptor. That the inhibitory action of Cbl is dominant in the long term was demonstrated in mice with disrupted c-Cbl, resulting in enhanced thymic positive selection (Murphy et al. 1998). Mice without Cbl-b displayed increased susceptibility for autoimmune diseases (Bachmaier et al. 2000).

Downstream of the T cell (TCR), B cell (BCR) and immunoglobulin receptors (FcRs) so-called immunoreceptor tyrosine-based activation motifs (ITAM) serve as initiation points for protein tyrosine kinase cascades. After the cross-linking of immunoreceptors, src-like membrane-anchored protein tyrosine kinases (Fyn, Blk, Lck) are concentrated, and locally phosphorylate ITAMs. ITAMs then recruit SH2 domain-containing protein tyrosine kinases including ZAP-70 and Syk. This enforced phosphorylation activity generates further PI3K docking sites on transmembrane adapter proteins (TRAPs) such as CD19, CD28, T cell receptor interacting molecule (TRIM), NTAL (also called LAB/Lat2) and linker for activation of T cells (LAT). Mice without LAT cannot generate T cells beyond the CD4$^-$/CD8$^-$ stage (Zhang et al. 1999; Roncagalli et al. 2010; Fuller et al. 2011). As for the growth factors mentioned above, soluble adapter proteins [Grb2, Gab and the B cell adaptor for phosphoinositide 3-kinase (BCAP)] contribute to PI3K activation. Deletions of BCAP (Yamazaki et al. 2002) or CD19 (Rickert et al. 1995) both lead to severe B cell phenotypes (Simeoni et al. 2004).

Toll-like receptor (TLR)-triggered responses are essential in host defense. The activation of TLRs by various ligands has been reported to activate PI3K. Elevated PI3K activity was mostly associated with an attenuation of the NFκB (nuclear factor kappa-light-chain-enhancer of activated B cells) pathway and cytokine production at multiple levels (Fukao and Koyasu 2003; Hazeki et al. 2007). Depending on the stimuli and TLR receptor targeted, PI3K mediated pro- and anti-inflammatory effects (for a review see (Fukao and Koyasu 2003)). The activation of class I PI3Ks can take place via the interaction of p85 with aYxxM sequence on either TLR1, TLR2, or TLR6 (Li et al. 2010) or by p85 binding to Mal (MyD88-adaptor-like; also named Toll—IL-1 receptor domain-containing adaptor protein, TIRAP; Santos-Sierra et al. 2009). When the PI3K downstream target 3-phosphoinositide-dependent protein kinase-1 (PDK) was cell-specifically targeted in the myeloid cell linage, macrophages without PDK1 became more susceptible to lipopolysaccharide (LPS) stimulation via TLR4, and activation by Pam3CysSerLys4 (Pam3CSK4), a potent TLR2 agonist acting through TLR2/TLR1. This resulted in increased production of tumor necrosis factor-α (TNF-α) and interleukin-6 (IL-6) and a dramatically increased sensitivity to LPS-induced septic shock (Chaurasia et al. 2010).

5.1.5 Class IB PI3K: PI3Kγ

The PIK3cg gene encodes the only PI3K class IB member, PI3Kγ (Stoyanov et al. 1995; Stephens et al. 1997), which associates with a p101 (PIKr5; Stephens et al. 1997) or a p84/p87^{PIKAP} adapter protein (PIKr6; Suire et al. 2005; Voigt et al. 2006; Bohnacker et al. 2009). PI3Kγ and its adapter subunits are highly expressed in leukocytes, and at lower levels in smooth muscle cells, endothelia and cardiomyocytes (Wymann and Marone 2005; Wymann et al. 2003b; Patrucco et al. 2004; Vecchione et al. 2005; Alloatti et al. 2005; Okkenhaug and Vanhaesebroeck 2003; Ghigo et al. 2010). The PI3Kγ complex is translocated and activated by βγ subunits (Gβγ) of trimeric G proteins. PI3Kγ is expressed at high levels in white blood cells throughout the hematopoietic system, and mainly relays signals downstream of G protein-coupled receptors (GPCRs). Thus linked to a plethora of chemokine and other receptors of inflammatory mediators, PI3Kγ mediates processes in inflammatory and allergic reactions (Wymann et al. 2003b; Ghigo et al. 2010; Deane and Fruman 2004). Although lower PI3Kγ expression was detected in cardiomyocytes, vascular smooth muscle and endothelia, PI3Kγ plays a role in the control of vascular tone (Vecchione et al. 2005), heart contractility (Patrucco et al. 2004; Crackower et al. 2002; Oudit et al. 2004) and progress of atherosclerosis (Fougerat et al. 2008). The role of the p101 and p84 adapter subunits has not yet been fully explored, but they are required for a functional relay of GPCR signals to PI3Kγ. In this respect, the p101 adapter subunit sensitizes the PI3Kγ complex to Gβγ subunits (Stephens et al. 1997; Krugmann et al. 1999; Maier et al. 1999; Brock et al. 2003; Kurig et al. 2009), while p84 does not fulfill this function. Interestingly, p84-p110γ complexes require the interaction with activated Ras (see below), while p101 functions even when Ras activation is blocked (Kurig et al. 2009). As p101-p110γ and p84-p110γ complexes were found to generate functionally distinct pools of PtdIns(3,4,5)P$_3$ (Bohnacker et al. 2009), it is likely that the two adapter subunits moderate input and output signals of the respective PI3Kγ complexes.

Alternatively, Gβγ subunits were also shown to activate p110β. Here early studies demonstrated that Gβγ subunits and phosphorylated Tyr peptides activated p85-p110β complexes synergistically (Maier et al. 1999; Kurosu et al. 1997; Tang and Downes 1997). Cellular studies using overexpression of GPCRs, p110β, genetic and pharmacological tools proposed that p110β was the main PI3K downstream of some GPCRs. While this makes sense for receptors that do not couple exclusively to B. Pertussis toxin-sensitive G proteins, for example the receptors for lysophosphatidic acid (LPA), thrombin, the bradykinine and A1 adenosine receptors, the sphingosine-1-phosphate receptor and more (Roche et al. 1998; Graness et al. 1998; Kubo et al. 2005; Guillermet-Guibert et al. 2008), the data for some other GPCRs has to be reviewed critically for receptor transactivation and co-operation with protein tyrosine kinase activities (Guillermet-Guibert et al. 2008). As an example, there is conflicting data demonstrating one hand that complement fragment 5a (C5a) signals exclusively via p110γ in macrophages (Hirsch et al. 2000), while others claimed that p110β was required (Guillermet-Guibert et al. 2008).

5.1.6 Activation of Class I PI3Ks by Small GTPases

A potential common activator for class I PI3Ks is activated, GTP-loaded Ras, as all class I PI3Ks display a Ras-binding domain (Fig. 5.2). This interaction is well documented for p110α (Sjolander and Lapetina 1992; Sjolander et al. 1991; Rodriguez-Viciana et al. 1994, 1996) and p110γ (Rubio et al. 1997, 1999). For p110α, Ras-interactions were demonstrated to be relevant in Ras-driven tumor promotion (Gupta et al. 2007), and an intact Ras-binding domain was shown to be crucial for the activation of the NADPH oxidase in neutrophils by GPCR ligands (Suire et al. 2006). These findings were corroborated by the resolution of the crystal structure of a Ras-p110γ complex defining the interface of the two proteins in detail. Ras-induced conformational changes in p110γ show that Ras is not only a docking site for p110 at the membrane, but a potent activator (Pacold et al. 2000). The importance of Ras activation upstream of p110β is controversial (Kang et al. 2006; Rodriguez-Viciana et al. 2004; Marques et al. 2008), and p110β was suggested to interact with the small GTPase Rab5 localized on early endosomes (Christoforidis et al. 1999; Shin et al. 2005; Kurosu and Katada 2001; Ciraolo et al. 2008; Jia et al. 2008). The p110δ catalytic subunit interacts with TC21 (or RRas2 (Rodriguez-Viciana et al. 2004; Delgado et al. 2009)), and depends on the GTPase for a translocation to T- and B-cell receptors (Delgado et al. 2009).

5.1.7 Class II PI3Ks

Class II PI3Ks are large enzymes (170–200 kDa) and include the three family members PI3K-C2α, β, and γ, which all have a C-terminal extension containing a Phox homology (PX) and a C2 homology domain. The C-terminal C2 domain is Ca^{2+}-insensitive due to the lack of a conserved aspartate residue (Falasca and Maffucci 2007; for a new classification of C2 domains see (Zhang and Aravind 2010)). The founder of the class II PI3K family was the *drosophila* PI3K_68D (MacDougall et al. 1995).

In contrast to class I PI3Ks, no class II regulatory subunits were identified. Class II PI3Ks have been mapped to the trans-Golgi network and low-density microsomes, but their mode of action is still poorly defined. The PI3KC2α and β isoforms have been reported to associate with their N-terminal region with clathrin (Domin et al. 2000; Gaidarov et al. 2001, 2005; Wheeler and Domin 2006). PI3KC2α has been clearly attributed roles in clathrin assembly and clathrin-mediated, microtubule-dependent vesicular trafficking (Domin et al. 2000; Gaidarov et al. 2001, 2005; Zhao et al. 2007). Recent reports identified PI3KC2α as an essential factor in dynamin-independent endocytosis and fluid-phase endocytosis. As PI3KC2α and the PtdIns-3-*P* -binding Early Endosomal Antigen (EEA1) were recruited to cargo vesicles, it is likely that the relevant *in vivo* product involved in the process was PtdIns-3-*P* (Krag et al. 2010).

Several extracellular stimuli activate class II PI3Ks, such as growth factors like EGF, PDGF, insulin and SCF, chemokines (MCP-1), cytokines (leptin, TNF-α), and lysophosphatidic acid (LPA; Arcaro et al. 2000; Maffucci et al. 2005). Proline-rich regions in the N-terminus, and interactions with signaling adapters like Grb2 were proposed to mediate interaction with growth factor receptors (Wheeler and Domin 2006). For PI3KC2β it was shown that this enzyme can promote LPA-induced cell migration of ovarian and cervical cancer cells (Maffucci et al. 2005; for a review on class II PI3K in cancer see (Traer et al. 2006)). Other results point to a role of class II PI3Ks in insulin signaling (Cui et al. 2011; Falasca et al. 2007; Dominguez et al. 2011), which is supported by the finding that class II regulates exocytosis of insulin granules in pancreatic beta cells (Dominguez et al. 2011). Interestingly, a polymorphism in the PIK3C2G gene (encoding PI3KC2γ could be associated with type 2 diabetes in a Japanese population (Daimon et al. 2008)), and the nematode *C. elegans* accumulates fat when its only class II PI3K gene (*F39B1.1*) product is down-regulated. So far, these findings were not duplicated in mice where the *PIK3C2A* (PI3KC2α) and *PIK3C2B* (PI3KC2β) loci were targeted: in the latter mice, fat and body mass were actually significantly reduced. It must be said, however, that the mice lacking functional PI3KC2β were mainly investigated for epithelial differentiation (Harada et al. 2005), and that the mice with a modified *PIK3C2A* locus displayed a defect in renal function and still expressed a trace of a truncated PI3KC2α protein lacking the C-terminal PX and C2 domains (Harris et al. 2011).

The available data connects class II PI3Ks to multiple signaling events like the activation of MAPK (but not PKB/Akt (Cui et al. 2011)), activation of Ca^{2+}-triggered potassium channels (KCa3.1; Srivastava et al. 2009), the regulation of Rho (Wang et al. 2006), clathrin coated vesicular movement on microtubules (Zhao et al. 2007), exocytosis (Meunier et al. 2005) and endocytotic events mentioned above. For class II PI3Ks it is presently not possible to delineate connected signaling pathways, and to define a predictive network linked to general signaling outputs.

5.1.8 Class III PI3Ks

Class III PI3K: The *Saccharomyces cerevisiae* Vps34 protein (Vps34p, vacuolar protein sorting mutant 34) is the prototype of class III PI3K and plays an essential role in vesicular and protein trafficking from the Golgi to the yeast vacuole, which is the yeast equivalent to lysosomes in mammals (Schu et al. 1993). In yeast, Vps34p binds to the N-terminally myristoylated Ser/Thr kinase Vps15p. It has been shown that a functional Vps15p kinase is needed for the activation and recruitment of Vps34p to Golgi membranes (Herman et al. 1992). The Vps15p orthologue in mammals is p150 (also referred to as hVps15), which translocates hVps34p to Rab5-positive early endosomes and Rab-7 positive late endosomes. Therefore, hVps34p is also central to endocytosis, vesicular trafficking (Christoforidis et al. 1999; Murray et al. 2002), and phagocytic uptake of bacteria (Sun et al. 2008).

Moreover, Vps34p interacts directly with Beclin-1 (yeast Atg6/Vps30p) and Atg14L to form a complex I, and alternative complexes containing UVRAG or UVRAG and Rubicon at the place of Atg14L (Funderburk et al. 2010; Matsunaga et al. 2010), which all play an important role in autophagy (Simonsen and Tooze 2009). Recently, it was proposed that the Vps34p-hVps15p-Beclin-1-Atg14L complex would also involve p110β as an element to regulate the autophagy (Dou et al. 2010). In yeast, deletion of Atg14 leads to defects in autophagy, while own regulation of cellular Beclin-1 affects autophagy and vesicular trafficking minimally (Kihara et al. 2001). In contrast, mice heterozygous for Beclin-1 display a decreased rate of autophagy and enhanced tumor formation (Qu et al. 2003; Yue et al. 2003). Recently, Vps34p was linked to the induction of autophagy in nutrient, amino acid, as well as energy (glucose) -deprived cells, and a role for Vps34p in the amino acid-induced activation of mTOR was proposed (Dann and Thomas 2006; Gulati and Thomas 2007; Nobukuni et al. 2007). The latter connection is controversial, and the deletion of *VPS34* in fruit flies did not affect TOR activity (Juhasz et al. 2008). In mice, Vps34 (encoded by PIK3C3) is required in early embryogenesis, and effects on mTOR activation were documented (Zhou and Wang 2010). Further in-depth reviews on class III PI3Ks can found in (Backer 2008; Simonsen and Tooze 2009; Funderburk et al. 2010; Backer 2010).

5.1.9 Downstream of Class I PI3Ks

The class I PI3K product PtdIns(3,4,5)P_3 is produced at the plasma membrane, where it serves as a docking site for proteins with PtdIns(3,4,5)P_3-specific PH domains (Ferguson et al. 2000). Members of the protein kinase B family (PKBα,β,γ/Akt1,2,3) are the most prominent representants of these signaling molecules, and link PI3Ks to the control metabolic activity, growth, proliferation and cellular survival pathways (Fig. 5.4). When PKB/Akt is recruited the plasma membrane, it is phosphorylated on Thr308 (numbers refer to PKBα/Akt1) by PDK1. In this process, binding of PDK1 itself to PtdIns(3,4,5)P_3 is crucial, as cells harboring a PDK1 with a non-functional PH domain cannot efficiently trigger PKB/Aktphosphorylation (McManus et al. 2004). To gain full activity, a second phosphorylation in the C-terminal, hydrophobic motif of PKB/Akt (Ser 473) by activities classified as "PDK2s" is required (Yang et al. 2002; Biondi and Nebreda 2003). A number of kinases have been shown to classify as PDK2 activities, such as mitogen-activated protein kinase-activated kinase 2 (MAPKAP-2), integrin-linked kinase (ILK), DNA-dependent protein kinase (DNA-PK$_{cs}$; Feng et al. 2004; Hanada et al. 2004; Bozulic and Hemmings 2009) and protein kinase Cβ (PKCβ; Kawakami et al. 2004). Finally it has been shown that the TOR complex 2 (TORC2, the [Rictor]-TOR complex [rapamycin-insensitive companion of TOR]) displays major PDK2 activity (Sarbassov et al. 2005, for reviews see (Bozulic and Hemmings 2009; Polak and Hall 2006; Manning and Cantley 2007; Zoncu et al. 2011)). PKB/Akt-mediated phosphorylations regulate many downstream targets, both positively and negatively (see Fig. 5.4). Besides the modulation of PKB/Akt activity, the phosphorylation of Ser 473 seems also to be required for an efficient phosphorylation of N-terminal residues of the PKB/Akt substrates FOXO1/3A/4 (Jacinto et al. 2006). The hydrophobic motif phosphorylation might thus direct the PKB/Akt substrate selectivity (for a review see (Manning and Cantley 2007)).

The activation of the PI3K/PKB pathway is counteracted by two phosphoinositide phosphatases: (i) the 3-phosphoinositide phosphatase and tensin homolog deleted in chromosome ten (PTEN) regenerates PtdIns(4,5)P_2 from PtdIns(3,4,5)P_3 (Stambolic et al. 1998). PTEN is often mutated in tumors, leading to the accumulation of PtdIns(3,4,5)P_3 and a constitutive activation of the PI3K pathway (Liaw et al. 1997; Sansal and Sellers 2004; Hobert and Eng 2009; Farooq et al. 2010; Vivanco and Sawyers 2002; Wymann and Marone 2005; Cully et al. 2006). (ii) the SH2-domain-containing inositol phosphatase (SHIP), which has a 5-phosphoinositide phosphatase activity, generates PtdIns(3,4)P_2 from PtdIns(3,4,5)P_3 (Majerus et al. 1999; Kisseleva et al. 2000; Rohrschneider et al. 2000; Kalesnikoff et al. 2003). Besides the above lipid phosphatases, the protein phosphatase PHLPP can dampen PKB/Akt activation by the removal of the phosphate group at Ser 473 (Brognard et al. 2007; Gao et al. 2005; Mendoza and Blenis 2007).

5.1.10 Control of Cellular Growth, Transcription, and Translation

PKB/Akt controls cell growth via the nutrient sensor mTOR: PKB/Akt phosphorylates and thus inhibits TSC2 (tuberin), which constitutes together with TSC1 (hamartin) the tuberous sclerosis complex (Pan et al. 2004). In its active form, the TSC1/2 complex prevents the exchange of GDP for GTP on the GTPaseRheb. As the activation of TOR requires GTP-loaded Rheb, the PKB/Akt-mediated phosphorylation of TSC2 therefore initiates the activation of the TOR complex 1 (TORC1, or [Raptor]-TOR complex [regulatory-associated protein of TOR]). Active TORC1 phosphorylates and blocks 4E-BP1, which releases translation of mRNAs with 5′-polypyrimidin regions, which is supported by the parallel, TORC1-mediated phosphorylation of p70^{S6K} (p70 S6 kinase) on Thr 389. S6K targets the ribosomal protein S6 (Dufner and Thomas 1999; Garami et al. 2003; Dann et al. 2007). Furthermore, TORC1 can be negatively regulated by PRAS40—controlling TORC1 substrate access (Van der Haar et al. 2007; Wang et al. 2007; Fonseca et al. 2007), or by an allosteric inhibition caused by the binding of the macrolide Rapamycin/FKBP12 complex to the FRB (FKBP/Rapamycin-binding) domain of TOR (Choi et al. 1996). When TORC1 signaling is permitted, transcription and translation are elevated, and

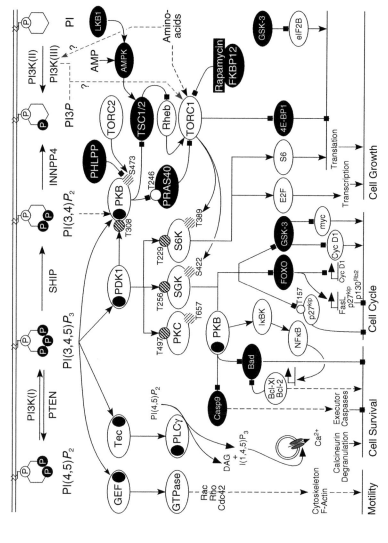

Fig. 5.4 Roadmap from PtdIns(3,4,5)P_3 to cellular outputs relevant in cancer and inflammation. PtdIns(3,4,5)P_3 is produced from ATP and PtdIns(4,5)P_2 by an activated class I PI3K (see Fig. 5.3) at the plasma membrane. Signaling of PI3K is completely reversed by the 3-phosphoinositide phosphatase PTEN, and the 5-inositol phosphatase SHIP produces PtdIns(3,4)P_2 with a limited cell activating capacity. *Pointed arrows* and *white protein symbols* denote activating pathways, *squared arrows* and *black proteins* represent inhibitory outputs. Pleckstrin homology (PH) domains sensitive to PtdIns(3,4,5)P_3, are shown as *black*

Fig. 5.4 (Continued) *ovals*. PH domains dock enzymes like phosphoinositide-dependent kinase 1 (PDK1) and protein kinase B (PKB) at the plasma membrane (Franke et al. 1995; Burgering and Coffer 1995; Stokoe et al. 1997). PDK1 then phosphorylates protein kinase B (PKB/Akt) on Thr308 (Alessi et al. 1997; the residue numbers here refer to the human α isoforms). PDK1 has been also reported to phosphorylate a number of other AGC (protein kinases A, G, and C) Ser/Thr kinases within a conserved activation loop sequence (*hatched, outlined circles*), such as Serum- and glucocorticoid-induced protein kinase (SGK), protein kinase C (PKC; Dutil et al. 1998), cAMP-dependent kinase (PKA), Ribosomal S6 kinase (RSK), PKC-related protein kinase 2 (PRK2), and p70^{S6K} (S6K; Pullen et al. 1998; for a recent review on AGC kinases see (Pearce et al. 2010)). Activity of PKB is further increased by a phosphorylation within the C-terminal hydrophobic motif (Ser473 in PKBα/Akt1) by so-called PDK2 activities (Yang et al. 2002; Biondi and Nebreda 2003). A prominent example is the mTOR complex 2 (TORC2; Sarbassov et al. 2005). Other reported PDK2 activities like the catalytic subunit of DNA-dependent protein kinase (DNA-PKcs, Feng et al. 2004), mitogen-activated protein kinase-activated kinase 2 (MAPKAP-2), integrin-linked kinase (ILK), and protein kinase Cβ (PKCβ; Kawakami et al. 2004) are not depicted for clarity. Interestingly, it was recently found that the site on SGK1 (Ser422) that corresponds to Ser473 on PKB is phosphorylated by TORC1, and SGK1 can exert similar actions as PKB (Hong et al. 2008; Toker 2008). When fully activated, PKB signals to a plethora of downstream targets modulating multiple cellular outputs. PKB-mediated phosphorylation and inactivation of the GTPase activating protein (GAP) activity of TSC2 (tuberin) in the TSC1 (hamartin)/TSC2 complex (TSC1/2; for a review see (Pan et al. 2004), leads to GTP loading of the small GTPaseRheb and the subsequent activation of the mTOR complex 1 (TORC1; Garami et al. 2003; Tee et al. 2003; Inoki et al. 2003). Important negative regulators in the area are the PHdomain leucine-rich repeat protein phosphatase (PHLPP; Brognard et al. 2007; Gao et al. 2005; Mendoza and Blenis 2007) dephosphorylating phospho-Ser473 of PKB; PRAS40 (proline-rich Akt substrate of 40 kDa or Akt1S1) that interferes with TORC1 substrate-binding when not phosphorylated by PKB (Van der Haar et al. 2007; Fonseca et al. 2007); and the FKBP12-rapamycin complex exclusively inhibiting with TORC1 (Choi et al. 1996; Polak and Hall 2009; Sengupta et al. 2010). When cellular energy levels are low, AMP concentrations rise and activate AMPK (trimeric AMP-activated protein kinase), which enhances the inhibitory effect of the TSC complex on Rheb (Hardie 2005; Martin and Hall 2005). PDK1 and TORC1 activate p70 S6 kinase (p70^{S6K}, S6K), and when GSK-3β and 4EBP1 are inhibited by PKB and TORC1, respectively, transcriptional and translation restrictions are released to promote cellular growth. Class O forkhead transcription factors (FOXOs) are phosphorylated by PKB and serum and glucocorticoid-inducible kinase (SGK), and then interact with 14.3.3 proteins trapping them in the cytosol. In a coordinated fashion, GSK-3β is inhibited by phosphorylation, and stops the phosphorylation of cyclin D1 and its degradation by the proteasome. The entry into cell cycle is tightly controlled by CDK inhibitor p27^{kip}, tumor suppressors like p130^{Rb2}, cyclin D1 (Cyc D1), myc. The expression levels (where indicated with *transcription arrows*) and activity status of these molecules is controlled by forkhead transcription factors (mainly FOXO1, 3A, 4) and the indicated upstream kinases. Anti-apoptotic signals are triggered downstream of PKB/Akt, which phosphorylates and inactivates caspase 9 (Casp9) and the cell death inducer Bad. A route via NFκB leads to the upregulation of Bcl-XL and Bcl-2, which neutralize Bad and promote cell survival. In lymphocytes, a Tec family kinase/PLCγ axis controls the sustained elevation of Ca^{2+} levels. Ca^{2+}–calmodulin then activates calcineurin, a protein phosphatase activating NF-AT (nuclear factor of activated T cells; not shown). Via PtdIns(3,4,5)P_3-sensitive GEFs (guanine nucleotide exchange factors) including Tiam (T cell lymphoma invasion and metastasis), Vav, PIX (PAK-associated guanine nucleotide exchange factor) and P-Rex (also stimulated by Gβγ subunits), small the small GTPasesRac, Rho and Cdc42 are stimulated and promote cytoskeletal changes and cell migration protein and lipid biosynthesis drive an increase in cell mass required for the entry in cell cycle progression.

5.1.11 PI3K-mediated Control of Cell Cycle Progression

Entry into cell cycle progression is tightly controlled by the alternating expression of cyclins and their interaction with cyclin-dependent kinases (CDKs) or CDKs inhibitors. Activated PKB/Akt phosphorylates forkhead transcription factors of the class O (FOXOs) on three different sites (for a review see (Manning and Cantley 2007)). When phosphorylated, FOXOs bind to 14-3-3 proteins, which act as phospho-Ser and phosphor-Thr "receptors" and retain FOXOs in the cytosol (Burgering and Kops 2002). When in the nucleus, FOXOs repress many genes required for cell cycle entry, and a cytosolic retention of FOXOs thus releases the transcription of cyclin D1, while the transcription of the CDK inhibitor p27^{Kip1} is attenuated (Alvarez et al. 2001; Burgering and Medema 2003). Glycogen synthase kinase 3β (GSK3β) phosphorylates cyclin D1, triggering its targeting and degradation to the proteasome. When phosphorylated by PKB/Akt, GSK3β is inhibited, and cyclin D1 accumulates (Liang and Slingerland 2003). The resulting increase in cyclin D1 levels combined with the concerted reduction in p27^{Kip1} allows cells to transit from G1 to the S phase (Liang and Slingerland 2003; Foijer and te Riele 2006).

5.1.12 PI3Ks—Driving Cell Survival and Anti-apoptotic Signaling

The stimulation of the PI3K/PKB pathway branches into many anti-apoptotic events: PKB/Akt directly phosphorylates and inhibits caspase 9, a protease crucial for the initiation of the apoptotic cascade (Cardone et al. 1998; for a cross-species comparison see (Datta et al. 1999)). PKB/Akt also inactivates the *Bcl-2*-associated death promoter (BAD) by a phosphorylation at Ser 136, which then liberates the anti-apoptotic proteins Bcl-2 and Bcl-XL (del Peso et al. 1997; Datta et al. 1997). In another branch PKB/Akt phosphorylates IκB kinase (IκBK), which blocks the action of the inhibitor I-κB to release the transcription factor NF-κB. NF-κB is now free to translocate to the nucleus where it activates transcription of cell survival proteins such as Bcl-2 and Bcl-XL (Li and Verma 2002; Ozes et al. 1999; for reviews see (Schinzel et al. 2004; Kaufmann et al. 2004)). As mentioned above, FOXOs are retained in the cytosol by the action of PKB/Akt. As FOXOs promote transcription of FasL, PKB/Akt prevents ligand-induced apoptosis (Brunet et al. 1999). The connections of these pathways are further detailed in Fig. 5.4.

In the above signaling scenarios, the PtdIns(3,4,5)P_3 → PDK1 → PKB/Akt axis seems to play a central role in the control of cellular growth, the entry into the cell cycle and the initiation of survival pathways. When PtdIns(3,4,5)P_3 is elevated due to loss of PTEN, this disrupts embryonic development in mice (Di Cristofano et al.

1998) and fruit flies (Stocker et al. 2002). Viability could be restored in PTEN null flies by the manipulation of the *drosophila* PKB/Akt PH domain: when flies expressed PKB/Akt with a PH domain with a low affinity for PtdIns(3,4,5)P_3, "lethal levels of PtdIns(3,4,5)P_3" were tolerated, suggesting that PKB/Akt was the major hub for PtdIns(3,4,5)P_3-sensing in the fly (Stocker et al. 2002). In the human genome there are, however, > 275 PH domain-containing proteins encoded (source: SMART database at smart.embl-heidelberg.de). These are complemented with 48 proteins with PX domains (a domain first found in phagocyte NADPH oxidase cytosolic factors), 27 with FYVE (Zinc finger domain first found in Fab1, YOTB, Vac1 and EEA1) domains, 20 with ENTH (Epsin*N*-terminal homology) domains, ANTH, FERM and other lipid-binding domains (Varnai et al. 2005; Balla 2005; Takenawa and Itoh 2006). For many of these proteins interactions with poly-phosphorylated phosphoinositides have been reported, and some prominent signaling molecules bind PtdIns(3,4,5)P_3 through PH domains (see also Fig. 5.4): TEC family protein tyrosine kinases including Btk (Brutons's tyrosine kinase; Readinger et al. 2009; Mohamed et al. 2009), and PLCγ1 (phospholipase Cγ1; Rebecchi and Pentyala 2000; Maroun et al. 2003; Ji et al. 1997) play an important role in innate and acquired immunity; and β-ARK1/GRK2 (β-adrenergic receptor kinase1/G protein-coupled receptor kinase; Takenawa and Itoh 2006; Jaber et al. 1996) modulates GPCR signaling and cardiovascular functions. Structural data for PtdIns(3,4,5)P_3-bound PH domains is available (Ferguson et al. 2000; Milburn et al. 2003), and the PH domain protein families have been thoroughly analyzed across various species (Park et al. 2008). In spite of the existing diversity of PtdIns(3,4,5)P_3 selective PH domain containing proteins, PKB research is with > one third of all articles overrepresented in the PI3K literature (for a recent in-depth review on phospholipid-binding domains see (Lemmon 2008)).

5.1.13 PI3Ks—A Connection to Migration and Polarization

Small Rho family GTPases including Rac, Cdc42, TC10 and Rho play important roles in cell migration, polarization and cytoskeletal rearrangements. The activity of these Rho GTPases is regulated by GAPs (GTPase activating proteins), GDIs (guanine nucleotide dissociation inhibitors) and guanine nucleotide exchange factors (GEFs). While GAPs increase the rate of hydrolysis of the GTP bound to activatedGTPases to GDP, GDIs retain Rho GTPases in the cytosol. Activated GEFs promote the reloading of small GTPases with GTP, and thus enable their subsequent interaction with downstream effectors (Heasman and Ridley 2008). Rho GEFs can be divided in (i) Dbl-like proteins (> 70 members), which contain the catalytic Dbl homology (DH) domain and a pleckstrin homology domain, and (ii) Dock family proteins (11 members) with the active Dock homology region (DHR-2; also named Docker-ZH2 domain) and a DHR-1 mediating translocations (Cote and Vuori 2007).

A number of Dbl-like GEFs contain PtdIns(3,4,5)P_3-binding PH domains. Many of these GEFs, like Tiam (T-lymphoma invasion and metastasis inducing protein; a

GEF for Rac), PIX (PAK-associated guanine nucleotide exchange factor; reloading Rac and Cdc42), Vav (targeting RhoA, Rac, and Cdc42), and ARNO (Arf nucleotide binding site opener; activating Arf1,6) integrate the PI3K signal with the input of an upstream protein kinase, before they are fully active (for a review see (Wymann and Marone 2005)). In contrast, P-Rex binds to Gβγ subunits released from trimeric G proteins after GPCR activation and PtdIns(3,4,5)P_3 before it activates Rac (Welch et al. 2002). For Dock family proteins, Dock2 and Dock180 were shown to bind PtdIns(3,4,5)P_3 via their DHR-1 domain (Cote and Vuori 2007).

The regulation of Rho GTPases by GEFs is crucial in many physiological processes and disease contexts. While Ras proteins are activated by mutations (Diaz-Flores and Shannon 2007), deviations in Rho family GTPasesignaling is often caused by the overexpression of corresponding GEFs (Ellenbroek and Collard 2007; Vega and Ridley 2008). Rho GTPase activation promotes the dissemination of cancer cells and immune cells throughout tissues, driving metastasis and chronic inflammation. A subset of small GTPases such as RhoA, Rac and Cdc42 also activate transcriptional events (for a review see (Benitah et al. 2004)).

5.2 PI3Ks in Physiology and Disease

As outlined above, and illustrated in Figs. 5.3 and 5.4, PI3K relays growth factor, cytokine and G protein receptor-coupled signaling to a network balancing a cell's activities and cellular energy consumption. Downstream of PI3Ks, the TOR complexes 1 and 2 are important hubs integrating hormonal input, energy and nutrient availability. PtdIns(3,4,5)P_3-dependent kinase cascades originating for example from PDK1, PKB/Akt, TEC family kinases or the activation of PtdIns(3,4,5)P_3-sensitive GEFs contribute to the control of growth, cell cycle, survival and migration. In chronic inflammatory and autoimmune disease, an overshooting cytokine network triggers the activation of immune cells, while in cancer oncogenes are activated by mutations and epigenetic effects. Many PI3K-dependent pathways are shared in cancer and inflammation, but operate in different contexts and yield cell-specific outputs. Various cell types use specific class I PI3K isoforms, and an understanding of PI3K isoform selective signaling is a prerequisite to develop refined targeted therapies in cancer and inflammation.

5.2.1 PI3Ks in Innate and Acquired Immunity

In acquired immunity, T lymphocytes and B lymphocytes are activated by specific antigens exposed to them by antigen-presenting cells, such as dendritic cells (DCs) and activated macrophages. The development of T cell subsets (Th1, Th2, Th17, Treg, $CD8^+$ cytotoxic T cells, etc.) is fine tuned by cytokine signals and regulated by

PI3Ks (Deane and Fruman 2004; Okkenhaug and Fruman 2010). T helper cells (Th) are then required for B cell development and to raise a humoral immune response.

Mice lacking functional PI3Kγ (Patrucco et al. 2004; Hirsch et al. 2000; Li et al. 2000; Sasaki et al. 2000) and/or PI3Kδ (Okkenhaug et al. 2002; Clayton et al. 2002; Jou et al. 2002) are viable and fertile, and were extensively studied in inflammatory disease models. Mutant mice without the catalytic subunit of PI3Kδ (p110δ; Clayton et al. 2002; Jou et al. 2002), or mice expressing a catalytically inactive p110δ (with a D910A mutation; Okkenhaug et al. 2002) display impaired development of marginal zone B-cells and peritoneal B1-cells, and signals emerging from the B-cell receptor (BCR) are attenuated. In mice lacking the p110δ protein completely, T cell maturation in the thymus was normal, while mice with the catalytically inactive p110δ produced more naïve peripheral T-cells. Later it was reported that mice with inactive PI3Kδ have elevated counts of Foxp3$^+$ regulatory T-cells (Treg) in the thymus, but Foxp3$^+$ cell numbers were reduced in peripheral organs, likely modulated by a PI3K-FOXO1/3a connection. Interestingly, in spite of impaired BCR signaling and reduced IgM and IgG responses, mice with inactive p110δ increase IgE production, and have a tendency to develop autoimmunity (Oak et al. 2006; Ji et al. 2007; Patton et al. 2006; Merkenschlager and von Boehmer 2010). As Th2 responses are also reduced in mice without functional p110δ, elevated IgE levels are best explained by the negative regulatory effect of p110δ on the IgE class switch (Zhang et al. 2008; Omori et al. 2006), or mechanisms of IgE production that do not require cognate T cell help (McCoy et al. 2006).

In T cells without functional PI3Kγ, initial TCR signaling is not affected directly (Sasaki et al. 2000), but secondary signals, and the accumulation of 3-phosphorylated phosphoinositides at the immune synapse is impaired (Alcazar et al. 2007). As a result, T-cells of mice without functional PI3Kγ display significant developmental and signaling defects, yielding impaired thymocyte selection, reduced numbers of double-positive (CD4$^+$ CD8$^+$) cells and an altered CD4 to CD8 ratio (Rodriguez-Borlado et al. 2003), as shortened CD4$^+$ memory T-cell survival (Barber et al. 2006). When PI3Kγ and PI3Kδ were genetically targeted, double mutant mice displayed severe defects in thymocyte development, loss of thymus structure reducing the number of CD4$^+$/CD8$^+$ double positive cells, and a dramatic shift towards Th2 immune responses resulting in highly elevated IgE levels (Ji et al. 2007; Webb et al. 2005).

PI3Kγ has been shown to be instrumental in migration of neutrophils, macrophages (Hirsch et al. 2000; Li et al. 2000; Sasaki et al. 2000; Wymann et al. 2000; Jones et al. 2003) and dendritic cells (Del Prete et al. 2004) towards chemokines and other GPCR ligands. PI3Kγ-derived PtdIns(3,4,5)P_3 was thus dubbed "the compass of leukocytes" (Rickert et al. 2000; Servant et al. 2000; Wang et al. 2002). Detailed investigations of neutrophil migratory processes confirmed that PI3Kγ is key for migration, but rather for cell polarization and "stop and go" decisions than for path finding (Ferguson et al. 2007). That a PtdIns(3,4,5)P_3 gradient is required in the process was nicely demonstrated in neutrophils from mice lacking the lipid 5′-phosphatase SHIP (Nishio et al. 2007), or from mice expressing constitutively membrane targeted PI3Kγ (Costa et al. 2007). While loss of PTEN did not affect neutrophil migration (Nishio et al. 2007), PTEN was essential for the formation of gradients in PtdIns(3,4,5)P_3 in the amoebic form of *Dictyostelium*discoideum (Chen et al. 2007; Funamoto et al. 2002). In T cells is has been shown that the guanine nucleotide exchange factor DOCK2 (dedicator of cytokinesis 2) mediates the activation of Rac largely independent of PI3K (Nombela-Arrieta et al. 2004, 2007). In neutrophils DOCK2 translocation to the leading edge is PtdIns(3,4,5)P_3-dependent, but was recently suggested to be supported by phosphatidic acid generated by ligand-stimulated phospholipase D activity (Nishikimi et al. 2009; Kunisaki et al. 2006).

Simplified, one could deduce from the above that resting cells of the myeloid linage depend on PI3K for migration and adhesion, while manipulations of PI3Ks in lymphocytes modulate linage development. Loss of a single class PI3K modulates the output of both the innate and the acquired immune system, but did not lead to severe immune deficiencies in mice. PI3Kγ and PI3Kδ are thus considered as valuable targets in inflammatory, allergic and autoimmune disease.

5.2.2 PI3K in Inflammation and Allergy

Tissue resident cells including macrophages and mast cells initiate inflammation and allergy when triggered by pathogens or antigens. Cytokines and chemokines released by these cells activate endothelia in close-by blood vessels to recruit neutrophils, monocytes or lymphocytes to the site of inflammation. PI3Kγ is required for chemokine-dependent recruitment of neutrophils to tissues, and macrophages require PI3Kγ to fight peritoneal infections (Hirsch et al. 2000). In allergy, tissue mast cell concentrations are elevated, and migration of mast cells also depends on PI3Kγ (Kitaura et al. 2005). Invading pathogens are opsonized by triggering the complement cascade, and are decorated with specific antibodies, or interact with Toll-like receptors (TLRs). All these actions can lead to PI3K activation and the promotion of cytokine production (Ghigo et al. 2010; Wymann et al. 2000, Fig. 5.3).

5.2.3 Atherosclerosis and Cardiovascular Disease

Atherosclerosis is initiated by the excessive uptake of oxidized low-density lipoproteins by (LDL) by macrophages. These macrophages accumulate in the intima of blood vessels, and chronic lipid uptake turns them into foam cells. The disintegration of foam cells leads to the formation of fatty streaks and atherosclerotic plaques. Finally, rupture and repair of atherosclerotic plaques leads to stenosis and eventually to the closure of arteries by thrombosis, culminating in myocardial infarction and stroke (Lusis 2000; Glass and Witztum 2001). Atherosclerosis is an inflammatory disease (Hansson and Hermansson 2011), and mouse genetic data demonstrates that chemokine receptor signaling selectively recruits monocytes (Boring et al. 1998) and T cells (Heller et al. 2006; Braunersreuther et al. 2007a, 2007b; Damas et al. 2007) during the onset of atherosclerosis. These findings have initiated the search for novel drugs to treat atherosclerosis beyond statins (Opar 2007).

Oxidized LDL activates PI3K in macrophages, and the release of granulocyte/macrophage colony-stimulating factor (GM-CSF) promotes the on-site proliferation of macrophages (Biwa et al. 2000a, 2000b). Interestingly, oxidized LDL did not activate PI3K in macrophages derived from PI3Kγ null mice (Chang et al. 2007). Mice devoid of apo-lipoprotein E (apoE; Zhang et al. 1992; Plump et al. 1992; Nakashima et al. 1994) or the LDL receptor (LDLR; Ishibashi et al. 1993) rapidly develop atherosclerotic plaques, which were significantly reduced in mice without PI3Kγ (Fougerat et al. 2008; Chang et al. 2007). The attenuation of plaque formation observed in PI3Kγ null mice was also reproduced using AS605240 (Fougerat et al. 2008; Chang et al. 2007), a selective PI3Kγ inhibitor (Camps et al. 2005). It has been reported that the lack of PI3Kγ attenuates E-selectin-dependent neutrophil adhesion to endothelial cells, and a role of PI3Kγ in endothelial cells was proposed to control cell recruitment significantly (Puri et al. 2005). Bone marrow transplantation experiments could, however, demonstrate that the main role of PI3Kγ in atherosclerosis is associated with the hematopoietic compartment (Fougerat et al. 2008, 2009). Macrophages are without doubt the executers of atherosclerosis, but a role for type 1 helper T cells (Th1) in the acceleration of atherosclerotic lesions has been proposed (Song et al. 2001), while regulatory T cells counteract the formation of plaques (Ait-Oufella et al. 2006; Nilsson et al. 2009).

The formation of atherosclerotic lesions remains non-symptomatic for a long time, and only stenosis and complete occlusion of critical blood vessels is noticed. The frequency of plaque rupture is enhanced in patients with hypertension due to increase shear forces. Interestingly, PI3Kγ null mice are protected against angiotensin II-induced hypertension (Vecchione et al. 2005). Moreover, loss of PI3Kγ also protected mice from ADP-induced thromboembolic vascular occlusion, which is initiated by micro-coagulation of blood platelets (Hirsch et al. 2001). Pharmacologic experiments using p110β selective compounds (TGX-221) demonstrated subsequently the importance of PI3Kβ in platelet-mediated thrombosis triggered by ADP, collagen and integrin-dependent stimulation (Jackson et al. 2005). These data were confirmed in mice expressing a catalytically inactive form of the p110β catalytic subunit of PI3Kβ (Canobbio et al. 2009). PI3Kβ downstream of alpha(IIb)beta3 integrins, and PI3Kγ and PI3Kβ downstream of the ADP-binding P2Y12 receptor thus cooperate to maintain stable platelet aggregates (Cosemans et al. 2006).

PI3Ks also promote cardiac hypertrophy, which is a consequence of chronic hypertension in humans: when specifically expressed in the heart, constitutively activated PI3Kα caused an increase in heart and cardiomyocyte size. In these mutant mice, cardiac function and architecture as determined by echocardiography was not affected (Shioi et al. 2000). A more dramatic increase in heart size could even be achieved by the targeted expression of a permanently activated form of PKB/Akt. Interestingly, this phenotype could be counteracted by treatment with rapamycin, demonstrating that the PI3K/PKB/TOR pathway is an important regulator of cell and organ size (Shioi et al. 2002). Similar results were obtained by the inactivation of PTEN in cardiomyocytes: heart size was increased due to an increase cell size of cardiomyocytes, but additionally a reduced cardiac contractility was observed in hearts lacking PTEN. Surprisingly it was found that the combination of targeting PTEN and elimination of PI3Kγ protein in cardiomyocytes reconstituted contractility without reverting heart size (Crackower et al. 2002). Class I PI3Ks, and likely PI3Kα (Luo et al. 2005), seem therefore to control cardiomyocyte size, while PI3Kγ is linked to the regulation of contractile force. Unchallenged PI3Kγ knock-out mice do not display a cardiovascular phenotype, but when subjected to pressure overload by transverse aortic constriction (TAC), they rapidly suffered from fatal heart failure. As mice expressing a catalytically inactive PI3Kγ protein had no signs of apoptosis and fibrosis in the heart, it became clear that PI3Kγ had a function in the heart that is not linked to its lipid kinase activity. Finally it was determined that PI3Kγ functions as a scaffold for cAMPsignaling, as it interacts with phosphodiesterase 3B. If PI3Kγ protein is absent in the heart, cAMP rises and contractility under stress increases (Patrucco et al. 2004). Recently, it was found that cAMP regulation works in both directions, because cAMP-dependent kinase (PKA) can phosphorylate and inactivate PI3Kγ in the heart (Perino et al. 2011).

In summary, inhibition of PI3K in hypertension, hypertrophy and atherosclerosis appears to be beneficial. Recent pharmacological studies indeed demonstrated that the inhibition of PI3Kγ and PI3Kδ reduced infarct size caused by inflammatory processes initiated after ischemia/reperfusion damage. Mice treated with the dual-specific PI3Kγ/δ inhibitor TG100-115 displayed reduced inflammation and edema at infarct sites. Tissue repair processes and endothelial cell mitogenesis, which are required after myocardial infarction, were not affected by the compounds (Doukas et al. 2006, 2007). Clinical trials in patients suffering from acute myocardial infarction were concluded (Table 5.1; refer to Vol. 1, Chap. 6).

5.2.4 Allergic and Hypersensitivity Responses

Mast cell are primary effector cells in inflammation, allergic disease such as, asthma, rhinitis and atopic dermatitis. Mast cells bind IgE with a high affinity receptor (FcεRI). Crosslinking of FcεRI receptors tips the balance towards mast activation, as src-like protein tyrosine kinases (e.g. Lyn) phosphorylate immunoreceptor tyrosine-based motifs (ITAMs) on the FcεRI receptor's β and γ chains. Subsequently, the SH2-containing Syk protein tyrosine is recruited and promotes the phosphorylation of multiple tyrosines on membrane-anchored adapters such as LAT, NTAL/LAB (Rivera 2005), and Grb2-associated binder 2 (Gab2; Gu et al. 2001). Class IA PI3Ks are then translocated and activated to trigger Bruton's tyrosine kinase (Btk) and phospholipase Cγ (PLCγ) by providing PtdIns(3,4,5)P_3 as a docking site for the PH domains of Btk and PLCγ. Activation of PLCγ leads eventually to the release of Ca^{2+} from intracellular stores, which trigger store operated Ca^{2+} channels to finally cause the release of histamine-containing granules and the production of inflammatory mediators (Kraft and Kinet 2007; Kim et al. 2008).

Surprisingly, it has been found that mice lacking functional PI3Kγ (knock-outs or catalytically inactive K833R mutants) are protected in models of passive systemic or cutaneous anaphylaxis. This finding was corroborated in bone marrow-derived mast cells (BMMCs), which depend on PI3Kγ for a full-scale degranulation response when exposed to IgE/antigen complexes. It was then established that a release of adenosine triggered PI3Kγ activation, which synergizes with the protein tyrosine kinase cascade downstream of FcεRI receptors (Laffargue et al. 2002; Wymann et al. 2003a). Mice harboring a catalytically inactive PI3Kδ also displayed a partially attenuated response to IgE/antigen complexes, while the relay of stem cell factor signaling to PI3K was completely abrogated, suggesting that in mast cells PI3Kδ is the only class IA PI3K associating with the c-kit receptor (Ali et al. 2004, 2008). PI3Kγ and PI3Kδ are thus currently evaluated as therapeutic targets in allergic disease: although asthma models in the mouse have a somewhat limited predictive value for the clinical outcome in man (Stevenson and Birrell 2011), studies using the PI3Kδ-selective inhibitor IC87114 in ovalbumin challenged BALB/c mice attenuated a number of disease parameters like leukocyte recruitment, mucus secretion, and Th2-derived release of cytokines and IgE into lung cavities (Lee et al. 2006a, 2006b). Paradoxically, other studies showed that unspecific and ovalbumin-specific IgE levels increase due to PI3Kδ inhibition (Zhang et al. 2008; Omori et al. 2006). Dual inhibition of PI3Kγ and PI3Kγ was achieved using aerosols of TG100-115 in ovalbumin-challenged mice, and was efficiently reducing airway hyper-responsiveness (AHR) even in a semi-therapeutic setting where the drug was applied after the ovalbumin challenge (Doukas et al. 2009). Other studies pointed to a role of PI3Kγ in the chemokine-mediated and ovalbumin-induced leukocyte recruitment to the lung in response to ovalbumin sensitization (Thomas et al. 2005, 2009).

While patients with allergic asthma often respond to treatment with corticosteroids or $β_2$-adrenergic agonists acting as bronchodilators, chronic obstructive pulmonary disease (COPD) patients suffer from a progressive disease refractive to current treatment (Barnes 2008; Hansel and Barnes 2009). COPD is induced by smoking in >90% of all cases, and exposure to cigarette smoke or LPS are used in animal models to mimic the human disease driven by type 1 helper T cells (Th1). In a smoke exposure model, TG100-115 successfully attenuated inflammatory readouts and reversed the steroid resistance observed in these settings (Doukas et al. 2009). As similar results were obtained by genetic inactivation of PI3Kδ (Marwick et al. 2009), and using PI3Kδ-specific inhibitors (IC87114; To et al. 2010), resolution of COPD parameters might have been mediated by the inhibition of PI3Kδ even in the case of TG100-115. Because ARH and COPD models are currently discussed controversially (Stevenson and Birrell 2011), and reference molecules for COPD models are missing, definitive conclusions concerning the best PI3K isoform profiles require further studies.

5.2.5 Autoimmune Diseases: Rheumatoid Arthritis

Rheumatoid arthritis (RA) is a systemic autoimmune disease and affects about 1% of the world's population. RA has a gradual onset inflicting initially a limited number of synovial joints, where inflammation progresses to cause cartilage and bone erosion, culminating in joint destruction. RA can involve other tissues like skin, blood vessels, heart, lungs and muscles. The disease is initiated by CD4$^+$ memory T cells, which cross the synovial membrane. Subsequently, they release cytokines including IL-2 and interferon γ (IFN-γ), which in turn activate macrophages and fibroblasts, and trigger monocyte recruitment. A wave of pro-inflammatory cytokines such has TNF-α, IL-1 and IL-6 isthen released to set off chronic inflammation (Firestein 2003, 2006; Steiner 2007). In the final stage of the disease, T- and B-cells, dendritic cells, macrophages, mast cells, and hyperplastic synovial fibroblasts collaborate to maintain inflammation. A constant influx of high numbers of neutrophils into the joint endorses cartilage and bone destruction, and tissue repair and neovascularization in the synovial membrane promote the process. As mentioned above, PI3Kγ and PI3Kδ have non-redundant roles in T- and B-cell differentiation and function, and the lipid kinases are key to leukocyte and mast cell migration, and mast cell degranulation.

Mouse models for RA include active immunization models like collagen-induced arthritis (CIA, Stuart et al. 1984) and passive models utilizing auto-antibodies from immunized or auto-immune mice (anti-collagen II-IA: Terato et al. 1992; K/BxN serum model: Korganow et al. 1999; Schaller et al. 2001). CIA is initiated by the intradermal injection of type II collagen. Subsequently, features of the human disease like cell-infiltration into the synovial space, hyperplasia, pannus formation, and cartilage and bone erosion, can be observed. CIA needs functional T- and B-cells, whereas in the passive models full T- and B-cell function is dispensable. In contrary to mouse models, no specific auto-antigen has been identified in the human disease.

When PI3Kγ null mice were challenged in a passive (αCII-IA) RA model, mutant mice were protected from RA development, and showed minimal paw swelling, and bone and cartilage erosion. An orally available salt of the PI3Kγ-specific inhibitor AS605240 was effective in the CIA and αCII-IA model, and even had a therapeutic effect when added after the onset of the disease. Both, the genetic ablation of PI3Kγ and the pharmacological inhibition of the enzyme suppressed the recruitment of neutrophils to the joints, which is a hallmark of RA (Camps et al. 2005).

5.2.6 *Systemic Lupus Erythematosus*

Systemic lupus erythematosus (SLE) is a complex autoimmune disease correlating with polygenic genetic disposition (Harley et al. 2008; Fairhurst et al. 2006). The disease is initiated by autoreactive CD4$^+$ memory T-cells, which trigger polyclonal B-cell expansion, leading to hyper-gammaglobulimia. Anti-nuclear autoantibodies (ANAs) often precede the clinical manifestation of the disease by years. In the late stage of the disease, patients eventually develop glomerulonephritis due deposition of autoantibody complexes in the kidney, culminating in renal failure.

Inbred mice of the MRL strain, which are homozygous for the lymphoproliferation (lpr) mutation (the MRL-*lpr/lpr* model; Cohen and Eisenberg 1991; Singer et al.

1994) progress spontaneously towards a SLE-like autoimmune disease. The MRL-*lpr/lpr* mice show clinical features of the human disease like the generation of ANAs, increased numbers of autoreactive $CD4^+$ T-cells, and finally the accumulation of immunoglobulin complexes in kidney and salivary glands (Liu and Wakeland 2001).

That PI3K signaling could be relevant in the progress of SLE emerged when it was observed that the heterozygous deletion of PTEN, or the constitutive activation of class IA PI3K by the transgenic expression of a truncated regulatory subunit (Lck promoter—p65(PI3K) transgene), in T lymphocytes led to a SLE-related disease (Di Cristofano et al. 1999; Borlado et al. 2000). When PI3Kγ was deleted from Lck-p65 mice, these animals showed an attenuated form of SLE. Lymphoproliferation and T-cell infiltration was still imminent (Barber et al. 2006), but the survival of $CD4^+$ T-cells was impaired. This resulted in a reduced progress of nephritis and longer life span (Barber et al. 2005). The importance of PI3Kγ in the SLE progression was further underlined by the action of the PI3Kγ inhibitor AS605240 in MRL-*lpr/lpr*mice, where it reduced $CD4^+$ T-cells, autoantibody concentrations and kidney failure, thus increasing live span. In mice, PI3Kγ inhibition was better tolerated than dexamethasone used as a reference drug. It must be noted that glucocorticoids lead to a dramatic immunosuppression in rodents, which makes them very susceptible to infections (Chatham and Kimberly 2001). The comparison of dexamethasone and AS605240 indicated that PI3Kγ inhibitors might have a decent therapeutic window in SLE, without causing too severe side effects.

5.2.7 PI3Ks in Chronic Inflammation and Allergy—Preliminary Conclusions

Combining mechanistic, cellular and mouse model data, we have a good validation that PI3Ks are valuable drug targets in chronic inflammation, allergy and autoimmune disease. As mentioned above, PI3Kγ and PI3Kδ are required for B- and T-cell development and function, and PI3Kγ has a prominent role modulating chemotaxis and recruitment of myeloid cells. PI3Kβ and PI3Kγ are involved in platelet aggregation. A role for PI3Kγ has also been demonstrated in pancreatitis (Lupia et al. 2004). Other diseases like psoriasis (Lowes et al. 2007; Schon and Boehncke 2005), and multiple sclerosis (Hauser and Oksenberg 2006; Hemmer et al. 2002) rely on cellular networks that should also respond to PI3K inhibition. The effects of PI3Kγ on cardiovascular tone open avenues for preventive treatments in hypertension and cardiovascular disease (refer to Vol. 1, Chap. 6).

In spite of all these findings, there are not too many isoform-specific PI3K inhibitors available (see Table 5.1), and only one was transiently in clinical trials for anti-inflammatory actions (TG100-115). One reason for this might be that PI3K inhibitors in non-fatal disease have to meet higher safety standards, another that they have to be better and cheaper than existing medication, as they compete with steroids, non-steroidal anti-inflammatory drugs (NSAIDs), and recently developed biological targeted therapies (e.g. anti-TNF-α, IgE, IL neutralizing antibodies). Data of biotech companies has been presented in recent scientific meetings, demonstrating that there is a pipeline for isoform specific inhibitors.

5.3 The PI3K/mTOR Pathway in Cancer

Patients with metastatic solid tumors have often a very bad prognosis, and even under best possible standard care the survival after diagnosis is short. The development of imatinib/Gleevec and its success as BCR-abl kinase inhibitor for the treatment of chronic myelogenousleukemia raised hopes that targeted therapies with limited adverse effects could be achieved for other cancers. As shown in Fig. 5.3, many oncogenes activated or amplified in cancer feed into the PI3K pathway: ErbB2/Her2 is amplified in breast cancer, c-kit is mutated in gastrointestinal stromal tumors (GIST), VEGFR promotes angiogenesis in growing tumors, EGFR drives proliferation of non-small cell lung cancer of non-smokers, mutated Ras signals in lung cancer of smokers, and last but not least, antagonists of the PI3K pathway such as lipid phosphatases are frequently lost in cancer.

5.3.1 PtdIns(3,4,5)P_3 Rising—Loss of PTEN

The tumor suppressor PTEN counteracts and balances the action of PI3K by hydrolysis of PtdIns(3,4,5)P_3. In normal cells, PTEN levels and activity are tightly controlled by transcription factors, methylation, oxidation, phosphorylation, ubiquitination, micro RNAs and more (for reviews see (Leslie et al. 2008; Carracedo et al. 2011)). The importance of levelingPtdIns(3,4,5)P_3 was convincingly demonstrated in mice with a targeted PTEN locus: even *Pten* heterozygous mice developed multiple forms cancers, manifesting in the prostate, breast, uterus, and other organs (Di Cristofano et al. 1998; Suzuki et al. 1998; Podsypanina et al. 1999).

Many tumors attenuate expression of PTEN, which often occurs by methylation of the *PTEN* promoter, or by a process called "loss of heterozygosity". This involves the deletion of both PTEN alleles, and results in a complete loss of PTEN protein expression. Downregulation of PTEN by promoter methylation has been frequently detected in melanoma, prostate, breast, endometrial and colorectal cancer, as well as and leukemia (Khan et al. 2004; Goel et al. 2004; Stahl et al. 2004; Mirmohammadsadegh et al. 2006). Spontaneous mutations in *PTEN* have been identified in more than half of all melanoma, glioma, prostate, endometrial and ovarian cancers, while attenuation of PTEN is less frequent in breast cancer (Mirmohammadsadegh et al. 2006; Li et al. 1997; Cairns et al. 1997; Wu et al. 2003). Mutation of *PTEN* is a late step in tumor progression (discussed in more detail in (Wymann and Marone 2005)), and is usually detected in late stage or metastatic tumors (for reviews see (Vivanco and Sawyers 2002; Wymann and Marone 2005; Cully et al. 2006; Abraham 2004)). A reason for the late appearance of changes in PTEN in tumors is likely to be "oncogene-inducible senescence" (for a short review see (Braig and Schmitt 2006)). Oncogene-inducible senescence was observed before for Ras and BRaf, and has been described as a fail-safe mechanism preventing tumor growth after single oncogene mutations. For PTEN, oncogene-inducible senescence has been demonstrated in a conditional *PTEN*-deficient mouse model: deletion of both PTEN alleles in the prostate is surprisingly slow to generate prostate cancer. When p53 was targeted at the same time, however, aggressive and fatal prostate cancer developed rapidly (Chen et al. 2005), demonstrating that loss of PTEN is opposed by p53 tumor suppressor genes. Tumors thus profit from loss of PTEN only at a late stage and as "second hit" mutagenesis.

5.3.2 Mutations in PIK3CA (p110α)

The key role for PI3Ks in cancer progression was further underlined by frequently occurring mutations in the gene coding for the catalytic subunit of PI3Kα (PIK3CA) in human tumors. *PIK3CA* mutations cluster in two hotspots coding for the helical (see PI3Ka in Fig. 5.2; exon 9) and the catalytic domain (PI3Kc; exon 20; Samuels et al. 2004; Thomas et al. 2007; TCGA study 2008; Parsons et al. 2008; Stemke-Hale et al. 2008; for a review see (Bader et al. 2005)). When mutated, both sites yield a constitutive active enzyme with transforming capacity in fibroblasts. Cells expressing mutant p110α display also an increased invasive capacity. This implies that *PIK3CA* mutations promote tumor cell survival and metastasis (Kang et al. 2005; Ikenoue et al. 2005; Samuels et al. 2005). In human cancers, the helical domain residues Glu542 and Glu545 are usually mutated to Lysine, and the C-terminal kinase domain residue His1047 is converted to Arginine. A rational for the increased activity of mutant p110α was provided by crystallographic data obtained from a complex of p85 fragments bound to the N-terminus of p110α (Miled et al. 2007; Huang et al. 2007). As described above, the p85 regulatory subunit interacts tightly with p110α, stabilizes the p110α protein and inhibits PI3K activity at the same time. In normal wild type p110α, the N-terminal SH2 domain of p85 (nSH2) mediates its inhibition via contacts within the helical domain. When negative charges in the helical domain are inverted by the mutation to Lys (Glu545Lys), charge-charge interactions are disrupted and full PI3K activity is released. The activation of p110α by C-terminal mutations (His1047Arg) can be best understood in the context of the concept of the "regulatory square" recently proposed for the regulation of p110β activity (Zhang et al. 2011; Vogt 2011). Of a set of three α-helices forming a square around the catalytic groove of p110s, the "elbow" at the start of the last α-helix is in contact with the C-terminal SH2 domain of p85 (cSH2). In the presence of the cSH2, the C-terminal helix in p110 seems to be clamped into a conformation that constrains residues in the catalytic loop into an inactive conformation. In wild type p110β, a Leucine (Leu1043 in p110β) at the elbow position allows for high basal activity, and p110β activity could indeed be restrained when Leu1043 was exchanged for a Histidine (Zhang et al. 2011). The His1047 in the elbow region seems thus to restrict p110α activity inherently, while p110β and p110δ are constrained by the cSH2 of p85 (for reviews see (Backer 2010; Vogt 2011; Vadas et al. 2011)).

PIK3CA is amplified in various tumors, frequently in ovarian, cervix and lung cancers (Shayesteh et al. 1999; Zhang et al. 2002; Racz et al. 1999). Activating *PIK3CA* mutations were detected in solid tumors in breast, endometrial, colorectal, upper digestive tract, gastric, pancreas, brain, lung and hepatocellular carcinomas (Bader et al. 2005; Samuels and Ericson 2006; for a list of mutation frequencies see (Liu et al. 2009)).

5.3.3 Mutations in p85 Regulating Class IA PI3Ks

Truncated forms of p85 regulatory subunits have been shown earlier to constitutively activate class IA PI3Ks. A truncation mutant containing amino acids 1-571 of p85α fused to a fragment of the eph tyrosine kinase family (p65-PI3K) has been isolated from a mouse lymphoma model (Borlado et al. 2000; Chan et al. 2002; Jimenez et al. 1998; for mechanistic investigations see (Shekar et al. 2005, #40429; Backer 2010, #62337; Huang et al. 2008, #56273; Huang et al. 2007, #46673; Miled et al. 2007, #38727)), but was not identified in human cancers. A somewhat longer truncation mutant was isolated from a human lymphoma cell line (p76-PI3K; Jucker et al. 2002), and an infrequent incidence of p85α (*PIK3R1*) mutations were reported in ovarian and colon tumors (Philp et al. 2001), and breast cancer (Wood et al. 2007). The highest frequency of p85α mutations was detected in glioblastoma (TCGA study 2008; Parsons et al. 2008). Until recently, little was known concerning their capacity to activate p110α, and their relevance in tumor progression. A meta-analysis of p85α mutations was performed in (Jaiswal et al. 2009), where a number of mutants modulating contacts between the p85α-iSH2 and the p110 C2 domain were investigated. In particular Asn564Asp and Asp560Tyr mutants of p85α were effectively promoting fatal tumorigenesis in a BaF3 cell mouse model (Jaiswal et al. 2009). *PI3KR1* was also recently identified as a colon cancer oncogene in a transposon insertion screen (Starr et al. 2009). The attenuation of lipid kinase activity by p85-p110 interactions is therefore crucial in cellular hemostasis, and the discussed structural studies are key for the understanding how mutations in the class IA heterodimer releases constraints on lipid kinase in disease.

5.3.4 Downstream of PI3K: TOR

Cellular growth is an important parameter in tumor progression, and is regulated by the availability of energy and nutrients. A central hub to integrate nutrient, energy, but also hormonal inputs are the target of rapamycin (TOR) complexes. There are two target of rapamycin (TOR) complexes: (i) TORC1, where the TOR protein kinase is associated with Raptor and (ii) TORC2 is bound to Rictor. TORC1 is activated by downstream of PKB/Akt (schematically shown in Fig. 5.4, for a in depth review on TOR signaling see (Zoncu et al. 2011)) and the Ras/MAPK cascades (not discussed here), which phosphorylate and inactivate TSC2. An energy sensing pathway acting via AMPK (AMP-dependent protein kinase) can shut off TORC1 when AMP accumulates. Finally, amino acids regulate TORC1 association with RAG proteins, re-localizing the complex to late endosomes (Sancak et al. 2010; Sancak and Sabatini 2009). Activation of TORC1 initiates cap-dependent translation via phosphorylation of 4E-BP1 (eIF4E-binding protein 1), and phosphorylation of p70^{S6K} promotes translation of ribosomal proteins (Dufner and Thomas 1999) and ribosome biogenesis (Wullschleger et al. 2006).

Mutations or loss of heterozygosity in TSC components gives rise to the initially benign, autosomal dominant TSC syndrome (van Slegtenhorst et al. 1997), manifested by hamartomas in a variety of organs (Cheadle et al. 2000), and an elevated risk to develop renal carcinoma. The serine-threonine protein kinase LKB1 upstream of AMPK normally balances TORC1 activity (Woods et al. 2003). Loss of function of LKB1 causes the Peutz-Jeghers syndrome, a familial colorectal polyp disorder. Peutz-Jeghers syndrome patients have a high risk for cancers in various tissues (Boudeau et al. 2003). Some signaling molecules downstream of TOR were also used as diagnostic markers: elevation of eIF4E correlates with a bad prognosis in a variety of cancers (Bjornsti and Houghton 2004).

5.4 Pharmacological Targeting of PI3K/TOR Signaling

Drugs to inhibit signals emerging from mutated or up-regulated growth factor receptors are already on the market or in clinical trails, and include neutralizing antibodies and protein tyrosine kinase receptor inhibitors. Knowledge from targeted therapies interfering with for example EGFR, ErbB2/Her2, and VEGFR, have validated two strategies to attack tumor cells: a first one aiming to reverse tumor autonomous signaling, and a second one targeting tumor-induced angiogenesis. As it turns out, targeting PI3K and TOR contributes to both approaches.

5.4.1 Targeting TOR—Rapamycin and Derivatives

Besides their effects on immune cells, Rapamycin derivatives (rapalogs; see Table 5.1) act as anti-angiogenic drugs. Rapamycin considerably reduces the production of VEGF, and more importantly intercepts the action of this growth factor on vascular endothelia (Guba et al. 2002). The latter has been shown to be mediated by the inhibition of hypoxia-inducible factor 1α (HIF1α) expression (Lane et al. 2009). RAD001, CCI-779, AP23573 and other rapalogs are highly specific, allosteric inhibitors of TORC1. Rapalogs bind to FK506-binding protein 12 (FKBP12) with a K_d in the sub-nanomolar range (Banaszynski et al. 2005), and this rapalog/FKBP12 complex then tightly interacts with the FRB domain of the TOR kinase. The activity of TORC2 is not affected by rapalogs.

Presently, there are >700 clinical trials registered exploring the actions of these compounds in proliferative disease, mostly designed as studies combining rapalogs and best standard of care, chemotherapy, or other targeted therapies. RAD001/Everolimus and CCI-779/Temsirolimus made it recently to market, and are approved for the treatment of organ rejection and renal cell carcinoma (Motzer et al. 2008; Atkins et al. 2009; Dancey 2010; for more references see Table 5.1). The endpoint of the clinical studies was progression-free survival (PFS). Renal cell carcinoma patients receiving Everolimus had a median PFS of 4.9 month, while the PFS of patients receiving placebo was 1.9 month (Motzer et al. 2008; Atkins et al. 2009; FDA documentation at http://www.accessdata.fda.gov/drugsatfda_docs/label/2009/022334lbl.pdf). Encouraging results with rapalogs were also obtained in mantle cell lymphoma (Johnston et al. 2010b; see also the approval of Temsirolimus (Dancey 2010)), soft tissue and bone sarcoma (Blay 2011), and endometrial cancer (for reviews see (Hay 2005; Faivre et al. 2006; Guertin and Sabatini 2007)).

In a subset of tumor-derived cell lines and patient biopsies elevated levels of PKB/Akt Ser473 phosphorylation were detected after the inhibition of TORC1 using rapalogs (Sun et al. 2005; O'Reilly et al. 2006; O'Donnell et al. 2008; Tabernero et al. 2008). It has been established, that $p70^{S6K}$, activated downstream of TORC1, phosphorylates insulin receptor substrate-1 (IRS1), and blocks its interaction with the insulin receptor. This leads to reduced phosphorylation of IRS on YxxM motifs, and thus attenuates recruitment of p85/p110 complexes and class IA PI3K (Shaw and Cantley 2006; Harrington et al. 2005; Manning 2004). When the activity of $p70^{S6K}$ is blocked due to TORC1 inhibition, PKB/Akt is hyper-phosphorylated and eventually promotes cell survival and metastasis (see Fig. 5.4). Others have reported that the targeting of TORC1 can trigger a PI3K-dependent feedback loop activating MAPK in human cancer (Carracedo et al. 2008). All this raised concerns that feedback activation of these stimulatory pathways would diminish the success of rapalog-based therapies (Hay 2005; Shaw and Cantley 2006; Rosen and She 2006). In contrast, it has been recently demonstrated that high basal levels of PKB/Akt Ser473 phosphorylation correlate with a sensitivity to RAD001, and that a subsequent RAD001-induced increase in phospho-PKB/Akt does not correlate with cell viability after rapalogs exposure (Breuleux et al. 2009). It has already been demonstrated earlier, that cells expressing constitutively activated PKB/Akt (myr-PKB/Akt; *N*-terminally myristoylated) were sensitive to CCI-779 (Neshat et al. 2001). Similarly, it was found that transgenic expression of myr-PKB/Akt in endothelial cells induced a pathological, tumor-like form of angiogenesis in non-tumoral tissues, which could be reversed by rapamycin (Phung et al. 2006). Early notions that loss of PTEN was an indicator for sensitivity to rapalogs (Neshat et al. 2001), have been disputed recently (Yang et al. 2008; for in-depth reviews see (Faivre et al. 2006; Hay 2005; Dancey 2010)). In the future it will be interesting to compare the efficiency of rapalogs with the performance of molecules targeting PI3K and mTOR kinase activities in clinical settings.

5.4.2 Targeting PI3K and mTOR Kinase

Although wortmannin (IC_{50} = low nM, Arcaro and Wymann 1993; Yano et al. 1993; Wymann and Arcaro 1994) and LY294002 (IC_{50} = high µM, Vlahos et al. 1994) inhibit a broad range of PI3K and PIKK family enzymes, the two compounds were instrumental to dissect PI3K signaling and to initiate PI3K drug development (Marone et al. 2008; Wymann and Schneiter 2008). The development of drug-like PI3K inhibitors was complicated by the relative difficulty to set up relevant *in vitro* and cell-based high throughput assays. Phosphoinositides are (still) complicated substrates, and variations of assay conditions can influence PI3K activity considerably. Moreover, cellular PI3K activity is mostly detected indirectly by phosphorylated PKB/Akt, where the readout can be convoluted by feedback mechanisms. The direct detection of cellular PtdIns(3,4,5)P_3 is cumbersome (Dove and Michell 2009), and only recently non-radioactive, mass-spectroscopy-based methods became available to determine cellular PtdIns(3,4,5)P_3 (Clark et al. 2011; Kiefer et al. 2010; Pettitt et al. 2006; for a commentary see (Wymann and Wenk 2011)).

Early efforts to produce drug-like molecules from wortmannin have been initiated even before PI3K was identified as the inhibitor's target. Wander AG in Bern, Switzerland, was the first pharmaceutical company to unknowingly develop "PI3K inhibitors" as anti-inflammatory compounds (Baggiolini et al. 1987). Trials to separate toxicity and pharmacological action of wortmannin-derivatives were not successful before the target enzyme was identified (Arcaro and Wymann 1993; Yano et al. 1993; Wymann and Arcaro 1994; Thelen et al. 1994), but were taken up by others later: Ihle and colleagues (Ihle et al. 2004) modified the furan ring of wortmannin to slow down covalent reactions of wortmannin-derivatives (Wymann et al. 1996). The result of this work led to the development of the wortmannin-derivative PX-866, which displays reduced liver toxicity as compared to wortmannin, and is currently in clinical trials in solid tumors (see Table 5.1, Ihle et al. 2004; Williams et al. 2006).

Semaphore Pharmaceuticals modified LY294002, and linked it to a RGD peptide to yield SF1126. SF1126 displays an increased solubility and targets the LY294002-derivative to integrins on cancers cells (Garlich et al. 2008; Ozbay et al. 2010).

Recently, the design of small molecules targeting PI3K has been facilitated by the availability of an extensive collection of inhibitor/PI3K structures. The release of the first class I PI3K structure—the catalytic PI3Kγ subunit p110γ (Walker et al. 1999)—was soon followed by p110γ bound to ATP and a variety of PI3K inhibitors such as wortmannin, LY294002, and the less specific kinase inhibitors staurosporine and quercetin (Walker et al. 2000). The elucidation of the PI3Kγ structure bound to Ras provided further insight into the regulation of the lipid kinase, and its putative orientation in respect to the plasma membrane (Pacold et al. 2000). The elucidation of structures of partial p110α bound to p85 fragments (Miled et al. 2007; Huang et al. 2007) and p110β (Zhang et al. 2011) provided insight into the regulation of class IA PI3K activities (see above, for a review see (Vadas et al. 2011)). Finally, the resolution of crystal structures for p110δ and bound inhibitors clarified the dynamics of the various p110 isoform structures, and explained specific structural dynamics opening a specificity pocket in proximity to Met752 of p110δ, which is relevant for p110δ inhibitor isoform specificity (e.g. for the p110δ inhibitor IC87114 (Knight et al. 2006; Berndt et al. 2010)).

In the past years, PI3K inhibitors have been refined, and several molecules targeting all class I PI3K isoforms (pan-PI3K inhibitors) have entered clinical testing (see Table 5.1 for molecules and references). There is a plethora of data available documenting the preclinical action and efficacy of PI3K inhibitors, which was established in xenograft and in syngeneic mouse models (for reviews see (Marone et al. 2008; Liu et al. 2009; Engelman 2009)). In these models it was clearly documented that PI3K inhibition acts usually cytostatic, arrests the cell cycle of tumor cells in G1 and only exceptionally triggers apoptosis (in vitro). In the in vivo setting, pan-PI3K inhibition displays a strong anti-angiogenic effect, which results in PI3K-induced tumor cell death and impressive reduction in tumor size in mouse models. Importantly, the effects of pharmacological targeting of PI3Ks was matched in genetic models. In this respect, it was demonstrated that PI3Kα plays not only an important role in driving tumor cell growth (see mutations in p110α, above), but takes a central role in angiogenesis (Graupera et al. 2008). Interestingly, the ablation of PI3Kβ activity attenuated tumor growth in the mammary gland (Ciraolo et al. 2008) and prostate (Jia et al. 2008; in mutant mice. When prostate cancer tumor formation was induced by the loss of PTEN, is was only the inactivation of PI3Kβ that was efficiently preventing tumor growth, while the inactivation of PI3Kα remained without significant effect (Jia et al. 2008). These results were confirmed using inducible shRNA vectors targeting specific PI3K isoforms (Wee et al. 2008).

Although class I PI3Ks and their lipid product PtdIns(3,4,5)P_3 are central to metabolic control (Wymann et al. 2003b), PI3K inhibitors did not display excessive toxicity in mouse models. As constitutive, genetic inactivation of PI3Kα causes embryonic lethality (Graupera et al. 2008; Bi et al. 1999), and genetic inactivation of PI3Kβ effects male fertility (Ciraolo et al. 2010) and triggers late stage insulin resistance (Ciraolo et al. 2008; Jia et al. 2008), but pharmacological pan-PI3K inhibition only mildly lowers blood glucose levels, it is tempting to speculate that an intermediate restoration of PI3K signaling in non-tumor tissue due to a short half-life of PI3K inhibitors reduces toxicity.

Many of the early PI3K inhibitors and clinical candidates act on PI3K and TOR in parallel. This dual mode of action was predicted for wortmannin early on (Wymann et al. 1996), and is followed by compounds like PI-103 (Fan et al. 2006), BEZ235 (Maira et al. 2008; Marone et al. 2009), BGT226, PX-866 and other molecules listed in Table 5.1. As for the pan-PI3K inhibitors above, dual-PI3K/mTOR kinase inhibitors produce impressive tumor responses in vivo (Marone et al. 2009; Workman et al. 2010; Ihle and Powis 2010; Falasca 2010; Roock et al. 2011; for reviews see (Marone et al. 2008; Liu et al. 2009; Engelman 2009; Wong et al. 2010)). In some cases, strong, prolonged mTOR kinase inhibition induces a dose-dependent hyperphosphorylation of PKB/Akt (as observed for PI-103 in melanoma (Marone et al. 2009)). Like for the TORC1-mediated feedback loop discussed above for rapamycin, it might be crucial to better elucidate feedbacks in the PI3K/mTOR pathway to optimize the action of dual-PI3K/mTOR inhibitors.

5.4.3 Plasticity of Tumorsignaling Pathways—Resistance

Cancer progression is a complex process and involves signaling pathways distinct from PI3K. Feedback loops and cross-talk between signaling pathways can provide escape routes for cancer cells and lead to adaptive resistance. A better knowledge of the plasticity and dynamics of signaling pathways in a given patient's tumor maximizes the chances of successful targeted therapies. Imatinib (Gleevec) inactivating the constitutively activated Bcr-Abl kinase in chronic myelogenousleukemia (CML) patients is initially very efficient, because Bcr-Abl is initially the exclusive driver of CML cell proliferation. Only when Bcr-Abl is further mutated to prevent imatinib binding, or when leukemia cells acquire further oncogenic mutations, resistance to imatinib occurs.

PI3K integrates plenty of input signals from upstream receptors (Fig. 5.3) that are currently targeted with neutralizing anti-bodies or protein tyrosine kinase inhibitors. As such, breast cancer patients with tumors depending on HER2/ErbB2 are treated with trastuzumab (Herceptin), while non-small cell lung cancer patients with EGF receptor amplifications or mutations are treated with gefitinib (Iressa) or erlotinib (Tarceva). In these settings, PI3K inhibition is expected to be beneficial. Cells with a loss of PTEN or activating mutations in p110α have been shown to be sensitive to PI3K inhibition. When Ras is mutated in a tumor, such as lung cancer, the MAPK pathway and PI3K signaling are activated in parallel. Here, mouse models have demonstrated that PI3K inhibitors meet resistance, but are efficient when combined with Raf or MEK inhibitors (Wee et al. 2009; Grant 2008; Engelman et al. 2008; Downward 2008). That tumor cells develop resistance to PI3K inhibitors by the mutation of gatekeeper residues is presently considered to be unlikely, as mutational screens to generate inhibitor resistant PI3K did not produce inhibitor-resistant enzyme with relevant activity (Zunder et al. 2008).

5.5 Closing Remarks

Looking back two decades when PI3K, wortmannin, TOR, and PKB/Akt (at that time called RAC1) entered the picture, we came a long way: a plethora of input signals for PI3Ks have been identified, and the map downstream of PI3K is well filled and connected to important hubs signaling through PKB/Akt and TOR. In some instances, the available literature focuses still too much on "topical" molecules, and we thus lack a deeper understanding if specific isoforms or relatives of lipid and protein kinases and phosphatases have non-redundant physiologic functions. The same is true for the PI3K regulatory and adaptor subunits and their splice variants, where we lack precise mechanisms how they control PI3K isoform-specific signaling in time and space. The feedback of TOR and S6K attenuating the coupling of PI3K to the activation of the insulin receptor illustrates that the whole PI3K-PKB-TOR pathway is regulated in a highly complex manner. Present therapeutic strategies using ATP-binding site inhibitors to block PI3K and PIKK activities does not quite match the complexity of the signaling network. Allosteric inhibitors and compounds targeting specific signaling complexes could provide more specific tools. Future approaches will also require solid quantification and the establishment of phosphoinositide fluxes.

Acknowledgements I apologize for not citing numerous excellent original articles due to space restrictions.

References

Abraham RT (2004) PI 3-kinase related kinases: 'big' players in stress-induced signaling pathways. DNA Repair (Amst) 3:883–887

Ait-Oufella H, Salomon BL, Potteaux S, Robertson AK, Gourdy P, Zoll J, Merval R, Esposito B, Cohen JL, Fisson S, Flavell RA, Hansson GK, Klatzmann D, Tedgui A, Mallat Z (2006) Natural regulatory T cells control the development of atherosclerosis in mice. Nat Med 12:178–180

Albert S, Serova M, Dreyer C, Sablin MP, Faivre S, Raymond E (2010) New inhibitors of the mammalian target of rapamycin signaling pathway for cancer. Expert Opin Investig Drugs 19:919–930

Alcazar I, Marques M, Kumar A, Hirsch E, Wymann M, Carrera AC, Barber DF (2007) Phosphoinositide 3-kinase gamma participates in T cell receptor-induced T cell activation. J Exp Med 204:2977–2987

Alessi DR, James SR, Downes CP, Holmes AB, Gaffney PR, Reese CB, Cohen P (1997) Characterization of a 3-phosphoinositide-dependent protein kinase which phosphorylates and activates protein kinase Balpha. Curr Biol 7:261–269

Ali K, Bilancio A, Thomas M, Pearce W, Gilfillan AM, Tkaczyk C, Kuehn N, Gray A, Giddings J, Peskett E, Fox R, Bruce I, Walker C, Sawyer C, Okkenhaug K, Finan P, Vanhaesebroeck B (2004) Essential role for the p110delta phosphoinositide 3-kinase in the allergic response. Nature 431:1007–1011

Ali K, Camps M, Pearce WP, Ji H, Ruckle T, Kuehn N, Pasquali C, Chabert C, Rommel C, Vanhaesebroeck B (2008) Isoform-specific functions of phosphoinositide 3-kinases: p110 delta but not p110 gamma promotes optimal allergic responses in vivo. J Immunol 180:2538–2544

Alloatti G, Marcantoni A, Levi R, Gallo MP, Del Sorbo L, Patrucco E, Barberis L, Malan D, Azzolino O, Wymann M, Hirsch E, Montrucchio G (2005) Phosphoinositide 3-kinase gamma controls autonomic regulation of the mouse heart through Gi-independent downregulation of cAMP level. FEBS Lett 579:133–140

Altomare DA, Zhang L, Deng J, Di Cristofano A, Klein-Szanto AJ, Kumar R, Testa JR (2010) GSK690693 delays tumor onset and progression in genetically defined mouse models expressing activated Akt. Clin Cancer Res 16:486–496

Alvarez B, Martinez AC, Burgering BM, Carrera AC (2001) Forkhead transcription factors contribute to execution of the mitotic programme in mammals. Nature 413:744–747

Amato RJ, Jac J, Giessinger S, Saxena S, Willis JP (2009) A phase 2 study with a daily regimen of the oral mTOR inhibitor RAD001 (everolimus) in patients with metastatic clear cell renal cell cancer. Cancer 115:2438–2446

Anderson RA, Boronenkov IV, Doughman SD, Kunz J, Loijens JC (1999) Phosphatidylinositol phosphate kinases, a multifaceted family of signaling enzymes. J Biol Chem 274:9907–9910

Arcaro A, Wymann MP (1993) Wortmannin is a potent phosphatidylinositol 3-kinase inhibitor: the role of phosphatidylinositol 3,4,5-trisphosphate in neutrophil responses. Biochem J 296:297–301

Arcaro A, Zvelebil MJ, Wallasch C, Ullrich A, Waterfield MD, Domin J (2000) Class II phosphoinositide 3-kinases are downstream targets of activated polypeptide growth factor receptors. Mol Cell Biol 20:3817–3830

Arcaro A, Khanzada UK, Vanhaesebroeck B, Tetley TD, Waterfield MD, Seckl MJ (2002) Two distinct phosphoinositide 3-kinases mediate polypeptide growth factor-stimulated PKB activation. EMBO J 21:5097–5108

Argiris A, Cohen E, Karrison T, Esparaz B, Mauer A, Ansari R, Wong S, Lu Y, Pins M, Dancey J, Vokes E (2006) A phase II trial of perifosine, an oral alkylphospholipid, in recurrent or metastatic head and neck cancer. Cancer Biol Ther 5:766–770

Atkins MB, Hidalgo M, Stadler WM, Logan TF, Dutcher JP, Hudes GR, Park Y, Liou SH, Marshall B, Boni JP, Dukart G, Sherman ML (2004) Randomized phase II study of multiple dose levels of CCI-779, a novel mammalian target of rapamycin kinase inhibitor, in patients with advanced refractory renal cell carcinoma. J Clin Oncol 22:909–918

Atkins MB, Yasothan U, Kirkpatrick P (2009) Everolimus. Nat Rev Drug Discov 8:535–536

Aziz SA, Jilaveanu LB, Zito C, Camp RL, Rimm DL, Conrad P, Kluger HM (2010) Vertical targeting of the phosphatidylinositol-3 kinase pathway as a strategy for treating melanoma. Clin Cancer Res 16:6029–6039

Bachmaier K, Krawczyk C, Kozieradzki I, Kong YY, Sasaki T, Oliveira-dos-Santos A, Mariathasan S, Bouchard D, Wakeham A, Itie A, Le J, Ohashi PS, Sarosi I, Nishina H, Lipkowitz S, Penninger JM (2000) Negative regulation of lymphocyte activation and autoimmunity by the molecular adaptor Cbl-b. Nature 403:211–216

Backer JM (2008) The regulation and function of Class III PI3Ks: novel roles for Vps34. Biochem J 410:1–17

Backer JM (2010) The regulation of class IA PI 3-kinases by inter-subunit interactions. Curr Top Microbiol Immunol 346:87–114

Bader AG, Kang S, Zhao L, Vogt PK (2005) Oncogenic PI3K deregulates transcription and translation. Nat Rev Cancer 5:921–929

Baggiolini M, Dewald B, Schnyder J, Ruch W, Cooper PH, Payne TG (1987) Inhibition of the phagocytosis-induced respiratory burst by the fungal metabolite wortmannin and some analogues. Exp Cell Res 169:408–418

Bailey HH, Mahoney MR, Ettinger DS, Maples WJ, Fracasso PM, Traynor AM, Erlichman C, Okuno SH (2006) Phase II study of daily oral perifosine in patients with advanced soft tissue sarcoma. Cancer 107:2462–2467

Balla T (2005) Inositol-lipid binding motifs: signal integrators through protein-lipid and protein-protein interactions. J Cell Sci 118:2093–2104

Balla A, Balla T (2006) Phosphatidylinositol 4-kinases: old enzymes with emerging functions. Trends Cell Biol 16:351–361

Balla T, Varnai P (2002) Visualizing cellular phosphoinositide pools with GFP-fused protein-modules. Sci STKE 2002:PL3

Banaszynski LA, Liu CW, Wandless TJ (2005) Characterization of the FKBP.rapamycin.FRB ternary complex. J Am Chem Soc 127:4715–4721

Barber DF, Bartolome A, Hernandez C, Flores JM, Redondo C, Fernandez-Arias C, Camps M, Ruckle T, Schwarz MK, Rodriguez S, Martinez AC, Balomenos D, Rommel C, Carrera AC (2005) PI3Kgamma inhibition blocks glomerulonephritis and extends lifespan in a mouse model of systemic lupus. Nat Med 11:933–935

Barber DF, Bartolome A, Hernandez C, Flores JM, Fernandez-Arias C, Rodriguez-Borlado L, Hirsch E, Wymann M, Balomenos D, Carrera AC (2006) Class IB-phosphatidylinositol 3-kinase (PI3K) deficiency ameliorates IA-PI3K-induced systemic lupus but not T cell invasion. J Immunol 176:589–593

Barnes PJ (2008) Immunology of asthma and chronic obstructive pulmonary disease. Nat Rev Immunol 8:183–192

Barylko B, Gerber SH, Binns DD, Grichine N, Khvotchev M, Sudhof TC, Albanesi JP (2001) A novel family of phosphatidylinositol 4-kinases conserved from yeast to humans. J Biol Chem 276:7705–7708

Baselga J, Semiglazov V, Dam P van, Manikhas A, Bellet M, Mayordomo J, Campone M, Kubista E, Greil R, Bianchi G, Steinseifer J, Molloy B, Tokaji E, Gardner H, Phillips P, Stumm M, Lane HA, Dixon JM, Jonat W, Rugo HS (2009) Phase II randomized study of neoadjuvant everolimus plus letrozole compared with placebo plus letrozole in patients with estrogen receptor-positive breast cancer. J Clin Oncol 27:2630–2637

Batra-Safferling R, Granzin J, Modder S, Hoffmann S, Willbold D (2010) Structural studies of the phosphatidylinositol 3-kinase (PI3K) SH3 domain in complex with a peptide ligand: role of the anchor residue in ligand binding. Biol Chem 391:33–42

Bellacosa A, Testa JR, Staal SP, Tsichlis PN (1991) A retroviral oncogene, akt, encoding a serine-threonine kinase containing an SH2-like region. Science 254:274–277

Benitah SA, Valeron PF, Van Aelst L, Marshall CJ, Lacal JC (2004) Rho GTPases in human cancer: an unresolved link to upstream and downstream transcriptional regulation. Biochim Biophys Acta 1705:121–132

Berger P, Bonneick S, Willi S, Wymann M, Suter U (2002) Loss of phosphatase activity in myotubularin-related protein 2 is associated with Charcot-Marie-Tooth disease type 4B1. Hum Mol Genet 11:1569–1579

Berndt A, Miller S, Williams O, Le DD, Houseman BT, Pacold JI, Gorrec F, Hon WC, Liu Y, Rommel C, Gaillard P, Ruckle T, Schwarz MK, Shokat KM, Shaw JP, Williams RL (2010) The p110delta structure: mechanisms for selectivity and potency of new PI(3)K inhibitors. Nat Chem Biol 6:117–124

Bi L, Okabe I, Bernard DJ, Wynshaw-Boris A, Nussbaum RL (1999) Proliferative defect and embryonic lethality in mice homozygous for a deletion in the p110alpha subunit of phosphoinositide 3-kinase. J Biol Chem 274:10963–10968

Biondi RM, Nebreda AR (2003) Signalling specificity of Ser/Thr protein kinases through docking-site-mediated interactions. Biochem J 372:1–13

Biwa T, Sakai M, Matsumura T, Kobori S, Kaneko K, Miyazaki A, Hakamata H, Horiuchi S, Shichiri M (2000a) Sites of action of protein kinase C and phosphatidylinositol 3-kinase are distinct in oxidized low density lipoprotein-induced macrophage proliferation. J Biol Chem 275:5810–5816

Biwa T, Sakai M, Shichiri M, Horiuchi S (2000b) Granulocyte/macrophage colony-stimulating factor plays an essential role in oxidized low density lipoprotein-induced macrophage proliferation. J Atheroscler Thromb 7:14–20

Bjornsti MA, Houghton PJ (2004) The TOR pathway: a target for cancer therapy. Nat Rev Cancer 4:335–348

Blank U, Launay P, Benhamou M, Monteiro RC (2009) Inhibitory ITAMs as novel regulators of immunity. Immunol Rev 232:59–71

Blay JY (2011) Updating progress in sarcoma therapy with mTOR inhibitors. Ann Oncol 22:280–287

Bohnacker T, Marone R, Collmann E, Calvez R, Hirsch E, Wymann MP (2009) PI3Kgamma adaptor subunits define coupling to degranulation and cell motility by distinct PtdIns(3,4,5)P3 pools in mast cells. Sci Signal 2:ra27

Bondeva T, Pirola L, Bulgarelli-Leva G, Rubio I, Wetzker R, Wymann MP (1998) Bifurcation of lipid and protein kinase signals of PI3Kgamma to the protein kinases PKB and MAPK. Science 282:293–296

Boring L, Gosling J, Cleary M, Charo IF (1998) Decreased lesion formation in CCR2-/- mice reveals a role for chemokines in the initiation of atherosclerosis. Nature 394:894–897

Borlado LR, Redondo C, Alvarez B, Jimenez C, Criado LM, Flores J, Marcos MA, Martinez AC, Balomenos D, Carrera AC (2000) Increased phosphoinositide 3-kinase activity induces a lymphoproliferative disorder and contributes to tumor generation in vivo. FASEB J 14:895–903

Boudeau J, Sapkota G, Alessi DR (2003) LKB1, a protein kinase regulating cell proliferation and polarity. FEBS Lett 546:159–165

Bozulic L, Hemmings BA (2009) PIKKing on PKB: regulation of PKB activity by phosphorylation. Curr Opin Cell Biol 21:256–261

Brachmann SM, Hofmann I, Schnell C, Fritsch C, Wee S, Lane H, Wang S, Garcia-Echeverria C, Maira SM (2009) Specific apoptosis induction by the dual PI3K/mTor inhibitor NVP-BEZ235 in HER2 amplified and PIK3CA mutant breast cancer cells. Proc Natl Acad Sci USA 106:22299–22304

Braig M, Schmitt CA (2006) Oncogene-induced senescence: putting the brakes on tumor development. Cancer Res 66:2881–2884

Braunersreuther V, Mach F, Steffens S (2007a) The specific role of chemokines in atherosclerosis. Thromb Haemost 97:714–721

Braunersreuther V, Zernecke A, Arnaud C, Liehn EA, Steffens S, Shagdarsuren E, Bidzhekov K, Burger F, Pelli G, Luckow B, Mach F, Weber C (2007b) Ccr5 but not Ccr1 deficiency reduces development of diet-induced atherosclerosis in mice. Arterioscler Thromb Vasc Biol 27:373–379

Breuleux M, Klopfenstein M, Stephan C, Doughty CA, Barys L, Maira SM, Kwiatkowski D, Lane HA (2009) Increased AKT S473 phosphorylation after mTORC1 inhibition is rictor dependent and does not predict tumor cell response to PI3K/mTOR inhibition. Mol Cancer Ther 8:742–753

Brock C, Schaefer M, Reusch HP, Czupalla C, Michalke M, Spicher K, Schultz G, Nurnberg B (2003) Roles of G beta gamma in membrane recruitment and activation of p110 gamma/p101 phosphoinositide 3-kinase gamma. J Cell Biol 160:89–99

Brognard J, Sierecki E, Gao T, Newton AC (2007) PHLPP and a second isoform, PHLPP2, differentially attenuate the amplitude of Akt signaling by regulating distinct Akt isoforms. Mol Cell 25:917–931

Brunet A, Bonni A, Zigmond MJ, Lin MZ, Juo P, Hu LS, Anderson MJ, Arden KC, Blenis J, Greenberg ME (1999) Akt promotes cell survival by phosphorylating and inhibiting a Forkhead transcription factor. Cell 96:857–868

Brunn GJ, Williams J, Sabers C, Wiederrecht G, Lawrence JC Jr, Abraham RT (1996) Direct inhibition of the signaling functions of the mammalian target of rapamycin by the phosphoinositide 3-kinase inhibitors, wortmannin and LY294002. EMBO J 15:5256–5267

Bunney TD, Katan M (2010) Phosphoinositide signalling in cancer: beyond PI3K and PTEN. Nat Rev Cancer 10:342–352

Buonamici S, Williams J, Morrissey M, Wang A, Guo R, Vattay A, Hsiao K, Yuan J, Green J, Ospina B, Yu Q, Ostrom L, Fordjour P, Anderson DL, Monahan JE, Kelleher JF, Peukert S, Pan S, Wu X, Maira SM, Garcia-Echeverria C, Briggs KJ, Watkins DN, Yao YM, Lengauer C, Warmuth M, Sellers WR, Dorsch M (2010) Interfering with resistance to smoothened antagonists by inhibition of the PI3K pathway in medulloblastoma. Sci Transl Med 2:51ra70

Burgering BM, Coffer PJ (1995) Protein kinase B (c-Akt) in phosphatidylinositol-3-OH kinase signal transduction. Nature 376:599–602

Burgering BM, Kops GJ (2002) Cell cycle and death control: long live Forkheads. Trends Biochem Sci 27:352–360

Burgering BM, Medema RH (2003) Decisions on life and death: FOXO Forkhead transcription factors are in command when PKB/Akt is off duty. J Leukoc Biol 73:689–701

Cairns P, Okami K, Halachmi S, Halachmi N, Esteller M, Herman JG, Jen J, Isaacs WB, Bova GS, Sidransky D (1997) Frequent inactivation of PTEN/MMAC1 in primary prostate cancer. Cancer Res 57:4997–5000

Camps M, Ruckle T, Ji H, Ardissone V, Rintelen F, Shaw J, Ferrandi C, Chabert C, Gillieron C, Francon B, Martin T, Gretener D, Perrin D, Leroy D, Vitte PA, Hirsch E, Wymann MP, Cirillo R, Schwarz MK, Rommel C (2005) Blockade of PI3Kgamma suppresses joint inflammation and damage in mouse models of rheumatoid arthritis. Nat Med 11:936–943

Canobbio I, Stefanini L, Cipolla L, Ciraolo E, Gruppi C, Balduini C, Hirsch E, Torti M (2009) Genetic evidence for a predominant role of PI3Kbeta catalytic activity in ITAM- and integrin-mediated signaling in platelets. Blood 114:2193–2196

Cantley LC (2002) The phosphoinositide 3-kinase pathway. Science 296:1655–1657

Cao C, Backer JM, Laporte J, Bedrick EJ, Wandinger-Ness A (2008) Sequential actions of myotubularin lipid phosphatases regulate endosomal PI(3)P and growth factor receptor trafficking. Mol Biol Cell 19:3334–3346

Carayol N, Vakana E, Sassano A, Kaur S, Goussetis DJ, Glaser H, Druker BJ, Donato NJ, Altman JK, Barr S, Platanias LC (2010) Critical roles for mTORC2- and rapamycin-insensitive mTORC1-complexes in growth and survival of BCR-ABL-expressing leukemic cells. Proc Natl Acad Sci USA 107:12469–12474

Cardone MH, Roy N, Stennicke HR, Salvesen GS, Franke TF, Stanbridge E, Frisch S, Reed JC (1998) Regulation of cell death protease caspase-9 by phosphorylation. Science 282:1318–1321

Carol H, Morton CL, Gorlick R, Kolb EA, Keir ST, Reynolds CP, Kang MH, Maris JM, Billups C, Smith MA, Houghton PJ, Lock RB (2010) Initial testing (stage 1) of the Akt inhibitor GSK690693 by the pediatric preclinical testing program. Pediatr Blood Cancer 55:1329–1337

Carracedo A, Ma L, Teruya-Feldstein J, Rojo F, Salmena L, Alimonti A, Egia A, Sasaki AT, Thomas G, Kozma SC, Papa A, Nardella C, Cantley LC, Baselga J, Pandolfi PP (2008) Inhibition of mTORC1 leads to MAPK pathway activation through a PI3K-dependent feedback loop in human cancer. J Clin Invest 118:3065–3074

Carracedo A, Alimonti A, Pandolfi PP (2011) PTEN level in tumor suppression: how much is too little? Cancer Res 71:629–633

Chan DW, Son SC, Block W, Ye R, Khanna KK, Wold MS, Douglas P, Goodarzi AA, Pelley J, Taya Y, Lavin MF, Lees-Miller SP (2000) Purification and characterization of ATM from human placenta. A manganese-dependent, wortmannin-sensitive serine/threonine protein kinase. J Biol Chem 275:7803–7810

Chan TO, Rodeck U, Chan AM, Kimmelman AC, Rittenhouse SE, Panayotou G, Tsichlis PN (2002) Small GTPases and tyrosine kinases coregulate a molecular switch in the phosphoinositide 3-kinase regulatory subunit. Cancer Cell 1:181–191

Chang JD, Sukhova GK, Libby P, Schvartz E, Lichtenstein AH, Field SJ, Kennedy C, Madhavarapu S, Luo J, Wu D, Cantley LC (2007) Deletion of the phosphoinositide 3-kinase p110gamma gene attenuates murine atherosclerosis. Proc Natl Acad Sci USA 104:8077–8082

Chatham WW, Kimberly RP (2001) Treatment of lupus with corticosteroids. Lupus 10:140–147

Chaurasia B, Mauer J, Koch L, Goldau J, Kock AS, Bruning JC (2010) Phosphoinositide-dependent kinase 1 provides negative feedback inhibition to Toll-like receptor-mediated NF-kappaB activation in macrophages. Mol Cell Biol 30:4354–4366

Cheadle JP, Reeve MP, Sampson JR, Kwiatkowski DJ (2000) Molecular genetic advances in tuberous sclerosis. Hum Genet 107:97–114

Chee KG, Longmate J, Quinn DI, Chatta G, Pinski J, Twardowski P, Pan CX, Cambio A, Evans CP, Gandara DR, Lara PN J. (2007) The AKT inhibitor perifosine in biochemically recurrent prostate cancer: a phase II California/Pittsburgh cancer consortium trial. Clin Genitourin Cancer 5:433–437

Chen Z, Trotman LC, Shaffer D, Lin HK, Dotan ZA, Niki M, Koutcher JA, Scher HI, Ludwig T, Gerald W, Cordon-Cardo C, Pandolfi PP (2005) Crucial role of p53-dependent cellular senescence in suppression of Pten-deficient tumorigenesis. Nature 436:725–730

Chen L, Iijima M, Tang M, Landree MA, Huang YE, Xiong Y, Iglesias PA, Devreotes PN (2007) PLA2 and PI3K/PTEN pathways act in parallel to mediate chemotaxis. Dev Cell 12:603–614

Cheng H, Bagrodia S, Bailey S, Edwards M, Hoffman J, Hu Q, Kania R, Knighton DR, Marx MA, Ninkovic S, Sun S, Zhang E (2010) Discovery of the highly potent PI3K/mTOR dual inhibitor PF-04691502 through structure based drug design. Med Chem Commun 1:139–144

Choi J, Chen J, Schreiber SL, Clardy J (1996) Structure of the FKBP12-rapamycin complex interacting with the binding domain of human FRAP. Science 273:239–242

Chresta CM, Davies BR, Hickson I, Harding T, Cosulich S, Critchlow SE, Vincent JP, Ellston R, Jones D, Sini P, James D, Howard Z, Dudley P, Hughes G, Smith L, Maguire S, Hummersone M, Malagu K, Menear K, Jenkins R, Jacobsen M, Smith GC, Guichard S, Pass M (2010) AZD8055 is a potent, selective, and orally bioavailable ATP-competitive mammalian target of rapamycin kinase inhibitor with in vitro and in vivo antitumor activity. Cancer Res 70:288–298

Christoforidis S, Miaczynska M, Ashman K, Wilm M, Zhao L, Yip SC, Waterfield MD, Backer JM, Zerial M (1999) Phosphatidylinositol-3-OH kinases are Rab5 effectors. Nat Cell Biol 1:249–252

Cimprich KA, Cortez D (2008) ATR: an essential regulator of genome integrity. Nat Rev Mol Cell Biol 9:616–627

Ciraolo E, Iezzi M, Marone R, Marengo S, Curcio C, Costa C, Azzolino O, Gonella C, Rubinetto C, Wu H, Dastru W, Martin EL, Silengo L, Altruda F, Turco E, Lanzetti L, Musiani P, Ruckle T, Rommel C, Backer JM, Forni G, Wymann MP, Hirsch E (2008) Phosphoinositide 3-kinase p110beta activity: key role in metabolism and mammary gland cancer but not development. Sci Signal 1:ra3

Ciraolo E, Morello F, Hobbs RM, Wolf F, Marone R, Iezzi M, Lu X, Mengozzi G, Altruda F, Sorba G, Guan K, Pandolfi PP, Wymann MP, Hirsch E (2010) Essential role of the p110beta subunit of phosphoinositide 3-OH kinase in male fertility. Mol Biol Cell 21:704–711

Clark J, Anderson KE, Juvin V, Smith TS, Karpe F, Wakelam MJ, Stephens LR, Hawkins PT (2011) Quantification of PtdInsP(3) molecular species in cells and tissues by mass spectrometry. Nat Methods 8:267–272

Clayton E, Bardi G, Bell SE, Chantry D, Downes CP, Gray A, Humphries LA, Rawlings D, Reynolds H, Vigorito E, Turner M (2002) A crucial role for the p110delta subunit of phosphatidylinositol 3-kinase in B cell development and activation. J Exp Med 196:753–763

Cohen PL, Eisenberg RA (1991) Lpr and gld: single gene models of systemic autoimmunity and lymphoproliferative disease. Annu Rev Immunol 9:243–269

Cosemans JM, Munnix IC, Wetzker R, Heller R, Jackson SP, Heemskerk JW (2006) Continuous signaling via PI3K isoforms beta and gamma is required for platelet ADP receptor function in dynamic thrombus stabilization. Blood 108:3045–3052

Costa C, Barberis L, Ambrogio C, Manazza AD, Patrucco E, Azzolino O, Neilsen PO, Ciraolo E, Altruda F, Prestwich GD, Chiarle R, Wymann M, Ridley A, Hirsch E (2007) Negative feedback regulation of Rac in leukocytes from mice expressing a constitutively active phosphatidylinositol 3-kinase gamma. Proc Natl Acad Sci USA 104:14354–14359

Cote JF, Vuori K (2007) GEF what? Dock180 and related proteins help Rac to polarize cells in new ways. Trends Cell Biol 17:383–393

Crackower MA, Oudit GY, Kozieradzki I, Sarao R, Sun H, Sasaki T, Hirsch E, Suzuki A, Shioi T, Irie-Sasaki J, Sah R, Cheng HY, Rybin VO, Lembo G, Fratta L, Oliveira-dos-Santos AJ, Benovic JL, Kahn CR, Izumo S, Steinberg SF, Wymann MP, Backx PH, Penninger JM (2002) Regulation of myocardial contractility and cell size by distinct PI3K-PTEN signaling pathways. Cell 110:737–749

Cui ZG, Hong NY, Kang HK, Lee DH, Lee YK, Park DB (2011) The alpha-isoform of class II phosphoinositide 3-kinase is necessary for the activation of ERK but not Akt/PKB. Mol Cell Biochem 346:95–101

Cully M, You H, Levine AJ, Mak TW (2006) Beyond PTEN mutations: the PI3K pathway as an integrator of multiple inputs during tumorigenesis. Nat Rev Cancer 6:184–192

Daimon M, Sato H, Oizumi T, Toriyama S, Saito T, Karasawa S, Jimbu Y, Wada K, Kameda W, Susa S, Yamaguchi H, Emi M, Muramatsu M, Kubota I, Kawata S, Kato T (2008) Association of the PIK3C2G gene polymorphisms with type 2 DM in a Japanese population. Biochem Biophys Res Commun 365:466–471

Damas JK, Smith C, Oie E, Fevang B, Halvorsen B, Waehre T, Boullier A, Breland U, Yndestad A, Ovchinnikova O, Robertson AK, Sandberg WJ, Kjekshus J, Tasken K, Froland SS, Gullestad L, Hansson GK, Quehenberger O, Aukrust P (2007) Enhanced expression of the homeostatic chemokines CCL19 and CCL21 in clinical and experimental atherosclerosis: possible pathogenic role in plaque destabilization. Arterioscler Thromb Vasc Biol 27:614–620

Dancey J (2010) mTOR signaling and drug development in cancer. Nat Rev Clin Oncol 7:209–219
Dann SG, Thomas G (2006) The amino acid sensitive TOR pathway from yeast to mammals. FEBS Lett 580:2821–2829
Dann SG, Selvaraj A, Thomas G (2007) mTOR Complex1-S6K1 signaling: at the crossroads of obesity, diabetes and cancer. Trends Mol Med 13:252–259
Datta SR, Dudek H, Tao X, Masters S, Fu H, Gotoh Y, Greenberg ME (1997) Akt phosphorylation of BAD couples survival signals to the cell-intrinsic death machinery. Cell 91:231–241
Datta SR, Brunet A, Greenberg ME (1999) Cellular survival: a play in three Akts. Genes Dev 13:2905–2927
Davies BR, Dudley P, Cosulich S, Luke R, Thompson N, Collins J, McHardy T, Garrett M, Ogilvie D (2009) CCT129254 (AT11854) is a well tolerated, orally bioavailable inhibitor of AKT/PKB with pharmacodynamic and antitumor activity in a range of xenograft models. Mol Cancer Ther 8:C208
Deane JA, Fruman DA (2004) Phosphoinositide 3-kinase: diverse roles in immune cell activation. Annu Rev Immunol 22:563–598
Dehnhardt CM, Venkatesan AM, Delos Santos E, Chen Z, Santos O, Ayral-Kaloustian S, Brooijmans N, Mallon R, Hollander I, Feldberg L, Lucas J, Chaudhary I, Yu K, Gibbons J, Abraham R, Mansour TS (2010) Lead optimization of N-3-Substituted 7-morpholinotriazolopyrimidines as dual phosphoinositide 3-kinase/mammalian target of rapamycin inhibitors: Discovery of PKI-402. J Med Chem 53:798–810
Del Prete A, Vermi W, Dander E, Otero K, Barberis L, Luini W, Bernasconi S, Sironi M, Santoro A, Garlanda C, Facchetti F, Wymann MP, Vecchi A, Hirsch E, Mantovani A, Sozzani S (2004) Defective dendritic cell migration and activation of adaptive immunity in PI3Kgamma-deficient mice. EMBO J 23:3505–3515
Delgado P, Cubelos B, Calleja E, Martinez-Martin N, Cipres A, Merida I, Bellas C, Bustelo XR, Alarcon B (2009) Essential function for the GTPase TC21 in homeostatic antigen receptor signaling. Nat Immunol 10:880–888
Di Cristofano A, Pesce B, Cordon-Cardo C, Pandolfi PP (1998) Pten is essential for embryonic development and tumour suppression. Nat Genet 19:348–355
Di Cristofano A, Kotsi P, Peng YF, Cordon-Cardo C, Elkon KB, Pandolfi PP (1999) Impaired Fas response and autoimmunity in Pten +/- mice. Science 285:2122–2125
Di Paolo G, De Camilli P (2006) Phosphoinositides in cell regulation and membrane dynamics. Nature 443:651–657
Diaz R, Nguewa PA, Diaz-Gonzalez JA, Hamel E, Gonzalez-Moreno O, Catena R, Serrano D, Redrado M, Sherris D, Calvo A (2009) The novel Akt inhibitor Palomid 529 (P529) enhances the effect of radiotherapy in prostate cancer. Br J Cancer 100:932–940
Diaz-Flores E, Shannon K (2007) Targeting oncogenic Ras. Genes Dev 21:1989–1992
Domin J, Gaidarov I, Smith ME, Keen JH, Waterfield MD (2000) The class II phosphoinositide 3-kinase PI3K-C2alpha is concentrated in the trans-Golgi network and present in clathrin-coated vesicles. J Biol Chem 275:11943–11950
Dominguez V, Raimondi C, Somanath S, Bugliani M, Loder MK, Edling CE, Divecha N, da Silva-Xavier G, Marselli L, Persaud SJ, Turner MD, Rutter GA, Marchetti P, Falasca M, Maffucci T (2011) Class II phosphoinositide 3-kinase regulates exocytosis of insulin granules in pancreatic beta cells. J Biol Chem 286:4216–4225
Dou Z, Chattopadhyay M, Pan JA, Guerriero JL, Jiang YP, Ballou LM, Yue Z, Lin RZ, Zong WX (2010) The class IA phosphatidylinositol 3-kinase p110-beta subunit is a positive regulator of autophagy. J Cell Biol 191:827–843
Doukas J, Wrasidlo W, Noronha G, Dneprovskaia E, Fine R, Weis S, Hood J, Demaria A, Soll R, Cheresh D (2006) Phosphoinositide 3-kinase gamma/delta inhibition limits infarct size after myocardial ischemia/reperfusion injury. Proc Natl Acad Sci USA 103:19866–19871
Doukas J, Wrasidlo W, Noronha G, Dneprovskaia E, Hood J, Soll R (2007) Isoform-selective PI3K inhibitors as novel therapeutics for the treatment of acute myocardial infarction. Biochem Soc Trans 35:204–206

Doukas J, Eide L, Stebbins K, Racanelli-Layton A, Dellamary L, Martin M, Dneprovskaia E, Noronha G, Soll R, Wrasidlo W, Acevedo LM, Cheresh DA (2009) Aerosolized phosphoinositide 3-kinase gamma/delta inhibitor TG100–115 [3-[2,4-diamino-6-(3-hydroxyphenyl)pteridin-7-yl]phenol] as a therapeutic candidate for asthma and chronic obstructive pulmonary disease. J Pharmacol Exp Ther 328:758–765

Dove SK, Johnson ZE (2007) Our FABulous VACation: a decade of phosphatidylinositol 3,5-bisphosphate. Biochem Soc Symp 129–139

Dove SK, Michell RH (2009) Inositol lipid-dependent functions in Saccharomyces cerevisiae: analysis of phosphatidylinositol phosphates. Methods Mol Biol 462:59–74

Dove SK, Dong K, Kobayashi T, Williams FK, Michell RH (2009) Phosphatidylinositol 3,5-bisphosphate and Fab1p/PIKfyve underPPIn endo-lysosome function. Biochem J 419:1–13

Downward J (2008) Targeting RAS and PI3K in lung cancer. Nat Med 14:1315–1316

Dufner A, Thomas G (1999) Ribosomal S6 kinase signaling and the control of translation. Exp Cell Res 253:100–109

Durocher D, Jackson SP (2001) DNA-PK, ATM and ATR as sensors of DNA damage: variations on a theme? Curr Opin Cell Biol 13:225–231

Dutil EM, Toker A, Newton AC (1998) Regulation of conventional protein kinase C isozymes by phosphoinositide-dependent kinase 1 (PDK-1). Curr Biol 8:1366–1375

Edling CE, Selvaggi F, Buus R, Maffucci T, Di Sebastiano P, Friess H, Innocenti P, Kocher HM, Falasca M (2010) Key role of phosphoinositide 3-kinase class IB in pancreatic cancer. Clin Cancer Res 16:4928–4937

Ellard SL, Clemons M, Gelmon KA, Norris B, Kennecke H, Chia S, Pritchard K, Eisen A, Vandenberg T, Taylor M, Sauerbrei E, Mishaeli M, Huntsman D, Walsh W, Olivo M, McIntosh L, Seymour L (2009) Randomized phase II study comparing two schedules of everolimus in patients with recurrent/metastatic breast cancer: NCIC Clinical Trials Group IND.163. J Clin Oncol 27:4536–4541

Ellenbroek SI, Collard JG (2007) Rho GTPases: functions and association with cancer. Clin Exp Metastasis 24:657–672

Elrod HA, Lin YD, Yue P, Wang X, Lonial S, Khuri FR, Sun SY (2007) The alkylphospholipid perifosine induces apoptosis of human lung cancer cells requiring inhibition of Akt and activation of the extrinsic apoptotic pathway. Mol Cancer Ther 6:2029–2038

Eng C (1998) Genetics of Cowden syndrome: through the looking glass of oncology. Int J Oncol 12:701–710

Engelman JA (2009) Targeting PI3K signalling in cancer: opportunities, challenges and limitations. Nat Rev Cancer 9:550–562

Engelman JA, Luo J, Cantley LC (2006) The evolution of phosphatidylinositol 3-kinases as regulators of growth and metabolism. Nat Rev Genet 7:606–619

Engelman JA, Chen L, Tan X, Crosby K, Guimaraes AR, Upadhyay R, Maira M, McNamara K, Perera SA, Song Y, Chirieac LR, Kaur R, Lightbown A, Simendinger J, Li T, Padera RF, Garcia-Echeverria C, Weissleder R, Mahmood U, Cantley LC, Wong KK (2008) Effective use of PI3K and MEK inhibitors to treat mutant Kras G12D and PIK3CA H1047R murine lung cancers. Nat Med 14:1351–1356

Ernst DS, Eisenhauer E, Wainman N, Davis M, Lohmann R, Baetz T, Belanger K, Smylie M (2005) Phase II study of perifosine in previously untreated patients with metastatic melanoma. Invest New Drugs 23:569–575

Fairhurst AM, Wandstrat AE, Wakeland EK (2006) Systemic lupus erythematosus: multiple immunological phenotypes in a complex genetic disease. Adv Immunol 92:1–69

Faivre S, Kroemer G, Raymond E (2006) Current development of mTOR inhibitors as anticancer agents. Nat Rev Drug Discov 5:671–688

Falasca M (2010) PI3K/Akt signalling pathway specific inhibitors: a novel strategy to sensitize cancer cells to anti-cancer drugs. Curr Pharm Des 16:1410–1416

Falasca M, Maffucci T (2007) Role of class II phosphoinositide 3-kinase in cell signalling. Biochem Soc Trans 35:211–214

Falasca M, Hughes WE, Dominguez V, Sala G, Fostira F, Fang MQ, Cazzolli R, Shepherd PR, James DE, Maffucci T (2007) The role of phosphoinositide 3-kinase C2alpha in insulin signaling. J Biol Chem 282:28226–28236

Fan QW, Knight ZA, Goldenberg DD, Yu W, Mostov KE, Stokoe D, Shokat KM, Weiss WA (2006) A dual PI3 kinase/mTOR inhibitor reveals emergent efficacy in glioma. Cancer Cell 9:341–349

Farooq A, Walker LJ, Bowling J, Audisio RA (2010) Cowden syndrome. Cancer Treat Rev 36:577–583

Feldman ME, Apsel B, Uotila A, Loewith R, Knight ZA, Ruggero D, Shokat KM (2009) Active-site inhibitors of mTOR target rapamycin-resistant outputs of mTORC1 and mTORC2. PLoS Biol 7:e38

Feng J, Park J, Cron P, Hess D, Hemmings BA (2004) Identification of a PKB/Akt hydrophobic motif Ser-473 kinase as DNA-dependent protein kinase. J Biol Chem 279:41189–41196

Ferguson KM, Kavran JM, Sankaran VG, Fournier E, Isakoff SJ, Skolnik EY, Lemmon MA (2000) Structural basis for discrimination of 3-phosphoinositides by pleckstrin homology domains. Mol Cell 6:373–384

Ferguson GJ, Milne L, Kulkarni S, Sasaki T, Walker S, Andrews S, Crabbe T, Finan P, Jones G, Jackson S, Camps M, Rommel C, Wymann M, Hirsch E, Hawkins P, Stephens L (2007) PI(3)Kgamma has an important context-dependent role in neutrophil chemokinesis. Nat Cell Biol 9:86–91

Firestein GS (2003) Evolving concepts of rheumatoid arthritis. Nature 423:356–361

Firestein GS (2006) Inhibiting inflammation in rheumatoid arthritis. N Engl J Med 354:80–82

Foijer F, te Riele H (2006) Check, double check: the G2 barrier to cancer. Cell Cycle 5:831–836

Folkes AJ, Ahmadi K, Alderton WK, Alix S, Baker SJ, Box G, Chuckowree IS, Clarke PA, Depledge P, Eccles SA, Friedman LS, Hayes A, Hancox TC, Kugendradas A, Lensun L, Moore P, Olivero AG, Pang J, Patel S, Pergl-Wilson GH, Raynaud FI, Robson A, Saghir N, Salphati L, Sohal S, Ultsch MH, Valenti M, Wallweber HJ, Wan NC, Wiesmann C, Workman P, Zhyvoloup A, Zvelebil MJ, Shuttleworth SJ (2008) The identification of 2-(1H-Indazol-4-yl)-6-(4-methanesulfonyl-piperazin-1-ylmethyl)-4-morpholin -4-yl-thieno[3,2-d]pyrimidine (GDC-0941) as a potent, selective, orally bioavailable inhibitor of class I PI3 kinase for the treatment of cancer. J Med Chem 51(18):5522–5532

Fonseca BD, Smith EM, Lee VH, Mackintosh C, Proud CG (2007) PRAS40 is a target for mammalian target of rapamycin complex 1 and is required for signaling downstream of this complex. J Biol Chem 282:24514–24524

Fougerat A, Gayral S, Gourdy P, Schambourg A, Ruckle T, Schwarz MK, Rommel C, Hirsch E, Arnal JF, Salles JP, Perret B, Breton-Douillon M, Wymann MP, Laffargue M (2008) Genetic and pharmacological targeting of phosphoinositide 3-kinase-gamma reduces atherosclerosis and favors plaque stability by modulating inflammatory processes. Circulation 117:1310–1317

Fougerat A, Gayral S, Malet N, Briand-Mesange F, Breton-Douillon M, Laffargue M (2009) Phosphoinositide 3-kinases and their role in inflammation: potential clinical targets in atherosclerosis? Clin Sci (Lond) 116:791–804

Franke TF, Yang SI, Chan TO, Datta K, Kazlauskas A, Morrison DK, Kaplan DR, Tsichlis PN (1995) The protein kinase encoded by the Akt proto-oncogene is a target of the PDGF-activated phosphatidylinositol 3-kinase. Cell 81:727–736

Frias M de, Iglesias-Serret D, Cosialls AM, Coll-Mulet L, Santidrian AF, Gonzalez-Girones DM, la Banda E de, Pons G, Gil J (2009) Akt inhibitors induce apoptosis in chronic lymphocytic leukemia cells. Haematologica 94:1698–1707

Fukao T, Koyasu S (2003) PI3K and negative regulation of TLR signaling. Trends Immunol 24:358–363

Fuller DM, Zhu M, Ou-Yang CW, Sullivan SA, Zhang W (2011) A tale of two TRAPs: LAT and LAB in the regulation of lymphocyte development, activation, and autoimmunity. Immunol Res 49:97–108

Funamoto S, Meili R, Lee S, Parry L, Firtel RA (2002) Spatial and temporal regulation of 3-phosphoinositides by PI 3-kinase and PTEN mediates chemotaxis. Cell 109:611–623

Funderburk SF, Wang QJ, Yue Z (2010) The Beclin 1-VPS34 complex—at the crossroads of autophagy and beyond. Trends Cell Biol 20:355–362

Gaidarov I, Smith ME, Domin J, Keen JH (2001) The class II phosphoinositide 3-kinase C2alpha is activated by clathrin and regulates clathrin-mediated membrane trafficking. Mol Cell 7:443–449

Gaidarov I, Zhao Y, Keen JH (2005) Individual phosphoinositide 3-kinase C2alpha domain activities independently regulate clathrin function. J Biol Chem 280:40766–40772

Galanis E, Buckner JC, Maurer MJ, Kreisberg JI, Ballman K, Boni J, Peralba JM, Jenkins RB, Dakhil SR, Morton RF, Jaeckle KA, Scheithauer BW, Dancey J, Hidalgo M, Walsh DJ (2005) Phase II trial of temsirolimus (CCI-779) in recurrent glioblastoma multiforme: a North Central Cancer Treatment Group Study. J Clin Oncol 23:5294–5304

Gao T, Furnari F, Newton AC (2005) PHLPP: a phosphatase that directly dephosphorylates Akt, promotes apoptosis, and suppresses tumor growth. Mol Cell 18:13–24

Garami A, Zwartkruis FJ, Nobukuni T, Joaquin M, Roccio M, Stocker H, Kozma SC, Hafen E, Bos JL, Thomas G (2003) Insulin activation of Rheb, a mediator of mTOR/S6K/4E-BP signaling, is inhibited by TSC1 and 2. Mol Cell 11:1457–1466

Garlich JR, De P, Dey N, Su JD, Peng X, Miller A, Murali R, Lu Y, Mills GB, Kundra V, Shu HK, Peng Q, Durden DL (2008) A vascular targeted pan phosphoinositide 3-kinase inhibitor prodrug, SF1126, with antitumor and antiangiogenic activity. Cancer Res 68:206–215

Ghigo A, Damilano F, Braccini L, Hirsch E (2010) PI3K inhibition in inflammation: Toward tailored therapies for specific diseases. Bioessays 32:185–196

Ghobrial IM, Gertz M, Laplant B, Camoriano J, Hayman S, Lacy M, Chuma S, Harris B, Leduc R, Rourke M, Ansell SM, Deangelo D, Dispenzieri A, Bergsagel L, Reeder C, Anderson KC, Richardson PG, Treon SP, Witzig TE (2010) Phase II trial of the oral mammalian target of rapamycin inhibitor everolimus in relapsed or refractory Waldenstrom macroglobulinemia. J Clin Oncol 28:1408–1414

Glass CK, Witztum JL (2001) Atherosclerosis. The road ahead. Cell 104:503–516

Goel A, Arnold CN, Niedzwiecki D, Carethers JM, Dowell JM, Wasserman L, Compton C, Mayer RJ, Bertagnolli MM, Boland CR (2004) Frequent inactivation of PTEN by promoter hypermethylation in microsatellite instability-high sporadic colorectal cancers. Cancer Res 64:3014–3021

Graham TR, Burd CG (2011) Coordination of Golgi functions by phosphatidylinositol 4-kinases. Trends Cell Biol 21:113–121

Graness A, Adomeit A, Heinze R, Wetzker R, Liebmann C (1998) A novel mitogenic signaling pathway of bradykinin in the human colon carcinoma cell line SW-480 involves sequential activation of a Gq/11 protein, phosphatidylinositol 3-kinase beta, and protein kinase Cepsilon. J Biol Chem 273:32016–32022

Grant S (2008) Cotargeting survival signaling pathways in cancer. J Clin Invest 118:3003–3006

Graupera M, Guillermet-Guibert J, Foukas LC, Phng LK, Cain RJ, Salpekar A, Pearce W, Meek S, Millan J, Cutillas PR, Smith AJ, Ridley AJ, Ruhrberg C, Gerhardt H, Vanhaesebroeck B (2008) Angiogenesis selectively requires the p110alpha isoform of PI3K to control endothelial cell migration. Nature 453:662–666

Gridelli C, Rossi A, Morgillo F, Bareschino MA, Maione P, Di Maio M, Ciardiello F (2007) A randomized phase II study of pemetrexed or RAD001 as second-line treatment of advanced non-small-cell lung cancer in elderly patients: treatment rationale and protocol dynamics. Clin Lung Cancer 8:568–571

Gu H, Saito K, Klaman LD, Shen J, Fleming T, Wang Y, Pratt JC, Lin G, Lim B, Kinet JP, Neel BG (2001) Essential role for Gab2 in the allergic response. Nature 412:186–190

Guba M, Breitenbuch P von, Steinbauer M, Koehl G, Flegel S, Hornung M, Bruns CJ, Zuelke C, Farkas S, Anthuber M, Jauch KW, Geissler EK (2002) Rapamycin inhibits primary and metastatic tumor growth by antiangiogenesis: involvement of vascular endothelial growth factor. Nat Med 8:128–135

Guertin DA, Sabatini DM (2007) Defining the role of mTOR in cancer. Cancer Cell 12:9–22

Guillermet-Guibert J, Bjorklof K, Salpekar A, Gonella C, Ramadani F, Bilancio A, Meek S, Smith AJ, Okkenhaug K, Vanhaesebroeck B (2008) The p110beta isoform of phosphoinositide 3-kinase signals downstream of G protein-coupled receptors and is functionally redundant with p110gamma. Proc Natl Acad Sci USA 105:8292–8297

Gulati P, Thomas G (2007) Nutrient sensing in the mTOR/S6K1 signalling pathway. Biochem Soc Trans 35:236–238

Gupta S, Ramjaun AR, Haiko P, Wang Y, Warne PH, Nicke B, Nye E, Stamp G, Alitalo K, Downward J (2007) Binding of ras to phosphoinositide 3-kinase p110alpha is required for ras-driven tumorigenesis in mice. Cell 129:957–968

Hamerman JA, Ni M, Killebrew JR, Chu CL, Lowell CA (2009) The expanding roles of ITAM adapters FcRgamma and DAP12 in myeloid cells. Immunol Rev 232:42–58

Han EK, Leverson JD, McGonigal T, Shah OJ, Woods KW, Hunter T, Giranda VL, Luo Y (2007) Akt inhibitor A-443654 induces rapid Akt Ser-473 phosphorylation independent of mTORC1 inhibition. Oncogene 26:5655–5661

Hanada M, Feng J, Hemmings BA (2004) Structure, regulation and function of PKB/AKT—a major therapeutic target. Biochim Biophys Acta 1697:3–16

Hansel TT, Barnes PJ (2009) New drugs for exacerbations of chronic obstructive pulmonary disease. Lancet 374:744–755

Hansson GK, Hermansson A (2011) The immune system in atherosclerosis. Nat Immunol 12:204–212

Harada K, Truong AB, Cai T, Khavari PA (2005) The class II phosphoinositide 3-kinase C2beta is not essential for epidermal differentiation. Mol Cell Biol 25:11122–11130

Hardie DG (2005) New roles for the LKB1 → AMPK pathway. Curr Opin Cell Biol 17:167–173

Harley JB, Alarcon-Riquelme ME, Criswell LA, Jacob CO, Kimberly RP, Moser KL, Tsao BP, Vyse TJ, Langefeld CD, Nath SK, Guthridge JM, Cobb BL, Mirel DB, Marion MC, Williams AH, Divers J, Wang W, Frank SG, Namjou B, Gabriel SB, Lee AT, Gregersen PK, Behrens TW, Taylor KE, Fernando M, Zidovetzki R, Gaffney PM, Edberg JC, Rioux JD, Ojwang JO, James JA, Merrill JT, Gilkeson GS, Seldin MF, Yin H, Baechler EC, Li QZ, Wakeland EK, Bruner GR, Kaufman KM, Kelly JA (2008) Genome-wide association scan in women with systemic lupus erythematosus identifies susceptibility variants in ITGAM, PXK, KIAA1542 and other loci. Nat Genet 40:204–210

Harrington LS, Findlay GM, Lamb RF (2005) Restraining PI3K: mTOR signalling goes back to the membrane. Trends Biochem Sci 30:35–42

Harris DP, Vogel P, Wims M, Moberg K, Humphries J, Jhaver KG, DaCosta CM, Shadoan MK, Xu N, Hansen GM, Balakrishnan S, Domin J, Powell DR, Oravecz T (2011) Requirement for class II phosphoinositide 3-kinase C2alpha in maintenance of glomerular structure and function. Mol Cell Biol 31:63–80

Hauser SL, Oksenberg JR (2006) The neurobiology of multiple sclerosis: genes, inflammation, and neurodegeneration. Neuron 52:61–76

Hay N (2005) The Akt-mTOR tango and its relevance to cancer. Cancer Cell 8:179–183

Hazeki K, Nigorikawa K, Hazeki O (2007) Role of phosphoinositide 3-kinase in innate immunity. Biol Pharm Bull 30:1617–1623

Heasman SJ, Ridley AJ (2008) Mammalian Rho GTPases: new insights into their functions from in vivo studies. Nat Rev Mol Cell Biol 9:690–701

Heerding DA, Rhodes N, Leber JD, Clark TJ, Keenan RM, Lafrance LV, Li M, Safonov IG, Takata DT, Venslavsky JW, Yamashita DS, Choudhry AE, Copeland RA, Lai Z, Schaber MD, Tummino PJ, Strum SL, Wood ER, Duckett DR, Eberwein D, Knick VB, Lansing TJ, McConnell RT, Zhang S, Minthorn EA, Concha NO, Warren GL, Kumar R (2008) Identification of 4-(2-(4-amino-1,2,5-oxadiazol-3-yl)-1-ethyl-7-{[(3S)-3-piperidinylmethyl]oxy}-1H-imidazo[4,5-c]pyridin-4-yl)-2-methyl-3-butyn-2-ol (GSK690693), a novel inhibitor of AKT kinase. J Med Chem 51:5663–5679

Heller EA, Liu E, Tager AM, Yuan Q, Lin AY, Ahluwalia N, Jones K, Koehn SL, Lok VM, Aikawa E, Moore KJ, Luster AD, Gerszten RE (2006) Chemokine CXCL10 promotes atherogenesis by modulating the local balance of effector and regulatory T cells. Circulation 113:2301–2312

Hemmer B, Archelos JJ, Hartung HP (2002) New concepts in the immunopathogenesis of multiple sclerosis. Nat Rev Neurosci 3:291–301

Herman PK, Stack JH, Emr SD (1992) An essential role for a protein and lipid kinase complex in secretory protein sorting. Trends Cell Biol 2:363–368

Herman SE, Gordon AL, Wagner AJ, Heerema NA, Zhao W, Flynn JM, Jones J, Andritsos L, Puri KD, Lannutti BJ, Giese NA, Zhang X, Wei L, Byrd JC, Johnson AJ (2010) Phosphatidylinositol 3-kinase-delta inhibitor CAL-101 shows promising preclinical activity in chronic lymphocytic leukemia by antagonizing intrinsic and extrinsic cellular survival signals. Blood 116:2078–2088

Hess G, Herbrecht R, Romaguera J, Verhoef G, Crump M, Gisselbrecht C, Laurell A, Offner F, Strahs A, Berkenblit A, Hanushevsky O, Clancy J, Hewes B, Moore L, Coiffier B (2009) Phase III study to evaluate temsirolimus compared with investigator's choice therapy for the treatment of relapsed or refractory mantle cell lymphoma. J Clin Oncol 27:3822–3829

Hiom K (2005) DNA repair: how to PIKK a partner. Curr Biol 15:R473–R475

Hirsch E, Katanaev VL, Garlanda C, Azzolino O, Pirola L, Silengo L, Sozzani S, Mantovani A, Altruda F, Wymann MP (2000) Central role for G protein-coupled phosphoinositide 3-kinase gamma in inflammation. Science 287:1049–1053

Hirsch E, Bosco O, Tropel P, Laffargue M, Calvez R, Altruda F, Wymann M, Montrucchio G (2001) Resistance to thromboembolism in PI3Kgamma-deficient mice. FASEB J 15:2019–2021

Hoang B, Frost P, Shi Y, Belanger E, Benavides A, Pezeshkpour G, Cappia S, Guglielmelli T, Gera J, Lichtenstein A (2010) Targeting TORC2 in multiple myeloma with a new mTOR kinase inhibitor. Blood 116:4560–4568

Hobert JA, Eng C (2009) PTEN hamartoma tumor syndrome: an overview. Genet Med 11:687–694

Hoedemaeker FJ, Siegal G, Roe SM, Driscoll PC, Abrahams JP (1999) Crystal structure of the C-terminal SH2 domain of the p85alpha regulatory subunit of phosphoinositide 3-kinase: an SH2 domain mimicking its own substrate. J Mol Biol 292:763–770

Hong F, Larrea MD, Doughty C, Kwiatkowski DJ, Squillace R, Slingerland JM (2008) mTOR-raptor binds and activates SGK1 to regulate p27 phosphorylation. Mol Cell 30:701–711

Huang CH, Mandelker D, Schmidt-Kittler O, Samuels Y, Velculescu VE, Kinzler KW, Vogelstein B, Gabelli SB, Amzel LM (2007) The structure of a human p110alpha/p85alpha complex elucidates the effects of oncogenic PI3Kalpha mutations. Science 318:1744–1748

Huang CH, Mandelker D, Gabelli SB, Amzel LM (2008) Insights into the oncogenic effects of PIK3CA mutations from the structure of p110alpha/p85alpha. Cell Cycle 7:1151–1156

Hudes G, Carducci M, Tomczak P, Dutcher J, Figlin R, Kapoor A, Staroslawska E, Sosman J, McDermott D, Bodrogi I, Kovacevic Z, Lesovoy V, Schmidt-Wolf IG, Barbarash O, Gokmen E, O'Toole T, Lustgarten S, Moore L, Motzer RJ (2007) Temsirolimus, interferon alfa, or both for advanced renal-cell carcinoma. N Engl J Med 356:2271–2281

Ihle NT, Powis G (2010) Inhibitors of phosphatidylinositol-3-kinase in cancer therapy. Mol Aspects Med 31:135–144

Ihle NT, Williams R, Chow S, Chew W, Berggren MI, Paine-Murrieta G, Minion DJ, Halter RJ, Wipf P, Abraham R, Kirkpatrick L, Powis G (2004) Molecular pharmacology and antitumor activity of PX-866, a novel inhibitor of phosphoinositide-3-kinase signaling. Mol Cancer Ther 3:763–772

Ikeda H, Hideshima T, Fulciniti M, Perrone G, Miura N, Yasui H, Okawa Y, Kiziltepe T, Santo L, Vallet S, Cristea D, Calabrese E, Gorgun G, Raje NS, Richardson P, Munshi NC, Lannutti BJ, Puri KD, Giese NA, Anderson KC (2010) PI3K/p110{delta} is a novel therapeutic target in multiple myeloma. Blood 116:1460–1468

Ikenoue T, Kanai F, Hikiba Y, Obata T, Tanaka Y, Imamura J, Ohta M, Jazag A, Guleng B, Tateishi K, Asaoka Y, Matsumura M, Kawabe T, Omata M (2005) Functional analysis of PIK3CA gene mutations in human colorectal cancer. Cancer Res 65:4562–4567

Inoki K, Li Y, Xu T, Guan KL (2003) Rheb GTPase is a direct target of TSC2 GAP activity and regulates mTOR signaling. Genes Dev 17:1829–1834

Ishibashi S, Brown MS, Goldstein JL, Gerard RD, Hammer RE, Herz J (1993) Hypercholesterolemia in low density lipoprotein receptor knockout mice and its reversal by adenovirus-mediated gene delivery. J Clin Invest 92:883–893

Izzard RA, Jackson SP, Smith GC (1999) Competitive and noncompetitive inhibition of the DNA-dependent protein kinase. Cancer Res 59:2581–2586

Jaber M, Koch WJ, Rockman H, Smith B, Bond RA, Sulik KK, Ross JJ, Lefkowitz RJ, Caron MG, Giros B (1996) Essential role of beta-adrenergic receptor kinase 1 in cardiac development and function. Proc Natl Acad Sci USA 93:12974–12979

Jacinto E, Facchinetti V, Liu D, Soto N, Wei S, Jung SY, Huang Q, Qin J, Su B (2006) SIN1/MIP1 maintains rictor-mTOR complex integrity and regulates Akt phosphorylation and substrate specificity. Cell 127:125–137

Jackson SP, Schoenwaelder SM, Goncalves I, Nesbitt WS, Yap CL, Wright CE, Kenche V, Anderson KE, Dopheide SM, Yuan Y, Sturgeon SA, Prabaharan H, Thompson PE, Smith GD, Shepherd PR, Daniele N, Kulkarni S, Abbott B, Saylik D, Jones C, Lu L, Giuliano S, Hughan SC, Angus JA, Robertson AD, Salem HH (2005) PI 3-kinase p110beta: a new target for antithrombotic therapy. Nat Med 11:507–514

Jaiswal BS, Janakiraman V, Kljavin NM, Chaudhuri S, Stern HM, Wang W, Kan Z, Dbouk HA, Peters BA, Waring P, Dela Vega T, Kenski DM, Bowman KK, Lorenzo M, Li H, Wu J, Modrusan Z, Stinson J, Eby M, Yue P, Kaminker JS, De Sauvage FJ, Backer JM, Seshagiri S (2009) Somatic mutations in p85alpha promote tumorigenesis through class IA PI3K activation. Cancer Cell 16:463–474

Jessen K, Wang S, Kessler L, Guo X, Kucharski J, Staunton J, Lan L, Elia M, Stewart J, Brown J, Li L, Chan K, Martin M, Ren P, Rommel C, Liu Y (2009) Abstract B148: INK128 is a potent and selective TORC1/2 inhibitor with broad oral antitumor activity. Mol Cancer Ther 8 (Abstract B148)

Ji QS, Winnier GE, Niswender KD, Horstman D, Wisdom R, Magnuson MA, Carpenter G (1997) Essential role of the tyrosine kinase substrate phospholipase C-gamma1 in mammalian growth and development. Proc Natl Acad Sci USA 94:2999–3003

Ji H, Rintelen F, Waltzinger C, Bertschy Meier D, Bilancio A, Pearce W, Hirsch E, Wymann MP, Ruckle T, Camps M, Vanhaesebroeck B, Okkenhaug K, Rommel C (2007) Inactivation of PI3Kgamma and PI3Kdelta distorts T-cell development and causes multiple organ inflammation. Blood 110:2940–2947

Jia S, Liu Z, Zhang S, Liu P, Zhang L, Lee SH, Zhang J, Signoretti S, Loda M, Roberts TM, Zhao JJ (2008) Essential roles of PI(3)K-p110beta in cell growth, metabolism and tumorigenesis. Nature 454:776–779

Jimenez C, Jones DR, Rodriguez-Viciana P, Gonzalez-Garcia A, Leonardo E, Wennstrom S, Kobbe C von, Toran JL, R-Borlado L, Calvo V, Copin SG, Albar JP, Gaspar ML, Diez E, Marcos MA, Downward J, Martinez AC, Merida I, Carrera AC (1998) Identification and characterization of a new oncogene derived from the regulatory subunit of phosphoinositide 3-kinase. EMBO J 17:743–753

Johnston PB, Inwards DJ, Colgan JP, Laplant BR, Kabat BF, Habermann TM, Micallef IN, Porrata LF, Ansell SM, Reeder CB, Roy V, Witzig TE (2010a) A phase II trial of the oral mTOR inhibitor everolimus in relapsed Hodgkin lymphoma. Am J Hematol 85:320–324

Johnston PB, Yuan R, Cavalli F, Witzig TE (2010b) Targeted therapy in lymphoma. J Hematol Oncol 3:45

Jones GE, Prigmore E, Calvez R, Hogan C, Dunn GA, Hirsch E, Wymann MP, Ridley AJ (2003) Requirement for PI 3-kinase gamma in macrophage migration to MCP-1 and CSF-1. Exp Cell Res 290:120–131

Jou ST, Carpino N, Takahashi Y, Piekorz R, Chao JR, Carpino N, Wang D, Ihle JN (2002) Essential, nonredundant role for the phosphoinositide 3-kinase p110delta in signaling by the B-cell receptor complex. Mol Cell Biol 22:8580–8591

Jucker M, Sudel K, Horn S, Sickel M, Wegner W, Fiedler W, Feldman RA (2002) Expression of a mutated form of the p85alpha regulatory subunit of phosphatidylinositol 3-kinase in a Hodgkin's lymphoma-derived cell line (CO). Leukemia 16:894–901

Juhasz G, Hill JH, Yan Y, Sass M, Baehrecke EH, Backer JM, Neufeld TP (2008) The class III PI(3)K Vps34 promotes autophagy and endocytosis but not TOR signaling in Drosophila. J Cell Biol 181:655–666

Kalesnikoff J, Sly LM, Hughes MR, Buchse T, Rauh MJ, Cao LP, Lam V, Mui A, Huber M, Krystal G (2003) The role of SHIP in cytokine-induced signaling. Rev Physiol Biochem Pharmacol 149:87–103

Kang S, Bader AG, Vogt PK (2005) Phosphatidylinositol 3-kinase mutations identified in human cancer are oncogenic. Proc Natl Acad Sci USA 102:802–807

Kang S, Denley A, Vanhaesebroeck B, Vogt PK (2006) Oncogenic transformation induced by the p110beta, -gamma, and -delta isoforms of class I phosphoinositide 3-kinase. Proc Natl Acad Sci USA 103:1289–1294

Kapahi P, Chen D, Rogers AN, Katewa SD, Li PW, Thomas EL, Kockel L (2010) With TOR, less is more: a key role for the conserved nutrient-sensing TOR pathway in aging. Cell Metab 11:453–465

Kaufmann T, Schinzel A, Borner C (2004) Bcl-w(edding) with mitochondria. Trends Cell Biol 14:8–12

Kawakami Y, Nishimoto H, Kitaura J, Maeda-Yamamoto M, Kato RM, Littman DR, Leitges M, Rawlings DJ, Kawakami T (2004) Protein kinase C betaII regulates Akt phosphorylation on Ser-473 in a cell type- and stimulus-specific fashion. J Biol Chem 279:47720–47725

Khan S, Kumagai T, Vora J, Bose N, Sehgal I, Koeffler PH, Bose S (2004) PTEN promoter is methylated in a proportion of invasive breast cancers. Int J Cancer 112:407–410

Kiefer S, Rogger J, Melone A, Mertz AC, Koryakina A, Hamburger M, Kuenzi P (2010) Separation and detection of all phosphoinositide isomers by ESI-MS. J Pharm Biomed Anal 53:552–558

Kihara A, Kabeya Y, Ohsumi Y, Yoshimori T (2001) Beclin-phosphatidylinositol 3-kinase complex functions at the trans-Golgi network. EMBO Rep 2:330–335

Kim MS, Radinger M, Gilfillan AM (2008) The multiple roles of phosphoinositide 3-kinase in mast cell biology. Trends Immunol 29:493–501

Kisseleva MV, Wilson MP, Majerus PW (2000) The isolation and characterization of a cDNA encoding phospholipid-specific inositol polyphosphate 5-phosphatase. J Biol Chem 275:20110–20116

Kitaura J, Kinoshita T, Matsumoto M, Chung S, Kawakami Y, Leitges M, Wu D, Lowell CA, Kawakami T (2005) IgE- and IgE+Ag-mediated mast cell migration in an autocrine/paracrine fashion. Blood 105:3222–3229

Knight ZA, Gonzalez B, Feldman ME, Zunder ER, Goldenberg DD, Williams O, Loewith R, Stokoe D, Balla A, Toth B, Balla T, Weiss WA, Williams RL, Shokat KM (2006) A pharmacological map of the PI3-K family defines a role for p110alpha in insulin signaling. Cell 125:733–747

Knight SD, Adams ND, Burgess JL, Chaudhari AM, Darcy MG, Donatelli CA, Newlander KA, Parrish CA, Ridgers LH, Sarpong MA, Schmidt SJ, Van Aller G, Carson JD, Elkins PA, Diamond M, Gardiner CM, Garver E, Luo L, Raha K, Sung C-M, Tummino PJ, Auger KR, Dhanak D (2009) Identification of GSK2126458, a highly potent inhibitor pf phosphoinositide 3-kinase (PI3K) and the mammalian target of rapamycin (mTOR). Mol Cancer Ther 8:C62

Knowling M, Blackstein M, Tozer R, Bramwell V, Dancey J, Dore N, Matthews S, Eisenhauer E (2006) A phase II study of perifosine (D-21226) in patients with previously untreated metastatic or locally advanced soft tissue sarcoma: A National Cancer Institute of Canada Clinical Trials Group trial. Invest New Drugs 24:435–439

Kong D, Yamori T (2007) ZSTK474 is an ATP-competitive inhibitor of class I phosphatidylinositol 3 kinase isoforms. Cancer Sci 98:1638–1642

Kong D, Yamori T (2009) Advances in development of phosphatidylinositol 3-kinase inhibitors. Curr Med Chem 16:2839–2854

Kong D, Dan S, Yamazaki K, Yamori T (2010) Inhibition profiles of phosphatidylinositol 3-kinase inhibitors against PI3K superfamily and human cancer cell line panel JFCR39. Eur J Cancer 46:1111–1121

Korganow AS, Ji H, Mangialaio S, Duchatelle V, Pelanda R, Martin T, Degott C, Kikutani H, Rajewsky K, Pasquali JL, Benoist C, Mathis D (1999) From systemic T cell self-reactivity to organ-specific autoimmune disease via immunoglobulins. Immunity 10:451–461

Kraft S, Kinet JP (2007) New developments in FcepsilonRI regulation, function and inhibition. Nat Rev Immunol 7:365–378

Krag C, Malmberg EK, Salcini AE (2010) PI3KC2alpha, a class II PI3K, is required for dynamin-independent internalization pathways. J Cell Sci 123:4240–4250

Krugmann S, Hawkins PT, Pryer N, Braselmann S (1999) Characterizing the interactions between the two subunits of the p101/p110gamma phosphoinositide 3-kinase and their role in the activation of this enzyme by G beta gamma subunits. J Biol Chem 274:17152–17158

Krystal G, Damen JE, Helgason CD, Huber M, Hughes MR, Kalesnikoff J, Lam V, Rosten P, Ware MD, Yew S, Humphries RK (1999) SHIPs ahoy. Int J Biochem Cell Biol 31:1007–1010

Kubo H, Hazeki K, Takasuga S, Hazeki O (2005) Specific role for p85/p110beta in GTP-binding-protein-mediated activation of Akt. Biochem J 392:607–614

Kunisaki Y, Nishikimi A, Tanaka Y, Takii R, Noda M, Inayoshi A, Watanabe K, Sanematsu F, Sasazuki T, Sasaki T, Fukui Y (2006) DOCK2 is a Rac activator that regulates motility and polarity during neutrophil chemotaxis. J Cell Biol 174:647–652

Kurig B, Shymanets A, Bohnacker T, Prajwal, Brock C, Ahmadian MR, Schaefer M, Gohla A, Harteneck C, Wymann MP, Jeanclos E, Nurnberg B (2009) Ras is an indispensable coregulator of the class IB phosphoinositide 3-kinase p87/p110gamma. Proc Natl Acad Sci USA 106:20312–20317

Kurosu H, Katada T (2001) Association of phosphatidylinositol 3-kinase composed of p110beta-catalytic and p85-regulatory subunits with the small GTPase Rab5. J Biochem 130:73–78

Kurosu H, Maehama T, Okada T, Yamamoto T, Hoshino S, Fukui Y, Ui M, Hazeki O, Katada T (1997) Heterodimeric phosphoinositide 3-kinase consisting of p85 and p110beta is synergistically activated by the betagamma subunits of G proteins and phosphotyrosyl peptide. J Biol Chem 272:24252–24256

Laffargue M, Calvez R, Finan P, Trifilieff A, Barbier M, Altruda F, Hirsch E, Wymann MP (2002) Phosphoinositide 3-kinase gamma is an essential amplifier of mast cell function. Immunity 16:441–451

Laird AD, Sillman A, Sun B, Mengistab A, Chu F, Lee M, Cancilla B, Aggarwal SK, Bentzien F (2008) Evaluation of peripheral blood cells and hair as surrogate tissues for clinical trial pharmacodynamic assessment of XL147 and XL765, inhibitors of the PI3K signaling pathway. Eur J Cancer 6 (Abstract 89)

Lane HA, Wood JM, McSheehy PM, Allegrini PR, Boulay A, Brueggen J, Littlewood-Evans A, Maira SM, Martiny-Baron G, Schnell CR, Sini P, O'Reilly T (2009) mTOR inhibitor RAD001 (everolimus) has antiangiogenic/vascular properties distinct from a VEGFR tyrosine kinase inhibitor. Clin Cancer Res 15:1612–1622

Lannutti B, Kashishian A, Meadows SA, Steiner B, Ueno L, Webb HK, Puri KD, Ulrich RG, Vogt PK, Giese NA (2009) CAL-120, a novel dual p110β/p110δ phosphatidylinositol-3-kinase (PI3K) inhibitor, attenuates PI3K signaling and demonstrates potent in vivo antitumor activity against solid tumors. Mol Cancer Ther 8:B136

Lannutti BJ, Meadows SA, Herman SE, Kashishian A, Steiner B, Johnson AJ, Byrd JC, Tyner JW, Loriaux MM, Deininger M, Druker BJ, Puri KD, Ulrich RG, Giese NA (2011) CAL-101, a p110delta selective phosphatidylinositol-3-kinase inhibitor (PI3K) for the treatment of B-cell malignancies, inhibits PI3K signaling and cellular viability. Blood 117:591–594

Laplante M, Sabatini DM (2009) An emerging role of mTOR in lipid biosynthesis. Curr Biol 19:R1046–R1052

Laporte J, Blondeau F, Buj-Bello A, Mandel JL (2001) The myotubularin family: from genetic disease to phosphoinositide metabolism. Trends Genet 17:221–228

Lee KS, Lee HK, Hayflick JS, Lee YC, Puri KD (2006a) Inhibition of phosphoinositide 3-kinase delta attenuates allergic airway inflammation and hyperresponsiveness in murine asthma model. FASEB J 20:455–465

Lee KS, Park SJ, Kim SR, Min KH, Jin SM, Puri KD, Lee YC (2006b) Phosphoinositide 3-kinase-delta inhibitor reduces vascular permeability in a murine model of asthma. J Allergy Clin Immunol 118:403–409

Leevers SJ, Weinkove D, MacDougall LK, Hafen E, Waterfield MD (1996) The Drosophila phosphoinositide 3-kinase Dp110 promotes cell growth. EMBO J 15:6584–6594

Lemmon MA (2008) Membrane recognition by phospholipid-binding domains. Nat Rev Mol Cell Biol 9:99–111

Leslie NR, Batty IH, Maccario H, Davidson L, Downes CP (2008) Understanding PTEN regulation: PIP2, polarity and protein stability. Oncogene 27:5464–5476

Levy DS, Kahana JA, Kumar R (2009) AKT inhibitor, GSK690693, induces growth inhibition and apoptosis in acute lymphoblastic leukemia cell lines. Blood 113:1723–1729

Li Q, Verma IM (2002) NF-kappaB regulation in the immune system. Nat Rev Immunol 2:725–734

Li J, Yen C, Liaw D, Podsypanina K, Bose S, Wang SI, Puc J, Miliaresis C, Rodgers L, McCombie R, Bigner SH, Giovanella BC, Ittmann M, Tycko B, Hibshoosh H, Wigler MH, Parsons R (1997) PTEN, a putative protein tyrosine phosphatase gene mutated in human brain, breast, and prostate cancer. Science 275:1943–1947

Li Z, Jiang H, Xie W, Zhang Z, Smrcka AV, Wu D (2000) Roles of PLC-beta2 and -beta3 and PI3Kgamma in chemoattractant-mediated signal transduction. Science 287:1046–1049

Li X, Jiang S, Tapping RI (2010) Toll-like receptor signaling in cell proliferation and survival. Cytokine 49:1–9

Liang J, Slingerland JM (2003) Multiple roles of the PI3K/PKB (Akt) pathway in cell cycle progression. Cell Cycle 2:339–345

Liaw D, Marsh DJ, Li J, Dahia PL, Wang SI, Zheng Z, Bose S, Call KM, Tsou HC, Peacocke M, Eng C, Parsons R (1997) Germline mutations of the PTEN gene in Cowden disease, an inherited breast and thyroid cancer syndrome. Nat Genet 16:64–67

Ling K, Schill NJ, Wagoner MP, Sun Y, Anderson RA (2006) Movin' on up: the role of PtdIns(4,5)P(2) in cell migration. Trends Cell Biol 16:276–284

Liu K, Wakeland EK (2001) Delineation of the pathogenesis of systemic lupus erythematosus by using murine models. Adv Exp Med Biol 490:1–6

Liu Y, Shreder KR, Gai W, Corral S, Ferris DK, Rosenblum JS (2005) Wortmannin, a widely used phosphoinositide 3-kinase inhibitor, also potently inhibits mammalian polo-like kinase. Chem Biol 12:99–107

Liu X, Shi Y, Woods KW, Hessler P, Kroeger P, Wilsbacher J, Wang J, Wang JY, Li C, Li Q, Rosenberg SH, Giranda VL, Luo Y (2008) Akt inhibitor a-443654 interferes with mitotic progression by regulating aurora a kinase expression. Neoplasia 10:828–837

Liu P, Cheng H, Roberts TM, Zhao JJ (2009) Targeting the phosphoinositide 3-kinase pathway in cancer. Nat Rev Drug Discov 8:627–644

Lowes MA, Bowcock AM, Krueger JG (2007) Pathogenesis and therapy of psoriasis. Nature 445:866–873

Luo J, McMullen JR, Sobkiw CL, Zhang L, Dorfman AL, Sherwood MC, Logsdon MN, Horner JW, DePinho RA, Izumo S, Cantley LC (2005) Class IA phosphoinositide 3-kinase regulates heart size and physiological cardiac hypertrophy. Mol Cell Biol 25:9491–9502

Lupia E, Goffi A, De Giuli P, Azzolino O, Bosco O, Patrucco E, Vivaldo MC, Ricca M, Wymann MP, Hirsch E, Montrucchio G, Emanuelli G (2004) Ablation of phosphoinositide 3-kinase-gamma reduces the severity of acute pancreatitis. Am J Pathol 165:2003–2011

Lusis AJ (2000) Atherosclerosis. Nature 407:233–241

Lyons JF, Grimshaw KM, Woodhead SJ, Feltell RE, Reule M, Smyth T, Seavers LC, Harada I, Higgins J, Smith DM, Fazal L, Workman P (2007) AT13148, an orally bioavailable AKT kinase inhibitor with potent anti-tumour activity in both in vitro and in vivo models exhibiting AKT pathway deregulation. http://www.astex-therapeutics.com/event_pdfs/Astex%20AACR%20EORTC%20PKB%20Poster%20October%202007.pdf

Macara IG, Marinetti GV, Balduzzi PC (1984) Transforming protein of avian sarcoma virus UR2 is associated with phosphatidylinositol kinase activity: possible role in tumorigenesis. Proc Natl Acad Sci USA 81:2728–2732

MacDougall LK, Domin J, Waterfield MD (1995) A family of phosphoinositide 3-kinases in Drosophila identifies a new mediator of signal transduction. Curr Biol 5:1404–1415

Maffucci T, Cooke FT, Foster FM, Traer CJ, Fry MJ, Falasca M (2005) Class II phosphoinositide 3-kinase defines a novel signaling pathway in cell migration. J Cell Biol 169:789–799

Maier U, Babich A, Nurnberg B (1999) Roles of non-catalytic subunits in gbetagamma-induced activation of class I phosphoinositide 3-kinase isoforms beta and gamma. J Biol Chem 274:29311–29317

Maira SM, Stauffer F, Brueggen J, Furet P, Schnell C, Fritsch C, Brachmann S, Chene P, De Pover A, Schoemaker K, Fabbro D, Gabriel D, Simonen M, Murphy L, Finan P, Sellers W, Garcia-Echeverria C (2008) Identification and characterization of NVP-BEZ235, a new orally available dual phosphatidylinositol 3-kinase/mammalian target of rapamycin inhibitor with potent in vivo antitumor activity. Mol Cancer Ther 7:1851–1863

Majerus PW, Kisseleva MV, Norris FA (1999) The role of phosphatases in inositol signaling reactions. J Biol Chem 274:10669–10672

Mallon R, Hollander I, Feldberg L, Lucas J, Soloveva V, Venkatesan A, Dehnhardt C, Delos Santos E, Chen Z, Dos Santos O, Ayral-Kaloustian S, Gibbons J (2010) Antitumor efficacy profile of PKI-402, a dual phosphatidylinositol 3-kinase/mammalian target of rapamycin inhibitor. Mol Cancer Ther 9:976–984

Manning BD (2004) Balancing Akt with S6K: implications for both metabolic diseases and tumorigenesis. J Cell Biol 167:399–403

Manning BD, Cantley LC (2007) AKT/PKB signaling: navigating downstream. Cell 129:1261–1274

Marone R, Cmiljanovic V, Giese B, Wymann MP (2008) Targeting phosphoinositide 3-kinase: moving towards therapy. Biochim Biophys Acta 1784:159–185

Marone R, Erhart D, Mertz AC, Bohnacker T, Schnell C, Cmiljanovic V, Stauffer F, Garcia-Echeverria C, Giese B, Maira SM, Wymann MP (2009) Targeting melanoma with dual phosphoinositide 3-kinase/mammalian target of rapamycin inhibitors. Mol Cancer Res 7:601–613

Maroun CR, Naujokas MA, Park M (2003) Membrane targeting of Grb2-associated binder-1 (Gab1) scaffolding protein through Src myristoylation sequence substitutes for Gab1 pleckstrin homology domain and switches an epidermal growth factor response to an invasive morphogenic program. Mol Biol Cell 14:1691–1708

Marques M, Kumar A, Cortes I, Gonzalez-Garcia A, Hernandez C, Moreno-Ortiz MC, Carrera AC (2008) Phosphoinositide 3-kinases p110alpha and p110beta regulate cell cycle entry, exhibiting distinct activation kinetics in G1 phase. Mol Cell Biol 28:2803–2814

Martin DE, Hall MN (2005) The expanding TOR signaling network. Curr Opin Cell Biol 17:158–166

Marwick JA, Caramori G, Stevenson CS, Casolari P, Jazrawi E, Barnes PJ, Ito K, Adcock IM, Kirkham PA, Papi A (2009) Inhibition of PI3Kdelta restores glucocorticoid function in smoking-induced airway inflammation in mice. Am J Respir Crit Care Med 179:542–548

Matsunaga K, Morita E, Saitoh T, Akira S, Ktistakis NT, Izumi T, Noda T, Yoshimori T (2010) Autophagy requires endoplasmic reticulum targeting of the PI3-kinase complex via Atg14L. J Cell Biol 190:511–521

McCoy KD, Harris NL, Diener P, Hatak S, Odermatt B, Hangartner L, Senn BM, Marsland BJ, Geuking MB, Hengartner H, Macpherson AJ, Zinkernagel RM (2006) Natural IgE production in the absence of MHC Class II cognate help. Immunity 24:329–339

McManus EJ, Collins BJ, Ashby PR, Prescott AR, Murray-Tait V, Armit LJ, Arthur JS, Alessi DR (2004) The in vivo role of PtdIns(3,4,5)P3 binding to PDK1 PH domain defined by knockin mutation. EMBO J 23:2071–2082

Mendoza MC, Blenis J (2007) PHLPPing it off: phosphatases get in the Akt. Mol Cell 25:798–800

Merkenschlager M, Boehmer H von (2010) PI3 kinase signalling blocks Foxp3 expression by sequestering Foxo factors. J Exp Med 207:1347–1350

Meunier FA, Osborne SL, Hammond GR, Cooke FT, Parker PJ, Domin J, Schiavo G (2005) Phosphatidylinositol 3-kinase C2alpha is essential for ATP-dependent priming of neurosecretory granule exocytosis. Mol Biol Cell 16:4841–4851

Milburn CC, Deak M, Kelly SM, Price NC, Alessi DR, Van Aalten DM (2003) Binding of phosphatidylinositol 3,4,5-trisphosphate to the pleckstrin homology domain of protein kinase B induces a conformational change. Biochem J 375:531–538

Miled N, Yan Y, Hon WC, Perisic O, Zvelebil M, Inbar Y, Schneidman-Duhovny D, Wolfson HJ, Backer JM, Williams RL (2007) Mechanism of two classes of cancer mutations in the phosphoinositide 3-kinase catalytic subunit. Science 317:239–242

Miller N (2009) Abstract B146: XL388: a novel, selective, orally bioavailable mTORC1 and mTORC2 inhibitor that demonstrates pharmacodynamic and antitumor activity in multiple human cancer xenograft models. Mol Cancer Ther 8(12): B146. doi:10.1158/1535-7163.TARG-09-B146

Minogue S, Anderson JS, Waugh MG, Dos Santos M, Corless S, Cramer R, Hsuan JJ (2001) Cloning of a human type II phosphatidylinositol 4-kinase reveals a novel lipid kinase family. J Biol Chem 276:16635–16640

Mirmohammadsadegh A, Marini A, Nambiar S, Hassan M, Tannapfel A, Ruzicka T, Hengge UR (2006) Epigenetic silencing of the PTEN gene in melanoma. Cancer Res 66:6546–6552

Mohamed AJ, Yu L, Backesjo CM, Vargas L, Faryal R, Aints A, Christensson B, Berglof A, Vihinen M, Nore BF, Smith CI (2009) Bruton's tyrosine kinase (Btk): function, regulation, and transformation with special emphasis on the PH domain. Immunol Rev 228:58–73

Morris JZ, Tissenbaum HA, Ruvkun G (1996) A phosphatidylinositol-3-OH kinase family member regulating longevity and diapause in Caenorhabditis elegans. Nature 382:536–539

Motzer RJ, Escudier B, Oudard S, Hutson TE, Porta C, Bracarda S, Grunwald V, Thompson JA, Figlin RA, Hollaender N, Urbanowitz G, Berg WJ, Kay A, Lebwohl D, Ravaud A (2008) Efficacy of everolimus in advanced renal cell carcinoma: a double-blind, randomised, placebo-controlled phase III trial. Lancet 372:449–456

Mruk DD, Cheng CY (2010) The myotubularin family of lipid phosphatases in disease and in spermatogenesis. Biochem J 433:253–262

Murphy MA, Schnall RG, Venter DJ, Barnett L, Bertoncello I, Thien CB, Langdon WY, Bowtell DD (1998) Tissue hyperplasia and enhanced T-cell signalling via ZAP-70 in c-Cbl-deficient mice. Mol Cell Biol 18:4872–4882

Murray JT, Panaretou C, Stenmark H, Miaczynska M, Backer JM (2002) Role of Rab5 in the recruitment of hVps34/p150 to the early endosome. Traffic 3:416–427

Nakashima Y, Plump AS, Raines EW, Breslow JL, Ross R (1994) ApoE-deficient mice develop lesions of all phases of atherosclerosis throughout the arterial tree. Arterioscler Thromb 14:133–140

Neshat MS, Mellinghoff IK, Tran C, Stiles B, Thomas G, Petersen R, Frost P, Gibbons JJ, Wu H, Sawyers CL (2001) Enhanced sensitivity of PTEN-deficient tumors to inhibition of FRAP/mTOR. Proc Natl Acad Sci USA 98:10314–10319

Nezis IP, Sagona AP, Schink KO, Stenmark H (2010) Divide and ProsPer: the emerging role of PtdIns3P in cytokinesis. Trends Cell Biol 20:642–649

Nilsson J, Wigren M, Shah PK (2009) Regulatory T cells and the control of modified lipoprotein autoimmunity-driven atherosclerosis. Trends Cardiovasc Med 19:272–276

Nishida K, Hirano T (2003) The role of Gab family scaffolding adapter proteins in the signal transduction of cytokine and growth factor receptors. Cancer Sci 94:1029–1033

Nishikimi A, Fukuhara H, Su W, Hongu T, Takasuga S, Mihara H, Cao Q, Sanematsu F, Kanai M, Hasegawa H, Tanaka Y, Shibasaki M, Kanaho Y, Sasaki T, Frohman MA, Fukui Y (2009) Sequential regulation of DOCK2 dynamics by two phospholipids during neutrophil chemotaxis. Science 324:384–387

Nishio M, Watanabe K, Sasaki J, Taya C, Takasuga S, Iizuka R, Balla T, Yamazaki M, Watanabe H, Itoh R, Kuroda S, Horie Y, Forster I, Mak TW, Yonekawa H, Penninger JM, Kanaho Y, Suzuki A, Sasaki T (2007) Control of cell polarity and motility by the PtdIns(3,4,5)P3 phosphatase SHIP1. Nat Cell Biol 9:36–44

Nobukuni T, Kozma SC, Thomas G (2007) hvps34, an ancient player, enters a growing game: mTOR Complex1/S6K1 signaling. Curr Opin Cell Biol 19:135–141

Nolte RT, Eck MJ, Schlessinger J, Shoelson SE, Harrison SC (1996) Crystal structure of the PI 3-kinase p85 amino-terminal SH2 domain and its phosphopeptide complexes. Nat Struct Biol 3:364–374

Nombela-Arrieta C, Lacalle RA, Montoya MC, Kunisaki Y, Megias D, Marques M, Carrera AC, Manes S, Fukui Y, Martinez-A C, Stein JV (2004) Differential requirements for DOCK2 and phosphoinositide-3-kinase gamma during T and B lymphocyte homing. Immunity 21:429–441

Nombela-Arrieta C, Mempel TR, Soriano SF, Mazo I, Wymann MP, Hirsch E, Martinez-A C, Fukui Y, Von Andrian UH, Stein JV (2007) A central role for DOCK2 during interstitial lymphocyte motility and sphingosine-1-phosphate-mediated egress. J Exp Med 204:497–510

Oak JS, Deane JA, Kharas MG, Luo J, Lane TE, Cantley LC, Fruman DA (2006) Sjogren's syndrome-like disease in mice with T cells lacking class 1A phosphoinositide-3-kinase. Proc Natl Acad Sci USA 103:16882–16887

O'Donnell A, Faivre S, Burris HA 3rd, Rea D, Papadimitrakopoulou V, Shand N, Lane HA, Hazell K, Zoellner U, Kovarik JM, Brock C, Jones S, Raymond E, Judson I (2008) Phase I pharmacokinetic and pharmacodynamic study of the oral mammalian target of rapamycin inhibitor everolimus in patients with advanced solid tumors. J Clin Oncol 26:1588–1595

Okkenhaug K, Fruman DA (2010) PI3Ks in lymphocyte signaling and development. Curr Top Microbiol Immunol 346:57–85

Okkenhaug K, Vanhaesebroeck B (2003) PI3K in lymphocyte development, differentiation and activation. Nat Rev Immunol 3:317–330

Okkenhaug K, Bilancio A, Farjot G, Priddle H, Sancho S, Peskett E, Pearce W, Meek SE, Salpekar A, Waterfield MD, Smith AJ, Vanhaesebroeck B (2002) Impaired B and T cell antigen receptor signaling in p110delta PI 3-kinase mutant mice. Science 297:1031–1034

Oliveira V, Romanow WJ, Geisen C, Otterness DM, Mercurio F, Wang HG, Dalton WS, Abraham RT (2008) A protective role for the human SMG-1 kinase against tumor necrosis factor-alpha-induced apoptosis. J Biol Chem 283:13174–13184

Omori SA, Cato MH, Anzelon-Mills A, Puri KD, Shapiro-Shelef M, Calame K, Rickert RC (2006) Regulation of class-switch recombination and plasma cell differentiation by phosphatidylinositol 3-kinase signaling. Immunity 25:545–557

Opar A (2007) Where now for new drugs for atherosclerosis? Nat Rev Drug Discov 6:334–335

O'Reilly KE, Rojo F, She QB, Solit D, Mills GB, Smith D, Lane H, Hofmann F, Hicklin DJ, Ludwig DL, Baselga J, Rosen N (2006) mTOR inhibition induces upstream receptor tyrosine kinase signaling and activates Akt. Cancer Res 66:1500–1508

Oudit GY, Sun H, Kerfant BG, Crackower MA, Penninger JM, Backx PH (2004) The role of phosphoinositide-3 kinase and PTEN in cardiovascular physiology and disease. J Mol Cell Cardiol 37:449–471

Ozbay T, Durden DL, Liu T, O'Regan RM, Nahta R (2010) In vitro evaluation of pan-PI3-kinase inhibitor SF1126 in trastuzumab-sensitive and trastuzumab-resistant HER2-over-expressing breast cancer cells. Cancer Chemother Pharmacol 65:697–706

Ozes ON, Mayo LD, Gustin JA, Pfeffer SR, Pfeffer LM, Donner DB (1999) NF-kappaB activation by tumour necrosis factor requires the Akt serine-threonine kinase. Nature 401:82–85

Pacold ME, Suire S, Perisic O, Lara-Gonzalez S, Davis CT, Walker EH, Hawkins PT, Stephens L, Eccleston JF, Williams RL (2000) Crystal structure and functional analysis of Ras binding to its effector phosphoinositide 3-kinase gamma. Cell 103:931–943

Pan D, Dong J, Zhang Y, Gao X (2004) Tuberous sclerosis complex: from Drosophila to human disease. Trends Cell Biol 14:78–85

Panaretou C, Domin J, Cockcroft S, Waterfield MD (1997) Characterization of p150, an adaptor protein for the human phosphatidylinositol (PtdIns) 3-kinase. Substrate presentation by phosphatidylinositol transfer protein to the p150.Ptdins 3-kinase complex. J Biol Chem 272:2477–2485

Park WS, Heo WD, Whalen JH, O'Rourke NA, Bryan HM, Meyer T, Teruel MN (2008) Comprehensive identification of PIP3-regulated PH domains from C. elegans to H. sapiens by model prediction and live imaging. Mol Cell 30:381–392

Parsons DW, Jones S, Zhang X, Lin JC, Leary RJ, Angenendt P, Mankoo P, Carter H, Siu IM, Gallia GL, Olivi A, McLendon R, Rasheed BA, Keir S, Nikolskaya T, Nikolsky Y, Busam DA, Tekleab H, Diaz LA Jr, Hartigan J, Smith DR, Strausberg RL, Marie SK, Shinjo SM, Yan H, Riggins GJ, Bigner DD, Karchin R, Papadopoulos N, Parmigiani G, Vogelstein B, Velculescu VE, Kinzler KW (2008) An integrated genomic analysis of human glioblastoma multiforme. Science 321:1807–1812

Patrucco E, Notte A, Barberis L, Selvetella G, Maffei A, Brancaccio M, Marengo S, Russo G, Azzolino O, Rybalkin SD, Silengo L, Altruda F, Wetzker R, Wymann MP, Lembo G, Hirsch E (2004) PI3Kgamma modulates the cardiac response to chronic pressure overload by distinct kinase-dependent and -independent effects. Cell 118:375–387

Patton DT, Garden OA, Pearce WP, Clough LE, Monk CR, Leung E, Rowan WC, Sancho S, Walker LS, Vanhaesebroeck B, Okkenhaug K (2006) Cutting edge: the phosphoinositide 3-kinase p110 delta is critical for the function of CD4 + CD25 + Foxp3 + regulatory T cells. J Immunol 177:6598–6602

Pearce LR, Komander D, Alessi DR (2010) The nuts and bolts of AGC protein kinases. Nat Rev Mol Cell Biol 11:9–22

Perino A, Ghigo A, Ferrero E, Morello F, Santulli G, Baillie GS, Damilano F, Dunlop AJ, Pawson C, Walser R, Levi R, Altruda F, Silengo L, Langeberg LK, Neubauer G, Heymans S, Lembo G, Wymann MP, Wetzker R, Houslay MD, Iaccarino G, Scott JD, Hirsch E (2011) Integrating Cardiac PIP(3) and cAMP Signaling through a PKA Anchoring Function of p110gamma. Mol Cell 42:84–95

del Peso L, Gonzalez-Garcia M, Page C, Herrera R, Nunez G (1997) Interleukin-3-induced phosphorylation of BAD through the protein kinase Akt. Science 278:687–689

Pettitt TR, Dove SK, Lubben A, Calaminus SD, Wakelam MJ (2006) Analysis of intact phosphoinositides in biological samples. J Lipid Res 47:1588–1596

Philp AJ, Campbell IG, Leet C, Vincan E, Rockman SP, Whitehead RH, Thomas RJ, Phillips WA (2001) The phosphatidylinositol 3′-kinase p85alpha gene is an oncogene in human ovarian and colon tumors. Cancer Res 61:7426–7429

Phung TL, Ziv K, Dabydeen D, Eyiah-Mensah G, Riveros M, Perruzzi C, Sun J, Monahan-Earley RA, Shiojima I, Nagy JA, Lin MI, Walsh K, Dvorak AM, Briscoe DM, Neeman M, Sessa WC, Dvorak HF, Benjamin LE (2006) Pathological angiogenesis is induced by sustained Akt signaling and inhibited by rapamycin. Cancer Cell 10:159–170

Pirola L, Zvelebil MJ, Bulgarelli-Leva G, Van Obberghen E, Waterfield MD, Wymann MP (2001) Activation loop sequences confer substrate specificity to phosphoinositide 3-kinase alpha (PI3Kalpha). Functions of lipid kinase-deficient PI3Kalpha in signaling. J Biol Chem 276:21544–21554

Plump AS, Smith JD, Hayek T, Aalto-Setala K, Walsh A, Verstuyft JG, Rubin EM, Breslow JL (1992) Severe hypercholesterolemia and atherosclerosis in apolipoprotein E-deficient mice created by homologous recombination in ES cells. Cell 71:343–353

Podsypanina K, Ellenson LH, Nemes A, Gu J, Tamura M, Yamada KM, Cordon-Cardo C, Catoretti G, Fisher PE, Parsons R (1999) Mutation of Pten/Mmac1 in mice causes neoplasia in multiple organ systems. Proc Natl Acad Sci USA 96:1563–1568

Polak P, Hall MN (2006) mTORC2 Caught in a SINful Akt. Dev Cell 11:433–434

Polak P, Hall MN (2009) mTOR and the control of whole body metabolism. Curr Opin Cell Biol 21:209–218

Pomel V, Klicic J, Covini D, Church DD, Shaw JP, Roulin K, Burgat-Charvillon F, Valognes D, Camps M, Chabert C, Gillieron C, Francon B, Perrin D, Leroy D, Gretener D, Nichols A, Vitte PA, Carboni S, Rommel C, Schwarz MK, Ruckle T (2006) Furan-2-ylmethylene thiazolidinediones as novel, potent, and selective inhibitors of phosphoinositide 3-kinase gamma. J Med Chem 49:3857–3871

Posadas EM, Gulley J, Arlen PM, Trout A, Parnes HL, Wright J, Lee MJ, Chung EJ, Trepel JB, Sparreboom A, Chen C, Jones E, Steinberg SM, Daniels A, Figg WD, Dahut WL (2005) A phase II study of perifosine in androgen independent prostate cancer. Cancer Biol Ther 4:1133–1137

Pratt JC, Igras VE, Maeda H, Baksh S, Gelfand EW, Burakoff SJ, Neel BG, Gu H (2000) Cutting edge: gab2 mediates an inhibitory phosphatidylinositol 3′-kinase pathway in T cell antigen receptor signaling. J Immunol 165:4158–4163

Pullen N, Dennis PB, Andjelkovic M, Dufner A, Kozma SC, Hemmings BA, Thomas G (1998) Phosphorylation and activation of p70s6k by PDK1. Science 279:707–710

Puri KD, Doggett TA, Huang CY, Douangpanya J, Hayflick JS, Turner M, Penninger J, Diacovo TG (2005) The role of endothelial PI3Kgamma activity in neutrophil trafficking. Blood 106:150–157

Qu X, Yu J, Bhagat G, Furuya N, Hibshoosh H, Troxel A, Rosen J, Eskelinen EL, Mizushima N, Ohsumi Y, Cattoretti G, Levine B (2003) Promotion of tumorigenesis by heterozygous disruption of the beclin 1 autophagy gene. J Clin Invest 112:1809–1820

Racz A, Brass N, Heckel D, Pahl S, Remberger K, Meese E (1999) Expression analysis of genes at 3q26-q27 involved in frequent amplification in squamous cell lung carcinoma. Eur J Cancer 35:641–646

Raynaud FI, Eccles SA, Patel S, Alix S, Box G, Chuckowree I, Folkes A, Gowan S, De Haven Brandon A, Di Stefano F, Hayes A, Henley AT, Lensun L, Pergl-Wilson G, Robson A, Saghir N, Zhyvoloup A, McDonald E, Sheldrake P, Shuttleworth S, Valenti M, Wan NC, Clarke PA, Workman P (2009) Biological properties of potent inhibitors of class I phosphatidylinositide 3-kinases: from PI-103 through PI-540, PI-620 to the oral agent GDC-0941. Mol Cancer Ther 8:1725–1738

Readinger JA, Mueller KL, Venegas AM, Horai R, Schwartzberg PL (2009) Tec kinases regulate T-lymphocyte development and function: new insights into the roles of Itk and Rlk/Txk. Immunol Rev 228:93–114

Rebecchi MJ, Pentyala SN (2000) Structure, function, and control of phosphoinositide-specific phospholipase C. Physiol Rev 80:1291–1335

Rhodes N, Heerding DA, Duckett DR, Eberwein DJ, Knick VB, Lansing TJ, McConnell RT, Gilmer TM, Zhang SY, Robell K, Kahana JA, Geske RS, Kleymenova EV, Choudhry AE, Lai Z, Leber JD, Minthorn EA, Strum SL, Wood ER, Huang PS, Copeland RA, Kumar R (2008) Characterization of an Akt kinase inhibitor with potent pharmacodynamic and antitumor activity. Cancer Res 68:2366–2374

Rickert RC, Rajewsky K, Roes J (1995) Impairment of T-cell-dependent B-cell responses and B-1 cell development in CD19-deficient mice. Nature 376:352–355

Rickert P, Weiner OD, Wang F, Bourne HR, Servant G (2000) Leukocytes navigate by compass: roles of PI3Kgamma and its lipid products. Trends Cell Biol 10:466–473

Rivera J (2005) NTAL/LAB and LAT: a balancing act in mast-cell activation and function. Trends Immunol 26:119–122

Rizzieri DA, Feldman E, Dipersio JF, Gabrail N, Stock W, Strair R, Rivera VM, Albitar M, Bedrosian CL, Giles FJ (2008) A phase 2 clinical trial of deforolimus (AP23573, MK-8669), a novel mammalian target of rapamycin inhibitor, in patients with relapsed or refractory hematologic malignancies. Clin Cancer Res 14:2756–2762

Roche S, Downward J, Raynal P, Courtneidge SA (1998) A function for phosphatidylinositol 3-kinase beta (p85alpha-p110beta) in fibroblasts during mitogenesis: requirement for insulin- and lysophosphatidic acid-mediated signal transduction. Mol Cell Biol 18:7119–7129

Rodriguez-Borlado L, Barber DF, Hernandez C, Rodriguez-Marcos MA, Sanchez A, Hirsch E, Wymann M, Martinez-A C, Carrera AC (2003) Phosphatidylinositol 3-kinase regulates the CD4/CD8T cell differentiation ratio. J Immunol 170:4475–4482

Rodriguez-Viciana P, Warne PH, Dhand R, Vanhaesebroeck B, Gout I, Fry MJ, Waterfield MD, Downward J (1994) Phosphatidylinositol-3-OH kinase as a direct target of Ras. Nature 370:527–532

Rodriguez-Viciana P, Warne PH, Vanhaesebroeck B, Waterfield MD, Downward J (1996) Activation of phosphoinositide 3-kinase by interaction with Ras and by point mutation. EMBO J 15:2442–2451

Rodriguez-Viciana P, Sabatier C, McCormick F (2004) Signaling specificity by Ras family GTPases is determined by the full spectrum of effectors they regulate. Mol Cell Biol 24:4943–4954

Roggo L, Bernard V, Kovacs AL, Rose AM, Savoy F, Zetka M, Wymann MP, Muller F (2002) Membrane transport in Caenorhabditis elegans: an essential role for VPS34 at the nuclear membrane. EMBO J 21:1673–1683

Rohrschneider LR, Fuller JF, Wolf I, Liu Y, Lucas DM (2000) Structure, function, and biology of SHIP proteins. Genes Dev 14:505–520

Rolli V, Gallwitz M, Wossning T, Flemming A, Schamel WW, Zurn C, Reth M (2002) Amplification of B cell antigen receptor signaling by a Syk/ITAM positive feedback loop. Mol Cell 10:1057–1069

Roncagalli R, Mingueneau M, Gregoire C, Malissen M, Malissen B (2010) LAT signaling pathology: an "autoimmune" condition without T cell self-reactivity. Trends Immunol 31:253–259

Roock WD, Vriendt VD, Normanno N, Ciardiello F, Tejpar S (2011) KRAS, BRAF, PIK3CA, and PTEN mutations: implications for targeted therapies in metastatic colorectal cancer. Lancet Oncol 12:594–603

Rosen N, She QB (2006) AKT and cancer—is it all mTOR? Cancer Cell 10:254–256

Rubio I, Rodriguez-Viciana P, Downward J, Wetzker R (1997) Interaction of Ras with phosphoinositide 3-kinase gamma. Biochem J 326:891–895

Rubio I, Wittig U, Meyer C, Heinze R, Kadereit D, Waldmann H, Downward J, Wetzker R (1999) Farnesylation of Ras is important for the interaction with phosphoinositide 3-kinase gamma. Eur J Biochem 266:70–82

Ryan CW, Vuky J, Chan JS, Chen Z, Beer TM, Nauman D (2011) A phase II study of everolimus in combination with imatinib for previously treated advanced renal carcinoma. Invest New Drugs 29:374–379

Sagona AP, Nezis IP, Pedersen NM, Liestol K, Poulton J, Rusten TE, Skotheim RI, Raiborg C, Stenmark H (2010) PtdIns(3)P controls cytokinesis through KIF13A-mediated recruitment of FYVE-CENT to the midbody. Nat Cell Biol 12:362–371

Samuels Y, Ericson K (2006) Oncogenic PI3K and its role in cancer. Curr Opin Oncol 18:77–82

Samuels Y, Wang Z, Bardelli A, Silliman N, Ptak J, Szabo S, Yan H, Gazdar A, Powell SM, Riggins GJ, Willson JK, Markowitz S, Kinzler KW, Vogelstein B, Velculescu VE (2004) High frequency of mutations of the PIK3CA gene in human cancers. Science 304:554

Samuels Y, Diaz LA Jr, Schmidt-Kittler O, Cummins JM, Delong L, Cheong I, Rago C, Huso DL, Lengauer C, Kinzler KW, Vogelstein B, Velculescu VE (2005) Mutant PIK3CA promotes cell growth and invasion of human cancer cells. Cancer Cell 7:561–573

Sancak Y, Sabatini DM (2009) Rag proteins regulate amino-acid-induced mTORC1 signalling. Biochem Soc Trans 37:289–290

Sancak Y, Bar-Peled L, Zoncu R, Markhard AL, Nada S, Sabatini DM (2010) Ragulator-Rag complex targets mTORC1 to the lysosomal surface and is necessary for its activation by amino acids. Cell 141:290–303

Sansal I, Sellers WR (2004) The biology and clinical relevance of the PTEN tumor suppressor pathway. J Clin Oncol 22:2954–2963

Santos-Sierra S, Deshmukh SD, Kalnitski J, Kuenzi P, Wymann MP, Golenbock DT, Henneke P (2009) Mal connects TLR2 to PI3Kinase activation and phagocyte polarization. EMBO J 28:2018–2027

Sarbassov DD, Guertin DA, Ali SM, Sabatini DM (2005) Phosphorylation and regulation of Akt/PKB by the rictor-mTOR complex. Science 307:1098–1101

Sarkaria JN, Tibbetts RS, Busby EC, Kennedy AP, Hill DE, Abraham RT (1998) Inhibition of phosphoinositide 3-kinase related kinases by the radiosensitizing agent wortmannin. Cancer Res 58:4375–4382

Sarkaria JN, Galanis E, Wu W, Dietz AB, Kaufmann TJ, Gustafson MP, Brown PD, Uhm JH, Rao RD, Doyle L, Giannini C, Jaeckle KA, Buckner JC (2010) Combination of temsirolimus (CCI-779) with chemoradiation in newly diagnosed glioblastoma multiforme (GBM) (NCCTG trial N027D) is associated with increased infectious risks. Clin Cancer Res 16:5573–5580

Sasaki T, Irie-Sasaki J, Jones RG, Oliveira-dos-Santos AJ, Stanford WL, Bolon B, Wakeham A, Itie A, Bouchard D, Kozieradzki I, Joza N, Mak TW, Ohashi PS, Suzuki A, Penninger JM (2000) Function of PI3Kgamma in thymocyte development, T cell activation, and neutrophil migration. Science 287:1040–1046

Sasaki T, Takasuga S, Sasaki J, Kofuji S, Eguchi S, Yamazaki M, Suzuki A (2009) Mammalian phosphoinositide kinases and phosphatases. Prog Lipid Res 48:307–343

Sattler M, Mohi MG, Pride YB, Quinnan LR, Malouf NA, Podar K, Gesbert F, Iwasaki H, Li S, Van Etten RA, Gu H, Griffin JD, Neel BG (2002) Critical role for Gab2 in transformation by BCR/ABL. Cancer Cell 1:479–492

Schaller M, Burton DR, Ditzel HJ (2001) Autoantibodies to GPI in rheumatoid arthritis: linkage between an animal model and human disease. Nat Immunol 2:746–753

Schauder C, Ma LC, Krug RM, Montelione GT, Guan R (2010) Structure of the iSH2 domain of human phosphatidylinositol 3-kinase p85beta subunit reveals conformational plasticity in the interhelical turn region. Acta Crystallogr Sect F Struct Biol Cryst Commun 66:1567–1571

Schinzel A, Kaufmann T, Borner C (2004) Bcl-2 family members: integrators of survival and death signals in physiology and pathology [corrected]. Biochim Biophys Acta 1644:95–105

Schon MP, Boehncke WH (2005) Psoriasis. N Engl J Med 352:1899–1912

Schu PV, Takegawa K, Fry MJ, Stack JH, Waterfield MD, Emr SD (1993) Phosphatidylinositol 3-kinase encoded by yeast VPS34 gene essential for protein sorting. Science 260:88–91

Sengupta S, Peterson TR, Sabatini DM (2010) Regulation of the mTOR complex 1 pathway by nutrients, growth factors, and stress. Mol Cell 40:310–322

Serra V, Markman B, Scaltriti M, Eichhorn PJ, Valero V, Guzman M, Botero ML, Llonch E, Atzori F, Di Cosimo S, Maira M, Garcia-Echeverria C, Parra JL, Arribas J, Baselga J (2008) NVP-BEZ235, a dual PI3K/mTOR inhibitor, prevents PI3K signaling and inhibits the growth of cancer cells with activating PI3K mutations. Cancer Res 68:8022–8030

Servant G, Weiner OD, Herzmark P, Balla T, Sedat JW, Bourne HR (2000) Polarization of chemoattractant receptor signaling during neutrophil chemotaxis. Science 287:1037–1040

Sessa C, Tosi D, Vigano L, Albanell J, Hess D, Maur M, Cresta S, Locatelli A, Angst R, Rojo F, Coceani N, Rivera VM, Berk L, Haluska F, Gianni L (2010) Phase Ib study of weekly mammalian target of rapamycin inhibitor ridaforolimus (AP23573; MK-8669) with weekly paclitaxel. Ann Oncol 21:1315–1322

Shapiro G, Kwak E, Baselga J, Rodon J, Scheffold C, Laird AD, Bedell C, Edelman G (2009) Phase I dose-escalation study of XL147, a PI3K inhibitor administered orally to patients with solid tumors. J Clin Oncol 27:15s (Abstract 3500)

Shaw RJ, Cantley LC (2006) Ras, PI(3)K and mTOR signalling controls tumour cell growth. Nature 441:424–430

Shayesteh L, Lu Y, Kuo WL, Baldocchi R, Godfrey T, Collins C, Pinkel D, Powell B, Mills GB, Gray JW (1999) PIK3CA is implicated as an oncogene in ovarian cancer. Nat Genet 21:99–102

Shekar SC, Wu H, Fu Z, Yip SC, Nagajyothi CSM, Girvin ME, Backer JM (2005) Mechanism of constitutive phosphoinositide 3-kinase activation by oncogenic mutants of the p85 regulatory subunit. J Biol Chem 280:27850–27855

Shiloh Y (2003) ATM and related protein kinases: safeguarding genome integrity. Nat Rev Cancer 3:155–168

Shin HW, Hayashi M, Christoforidis S, Lacas-Gervais S, Hoepfner S, Wenk MR, Modregger J, Uttenweiler-Joseph S, Wilm M, Nystuen A, Frankel WN, Solimena M, De Camilli P, Zerial M (2005) An enzymatic cascade of Rab5 effectors regulates phosphoinositide turnover in the endocytic pathway. J Cell Biol 170:607–618

Shioi T, Kang PM, Douglas PS, Hampe J, Yballe CM, Lawitts J, Cantley LC, Izumo S (2000) The conserved phosphoinositide 3-kinase pathway determines heart size in mice. EMBO J 19:2537–2548

Shioi T, McMullen JR, Kang PM, Douglas PS, Obata T, Franke TF, Cantley LC, Izumo S (2002) Akt/protein kinase B promotes organ growth in transgenic mice. Mol Cell Biol 22:2799–2809

Simeoni L, Kliche S, Lindquist J, Schraven B (2004) Adaptors and linkers in T and B cells. Curr Opin Immunol 16:304–313

Simonsen A, Tooze SA (2009) Coordination of membrane events during autophagy by multiple class III PI3-kinase complexes. J Cell Biol 186:773–782

Simonsen A, Wurmser AE, Emr SD, Stenmark H (2001) The role of phosphoinositides in membrane transport. Curr Opin Cell Biol 13:485–492

Singer GG, Carrera AC, Marshak-Rothstein A, Martinez C, Abbas AK (1994) Apoptosis, Fas and systemic autoimmunity: the MRL-lpr/lpr model. Curr Opin Immunol 6:913–920

Sjolander A, Lapetina EG (1992) Agonist-induced association of the p21ras GTPase-activating protein with phosphatidylinositol 3-kinase. Biochem Biophys Res Commun 189:1503–1508

Sjolander A, Yamamoto K, Huber BE, Lapetina EG (1991) Association of p21ras with phosphatidylinositol 3-kinase. Proc Natl Acad Sci USA 88:7908–7912

Slegtenhorst M van, Hoogt R de, Hermans C, Nellist M, Janssen B, Verhoef S, Lindhout D, Ouweland A Van Den, Halley D, Young J, Burley M, Jeremiah S, Woodward K, Nahmias J, Fox M, Ekong R, Osborne J, Wolfe J, Povey S, Snell RG, Cheadle JP, Jones AC, Tachataki M, Ravine D, Sampson JR, Reeve MP, Richardson P, Wilmer F, Munro C, Hawkins TL, Sepp T, Ali JB, Ward S, Green AJ, Yates JR, Kwiatkowska J, Henske EP, Short MP, Haines JH, Jozwiak S, Kwiatkowski DJ (1997) Identification of the tuberous sclerosis gene TSC1 on chromosome 9q34. Science 277:805–808

Song L, Leung C, Schindler C (2001) Lymphocytes are important in early atherosclerosis. J Clin Invest 108:251–259

Srivastava S, Di L, Zhdanova O, Li Z, Vardhana S, Wan Q, Yan Y, Varma R, Backer J, Wulff H, Dustin ML, Skolnik EY (2009) The class II phosphatidylinositol 3 kinase C2beta is required for the activation of the K+ channel KCa3.1 and CD4T-cells. Mol Biol Cell 20:3783–3791

Stack JH, Horazdovsky B, Emr SD (1995) Receptor-mediated protein sorting to the vacuole in yeast: roles for a protein kinase, a lipid kinase and GTP-binding proteins. Annu Rev Cell Dev Biol 11:1–33

Stahl JM, Sharma A, Cheung M, Zimmerman M, Cheng JQ, Bosenberg MW, Kester M, Sandirasegarane L, Robertson GP (2004) Deregulated Akt3 activity promotes development of malignant melanoma. Cancer Res 64:7002–7010

Stambolic V, Suzuki A, la Pompa JL de, Brothers GM, Mirtsos C, Sasaki T, Ruland J, Penninger JM, Siderovski DP, Mak TW (1998) Negative regulation of PKB/Akt-dependent cell survival by the tumor suppressor PTEN. Cell 95:29–39

Starr TK, Allaei R, Silverstein KA, Staggs RA, Sarver AL, Bergemann TL, Gupta M, O'Sullivan MG, Matise I, Dupuy AJ, Collier LS, Powers S, Oberg AL, Asmann YW, Thibodeau SN, Tessarollo L, Copeland NG, Jenkins NA, Cormier RT, Largaespada DA (2009) A transposon-based genetic screen in mice identifies genes altered in colorectal cancer. Science 323:1747–1750

Steiner G (2007) Auto-antibodies and autoreactive T-cells in rheumatoid arthritis: pathogenetic players and diagnostic tools. Clin Rev Allergy Immunol 32:23–36

Stemke-Hale K, Gonzalez-Angulo AM, Lluch A, Neve RM, Kuo WL, Davies M, Carey M, Hu Z, Guan Y, Sahin A, Symmans WF, Pusztai L, Nolden LK, Horlings H, Berns K, Hung MC, Van de Vijver MJ, Valero V, Gray JW, Bernards R, Mills GB, Hennessy BT (2008) An integrative genomic and proteomic analysis of PIK3CA, PTEN, and AKT mutations in breast cancer. Cancer Res 68:6084–6091

Stephens LR, Jackson TR, Hawkins PT (1993) Agonist-stimulated synthesis of phosphatidylinositol(3,4,5)-trisphosphate: a new intracellular signalling system? Biochim Biophys Acta 1179:27–75

Stephens LR, Eguinoa A, Erdjument-Bromage H, Lui M, Cooke F, Coadwell J, Smrcka AS, Thelen M, Cadwallader K, Tempst P, Hawkins PT (1997) The G beta gamma sensitivity of a PI3K is dependent upon a tightly associated adaptor, p101. Cell 89:105–114

Stevenson CS, Birrell MA (2011) Moving towards a new generation of animal models for asthma and COPD with improved clinical relevance. Pharmacol Ther 130:93–105

Stocker H, Andjelkovic M, Oldham S, Laffargue M, Wymann MP, Hemmings BA, Hafen E (2002) Living with lethal PIP3 levels: viability of flies lacking PTEN restored by a PH domain mutation in Akt/PKB. Science 295:2088–2091

Stokoe D, Stephens LR, Copeland T, Gaffney PR, Reese CB, Painter GF, Holmes AB, McCormick F, Hawkins PT (1997) Dual role of phosphatidylinositol-3,4,5-trisphosphate in the activation of protein kinase B. Science 277:567–570

Stoyanov B, Volinia S, Hanck T, Rubio I, Loubtchenkov M, Malek D, Stoyanova S, Vanhaesebroeck B, Dhand R, Nurnberg B, Gierschik P, Seedorf K, Hsuan JJ, Waterfield MD, Wetzker R (1995) Cloning and characterization of a G protein-activated human phosphoinositide-3 kinase. Science 269:690–693

Stoyanova S, Bulgarelli-Leva G, Kirsch C, Hanck T, Klinger R, Wetzker R, Wymann MP (1997) Lipid kinase and protein kinase activities of G-protein-coupled phosphoinositide 3-kinase gamma: structure-activity analysis and interactions with wortmannin. Biochem J 324:489–495

Stuart JM, Townes AS, Kang AH (1984) Collagen autoimmune arthritis. Annu Rev Immunol 2:199–218

Sugimoto Y, Whitman M, Cantley LC, Erikson RL (1984) Evidence that the Rous sarcoma virus transforming gene product phosphorylates phosphatidylinositol and diacylglycerol. Proc Natl Acad Sci USA 81:2117–2121

Suire S, Coadwell J, Ferguson GJ, Davidson K, Hawkins P, Stephens L (2005) p84, a new Gbetagamma-activated regulatory subunit of the type IB phosphoinositide 3-kinase p110gamma. Curr Biol 15:566–570

Suire S, Condliffe AM, Ferguson GJ, Ellson CD, Guillou H, Davidson K, Welch H, Coadwell J, Turner M, Chilvers ER, Hawkins PT, Stephens L (2006) Gbetagammas and the Ras binding domain of p110gamma are both important regulators of PI3Kgamma signalling in neutrophils. Nat Cell Biol 8:1303–1309

Sun SY, Rosenberg LM, Wang X, Zhou Z, Yue P, Fu H, Khuri FR (2005) Activation of Akt and eIF4E survival pathways by rapamycin-mediated mammalian target of rapamycin inhibition. Cancer Res 65:7052–7058

Sun Q, Fan W, Chen K, Ding X, Chen S, Zhong Q (2008) Identification of Barkor as a mammalian autophagy-specific factor for Beclin 1 and class III phosphatidylinositol 3-kinase. Proc Natl Acad Sci USA 105:19211–19216

Suzuki A, De la Pompa JL, Stambolic V, Elia AJ, Sasaki T, Del Barco Barrantes I, Ho A, Wakeham A, Itie A, Khoo W, Fukumoto M, Mak TW (1998) High cancer susceptibility and embryonic lethality associated with mutation of the PTEN tumor suppressor gene in mice. Curr Biol 8:1169–1178

Tabernero J, Rojo F, Calvo E, Burris H, Judson I, Hazell K, Martinelli E, Ramon y Cajal S, Jones S, Vidal L, Shand N, Macarulla T, Ramos FJ, Dimitrijevic S, Zoellner U, Tang P, Stumm M, Lane HA, Lebwohl D, Baselga J (2008) Dose- and schedule-dependent inhibition of the mammalian target of rapamycin pathway with everolimus: a phase I tumor pharmacodynamic study in patients with advanced solid tumors. J Clin Oncol 26:1603–1610

Takenawa T, Itoh T (2006) Membrane targeting and remodeling through phosphoinositide-binding domains. IUBMB Life 58:296–303

Tang X, Downes CP (1997) Purification and characterization of Gbetagamma-responsive phosphoinositide 3-kinases from pig platelet cytosol. J Biol Chem 272:14193–14199

TCGA study (2008) Comprehensive genomic characterization defines human glioblastoma genes and core pathways. Nature 455:1061–1068

Tee AR, Manning BD, Roux PP, Cantley LC, Blenis J (2003) Tuberous sclerosis complex gene products, Tuberin and Hamartin, control mTOR signaling by acting as a GTPase-activating protein complex toward Rheb. Curr Biol 13:1259–1268

Terato K, Hasty KA, Reife RA, Cremer MA, Kang AH, Stuart JM (1992) Induction of arthritis with monoclonal antibodies to collagen. J Immunol 148:2103–2108

Thelen M, Wymann MP, Langen H (1994) Wortmannin binds specifically to 1-phosphatidylinositol 3-kinase while inhibiting guanine nucleotide-binding protein-coupled receptor signaling in neutrophil leukocytes. Proc Natl Acad Sci USA 91:4960–4964

Thomas MJ, Smith A, Head DH, Milne L, Nicholls A, Pearce W, Vanhaesebroeck B, Wymann MP, Hirsch E, Trifilieff A, Walker C, Finan P, Westwick J (2005) Airway inflammation: chemokine-induced neutrophilia and the class I phosphoinositide 3-kinases. Eur J Immunol 35:1283–1291

Thomas RK, Baker AC, Debiasi RM, Winckler W, Laframboise T, Lin WM, Wang M, Feng W, Zander T, MacConaill L, Lee JC, Nicoletti R, Hatton C, Goyette M, Girard L, Majmudar K, Ziaugra L, Wong KK, Gabriel S, Beroukhim R, Peyton M, Barretina J, Dutt A, Emery C, Greulich H, Shah K, Sasaki H, Gazdar A, Minna J, Armstrong SA, Mellinghoff IK, Hodi FS, Dranoff G, Mischel PS, Cloughesy TF, Nelson SF, Liau LM, Mertz K, Rubin MA, Moch H, Loda M, Catalona W, Fletcher J, Signoretti S, Kaye F, Anderson KC, Demetri GD, Dummer R, Wagner S, Herlyn M, Sellers WR, Meyerson M, Garraway LA (2007) High-throughput oncogene mutation profiling in human cancer. Nat Genet 39:347–351

Thomas M, Edwards MJ, Sawicka E, Duggan N, Hirsch E, Wymann MP, Owen C, Trifilieff A, Walker C, Westwick J, Finan P (2009) Essential role of phosphoinositide 3-kinase gamma in eosinophil chemotaxis within acute pulmonary inflammation. Immunology 126:413–422

To Y, Ito K, Kizawa Y, Failla M, Ito M, Kusama T, Elliott WM, Hogg JC, Adcock IM, Barnes PJ (2010) Targeting phosphoinositide-3-kinase-delta with theophylline reverses corticosteroid insensitivity in chronic obstructive pulmonary disease. Am J Respir Crit Care Med 182:897–904

Toker A (2008) mTOR and Akt signaling in cancer: SGK cycles in. Mol Cell 31:6–8

Tolcher AW, Yap TA, Fearen I, Taylor A, Carpenter C, Brunetto AT, Beeram M, Papadopoulos K, Yan L, Bono J de (2009) J Clin Oncol 27:15s (suppl; Abstract 3503)

Traer CJ, Foster FM, Abraham SM, Fry MJ (2006) Are class II phosphoinositide 3-kinases potential targets for anticancer therapies? Bull Cancer 93:E53–E58

Vadas O, Burke JE, Zhang X, Berndt A, Williams RL (2011) Brakes and accelerators: the mechanics of phosphoinositide 3-kinase regulation. Sci Signal (in press)

Vakana E, Sassano A, Platanias LC (2010) Induction of autophagy by dual mTORC1-mTORC2 inhibition in BCR-ABL-expressing leukemic cells. Autophagy 6:966–967

Van Der Haar E, Lee SI, Bandhakavi S, Griffin TJ, Kim DH (2007) Insulin signalling to mTOR mediated by the Akt/PKB substrate PRAS40. Nat Cell Biol 9:316–323

Van Ummersen L, Binger K, Volkman J, Marnocha R, Tutsch K, Kolesar J, Arzoomanian R, Alberti D, Wilding G (2004) A phase I trial of perifosine (NSC 639966) on a loading dose/maintenance dose schedule in patients with advanced cancer. Clin Cancer Res 10:7450–7456

Vanhaesebroeck B, Leevers SJ, Ahmadi K, Timms J, Katso R, Driscoll PC, Woscholski R, Parker PJ, Waterfield MD (2001) Synthesis and function of 3-phosphorylated inositol lipids. Annu Rev Biochem 70:535–602

Vanhaesebroeck B, Guillermet-Guibert J, Graupera M, Bilanges B (2010) The emerging mechanisms of isoform-specific PI3K signalling. Nat Rev Mol Cell Biol 11:329–341

Varnai P, Bondeva T, Tamas P, Toth B, Buday L, Hunyady L, Balla T (2005) Selective cellular effects of overexpressed pleckstrin-homology domains that recognize PtdIns(3,4,5)P3 suggest their interaction with protein binding partners. J Cell Sci 118:4879–4888

Vecchione C, Patrucco E, Marino G, Barberis L, Poulet R, Aretini A, Maffei A, Gentile MT, Storto M, Azzolino O, Brancaccio M, Colussi GL, Bettarini U, Altruda F, Silengo L, Tarone G, Wymann MP, Hirsch E, Lembo G (2005) Protection from angiotensin II-mediated vasculotoxic and hypertensive response in mice lacking PI3Kgamma. J Exp Med 201:1217–1228

Vega FM, Ridley AJ (2008) Rho GTPases in cancer cell biology. FEBS Lett 582:2093–2101

Venkatesan AM, Dehnhardt CM, Delos Santos E, Chen Z, Dos Santos O, Ayral-Kaloustian S, Khafizova G, Brooijmans N, Mallon R, Hollander I, Feldberg L, Lucas J, Yu K, Gibbons J, Abraham RT, Chaudhary I, Mansour TS (2010) Bis(morpholino-1,3,5-triazine) derivatives: potent adenosine 5′-triphosphate competitive phosphatidylinositol-3-kinase/mammalian target of rapamycin inhibitors: discovery of compound 26 (PKI-587), a highly efficacious dual inhibitor. J Med Chem 53:2636–2645

Vink SR, Schellens JH, Beijnen JH, Sindermann H, Engel J, Dubbelman R, Moppi G, Hillebrand MJ, Bartelink H, Verheij M (2006) Phase I and pharmacokinetic study of combined treatment with perifosine and radiation in patients with advanced solid tumours. Radiother Oncol 80:207–213

Vivanco I, Sawyers CL (2002) The phosphatidylinositol 3-Kinase AKT pathway in human cancer. Nat Rev Cancer 2:489–501

Vlahos CJ, Matter WF, Hui KY, Brown RF (1994) A specific inhibitor of phosphatidylinositol 3-kinase, 2-(4-morpholinyl)-8-phenyl-4H-1-benzopyran-4-one (LY294002). J Biol Chem 269:5241–5248

Vogt PK (2011) PI3K p110beta: more tightly controlled or constitutively active? Mol Cell 41:499–501

Voigt P, Brock C, Nurnberg B, Schaefer M (2005) Assigning functional domains within the p101 regulatory subunit of phosphoinositide 3-kinase gamma. J Biol Chem 280:5121–5127

Voigt P, Dorner MB, Schaefer M (2006) Characterization of p87PIKAP, a novel regulatory subunit of phosphoinositide 3-kinase gamma that is highly expressed in heart and interacts with PDE3B. J Biol Chem 281:9977–9986

Wagner AJ, Burris HA III, Bono JS de, Jayson GC, Bendell JC, Gomez-Roca C, Dolly S, Zee Y-K, Ware JA, Yan Y, Mazina K, Derynck M, Holden S, Soria J-C (2009) Pharmacokinetics and pharmacodynamic biomarkers for the dual PI3K/mTOR inhibitor GDC-0980: initial phase I evaluation. Mol Cancer Ther 8:B137

Walker EH, Perisic O, Ried C, Stephens L, Williams RL (1999) Structural insights into phosphoinositide 3-kinase catalysis and signalling. Nature 402:313–320

Walker EH, Pacold ME, Perisic O, Stephens L, Hawkins PT, Wymann MP, Williams RL (2000) Structural determinants of phosphoinositide 3-kinase inhibition by wortmannin, LY294002, quercetin, myricetin, and staurosporine. Mol Cell 6:909–919

Wallace E, Xu R, Josey J, Gross SD, Miknis G, Fischer J, De Meese L, Humphries M, Regal K, Fell B, Condroski K, Burkard M, DeWolf WE, Gloor S, Hastings G, Zuzack J, Winkler J, Koch K (n d) AR-mTOR-1: A potent, selective mTOR 1/2 kinase inhibitor for the treatment of malignancy. Download at: http://www.arraybiopharma.com/_documents/Publication/PubAttachment364.pdf

Wang F, Herzmark P, Weiner OD, Srinivasan S, Servant G, Bourne HR (2002) Lipid products of PI(3)Ks maintain persistent cell polarity and directed motility in neutrophils. Nat Cell Biol 4:513–518

Wang Y, Yoshioka K, Azam MA, Takuwa N, Sakurada S, Kayaba Y, Sugimoto N, Inoki I, Kimura T, Kuwaki T, Takuwa Y (2006) Class II phosphoinositide 3-kinase alpha-isoform regulates Rho, myosin phosphatase and contraction in vascular smooth muscle. Biochem J 394:581–592

Wang L, Harris TE, Roth RA, Lawrence JC Jr (2007) PRAS40 regulates mTORC1 kinase activity by functioning as a direct inhibitor of substrate binding. J Biol Chem 282:20036–20044

Webb LM, Vigorito E, Wymann MP, Hirsch E, Turner M (2005) Cutting edge: T cell development requires the combined activities of the p110gamma and p110delta catalytic isoforms of phosphatidylinositol 3-kinase. J Immunol 175:2783–2787

Wee S, Wiederschain D, Maira SM, Loo A, Miller C, DeBeaumont R, Stegmeier F, Yao YM, Lengauer C (2008) PTEN-deficient cancers depend on PIK3CB. Proc Natl Acad Sci USA 105:13057–13062

Wee S, Jagani Z, Xiang KX, Loo A, Dorsch M, Yao YM, Sellers WR, Lengauer C, Stegmeier F (2009) PI3K pathway activation mediates resistance to MEK inhibitors in KRAS mutant cancers. Cancer Res 69:4286–4293

Welch HC, Coadwell WJ, Ellson CD, Ferguson GJ, Andrews SR, Erdjument-Bromage H, Tempst P, Hawkins PT, Stephens LR (2002) P-Rex1, a PtdIns(3,4,5)P3- and Gbetagamma-regulated guanine-nucleotide exchange factor for Rac. Cell 108:809–821

Werner M, Hobeika E, Jumaa H (2010) Role of PI3K in the generation and survival of B cells. Immunol Rev 237:55–71

Wheeler M, Domin J (2006) The N-terminus of phosphoinositide 3-kinase-C2beta regulates lipid kinase activity and binding to clathrin. J Cell Physiol 206:586–593

Whitman M, Kaplan DR, Schaffhausen B, Cantley L, Roberts TM (1985) Association of phosphatidylinositol kinase activity with polyoma middle-T competent for transformation. Nature 315:239–242

Williams R, Baker AF, Ihle NT, Winkler AR, Kirkpatrick L, Powis G (2006) The skin and hair as surrogate tissues for measuring the target effect of inhibitors of phosphoinositide-3-kinase signaling. Cancer Chemother Pharmacol 58:444–450

Wilson BS, Pfeiffer JR, Surviladze Z, Gaudet EA, Oliver JM (2001) High resolution mapping of mast cell membranes reveals primary and secondary domains of Fc(epsilon)RI and LAT. J Cell Biol 154:645–658

Witzig TE, Geyer SM, Ghobrial I, Inwards DJ, Fonseca R, Kurtin P, Ansell SM, Luyun R, Flynn PJ, Morton RF, Dakhil SR, Gross H, Kaufmann SH (2005) Phase II trial of single-agent temsirolimus (CCI-779) for relapsed mantle cell lymphoma. J Clin Oncol 23:5347–5356

Witzig TE, Reeder CB, Laplant BR, Gupta M, Johnston PB, Micallef IN, Porrata LF, Ansell SM, Colgan JP, Jacobsen ED, Ghobrial IM, Habermann TM (2011) A phase II trial of the oral mTOR inhibitor everolimus in relapsed aggressive lymphoma. Leukemia 25:341–347

Wong KK, Engelman JA, Cantley LC (2010) Targeting the PI3K signaling pathway in cancer. Curr Opin Genet Dev 20:87–90

Wood LD, Parsons DW, Jones S, Lin J, Sjoblom T, Leary RJ, Shen D, Boca SM, Barber T, Ptak J, Silliman N, Szabo S, Dezso Z, Ustyanksky V, Nikolskaya T, Nikolsky Y, Karchin R, Wilson PA, Kaminker JS, Zhang Z, Croshaw R, Willis J, Dawson D, Shipitsin M, Willson JK, Sukumar S, Polyak K, Park BH, Pethiyagoda CL, Pant PV, Ballinger DG, Sparks AB, Hartigan J, Smith DR, Suh E, Papadopoulos N, Buckhaults P, Markowitz SD, Parmigiani G, Kinzler KW, Velculescu VE, Vogelstein B (2007) The genomic landscapes of human breast and colorectal cancers. Science 318:1108–1113

Woods A, Johnstone SR, Dickerson K, Leiper FC, Fryer LG, Neumann D, Schlattner U, Wallimann T, Carlson M, Carling D (2003) LKB1 is the upstream kinase in the AMP-activated protein kinase cascade. Curr Biol 13:2004–2008

Workman P, Clarke PA, Raynaud FI, Montfort RL van (2010) Drugging the PI3 kinome: from chemical tools to drugs in the clinic. Cancer Res 70:2146–2157

Woscholski R, Kodaki T, McKinnon M, Waterfield MD, Parker PJ (1994) A comparison of demethoxyviridin and wortmannin as inhibitors of phosphatidylinositol 3-kinase. FEBS Lett 342:109–114

Wu H, Goel V, Haluska FG (2003) PTEN signaling pathways in melanoma. Oncogene 22:3113–3122

Wullschleger S, Loewith R, Hall MN (2006) TOR signaling in growth and metabolism. Cell 124:471–484

Wymann M, Arcaro A (1994) Platelet-derived growth factor-induced phosphatidylinositol 3-kinase activation mediates actin rearrangements in fibroblasts. Biochem J 298(Pt 3):517–520

Wymann MP, Pirola L (1998) Structure and function of phosphoinositide 3-kinases. Biochim Biophys Acta 1436:127–150

Wymann MP, Marone R (2005) Phosphoinositide 3-kinase in disease: timing, location, and scaffolding. Curr Opin Cell Biol 17:141–149

Wymann MP, Schneiter R (2008) Lipid signalling in disease. Nat Rev Mol Cell Biol 9:162–176

Wymann MP, Wenk MR (2011) Neutral not a loss: phosphoinositides beyond the head group. Nat Methods 8:219–220

Wymann MP, Bulgarelli-Leva G, Zvelebil MJ, Pirola L, Vanhaesebroeck B, Waterfield MD, Panayotou G (1996) Wortmannin inactivates phosphoinositide 3-kinase by covalent modification of Lys-802, a residue involved in the phosphate transfer reaction. Mol Cell Biol 16:1722–1733

Wymann MP, Sozzani S, Altruda F, Mantovani A, Hirsch E (2000) Lipids on the move: phosphoinositide 3-kinases in leukocyte function. Immunol Today 21:260–264

Wymann MP, Bjorklof K, Calvez R, Finan P, Thomast M, Trifilieff A, Barbier M, Altruda F, Hirsch E, Laffargue M (2003a) Phosphoinositide 3-kinase gamma: a key modulator in inflammation and allergy. Biochem Soc Trans 31:275–280

Wymann MP, Zvelebil M, Laffargue M (2003b) Phosphoinositide 3-kinase signalling—which way to target? Trends Pharmacol Sci 24:366–376

Xue Q, Hopkins B, Perruzzi C, Udayakumar D, Sherris D, Benjamin LE (2008) Palomid 529, a novel small-molecule drug, is a TORC1/TORC2 inhibitor that reduces tumor growth, tumor angiogenesis, and vascular permeability. Cancer Res 68:9551–9557

Yaguchi S, Fukui Y, Koshimizu I, Yoshimi H, Matsuno T, Gouda H, Hirono S, Yamazaki K, Yamori T (2006) Antitumor activity of ZSTK474, a new phosphatidylinositol 3-kinase inhibitor. J Natl Cancer Inst 98:545–556

Yamashita A, Kashima I, Ohno S (2005) The role of SMG-1 in nonsense-mediated mRNA decay. Biochim Biophys Acta 1754:305–315

Yamazaki T, Takeda K, Gotoh K, Takeshima H, Akira S, Kurosaki T (2002) Essential immunoregulatory role for BCAP in B cell development and function. J Exp Med 195:535–545

Yang J, Cron P, Good VM, Thompson V, Hemmings BA, Barford D (2002) Crystal structure of an activated Akt/protein kinase B ternary complex with GSK3-peptide and AMP-PNP. Nat Struct Biol 9:940–944

Yang L, Clarke MJ, Carlson BL, Mladek AC, Schroeder MA, Decker P, Wu W, Kitange GJ, Grogan PT, Goble JM, Uhm J, Galanis E, Giannini C, Lane HA, James CD, Sarkaria JN (2008) PTEN loss does not predict for response to RAD001 (Everolimus) in a glioblastoma orthotopic xenograft test panel. Clin Cancer Res 14:3993–4001

Yano H, Nakanishi S, Kimura K, Hanai N, Saitoh Y, Fukui Y, Nonomura Y, Matsuda Y (1993) Inhibition of histamine secretion by wortmannin through the blockade of phosphatidylinositol 3-kinase in RBL-2H3 cells. J Biol Chem 268:25846–25856

Yao JC, Lombard-Bohas C, Baudin E, Kvols LK, Rougier P, Ruszniewski P, Hoosen S, St Peter J, Haas T, Lebwohl D, Van Cutsem E, Kulke MH, Hobday TJ, O'Dorisio TM, Shah MH, Cadiot G, Luppi G, Posey JA, Wiedenmann B (2010) Daily oral everolimus activity in patients with metastatic pancreatic neuroendocrine tumors after failure of cytotoxic chemotherapy: a phase II trial. J Clin Oncol 28:69–76

Yue Z, Jin S, Yang C, Levine AJ, Heintz N (2003) Beclin 1, an autophagy gene essential for early embryonic development, is a haploinsufficient tumor suppressor. Proc Natl Acad Sci USA 100:15077–15082

Zhang D, Aravind L (2010) Identification of novel families and classification of the C2 domain superfamily elucidate the origin and evolution of membrane targeting activities in eukaryotes. Gene 469:18–30

Zhang S, Yu D (2010) PI(3)king apart PTEN's role in cancer. Clin Cancer Res 16:4325–4330

Zhang SH, Reddick RL, Piedrahita JA, Maeda N (1992) Spontaneous hypercholesterolemia and arterial lesions in mice lacking apolipoprotein E. Science 258:468–471

Zhang W, Sommers CL, Burshtyn DN, Stebbins CC, DeJarnette JB, Trible RP, Grinberg A, Tsay HC, Jacobs HM, Kessler CM, Long EO, Love PE, Samelson LE (1999) Essential role of LAT in T cell development. Immunity 10:323–332

Zhang A, Maner S, Betz R, Angstrom T, Stendahl U, Bergman F, Zetterberg A, Wallin KL (2002) Genetic alterations in cervical carcinomas: frequent low-level amplifications of oncogenes are associated with human papillomavirus infection. Int J Cancer 101:427–433

Zhang Y, Gao X, Saucedo LJ, Ru B, Edgar BA, Pan D (2003) Rheb is a direct target of the tuberous sclerosis tumour suppressor proteins. Nat Cell Biol 5:578–581

Zhang TT, Okkenhaug K, Nashed BF, Puri KD, Knight ZA, Shokat KM, Vanhaesebroeck B, Marshall AJ (2008) Genetic or pharmaceutical blockade of p110delta phosphoinositide 3-kinase enhances IgE production. J Allergy Clin Immunol 122:811–819e2

Zhang TT, Li H, Cheung SM, Costantini JL, Hou S, Al-Alwan M, Marshall AJ (2009) Phosphoinositide 3-kinase-regulated adapters in lymphocyte activation. Immunol Rev 232:255–272

Zhang X, Vadas O, Perisic O, Anderson KE, Clark J, Hawkins PT, Stephens LR, Williams RL (2011) Structure of lipid kinase p110beta/p85beta elucidates an unusual SH2-domain-mediated inhibitory mechanism. Mol Cell 41:567–578

Zhao JJ, Liu Z, Wang L, Shin E, Loda MF, Roberts TM (2005) The oncogenic properties of mutant p110alpha and p110beta phosphatidylinositol 3-kinases in human mammary epithelial cells. Proc Natl Acad Sci USA 102:18443–18448

Zhao Y, Gaidarov I, Keen JH (2007) Phosphoinositide 3-kinase C2alpha links clathrin to microtubule-dependent movement. J Biol Chem 282:1249–1256

Zhou X, Wang F (2010) Effects of neuronal PIK3C3/Vps34 deletion on autophagy and beyond. Autophagy 6:798–799

Zhuang J, Hawkins SF, Glenn MA, Lin K, Johnson GG, Carter A, Cawley JC, Pettitt AR (2010) Akt is activated in chronic lymphocytic leukemia cells and delivers a pro-survival signal: the therapeutic potential of Akt inhibition. Haematologica 95:110–118

Zoncu R, Efeyan A, Sabatini DM (2011) mTOR: from growth signal integration to cancer, diabetes and ageing. Nat Rev Mol Cell Biol 12:21–35

Zunder ER, Knight ZA, Houseman BT, Apsel B, Shokat KM (2008) Discovery of drug-resistant and drug-sensitizing mutations in the oncogenic PI3K isoform p110 alpha. Cancer Cell 14:180–192

Chapter 6
Phosphoinositide 3-Kinases in Health and Disease

Alessandra Ghigo, Fulvio Morello, Alessia Perino and Emilio Hirsch

Abstract In the last decade, the availability of genetically modified animals has revealed interesting roles for phosphoinositide 3-kinases (PI3Ks) as signaling platforms orchestrating multiple cellular responses, both in health and pathology. By acting downstream distinct receptor types, PI3Ks nucleate complex signaling assemblies controlling several biological process, ranging from cell proliferation and survival to immunity, cancer, metabolism and cardiovascular control. While the involvement of these kinases in modulating immune reactions and neoplastic transformation has long been accepted, recent progress from our group and others has highlighted new and unforeseen roles of PI3Ks in controlling cardiovascular function. Hence, the view is emerging that pharmacological targeting of distinct PI3K isoforms could be successful in treating disorders such as myocardial infarction and heart failure, besides inflammatory diseases and cancer. Currently, PI3Ks represent attractive drug targets for companies interested in the development of novel and safe treatments for such diseases. Numerous hit and lead compounds are now becoming available and, for some of them, clinical trials can be envisaged in the near future. In the following sections, we will outline the impact of specific PI3K isoforms in regulating different cellular contexts, including immunity, metabolism, cancer and cardiovascular system, both in physiological and disease conditions.

Keywords Cancer · Immunity · Inflammation · Glucose metabolism · Heart failure

6.1 Class I PI3Ks in Cancer

The PI3K pathway participates in several processes of cancer biology including cell transformation, proliferation, survival, motility and angiogenesis. In cancer cells, hyperactivity of PI3K signaling results from (i) gain-of-function mutations of a class

---

E. Hirsch (✉) · A. Ghigo · F. Morello · A. Perino
Molecular Biotechnology Center, Department of Genetics, Biology and Biochemistry,
University of Torino, Torino, Italy
e-mail: emilio.hirsch@unito.it

I PI3K, and/or (ii) abnormal activation upstream (e.g. tyrosine kinase receptors) or downstream PI3Ks (e.g. AKT and PTEN). In particular, the gene encoding for p110α (PIK3CA) is one of the most frequently mutated oncogenes in human tumors (Gymnopoulos et al. 2007), as somatic mutations of the PIK3CA gene have been reported in several cancer types including colon, ovary, breast, brain, liver, stomach, endometrial and lung cancer (Samuels et al. 2004). Three hot-spot mutations (E542K, E545K and H1047R) represent 80% of all PIK3CA mutations found in tumors and map two distinct domains of the p110α protein (Zhao and Vogt 2008). E542K and E545K mutations are situated within the helical domain, while H1047R lies in the kinase domain. Mutations in the helical domain seem to alter the binding of p110α to the regulatory subunit p85 and possibly interfere with the inhibitory action of p85 on p110α, thus mimicking an activation state by tyrosine kinase receptors. The H1047 is the most frequent mutation and occurs at the end of the activation loop of p110α, where it appears to directly influence the interaction between p110α and PIP_2 (Chaussade et al. 2009).

PIK3CA mutations either participate in the initiation of tumorigenesis or sustain cell growth in advanced tumors. When expressed in chicken embryo fibroblasts, E542K, E545K and H1047R p110α mutants induce oncogenic transformation with high efficiency (Bader et al. 2006). Although the expression of p110α mutants *in vitro* results in constitutive activation of Akt even in the absence of growth factors, the impact of p110α mutation on Akt *in vivo* is variable (Morrow et al. 2005; Vasudevan et al. 2009). Importantly, gain-of-function mutations in PIK3CA genes often coexist with additional alterations in the PI3K pathway in several types of tumors. For instance, mutated p110α has been associated with mutations of PTEN and K-Ras (Silvestris et al. 2009; Velasco et al. 2006) or with ERBB2 over-expression (Bachman et al. 2004; Saal et al. 2005). Co-occurrence of p110α gain-of-function with other specific oncogenic alterations in the PI3K pathway suggests that the mutational status of p110α and p85α may have different consequences on tumor formation and progression depending on tissue specificity and cell type.

Somatic mutations have also been reported in the regulatory subunit p85α (PIK3R1 gene), even though they are less frequent compared to mutations of p110α. On the contrary, mutations in the genes encoding other PI3K regulatory subunits (p85β—PIK3R2, and p55γ—PIK3R3), are rare events, which suggests an isoform-specific role for p85α in cancer. Most p85α mutations (i.e. D560Y, N564D, QYL579 deletion, DS459delN and DKRMNS560del) cluster in the two SH2 domains and in the inter-SH2 domain (Berenjeno and Vanhaesebroeck 2009; Jaiswal et al. 2009). Analysis of this region has revealed that these mutants, while retaining the ability to bind the p110 catalytic subunit, loose their inhibitory activity on p110. In addition, p85α mutations lead to unspecific activation of all class IA p110 isoforms. As p110β and p110δ have been found over-expressed in certain human cancers, the co-expression of mutated p85α in these tumors may thus further enhance the tumorigenic activity of PI3Ks.

To date, no genetic alterations have been found in the genes encoding for p110β, γ and δ. Conversely, increased expression of p110β and p110δ has been identified in glioblastomas (Knobbe and Reifenberger 2003), colon and bladder tumors (Benistant et al. 2000). Indeed, over-expression of wild-type p110β, δ and γ is sufficient to induce an oncogenic phenotype in cultured cells (Kang et al. 2006). Moreover, expression of myristoylated p110β induces the development of prostate intraepithelial tumors in mice (Lee et al. 2010) and expression of myristoylated p110γ induces a constitutive activation of Akt in Rat1 fibroblasts (Link et al. 2005). In line with this finding, p110γ has been found over-expressed in pancreatic cancer, where it is required for cell proliferation, as shown by reduced cell growth in the lack of p110γ lipid kinase activity (Edling et al. 2010). Taken together, these findings suggest that contrary to p110α, p110β, p110γ and 110δ explicate their oncogenic potential as wild-type proteins.

p110β has emerged as an interesting target in tumors such as breast and prostate cancers (Carvalho et al. 2010; Hill et al. 2010). In a mouse model of breast cancer driven by hyperactivation of the HER-2 signaling pathway, the absence of p110β lipid kinase activity strongly delays the appearance of the first tumor and reduces tumor growth *in vitro* even in a context of PTEN down-regulation (Ciraolo et al. 2008). Similarly, ablation of p110β blocks PTEN loss-driven tumorigenesis in the prostate. In this model, PTEN-mediated transformation appears to strictly depend on p110β, since prostate-specific knockout of p110α fails to affect tumor formation (Jia et al. 2008). These observations demonstrate the existence of a link between PTEN loss and p110β signaling. As a matter of fact, in the absence of p110α mutations, cancer cells harboring PTEN-null alleles depend on p110β lipid kinase activity, as treatment with p110β-selective inhibitors can block cell growth (Wee et al. 2008).

The role of PI3Ks in cancer has been extensively described in the last decade. Since our group has mainly focused on the area of non-oncological diseases, for broader description of PI3Ks in cancer we refer to dedicated reviews of expert authors (Engelman et al. 2006; Wong et al. 2010).

6.2 PI3Ks in Immunity and Inflammation

Protection against pathogens is achieved through concerted and synergic actions of a variety of cell types, which constitute the innate and adaptive immune responses. Innate immunity has evolved to rapidly recognize and eliminate non-self molecules through specialized phagocytic/effector cells, neutrophils and macrophages, which cooperate by providing the first line of antimicrobial defense. Neutrophils are the first to infiltrate inflamed tissues, where they carry out their defensive strategy based on bulky production of reactive oxygen species. Macrophages participate in the later phases of the inflammatory response. These cells are characterized by a high phagocytic activity and represent the principal scavengers of the immune system. Moreover, they secrete pro-inflammatory cytokines, which in turn boost the host defense. Other cell types, including mast cells and eosinophils, participate in the response to parasites, by releasing important mediators. Adaptive immunity constitutes a more sophisticated mechanism of defense, whose key feature is represented by specific antigen recognition of the invaders by T and B lymphocytes, which clonally

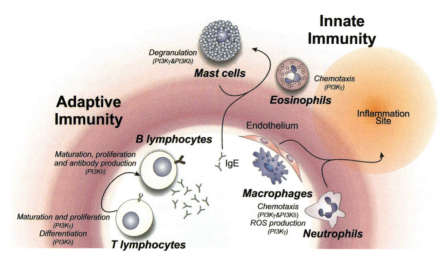

Fig. 6.1 PI3Ks control multiple aspects of innate and adaptive immunity. Class I PI3Kδ and PI3Kγ function as master regulators of distinct cell types, orchestrating innate and adaptive immune responses. In neutrophils and macrophages, both isoforms are required for correct directional movement, while ROS production is mainly controlled by PI3Kγ. The functional interaction between PI3Kδ and PI3Kγ is also relevant for mast cell degranulation. On the other hand, only PI3Kγ contributes to eosinophil migration. Within adaptive immunity, PI3Kδ and PI3Kγ cooperate in regulating distinct functions of T lymphocytes. While PI3Kγ is essential for thymocyte maturation and T cell proliferation, PI3Kδ mainly participates to T cell differentiation. Instead, B cell function is uniquely controlled by PI3Kδ

express a large repertoire of antigen receptors. Upon antigenic recognition, lymphocytes maintain a memory of this event, thus mounting a more rapid and efficient response upon subsequent exposures to the same agent.

Several mechanisms have evolved to allow innate and adaptive immunity to recognize and eliminate foreign antigens, while maintaining tolerance to the self. Among these, the PI3K signaling pathway is a major example of a mechanism that requires fine tuning in order to ensure proper defense without developing excessive inflammation and autoimmunity. In this context, PI3Ks are activated by stimulation through a variety of receptor types, including antigen receptors, co-stimulatory receptors and certain cytokine receptors. Hence, PI3Ks profoundly impact on the pathophysiological regulation of innate and adaptive immunity. In the last decades, studies based on pharmacological and genetic inhibition of PI3Ks have highlighted the role of distinct PI3K isoforms in regulating specific aspects of immune responses (Fig. 6.1). Although immune cells express all class I PI3Ks, a prominent role is hold by PI3Kδ and PI3Kγ. Indeed, PI3Kδ and PI3Kγ mutant mice show relevant phenotypes in their immune responses (Clayton et al. 2002; Hirsch et al. 2000; Jou et al. 2002; Li et al. 2000; Patrucco et al. 2004; Sasaki et al. 2000).

In the following sections, we will summarize the role of PI3Ks in innate and adaptive immunity, with a main focus on PI3Kδ and PI3Kγ, both in physiological and disease conditions.

6.2.1 PI3Ks in Neutrophils and Macrophages

Neutrophils and macrophages provide the first defensive barrier against the invasion by pathogens and microbial agents. Recruitment of these cells to the site of infection is initiated by sensing of specific inflammatory signals (chemokines and cytokines) released by the inflamed tissue. This directional cell migration is known as chemotaxis. The importance of PI3Ks in chemotaxis has been firstly uncovered by means of pan-PI3K inhibition with wortmannin (Okada et al. 1994). In the last 10 years, the availability of isoform-selective compounds and genetic modified animals has allowed for further dissection of the specific contribution of distinct PI3K isoforms. Amongst class I PI3Ks, PI3Kδ and PI3Kγ have clearly emerged as the main determinants of leukocyte migration, both *in vitro* and *in vivo*.

Neutrophils and macrophages from mice lacking PI3Kγ (PI3Kγ$^{-/-}$) show impaired *in vitro* migration in response to different G-protein coupled receptor (GPCR)-related stimuli such as fMLP, C5a, RANTES, IL-8 (Hirsch et al. 2000; Li et al. 2000; Patrucco et al. 2004; Sasaki et al. 2000). In agreement, PI3Kγ$^{-/-}$ animals display reduced number of infiltrating cells in a peritonitis model (Hirsch et al. 2000; Li et al. 2000; Sasaki et al. 2000). This phenotype can be explained by the inability of PI3Kγ$^{-/-}$ cells to correctly assemble and activate their molecular machinery controlling cell polarization, which represents an essential prerequisite for directional movement. Knock-in mice expressing a membrane-targeted PI3Kγ enzyme (characterized by delocalized production of PIP$_3$) recapitulate the phenotype of PI3Kγ$^{-/-}$ mice, demonstrating the importance of controlled spatial and temporal production of PIP$_3$ in leukocyte migration (Costa et al. 2007). Thus, PI3Kγ catalytic function is crucial for proper PIP$_3$ generation, and, in turn, for the regulation of Rac activity and cytoskeleton rearrangement at the leading edge (Barberis et al. 2009; Costa et al. 2007; Ferguson et al. 2007). In addition to PI3Kγ, PI3Kδ contributes to the fine modulation of directional movement, as the PI3Kδ-specific inhibitor IC87114 significantly dampens migration of neutrophils *in vitro* (Puri et al. 2004; Sadhu et al. 2003), as well as recruitment of inflammatory cells in a *in vivo* model of pulmonary inflammation (Puri et al. 2004).

The recruitment of inflammatory cells at the site of infection is a multistep process, including a first event of selectin-mediated "capture" of circulating leukocytes and subsequent "rolling" on the vascular endothelium, followed by integrin-mediated firm adhesion and extravasation. Both PI3Kγ and PI3Kδ are expressed in endothelial cells, where they regulate the complex interplay between leukocytes and the inflamed, sticky and leaky endothelium. Selective blockade of PI3Kγ in endothelial cells has been shown to reduce selectin-mediated attachment of neutrophils and to increase their rolling velocity (Puri et al. 2005). Similarly, endothelial PI3Kδ plays a central role in neutrophil adhesion and subsequent transendothelial migration in response to tumor necrosis factor α (TNF α) and leukotriene B4 (LTB4) (Puri et al. 2004). Consistent with an essential role for both isoforms in leukocyte migration, double knock-out PI3Kγ$^{-/-}$δ$^{-/-}$ mice display a more dramatic phenotype than single mutants (Puri et al. 2005). Nonetheless, PI3Kγ and PI3Kδ do not play overlapping roles, as they regulate temporally distinct events. Neutrophil emigration toward CXCL2 or CXCL1 is severely impaired in PI3Kγ$^{-/-}$ mice at an early time (first 90 min), but more prolonged responses are almost entirely PI3Kγ-independent and largely dependent on PI3Kδ (Liu et al. 2007).

After recruitment to the inflammation site, neutrophils and macrophages exert their antimicrobial function by producing and secreting reactive oxygen species (ROS), an event known as respiratory burst. In the absence of PI3Kγ, ROS production evoked by cytokine-primed neutrophils in response to fMLP is significantly reduced (Hirsch et al. 2000; Li et al. 2000; Sasaki et al. 2000). Pharmacological inhibition of PI3Kγ with selective inhibitors further demonstrates that in TNFα-primed human neutrophils PI3Kγ is needed to initiate the first phase of a temporally biphasic pathway of ROS production induced by fMLP. Instead, PI3Kδ and at least in part PI3Kα and PI3Kβ, are necessary for the subsequent amplification phase (Puri et al. 2004; Sadhu et al. 2003). Although the second phase of ROS production is mediated by PI3Kδ, both phases actually depend entirely on the first phase of ROS production, which is regulated exclusively by PI3Kγ (Condliffe et al. 2005).

6.2.2 PI3Ks in Mast Cells and Eosinophils

In their action against parasites and infections, neutrophils and macrophages are assisted by mast cells. On the other hand, aberrant activation of mast cells causes different allergic diseases. Mast cells are characterized by large intracytoplasmic granules, containing hystamin and heparin, which are rapidly released following cell activation, a process known as degranulation. In allergic reactions, mast cells remain inactive until an allergen binds to a special set of immunoglubulins of the IgE type, which are tightly associated to the IgE high affinity receptor (FCeRI) at the plasma membrane. At molecular level, allergen stimulation, through IgE binding, triggers the activation of the protein tyrosine kinase Lyn and recruitment of Syk, resulting in the phosphorylation of immunoreceptor tyrosine-based activation motifs (ITAMs). These phosphorylated motifs provide a docking sites for the SH2 domains of class IA PI3Ks adaptor subunits. The subsequent PIP$_3$ production is then essential to activate Bruton's tyrosine kinase (Btk) and subsequently phospholipase C-γ (PLCγ). These signaling pathways cause the opening of plasma membrane calcium channels and granules release (Rommel et al. 2007).

The first indication of the involvement of PI3Ks in mast cells came from the use of non-selective PI3K inhibitors. Indeed, pan-PI3K inhibitors like LY294002 and wortmannin impair mast cell degranulation (Tkaczyk et al. 2003). Class I PI3Kδ and PI3Kγ appear to be the main isoforms involved in this process. Treatment with the PI3Kδ-selective inhibitor IC87114 or genetic inactivation of PI3Kδ activity (PI3Kδ^{D910A}) dampen mast cell activity (Ali et al. 2004, 2008). Accordingly, PI3Kδ^{D910A} mice are protected from passive cutaneous anaphylaxis induced by IgE- and antigen-injection (Ali et al. 2004, 2008). Similarly, blockade of the GPCR-coupled PI3Kγ activity reduces mast cell degranulation and PI3Kγ$^{-/-}$ animals are resistant to passive systemic anaphylaxis (Laffargue et al. 2002). Recent findings suggest that only a restricted pool of PI3Kγ is committed to the modulation of this process. Indeed, the mast cell phenotype of PI3K$\gamma^{-/-}$ animals is completed rescued by the co-expression of p110γ catalytic subunit together with the regulatory subunit p84/87, but not with p101 (Bohnacker et al. 2009). The current view of the complex interplay between PI3Kγ and PI3Kδ in regulating mast cell function proposes a complex epistatic interaction, with PI3Kδ acting earlier in response to IgE and PI3Kγ functioning later to maximize degranulation (Hirsch et al. 2006).

Eosinophils are recruited and activated in response to mast cell degranulation, thus functioning as effector cells in the allergic disease. They are typical infiltrating cells at sites of allergen-IgE reactions, where they produce a wide array of mediators such as cytokines and ROS. PI3Ks have been shown to regulate eosinophil chemotaxis in response to different chemoattractants. IL-5-induced release of eosinophils from the bone marrow is severely impaired upon treatment with the pan-PI3K inhibitor wortmannin (Palframan et al. 1998). In addition, wortmannin decreases the number of eosinophils in the brochoalveolar lavage (BAL) of ovalbumin (OVA)-challenged animals (Tigani et al. 2001). Wortmannin and LY294002 have also been found to inhibit platelet-activating factor (PAF)-induced eosinophil chemotaxis and respiratory burst, but not eotaxin-induced migration (Mishra et al. 2005). Furthermore, intra-tracheal administration of PI3K inhibitors wortmannin or LY294002 could significantly attenuate inflammation symptoms and airway hyper-responsiveness, due to sensitization with OVA inhalation in a mouse model of asthma (Duan et al. 2005; Ezeamuzie et al. 2001).

The specific PI3K isoforms involved in regulating eosinophil chemotaxis are still unclear. OVA-sensitized PI3K$\gamma^{-/-}$ mice display reduced levels of allergen-induced eosinophilic airway inflammation and airway remodeling (Lim et al. 2009; Takeda et al. 2009), thus pointing to a crucial role of this class I isoform. However, other studies suggest that PI3Kγ mainly regulates the maintenance of eosinophilic inflammation *in vivo*, rather than the recruitment process, which seems to be modulated by other PI3Ks (Pinho et al. 2005). Further complexity comes from the unexpected finding that double mutant mice PI3K$\gamma^{KO}/\delta^{D910A}$ display marked eosinophilic inflammation in multiple mucosal organs, as well as increased amount of serum IgE, IL-4 and IL-5 levels (Ji et al. 2007).

6.2.3 PI3Ks in T Lymphocytes

T and B lymphocytes orchestrate a sophisticated mechanism of protection, featured by recognition of specific antigens and pathogens. T lymphocytes are involved in cell-mediated immunity and contribute to the control of humoral immunity, by exerting a strict control on the activity of B lymphocytes. The first suggestion of a key role of PI3Ks in regulating T lymphocyte function has come from studies with pan-PI3K inhibitors. Wortmannin impairs antigen (Ag)-induced IL-2 production, as well as Ag-induced CD4$^+$ T cells differentiation (Shi et al. 1997). In addition, wortmannin and LY294002 inhibit CD3-induced IL-2 synthesis and proliferation of CD8$^+$ T cells (Phu et al. 2001).

More recently, the development of isoform-selective inhibitors and genetically engineered animals has allowed to show that distinct PI3K isoforms control different processes of T lymphocyte function, including development, proliferation and migration. For instance, class I PI3Kγ and PI3Kδ are both required for proper T cell differentiation. Indeed, PI3Kγ$^{-/-}$ mice show a reduced number of peripheral T lymphocyte due to impaired maturation of thymocytes. In particular, ablation of PI3Kγ increases the ratio of double negative (CD4$^-$ CD8$^-$) on double positive (CD4$^+$ CD8$^+$) cells in thymus, blocking thymocyte development (Sasaki et al. 2000). This phenotype is worsened by the double ablation of PI3Kγ and PI3Kδ, leading to a dramatic increase in the number of double negative thymocytes. On the contrary, PI3Kδ inactivation (PI3Kδ^{D910A}) alone does not affect thymocyte development (Ji et al. 2007; Webb et al. 2005).

Both PI3Kδ and PI3Kγ are essential for the subsequent phase of proliferation. In naïve T cells, PI3Ks are engaged by the cross-linking of T-cell receptor (TCR), with or without co-stimulation by CD28, or by activation of the IL-2 receptor or chemokine receptors (Alcazar et al. 2007; Fruman and Cantley 2002). T cells lacking PI3Kγ show abnormal TCR-mediated signaling and reduced immunological synapse organization, as well as reduced proliferation (Sasaki et al. 2000). Similarly, knock-in mice expressing a kinase-inactive PI3Kδ display impaired antigen-specific T-cell responses and a reduction in T-cell activation and proliferation upon *in vitro* stimulation (Okkenhaug et al. 2002).

On the other hand, PI3Kδ is central for maturation of CD4$^+$ T cell and differentiation in distinct T cell subsets (Th1, Th2, Th17, Treg). PI3Kδ^{D910A} mice display reduction of both Th1 and Th2, *in vitro* and *in vivo* (Okkenhaug et al. 2006). Furthermore, PI3Kδ cooperates with SHIP to maintain the correct ratio of Th17 and Treg cells. SHIP1$^{-/-}$ mice show preferential differentiation in Treg compared to Th17 (Locke et al. 2009). On the other hand, in PI3Kδ^{D910A} mice peripheral Treg maturation is impaired (Ji et al. 2007; Liu et al. 2009; Oak et al. 2006; Patton et al. 2006). Reduction of Treg function, associated with increased B cell-mediated IgE production, renders PI3Kδ^{D910A} mice prone to autoimmunity (Ji et al. 2007; Oak et al. 2006). By contrast, the impaired Treg immunosuppressive function of PI3Kδ^{D910A} mice appears beneficial in the case of infection by the parasite *Leishmania major*. Indeed, the reduced Treg expansion of PI3Kδ^{D910A} mice seems to be responsible for a weakened Th1 response, thus preventing disease development (Liu et al. 2009).

The role of PI3Ks in the regulation of T cell migration is more controverse. In some circumstances, PI3Kγ signaling appears important for T cell chemotaxis in the mouse (Camps et al. 2005; Reif et al. 2004; Sasaki et al. 2000; Webb et al. 2005). In addition, treatment with the PI3Kγ specific inhibitor AS605240 has indicated that PI3Kγ plays a dominant role in the migratory response to CXCL12 (Smith et al. 2007) in primary human T lymphocytes. In contrast, migratory responses to a range of chemokines, including CXCL12, of T cells derived from mice expressing a catalytically-inactive form of PI3Kδ are largely unaffected (Reif et al. 2004), indicating that PI3Kγ is the predominant isoform involved in T cell migration. However, recent works have suggested that T cell migration principally depends on pathways involving the Rac guanine nucleotide exchange factor (GEF) DOCK2, rather than on PI3Kδ- and PI3Kγ-dependent signaling (Nombela-Arrieta et al. 2004). Instead, PI3Kδ is crucial for T lymphocyte trafficking, by regulating shedding and transcriptional shut off of the lymph node-homing receptor CD62L (L-selectin) (Sinclair et al. 2008).

Overall, these findings suggest that class I PI3Kδ and PI3Kγ cooperate in regulating T cell signaling, with PI3Kδ impacting more profoundly than PI3Kγ on regulation of cell-based immunity.

6.2.4 PI3Ks in B Lymphocytes

B lymphocytes represent the other major cellular component of the adaptive immune response. In contrast to what found in T cells, PI3Kγ does not play a significant role in B cells. Conversely, PI3Kδ has been shown to control different aspects of B lymphocyte function, including their maturation process and proliferation.

B cell development occurs through several stages. Immature B cells are produced in bone marrow, through Ig chain rearrangement of B cell progenitors (at pro-B and pre-B stages), followed by repertoire selection. These immature B cells then migrate to the spleen where, upon a further selection process, they differentiate into mature B lymphocytes by forming follicular (FO) and marginal zone (MZ) niches. Lack of PI3Kδ has been shown to affect the early steps of B cell maturation, as suggested by the increased pro B/pre B ratio. In agreement, the few immature B cells of PI3Kδ^{D910A} mice are unable to sustain FO and MZ pools (Clayton et al. 2002; Jou et al. 2002; Okkenhaug et al. 2002). Interestingly, B cell development is not further impaired when both PI3Kδ and PI3Kγ are eliminated and PI3Kγ$^{-/-}$ mice do not show any clear defect in B cell maturation, thus demonstrating that PI3Kγ does not contribute to regulation of B cell development (Webb et al. 2005).

Also B cell proliferation is strictly dependent on PI3Kδ signaling. B-cell proliferation in response to IgM stimulation and BAFF is decreased in cells expressing a catalytically-inactive form of PI3Kδ, whereas proliferation induced by IL-4, CD40 or LPS is only partially affected (Henley et al. 2008; Okkenhaug et al. 2002). In addition, PI3Kδ activity is indispensable for B-cell-receptor-induced DNA synthesis and proliferation, as well as IL-4-induced survival (Bilancio et al. 2006; Sujobert et al. 2005). PI3Kδ$^{-/-}$ and PI3Kδ^{D910A} B lymphocytes also display reduced antibody production upon T cell-dependent and independent stimulation (Clayton et al. 2002; Jou et al. 2002; Okkenhaug et al. 2002). More specifically, PI3Kδ^{D910A} mice show reduced IgM and IgG antibody responses (Okkenhaug et al. 2002). By contrast, IgE production is paradoxically increased by genetic or pharmacological inactivation of PI3Kδ, despite reduced Th2 responses (Zhang et al. 2008) due to the ability of PI3Kδ to modulate IgE switch (Omori et al. 2006; Zhang et al. 2008).

Similar to the case of T cells, migration of B cells is not regulated by PI3Ks. Rather, this process is mediated by DOCK2 and PI3K-independent Btk signaling (de Gorter et al. 2007; Nombela-Arrieta et al. 2004). Nonetheless, the complete response to CXCL13 is reduced in PI3Kδ^{D910A}, but not in PI3K$\gamma^{-/-}$ B lymphocytes, revealing a surprising and not well clarified role of PI3Kδ downstream G-protein coupled chemokine receptors (Nombela-Arrieta et al. 2004; Reif et al. 2004).

6.2.5 PI3Ks in Inflammatory and Autoimmune Diseases

Given the central role of class I PI3Kδ and PI3Kγ in the homeostasis of both innate and adaptive responses, these enzymes can also participate to the onset and/or progression of diseases characterized by deregulated activation of innate and/or adaptive immunity. Indeed, genetic inactivation of PI3Kγ or PI3Kδ modulates the susceptibility to specific diseases.

6.2.5.1 Rheumatoid Arthritis

Rheumatoid arthritis (RA) is an autoimmune disorder that affects the joints and is characterized by the progressive destruction of articular structures (Harris 1990). The pathogenesis of RA is not completely understood. Chemokines and other chemoattractants have been detected in the inflamed joints and are responsible for the local recruitment of leukocytes. Amongst these, neutrophils constitute the most abundant population and actively induce inflammatory response and tissue damage (Brennan and Feldmann 1996; Edwards and Hallett 1997; Szekanecz et al. 2003). As PI3Kγ is key in neutrophil chemotaxis, PI3Kγ deficiency is protective in different mouse models of RA. Camps et al. first showed that PI3K$\gamma^{-/-}$ animals are largely resistant to αCII-induced arthritis, where type II collagen-specific monoclonal antibodies are injected to initiate RA. Moreover, blockade of PI3Kγ by oral delivery of the isoform-selective compound AS605240 reproduces the protective effect of PI3K$\gamma^{-/-}$ mice in a model of collagen-induced arthritis, where typical features of the disease are triggered by intra-dermally injection of collagen II (Camps et al. 2005). In both cases, the protection correlates with defective neutrophil migration and thus to reduced accumulation of neutrophils in the joints. PI3Kγ inactivation is associated to a milder inflammatory arthritis also in an alternative mouse model of RA based on the transgenic overexpression of the human TNFα (Hayer et al. 2009). Interestingly, the genetic disruption of PI3Kγ reduces the severity of arthritis through both reduced invasion of leukocytes and reduced proliferation of synovial mesenchymal-derived fibroblasts. These findings challenge the concept of a leukocyte-restricted role of PI3Kγ in the pathogenesis of RA and suggest that the therapeutic potential of specific PI3Kγ inhibitors might be expanded to a broader spectrum of cell targets, thus yielding superior results in the potential treatment of RA.

In line with a cooperative role of PI3Kγ and PI3Kδ in regulating neutrophil function, also PI3K$\delta^{-/-}$ mice are protected from RA. Randis et al. [2008] have shown that administration of arthritogenic serum to PI3K$\delta^{-/-}$ mice results in a significant reduction of paw edema, similar to what observed in PI3Kγ$^{-/-}$ animals. A more pronounced protection is also observed in double PI3Kδ$^{-/-}$γ$^{-/-}$ mice, indicating the existence of a functional interaction between PI3Kγ and PI3Kδ in inflammatory arthritis. Accordingly, combined inhibition of PI3Kδ and PI3Kγ might represent an intriguing innovative treatment for RA.

6.2.5.2 Systemic Lupus Erythematosus

Systemic lupus erythematosus (SLE) is a chronic autoimmune disease characterized by deregulation of T-cell mediated B-cell activation, resulting in generalized B-cell expansion and hypergammaglobulinemia (Liu and Wakeland 2001). The involvement of PI3Ks in the pathogenesis of SLE was first uncovered by two works showing that increased PI3K activity, due to either PTEN ablation (PTEN$^{+/-}$) or overexpression of an activating form of the p85 regulatory subunit (p65^{PI3K}) in T lymphocytes, leads to a SLE-like phenotype (Borlado et al. 2000; Di Cristofano et al. 1999). Interestingly, the severity of the disease is attenuated in p65^{PI3K}Tg/PI3Kγ$^{-/-}$ animals, thus suggesting that PI3Kγ plays a crucial role in SLE (Barber et al. 2006). Blockade of PI3Kγ by the selective inhibitor AS605240 has also found effective in another mouse model of SLE, the MRL-lpr SLE-prone model (Barber et al. 2006). Taken together, these data encourage further study of selective PI3K inhibitors in the treatment of SLE.

6.2.5.3 Asthma

Asthma is a pulmonary disease characterized by bronchial hypersensitivity and involving a Th2 immune response mounted by CD4$^+$ T cells. Other inflammatory cells such as mast cells and eosinophils also play important roles. In particular, neutrophils contribute to the chronic evolution of the disease (Baraldo et al. 2007). The availability of pan-PI3K inhibitors first allowed to uncover the correlation between PI3K hyperactivity and the development of allergic conditions. Indeed, intratracheal administration of wortmannin or LY294002 significantly attenuates inflammation symptoms and airway hyperresponsiveness in a mouse model of asthma (Duan et al. 2005; Ezeamuzie et al. 2001).

PI3Kδ and PI3Kγ act as master regulators of mast cells and eosinophils, which constitute the principal mediators of allergy. Accordingly, PI3Kγ$^{-/-}$ animals are completely protected against systemic anaphylaxis (Laffargue et al. 2002). Similarly, PI3Kδ knock-in mice are partially resistant to passive cutaneous anaphylaxis induced by IgE- and antigen-injection (Ali et al. 2004, 2008). Furthermore, intratracheal administration of the PI3Kδ-selective inhibitor IC87114 significantly attenuates allergic airway inflammation and suppresses OVA-induced airway hyper-responsiveness to inhaled methacoline (Lee et al. 2006). Farghaly et al. have shown that the Th2 cytokine IL-13 fails to induce hyper-responsiveness in isolated tracheal rings from PI3Kδ^{D910A} mice. In this context, the reduced hyper-responsiveness may be attributed to a direct effect on airway structural cells rather than on infiltrating immune cells (Farghaly et al. 2008).

Overall, these studies have unveiled a key role for PI3Kδ, in addition to PI3Kγ, in the pathogenesis of allergic asthma. However, whether selective inhibition of PI3Kδ might be protective in allergy is still controversial. In OVA-immunized mice, blockade of PI3Kδ leads to a paradoxical increase of both total and OVA-specific IgE levels, despite diminished Th2 responses (Omori et al. 2006; Zhang et al. 2008). Hence, additional studies are needed to clarify the role of PI3Kδ in regulating allergen-mediated IgE production, as well as the resulting clinical implications in the treatment of allergic disease. As PI3Kγ and PI3Kδ cooperate in the onset and progression of allergic conditions, a combined inhibition of these isoforms appears as a reasonable therapeutic strategy (Doukas et al. 2009).

6.2.5.4 Chronic Obstructive Pulmonary Disease

PI3Kδ and PI3Kγ cooperate in the pathogenesis of chronic obstructive pulmonary disease (COPD). COPD is a common respiratory disease which, unlike allergic asthma, involves $CD8^+$ T cells releasing Th1-type cytokines, with the additional contribution of macrophages and neutrophils. In COPD, airflow limitation is progressive and may be steroid-resistant (Baraldo et al. 2007; Doherty 2004). The double selective inhibitor TG100-115, targeting both PI3Kδ and PI3Kγ, has been shown effective in controlling COPD in different mouse models. TG100-115 significantly reduces neutrophil accumulation as well as production of the classical Th1 cytokine TNFα in a LPS-induced model of COPD. Interestingly, TG100-115 is successful even in a steroid-resistant form of COPD induced in mice by cigarette smoke exposure (Doukas et al. 2009). Recent reports suggest that this beneficial effect is achieved through a selective involvement of PI3Kδ, as genetic inactivation of this isoform, and not of PI3Kγ, restores glucocorticoid responsiveness in smoke-induced airway inflammation (Marwick et al. 2009, 2010). Therefore, the complex interplay between PI3Kγ and PI3Kδ in the development of COPD needs further study. Currently, double selective compounds might represent the most promising response to the urgent need of new treatments for steroid-unresponsive inflammatory diseases.

6.3 PI3Ks in the Regulation of Glucose Metabolism and Insulin Sensitivity

The tight control of glucose metabolism represents a fundamental physiological mechanism regulated by a series of key metabolic hormones. The main player in this process is insulin, a peptide hormone secreted from pancreatic β-cells in response to elevated glucose concentrations. Insulin mainly acts by (i) inhibiting liver gluconeogenesis and glycogenolysis and (ii) stimulating glucose uptake in insulin-sensitive periferal tissues, mainly skeletal muscle and adipocytes. In addition, insulin affects

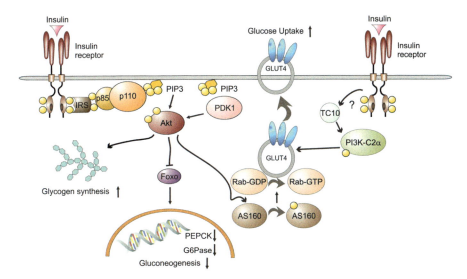

Fig. 6.2 PI3K function in glucose metabolism. Upon binding of insulin to its receptor (IR), the IRS adaptor proteins recruit class I PI3Ks, thus triggering PIP$_3$ production and PKB/Akt activation. PKB, in turn, mediates the translocation of the glucose transporter GLUT4 to the plasma membrane, which leads to glucose uptake from the extracellular space. Furthermore, Akt mediates the inhibition of FOXO transcription, thereby negatively regulating the expression of gluconeogenic enzymes such as PEPCK and G6Pase. Insulin also modulates glycogen synthesis through PKB-mediated inhibition of GSK3 and the consequent activation of glycogen synthase. On the other hand, class II PI3K-C2α, following its TC10-dependent activation, contributes to GLUT4 translocation. However, the exact mechanisms linking IR activation to PI3K-C2α are yet to be defined

lipid metabolism and stimulates the uptake of aminoacids, thus enhancing protein synthesis. Defects in insulin production and signaling underlie important pathological conditions such as diabetes mellitus and the metabolic syndrome. Type 2 diabetes represents a common disease and is characterized by impaired insulin-stimulated glucose uptake, increased hepatic glucose production and inadequate compensation by the pancreatic β-cells, ultimately leading to hyperglycemia (Kahn 1994).

The action of insulin is initiated by its binding to the insulin receptor (IR), a transmembrane glycoprotein with intrinsic protein tyrosine kinase activity. The adaptors IRS (insulin receptor substrate) are the first IR substrates that undergo tyrosine phosphorylation upon insulin stimulation (Myers and White 1993). Six IRS isoforms are expressed in mammals and play the role of linkers between the upstream tyrosine kinase and the downstream regulatory enzymes and adaptor molecules. PI3K was the first enzyme found to be associated with the IR/IRS signaling. Within this context, upon receptor activation, IRS serves as a docking site for the SH2 domains of the regulatory subunit of PI3K, which in turn recruit the p110 enzyme to the plasma membrane (Myers et al. 1992; Sun et al. 1993) (Fig. 6.2). Upon activation, the p110 catalytic subunit of PI3Ks produces the lipid second messenger PIP$_3$, which in turn activates downstream PH domain-containing effectors that control various metabolic processes such as glucose uptake, lipolysis inhibition, triglyceride formation, and glycogen synthesis. Within these processes, a key molecule activated following PIP_3 generation is PKB/Akt (Taniguchi et al. 2006). However, other phosphoinositide-activated kinases, such as atypical PKC, have also been implicated in insulin action (Farese et al. 2005).

Amongst the downstream effectors of IR-PI3Ks, Akt represents the central node in the insulin signaling network, stimulating blood glucose disposal and glycogen synthesis, and inhibiting gluconeogenesis. Three different isoforms of Akt, encoded by different genes, are found in mammals. The study of knock-out mice have identified the specific functions of each Akt isoform. Of the three isoforms, Akt2 is in particular the main regulator of glucose homeostasis *in vivo* (Cho et al. 2001). Indeed, the disruption of Akt2 in mice results in impaired glucose uptake and in a complete failure of insulin to suppress hepatic glucose output (Cho et al. 2001). Akt2 plays an important role also in humans in the regulation of glucose homeostasis, since a germline mutation in Akt2 correlates with development of type 2 diabetes (George et al. 2004).

In adipocytes and muscle cells, PI3K-dependent activation of Akt2 (and to a lesser extent of Akt1) controls insulin-mediated glucose uptake by stimulating GLUT4 translocation from an intracellular compartment to the plasma membrane (Bai et al. 2007; Stenkula et al. 2010). GLUT4 exocytosis is highly regulated by PIP_3 production, although the exact mechanism is still unclear. A number of Akt substrates, including the Rab-GAP AS160 (Akt substrate of 160 kDa) (Chen et al. 2011; Sano et al. 2003) and PI5-kinase (PIKfyve) (Berwick et al. 2004), are involved in this process. Other PI3K-dependent mediators of glucose uptake are atypical PKC (aPKC) isoforms that regulate the kinesin and Rab4-dependent GLUT4 exocytosis (Imamura et al. 2003). Further support for this model is given by the muscle-specific knock-out of the aPKC PKC-λ, which results in systemic insulin resistance and glucose intolerance (Farese et al. 2007).

The class IA PI3K pathway is also involved in insulin-mediated inhibition of hepatic gluconeogenesis (Agati et al. 1998; Kotani et al. 1999), a process indispensable in a starved condition and switched-off when external resources are available. In the latter condition, insulin negatively regulates gluconeogenesis by suppressing the expression of key gluconeogenic enzymes through the Akt-mediated phosphorylation of the transcription factor FoxO1. Phosphorylated FoxO1 is excluded from the nucleus and consequently fails to activate the transcription of genes required for the gluconeogenesis, such as phosphoenolpyruvate caboxykinase and glucose-6-phosphatase (Nakae et al. 2001).

Together with class IA PI3Ks, PI3Ks of other classes might be involved in insulin-mediated glucose disposal. Indeed, insulin stimulation triggers the production of PI3P, thus possibly involving either class II or III PI3K. PI3P binds the PX and FYVE domains of proteins involved in the control of vesicular trafficking. It is thus possible that PI3K other than class IA are involved in processes needed for plasma membrane fusion of cytoplasmic vesicles containing GLUT4 (Shepherd 2005). Interestingly, insulin-mediated PI3P production appears insensitive to the action of wortmannin, thus excluding a role for class III enzymes, while suggesting a role for the wortmannin-insensitive class II PI3K-C2 isoenzymes. In line with this view, PI3K-C2α is activated in response to insulin-dependent activation of the small GTPase TC10, thus triggering PI3P elevation necessary for full-scale GLUT4 plasma membrane translocation (Maffucci et al. 2003).

6.3.1 Class I PI3K Regulatory Subunits in Glucose Metabolism

The crucial role of PI3Ks in insulin signaling suggests that a dysfunction of the PI3K pathway may deeply affect insulin sensitivity. Indeed, several lines of evidence have indicated that PI3K signaling is compromised in the obese diabetic ob/ob mouse model (Folli et al. 1993; Kerouz et al. 1997), as well as in diabetic patients (Bjornholm et al. 1997; Goodyear et al. 1995).

Despite the key role of PI3K in the insulin signaling, gene deletion studies of adaptors of the p85 family have reported unexpected paradoxical effects. Indeed, mice lacking p85α develop hypoglycemia and increased insulin sensitivity (Fruman et al. 2000). This phenotype probably correlates with the up-regulation of the other *Pik3r1* splicing variants, p50α and p55α, in fat and muscle, leading to an elevation of PIP_3 levels and to facilitated GLUT4 translocation to the plasma membrane (Terauchi et al. 1999). A slight enhancement of insulin signaling is also present in mice lacking either p50α/p55α or p85β (Chen et al. 2004; Ueki et al. 2002). On the other hand, deletion of all *Pik3r1* gene products (p85, p55, and p50), results in perinatal lethality, associated with a substantial decrease in the expression and activity of class IA PI3K catalytic subunits. Nonetheless, *Pik3r1*-deficient mice are hypoglycemic and more insulin sensitive because of a more active glucose transport in insulin-responsive tissues (Fruman et al. 2000). Similar findings have been obtained with a hepatic deletion of *Pik3r1*, which causes improved insulin sensitivity in liver, muscle, and fat, but leads to a 60% decrease in total hepatic PI3K activity (Taniguchi et al. 2006). The phosphorylation of Akt downstream PI3K is enhanced as a consequence of reduced activity of the phosphatase PTEN, suggesting a role of p85 in PTEN regulation (Taniguchi et al. 2006). Indeed, PTEN is a potent negative regulator of the insulin signaling and loss of PTEN in adipose tissue results in increased insulin sensitivity and GLUT4 recruitment (Kurlawalla-Martinez et al. 2005). A recent study has shown that p85 directly binds and activates PTEN, which explains the paradox of increased insulin sensitivity in p85-deficient animals (Chagpar et al. 2010).

These studies indicate that a critical molecular balance between regulatory and catalytic subunits determines the optimal response of the PI3K pathway to insulin signaling (Ueki et al. 2002). In physiological conditions, the p85 regulatory subunit is more abundant than the catalytic p110 subunit. p85 monomers can thus inhibit PIP_3 production either by binding to phosphorylated IRS proteins (Ueki et al. 2002) or by altering subcellular localization of p110/p85 dimers (Inoue et al. 1998). Unbalanced p85 levels can compromise this regulatory mechanisms and lead to a paradoxical increase in PI3K activity. Finding of increased p85/55/50 protein expression, in mouse models of obesity as well as in type 2 diabetic patients further confirms this view (Bandyopadhyay et al. 2005) and suggests that p85 family members play a complex role in the regulation of PI3K-mediated insulin signaling.

6.3.2 Class I PI3K Catalytic Subunits in Glucose Metabolism

The specific role of single p110 isoforms in insulin signaling has recently begun to emerge, thanks to the development of selective pharmacological inhibitors and of mouse genetic models. Although the lethal phenotype of mice lacking either p110α or p110β hampers the study of their specific roles in insulin signaling, analysis of insulin-mediated responses in heterozygous animals suggests that both proteins might be involved. Heterozygous mice lacking either p110α or p110β show normal responses to insulin. However, mice heterozygous for both isoforms have decreased insulin sensitivity (Brachmann et al. 2005), suggesting that p110α and p110β play complementary roles in insulin signaling. Nonetheless, heterozygous mice expressing a catalytically inactive p110α become insulin resistant with aging, and develop glucose intolerance, hyperlipidemia, adiposity, as well as hyperglycemia and deregulate hepatic gluconeogenesis (Foukas et al. 2006). The use of isoform-selective p110 inhibitors has shown that p110α constitutes the major effector downstream of the IR, while p110β plays only a marginal role, by in providing a basal threshold of PIP_3 production that potentiates p110α activity (Knight et al. 2006). Indeed, mice expressing a kinase-dead p110β or carrying a liver-specific ablation of this enzyme develop only mild impaired insulin sensitivity and glucose homeostasis (Ciraolo et al. 2008; Jia et al. 2008). On the contrary, the disruption of p110α in liver results in impaired insulin action and glucose homeostasis which cannot be rescued by p110β (Sopasakis et al. 2010). These findings demonstrate that p110α is the primary PI3K isoform required for the metabolic actions of insulin in the liver. Instead, p110β plays a minor but not negligible role.

6.4 Class I PI3Ks in Cardiac Physiology and Disease

Three different class I PI3Ks are expressed in the myocardial tissue (p110α, p110β and p110γ). However, a growing amount of evidence indicates that different PI3K isoforms participate in distinct physiological and pathological processes within cardiomyocytes. In particular, class IA p110α mostly functions as a critical regulator of cardiomyocyte viability and growth downstream tyrosine kinase receptors such as IGF-I and insulin. Based on available studies, p110α activity can be generally considered as beneficial and protective for heart function. In contrast, class IB p110γ is uniquely controlled by GPCRs such as β-adrenergic receptors. Of note, robust activation of p110γ is essentially detected in the context of maladaptive remodeling and during the natural history of heart failure, where p110γ contributes to myocardial dysfunction and impaired contractility. Hence, pharmacological inhibition of p110γ has emerged as a novel potential therapeutic strategy in cardiac disease. Finally, our understanding of p110β function in the myocardium is still very initial and is awaiting for further and detailed study.

6.4.1 PI3Kα Regulates Cardiomyocyte Growth

Studies performed on gain and loss-of function models have consistently shown that PI3Kα activity represents a master switch leading to cardiomyocyte growth and survival (Fig. 6.3, upper panel). Systemic knockout of p110α is embryonic lethal without obvious cardiac abnormalities, but a general impairment in cell growth is observed (Bi et al. 1999). In line with this finding, over-expression of a myocardial-selective dominant-negative p110α results in smaller cardiomyocytes and reduced heart mass, without affecting tissue architecture and contractility (Crackower et al. 2002; McMullen et al. 2003; Shioi et al. 2000). On the other hand, over-expression of a cardiac-specific and constitutively active p110α leads to myocardial hypertrophy and increased heart/body weight, but without progression to hypertrophic cardiomyopathy or cardiac dysfunction (Shioi et al. 2000). These findings phenocopy the genetic loss of myocardial PTEN, which leads to PIP_3 accumulation (Crackower et al. 2002). In both over-expression and dominant negative models, the changes in p110α activity are paralleled by corresponding changes in Akt, Gsk3β and p70S6K activity, indicating that p110α boosts myocardial growth through the activation of canonical PIP_3-dependent anabolic pathways. Indeed, genetic manipulations of Akt lead to similar cardiac phenotypes: Akt1 knockout mice show reduced heart size, while different models of constitutively active Akt1 result in different degrees of cardiac hypertrophy (Condorelli et al. 2002; DeBosch et al. 2006; Matsui et al. 2002; McMullen et al. 2003).

Beyond cardiac development, the activity of p110α appears to control essentially physiological hypertrophy, as mice expressing a dominant-negative p110α are resistant to left ventricular hypertrophy induced by exercise training. Nonetheless, loss of p110α leaves the animals prone to compensatory hypertrophy following pressure overload by aortic constriction, indicating that p110α represents a master switch of physiological but not of pathological hypertrophy (McMullen et al. 2003). Importantly, cardiac p110α is functionally relevant in conditions of myocardial damage, where p110α protects cardiomyocytes from cell death and dysfunction caused by various pathological noxae (Fig. 6.3, lower panel). Indeed, it has been shown that in several heart disease models such as dilative myocardiopathy, myocardial infarction, chronic adrenergic stimulation and pressure overload, the loss of p110α is highly detrimental and accelerates adverse ventricular remodeling, while on the other hand constitutive p110α activity exerts cardioprotective effects (Lin et al. 2010; McMullen et al. 2007). Taken together, our understanding of myocardial PI3Kα suggests that any treatment targeting myocardial PI3Ks should leave p110α activity unchanged to circumvent a negative impact on cardiac function, especially in the context of cardiotoxicity (e.g. tumor chemotherapy) and heart failure.

Normal myocardium

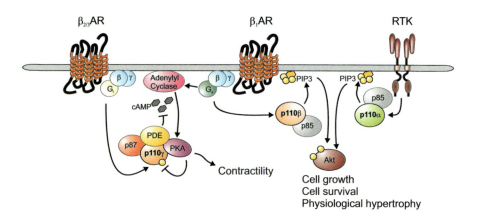

Failing myocardium

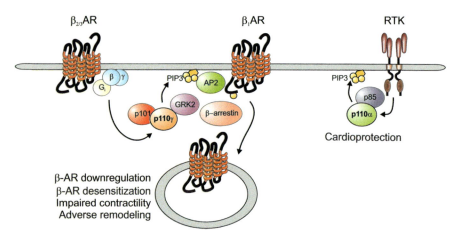

Fig. 6.3 PI3K signaling in normal and failing myocardium. *Upper panel*. In the myocardium, PI3Kα is activated by tyrosine kinase receptors (RTKs) such as IGF-I, EGF and insulin receptor. Downstream p110α, Akt and other classical PIP_3-dependent anabolic pathways promote normal myocardial development and growth. Moreover, PI3Kα is required for physiological hypertrophy. On the other hand, p110β represents the master PI3K isoform producing PIP_3 and activating Akt upon β-adrenergic stimulation. In physiological conditions, p110γ is expressed at low levels and contributes residually to PIP_3 production upon β-adrenergic stimulation. p110γ, along with the p87 regulatory subunit, is part of a macromolecular complex mediating the cAMP-dependent activation of PDEs, while the lipid kinase activity of p110γ is blunted by PKA phosphorylation. *Lower panel*. The natural history of heart failure is characterized by the progressive increase in p110γ lipid kinase activity, due to p110γ upregulation, regulatory isoform switch (from p87 to p101) and loss of inhibition by PKA. By interacting with GRK2, in this context p110γ leads to AP2 and β-arrestin mediated downregulation and desensitization of β-adrenergic receptors. In the presence of active myocardial damage, p110α contributes to cardiomyocyte survival and provides protection for adverse remodeling and failure

6.4.2 PI3Kγ Regulates Cardiac Contractility and Remodeling

PI3Kγ is expressed at high levels in hematopoietic cells, especially in leukocytes. However, p110γ is also present in the myocardium, including cardiomyocytes, endothelial cells and possibly myofibroblasts. Although p110γ levels are relatively low in cardiomyocytes, this class IB PI3K isoform has important functions both in heart physiology and disease. In particular, p110γ is a key player in the regulation of cardiac contractility, through combined kinase-dependent and kinase-independent molecular mechanisms interlacing cyclic AMP (cAMP) and PIP_3 signaling.

Hearts derived from p110γ knock-out mice are hyper-contractile compared to wild-type controls both in basal and stimulated conditions, due to functional activation of their contractile machinery (Crackower et al. 2002; Patrucco et al. 2004). This phenotype is caused by increased cellular levels of cAMP, a key second messenger controlling myocardial contractility in response to β-adrenergic stimulation. β-adrenergic receptors (β-ARs, which include $β_1$, $β_2$ and $β_3$ subtypes) are GPCRs physiologically activated by circulating or sympathetic nerve catecholamines (mostly epinephrine). All β-AR isoforms are coupled to G_s subunits, while $β_2$ and $β_3$ receptors also associate to G_i subunits (Brodde et al. 2006; Skeberdis et al. 2008). Catecholamine engagement of β-ARs leads to G_s-triggered stimulation of adenylyl cyclase, which in turn generates the second messenger cAMP. The main signaling effector of cAMP is protein kinase A (PKA), which is constituted by regulatory and catalytic subunits and whose function is compartmentalized and controlled by different proteins known as A-kinase anchor proteins (AKAPs) (Scott and Pawson 2009). In cardiomyocytes, PKA phosphorylates key players orchestrating myocardial contractility, such as phospholamban, the ryanodine receptor and troponin I (Chu et al. 2000; Marx et al. 2000; Stelzer et al. 2007).

Our group and others have shown that p110γ plays a key role in the regulation of cAMP signaling, as p110γ is required for the activation of cellular phosphodiesterases (PDEs) which hydrolyse cAMP to 5'-AMP and terminate signaling, such as PDE3s and PDE4s (Conti and Beavo 2007; Ghigo and Hirsch 2011; Kerfant et al. 2007; Patrucco et al. 2004; Perino et al. 2011). The regulation of PDE3B by p110γ is operated within a macromolecular complex which includes p110γ, the PI3K adaptor subunit p84/87 and PKA (Patrucco et al. 2004; Perino et al. 2011; Voigt et al. 2006). In this complex, p110γ functions as an AKAP, as it directly binds the RIIα subunit of PKA and thus allows the activation of PDE3B by PKA (Fig. 6.3, upper panel). Of note, this functional interaction does not require the kinase activity of p110γ, which instead operates as a scaffold (Hirsch et al. 2009; Perino et al. 2011). In mice lacking p110γ (and not in mice expressing a kinase-inactive p110γ), disassembly of this critical signaling complex leads to a major reduction in the phosphodiesterase activity, thus resulting in abnormal levels of myocardial cAMP. When p110γ-null mice are subjected to cardiac pressure overload, cAMP further raises uncontrolled and causes myocardial damage rapidly progressing towards overt cardiomyopathy and heart failure (Crackower et al. 2002; Patrucco et al. 2004). Similar to PDE3B, also PDE4 activity is controlled by p110γ, thus modulating cAMP signaling within cardiomyocyte sub-cellular compartments containing SR Ca^{2+} ATPase

(Kerfant et al. 2007). Taken together, these findings picture a scenario where PI3Kγ orchestrates a complex modulation of multiple phosphodiesterases, deeply affecting the intracellular compartmentalization of cAMP signaling in cardiomyocytes. Furthermore, we have recently shown that p110γ-associated PKA not only influences phosphodiesterase activity, but also modulates the lipid kinase activity of p110γ itself, as the proximity of PKA and p110γ within the same macromolecular complex allows active PKA to phosphorylate also p110γ on T1024. As a result of PKA phosphorylation, p110γ lipid kinase activity is reduced (Perino et al. 2011) (Fig. 6.3, upper panel).

A strict interplay exists between β-ARs and PI3Ks, as the stimulation of myocardial β-ARs not only mobilizes adenylyl cyclase and cAMP, but also PI3Kβ, PI3Kγ and thus PIP_3 (Rockman et al. 2002). By using mice expressing a kinase inactive p110γ or p110β, our group has recently shown *in vivo* that in physiological conditions, β-adrenergic stimulation of the myocardium essentially activates p110β, which is required to activate Akt. Instead, p110 γ activation only produces a minor fraction of β-adrenergic engaged PIP_3, as in the normal myocardium p110 γ is expressed at low levels and is strictly controlled by PKA-mediated inhibition (Perino et al. 2011). During the natural history of heart failure, the signaling scenario changes deeply, as p110γ levels and activity increase due to (i) up-regulation of p110γ, (ii) regulatory subunit switch (from p84/87 to p101) of p110γ and (iii) reduction of PKA-dependent inhibition of p110γ. As a result, β-adrenergic stimulation of the failing myocardium engages unbalanced p110γ activity, which contributes to adverse remodeling and left ventricular failure (Patrucco et al. 2004; Perino et al. 2011).

A key pathophysiological feature of heart failure is represented by uncontrolled catecholamine release, which produces a situation of chronic and abnormal stimulation of β-adrenergic receptors. Amongst the maladaptive effects of chronic β-adrenergic engagement is the progressive loss of the ability of β-ARs to transduce signals, a process called desensitization. Furthermore, continuous adrenergic stimulation culminates in β-AR endocytosis and cell surface downregulation. Due to desensitization and downregulation, adrenergic signaling is therefore progressively impaired in the failing myocardium, which loses tonic and phasic contractile responses to catecholamine stimulation (Bristow et al. 1982; Rockman et al. 2002). Both β-adrenergic desensitization and internalization processes depend on the phosphorylation of β-ARs, which permits the interaction between β-ARs and β-arrestins, cytoplasmic proteins that block coupling to G-proteins and start the internalization process. Once in early endosomes, phosphorylated β-ARs can be either dephosphorylated and recycled to the cell surface or subjected to degradation in late endosomes (Rockman et al. 2002). Nonetheless, β-ARs can induce maladaptive signaling pathways at the level of the early endosomes, before being degraded (Lefkowitz and Whalen 2004). Several studies have shown that p110γ and PIP_3 are key players in these detrimental processes. In particular, p110γ preferentially cooperates with G protein-coupled receptor kinase-2 (GRK-2, also known as β-adrenergic receptor kinase-1, β-ARK-1) via its PIK domain (Naga Prasad et al. 2001; Nienaber et al. 2003). GRK-2 constitutes a cytoplasmic complex with p110γ and, upon β-AR stimulation, translocates to the activated receptor, where p110γ is activated by G-proteins (Naga Prasad et al. 2000). Herein, PIP_3 produced by p110γ is then required for AP-2 adaptor recruitment at the plasma membrane, which leads to the consequent organization of clathrin-coated pits orchestrating the internalization of the activated receptor (Naga Prasad et al. 2002). *In vivo* studies have confirmed these mechanistic insights. In mice, over-expression of the PIK domain of PI3K reduces GRK-2-associated PI3K activity and consequently β-AR internalization, without affecting Akt or MAPK pathways (Naga Prasad et al. 2002; Nienaber et al. 2003). In pigs subjected to ventricular pacing-induced heart failure, overexpression of the PIK domain derived from failing hearts rapidly reverts contractility to normal levels (Perrino et al. 2005b). Furthermore, in a model of murine cardiomyopathy (calsequestrin over-expression), over-expression of a catalytically inactive p110γ is protective from β-adrenergic perturbation in heart failure, leading to reduced mortality (Perrino et al. 2005a). Our group has further shown that β-AR density remains unchanged after pressure overload in mice expressing a kinase-inactive p110γ and that in wild type mice suffering from pressure overload-induced heart failure, administration of a p110γ-specific pharmacological inhibitor can significantly improve both β-AR density and left ventricular contractility (Perino et al. 2011). Taken together, these findings picture a scenario where p110γ deregulation leads to maladaptive β-adrenergic perturbation during heart failure, by coupling to GRK-2 and AP-2 (Fig. 6.3, lower panel).

Our group has further shown that mice expressing a kinase-inactive p110γ are protected from adverse remodeling following pressure overload by transverse aortic constriction. By using bone marrow chimeras, Damilano et al. have provided evidence that interestingly, the detrimental effects of p110γ on the myocardium are multifactorial. In particular, cardiomyocyte p110γ plays a role in the long term deterioration of left ventricular contractility and diameter, as mice expressing a kinase-inactive p110γ in cardiomyocytes are protected from left ventricular dilation and failure at later time points (Damilano et al. 2011). Nonetheless, leukocyte p110γ is also key to myocardial infiltration by inflammatory cells, orchestrating local cytokine release and ultimately leading to cardiac fibrosis and diastolic dysfunction (Damilano et al. 2011). Indeed, the loss or inhibition of p110γ activity strongly reduces tissue inflammation in the context of cardiovascular diseases such as myocardial infarction and aortic atherosclerosis (Fougerat et al. 2008; Siragusa et al. 2010). These findings are in line with our broader understanding of PI3Kγ as a cornerstone signaling enzyme in the context of inflammatory and immune diseases.

In conclusion, these data indicate that p110γ signaling occupies a central spot in the molecular pathophysiology of heart failure, which is dominated by abnormal β-adrenergic stimulation and p110γ deregulation. In this context, p110γ escapes physiological feedback control mechanisms and orchestrates key aspects of myocardial damage and remodeling, such as β-adrenergic desensitization and downregulation, myocardial inflammation and fibrosis. As several proof-of-concept studies have suggested, p110γ inhibition by pharmacological inhibitors appears therefore as a promising strategy for the treatment of heart disease and calls for further translational and pharmacological studies.

6.5 Concluding Remarks

In the last decade, efforts from basic science have uncovered PI3K signaling as a fundamental biological process of eukaryotics. The high number of different isoforms grouped in the PI3K family might envisage a complex regulation of this signal transduction pathway, based on intricate interplays between distinct isoenzymes. Instead, there is growing evidence that specific cellular functions are peculiar to distinct PI3K isoforms. While the ubiquitous PI3Kα and PI3Kβ are mainly involved in glucose metabolism and cancer, the hematopoietic-restricted PI3Kδ and PI3Kγ play a major role in immunity, with specific functions of the two isoforms in selected subpopulations of immune cells. Furthermore, PI3Kα and PI3Kγ have emerged as master regulators of cardiac function, although further efforts are needed to clarify the role of other isoforms in this context. Overall, selective targeting of specific PI3K isoforms can be predicted to guarantee maximum of efficacy in the treatment of specific diseases, with minimal side effects. In some circumstances, however, different PI3K isoforms function in a synergistic manner in regulating specific biological process. Thus, some pathologic conditions might be more effectively treated with combined therapies targeting more than one selected PI3K. New isoforms-selective compounds are becoming now available, although their potency and isoform selectivity need further improvement before entering clinical trials. Only joint efforts of basic research on genetically engineered animals, pharmaceutical investigation and clinical trial will finally pave our way towards the clinical use of a PI3K inhibitor.

References

Agati JM, Yeagley D, Quinn PG (1998) Assessment of the roles of mitogen-activated protein kinase, phosphatidylinositol 3-kinase, protein kinase B, and protein kinase C in insulin inhibition of cAMP-induced phosphoenolpyruvate carboxykinase gene transcription. J Biol Chem 273:18751–18759

Alcazar I, Marques M, Kumar A, Hirsch E, Wymann M, Carrera AC, Barber DF (2007) Phosphoinositide 3-kinase gamma participates in T cell receptor-induced T cell activation. J Exp Med 204:2977–2987

Ali K, Bilancio A, Thomas M, Pearce W, Gilfillan AM, Tkaczyk C, Kuehn N, Gray A, Giddings J, Peskett E et al (2004) Essential role for the p110delta phosphoinositide 3-kinase in the allergic response. Nature 431:1007–1011

Ali K, Camps M, Pearce WP, Ji H, Ruckle T, Kuehn N, Pasquali C, Chabert C, Rommel C, Vanhaesebroeck B (2008) Isoform-specific functions of phosphoinositide 3-kinases: p110 delta but not p110 gamma promotes optimal allergic responses in vivo. J Immunol 180:2538–2544

Bachman KE, Argani P, Samuels Y, Silliman N, Ptak J, Szabo S, Konishi H, Karakas B, Blair BG, Lin C et al (2004) The PIK3CA gene is mutated with high frequency in human breast cancers. Cancer Biol Ther 3:772–775

Bader AG, Kang S, Vogt PK (2006) Cancer-specific mutations in PIK3CA are oncogenic in vivo. Proc Natl Acad Sci USA 103:1475–1479

Bai L, Wang Y, Fan J, Chen Y, Ji W, Qu A, Xu P, James DE, Xu T (2007) Dissecting multiple steps of GLUT4 trafficking and identifying the sites of insulin action. Cell Metab 5:47–57

Bandyopadhyay GK, Yu JG, Ofrecio J, Olefsky JM (2005) Increased p85/55/50 expression and decreased phosphotidylinositol 3-kinase activity in insulin-resistant human skeletal muscle. Diabetes 54:2351–2359

Baraldo S, Lokar Oliani K, Turato G, Zuin R, Saetta M (2007) The role of lymphocytes in the pathogenesis of asthma and COPD. Curr Med Chem 14:2250–2256

Barber DF, Bartolome A, Hernandez C, Flores JM, Fernandez-Arias C, Rodriguez-Borlado L, Hirsch E, Wymann M, Balomenos D, Carrera AC (2006) Class IB-phosphatidylinositol 3-kinase (PI3K) deficiency ameliorates IA-PI3K-induced systemic lupus but not T cell invasion. J Immunol 176:589–593

Barberis L, Pasquali C, Bertschy-Meier D, Cuccurullo A, Costa C, Ambrogio C, Vilbois F, Chiarle R, Wymann M, Altruda F et al (2009) Leukocyte transmigration is modulated by chemokine-mediated PI3Kgamma-dependent phosphorylation of vimentin. Eur J Immunol 39:1136–1146

Benistant C, Chapuis H, Roche S (2000) A specific function for phosphatidylinositol 3-kinase alpha (p85alpha-p110alpha) in cell survival and for phosphatidylinositol 3-kinase beta (p85alpha-p110beta) in de novo DNA synthesis of human colon carcinoma cells. Oncogene 19:5083–5090

Berenjeno IM, Vanhaesebroeck B (2009) PI3K regulatory subunits lose control in cancer. Cancer Cell 16:449–450

Berwick DC, Dell GC, Welsh GI, Heesom KJ, Hers I, Fletcher LM, Cooke FT, Tavare JM (2004) Protein kinase B phosphorylation of PIKfyve regulates the trafficking of GLUT4 vesicles. J Cell Sci 117:5985–5993

Bi L, Okabe I, Bernard DJ, Wynshaw-Boris A, Nussbaum RL (1999) Proliferative defect and embryonic lethality in mice homozygous for a deletion in the p110alpha subunit of phosphoinositide 3-kinase. J Biol Chem 274:10963–10968

Bilancio A, Okkenhaug K, Camps M, Emery JL, Ruckle T, Rommel C, Vanhaesebroeck B (2006) Key role of the p110delta isoform of PI3K in B-cell antigen and IL-4 receptor signaling: comparative analysis of genetic and pharmacologic interference with p110delta function in B cells. Blood 107:642–650

Bjornholm M, Kawano Y, Lehtihet M, Zierath JR (1997) Insulin receptor substrate-1 phosphorylation and phosphatidylinositol 3-kinase activity in skeletal muscle from NIDDM subjects after in vivo insulin stimulation. Diabetes 46:524–527

Bohnacker T, Marone R, Collmann E, Calvez R, Hirsch E, Wymann MP (2009) PI3Kgamma adaptor subunits define coupling to degranulation and cell motility by distinct PtdIns(3,4,5)P3 pools in mast cells. Sci Signal 2:ra27

Borlado LR, Redondo C, Alvarez B, Jimenez C, Criado LM, Flores J, Marcos MA, Martinez AC, Balomenos D, Carrera AC (2000) Increased phosphoinositide 3-kinase activity induces a lymphoproliferative disorder and contributes to tumor generation in vivo. Faseb J 14:895–903

Brachmann SM, Ueki K, Engelman JA, Kahn RC, Cantley LC (2005) Phosphoinositide 3-kinase catalytic subunit deletion and regulatory subunit deletion have opposite effects on insulin sensitivity in mice. Mol Cell Biol 25:1596–1607

Brennan FM, Feldmann M (1996) Cytokines in autoimmunity. Curr Opin Immunol 8:872–877

Bristow MR, Ginsburg R, Minobe W, Cubicciotti RS, Sageman WS, Lurie K, Billingham ME, Harrison DC, Stinson EB (1982) Decreased catecholamine sensitivity and beta-adrenergic-receptor density in failing human hearts. N Engl J Med 307:205–211

Brodde OE, Bruck H, Leineweber K (2006) Cardiac adrenoceptors: physiological and pathophysiological relevance. J Pharmacol Sci 100:323–337

Camps M, Ruckle T, Ji H, Ardissone V, Rintelen F, Shaw J, Ferrandi C, Chabert C, Gillieron C, Francon B et al (2005) Blockade of PI3Kgamma suppresses joint inflammation and damage in mouse models of rheumatoid arthritis. Nature medicine 11:936–943

Carvalho S, Milanezi F, Costa JL, Amendoeira I, Schmitt F (2010) PIKing the right isoform: the emergent role of the p110beta subunit in breast cancer. Virchows Arch 456:235–243

Chagpar RB, Links PH, Pastor MC, Furber LA, Hawrysh AD, Chamberlain MD, Anderson DH (2010) Direct positive regulation of PTEN by the p85 subunit of phosphatidylinositol 3-kinase. Proc Natl Acad Sci USA 107:5471–5476

Chaussade C, Cho K, Mawson C, Rewcastle GW, Shepherd PR (2009) Functional differences between two classes of oncogenic mutation in the PIK3CA gene. Biochem Biophys Res Commun 381:577–581

Chen D, Mauvais-Jarvis F, Bluher M, Fisher SJ, Jozsi A, Goodyear LJ, Ueki K, Kahn CR (2004) p50alpha/p55alpha phosphoinositide 3-kinase knockout mice exhibit enhanced insulin sensitivity. Mol Cell Biol 24:320–329

Chen S, Wasserman DH, MacKintosh C, Sakamoto K (2011) Mice with AS160/TBC1D4-Thr649Ala knockin mutation are glucose intolerant with reduced insulin sensitivity and altered GLUT4 trafficking. Cell Metab 13:68–79

Cho H, Mu J, Kim JK, Thorvaldsen JL, Chu Q, Crenshaw EB 3rd, Kaestner KH, Bartolomei MS, Shulman GI, Birnbaum MJ (2001) Insulin resistance and a diabetes mellitus-like syndrome in mice lacking the protein kinase Akt2 (PKB beta). Science 292:1728–1731

Chu G, Lester JW, Young KB, Luo W, Zhai J, Kranias EG (2000) A single site (Ser16) phosphorylation in phospholamban is sufficient in mediating its maximal cardiac responses to beta-agonists. J Biol Chem 275:38938–38943

Ciraolo E, Iezzi M, Marone R, Marengo S, Curcio C, Costa C, Azzolino O, Gonella C, Rubinetto C Wu H et al (2008) Phosphoinositide 3-kinase p110beta activity: key role in metabolism and mammary gland cancer but not development. Sci Signal 1:ra3

Clayton E, Bardi G, Bell SE, Chantry D, Downes CP, Gray A, Humphries LA, Rawlings D, Reynolds H, Vigorito E et al (2002) A crucial role for the p110delta subunit of phosphatidylinositol 3-kinase in B cell development and activation. J Exp Med 196:753–763

Condliffe AM, Davidson K, Anderson KE, Ellson CD, Crabbe T, Okkenhaug K, Vanhaesebroeck B, Turner M, Webb L, Wymann MP et al (2005) Sequential activation of class IB and class IA PI3K is important for the primed respiratory burst of human but not murine neutrophils. Blood 106:1432–1440

Condorelli G, Drusco A, Stassi G, Bellacosa A, Roncarati R, Iaccarino G, Russo MA, Gu Y, Dalton N, Chung C et al (2002) Akt induces enhanced myocardial contractility and cell size in vivo in transgenic mice. Proc Natl Acad Sci USA 99:12333–12338

Conti M, Beavo J (2007) Biochemistry and physiology of cyclic nucleotide phosphodiesterases: essential components in cyclic nucleotide signaling. Annu Rev Biochem 76:481–511

Costa C, Barberis L, Ambrogio C, Manazza AD, Patrucco E, Azzolino O, Neilsen PO, Ciraolo E, Altruda F, Prestwich GD et al (2007) Negative feedback regulation of Rac in leukocytes from mice expressing a constitutively active phosphatidylinositol 3-kinase gamma. Proc Natl Acad Sci USA 104:14354–14359

Crackower MA, Oudit GY, Kozieradzki I, Sarao R, Sun H, Sasaki T, Hirsch E, Suzuki A, Shioi T, Irie-Sasaki J et al (2002) Regulation of myocardial contractility and cell size by distinct PI3K-PTEN signaling pathways. Cell 110:737–749

Damilano F, Franco I, Perrino C, Schaefer K, Azzolino O, Carnevale D, Cifelli G, Carullo P, Ragona R, Ghigo A et al (2011) Distinct effects of leukocyte and cardiac phosphoinositide 3-kinase {gamma} activity in pressure overload-induced cardiac failure. Circulation 123:391–399

DeBosch B, Treskov I, Lupu TS, Weinheimer C, Kovacs A, Courtois M, Muslin AJ (2006) Akt1 is required for physiological cardiac growth. Circulation 113:2097–2104

De Gorter DJ, Beuling EA, Kersseboom R, Middendorp S, Van Gils JM, Hendriks RW, Pals ST, Spaargaren M (2007) Bruton's tyrosine kinase and phospholipase Cgamma2 mediate chemokine-controlled B cell migration and homing. Immunity 26:93–104

Di Cristofano A, Kotsi P, Peng YF, Cordon-Cardo C, Elkon KB, Pandolfi PP (1999) Impaired Fas response and autoimmunity in Pten +/- mice. Science 285:2122–2125

Doherty DE (2004) The pathophysiology of airway dysfunction. Am J Med 117(Suppl 12A):11S–23S

Doukas J, Eide L, Stebbins K, Racanelli-Layton A, Dellamary L, Martin M, Dneprovskaia E, Noronha G, Soll R, Wrasidlo W et al (2009) Aerosolized phosphoinositide 3-kinase gamma/

delta inhibitor TG100–115 [3-[2,4-diamino-6-(3-hydroxyphenyl)pteridin-7-yl]phenol] as a therapeutic candidate for asthma and chronic obstructive pulmonary disease. J Pharmacol Exp Ther 328:758–765

Duan W, Aguinaldo Datiles AM, Leung BP, Vlahos CJ, Wong WS (2005) An anti-inflammatory role for a phosphoinositide 3-kinase inhibitor LY294002 in a mouse asthma model. Int Immunopharmacol 5:495–502

Edling CE, Selvaggi F, Buus R, Maffucci T, Di Sebastiano P, Friess H, Innocenti P, Kocher HM, Falasca M (2010) Key role of phosphoinositide 3-kinase class IB in pancreatic cancer. Clin Cancer Res 16:4928–4937

Edwards SW, Hallett MB (1997) Seeing the wood for the trees: the forgotten role of neutrophils in rheumatoid arthritis. Immunology today 18:320–324

Engelman JA, Luo J, Cantley LC (2006) The evolution of phosphatidylinositol 3-kinases as regulators of growth and metabolism. Nat Rev Genet 7:606–619

Ezeamuzie CI, Sukumaran J, Philips E (2001) Effect of wortmannin on human eosinophil responses in vitro and on bronchial inflammation and airway hyperresponsiveness in Guinea pigs in vivo. Am J Respir Crit Care Med 164:1633–1639

Farese RV, Sajan MP, Standaert ML (2005) Atypical protein kinase C in insulin action and insulin resistance. Biochem Soc Trans 33:350–353

Farese RV, Sajan MP, Yang H, Li P, Mastorides S, Gower WR Jr, Nimal S, Choi CS, Kim S, Shulman GI et al (2007) Muscle-specific knockout of PKC-lambda impairs glucose transport and induces metabolic and diabetic syndromes. J Clin Invest 117:2289–2301

Farghaly HS, Blagbrough IS, Medina-Tato DA, Watson ML (2008) Interleukin 13 increases contractility of murine tracheal smooth muscle by a phosphoinositide 3-kinase p110delta-dependent mechanism. Molecular pharmacology 73:1530–1537

Ferguson GJ, Milne L, Kulkarni S, Sasaki T, Walker S, Andrews S, Crabbe T, Finan P, Jones G, Jackson S et al (2007) PI(3)Kgamma has an important context-dependent role in neutrophil chemokinesis. Nature Cell Biol 9:86–91

Folli F, Saad MJ, Backer JM, Kahn CR (1993) Regulation of phosphatidylinositol 3-kinase activity in liver and muscle of animal models of insulin-resistant and insulin-deficient diabetes mellitus. J Clin Invest 92:1787–1794

Fougerat A, Gayral S, Gourdy P, Schambourg A, Ruckle T, Schwarz MK, Rommel C, Hirsch E, Arnal JF, Salles JP et al (2008) Genetic and pharmacological targeting of phosphoinositide 3-kinase-gamma reduces atherosclerosis and favors plaque stability by modulating inflammatory processes. Circulation 117:1310–1317

Foukas LC, Claret M, Pearce W, Okkenhaug K, Meek S, Peskett E, Sancho S, Smith AJ, Withers DJ, Vanhaesebroeck B (2006) Critical role for the p110alpha phosphoinositide-3-OH kinase in growth and metabolic regulation. Nature 441:366–370

Fruman DA, Cantley LC (2002) Phosphoinositide 3-kinase in immunological systems. Sem Immunol 14:7–18

Fruman DA, Mauvais-Jarvis F, Pollard DA, Yballe CM, Brazil D, Bronson RT, Kahn CR, Cantley LC (2000) Hypoglycaemia, liver necrosis and perinatal death in mice lacking all isoforms of phosphoinositide 3-kinase p85 alpha. Nat Genet 26:379–382

George S, Rochford JJ, Wolfrum C, Gray SL, Schinner S, Wilson JC, Soos MA, Murgatroyd PR, Williams RM, Acerini CL et al (2004) A family with severe insulin resistance and diabetes due to a mutation in AKT2. Science 304:1325–1328

Ghigo A, Hirsch E (2011) PI3Kgamma mediates cardiac cAMP compartmentalization through scaffloding of distinct phosphodiesterases. In Heart Failure Winter Meeting (Les Diablerets)

Goodyear LJ, Giorgino F, Sherman LA, Carey J, Smith RJ, Dohm GL (1995) Insulin receptor phosphorylation, insulin receptor substrate-1 phosphorylation, and phosphatidylinositol 3-kinase activity are decreased in intact skeletal muscle strips from obese subjects. J Clin Invest 95:2195–2204

Gymnopoulos M, Elsliger MA, Vogt PK (2007) Rare cancer-specific mutations in PIK3CA show gain of function. Proc Natl Acad Sci USA 104:5569–5574

Harris ED Jr (1990) Rheumatoid arthritis. Pathophysiology and implications for therapy. N Engl J Med 322:1277–1289

Hayer S, Pundt N, Peters MA, Wunrau C, Kuhnel I, Neugebauer K, Strietholt S, Zwerina J, Korb A, Penninger J et al (2009) PI3K{gamma} regulates cartilage damage in chronic inflammatory arthritis. Faseb J 23:4288–4298

Henley T, Kovesdi D, Turner M (2008) B-cell responses to B-cell activation factor of the TNF family (BAFF) are impaired in the absence of PI3K delta. Eur J Immunol 38:3543–3548

Hill KM, Kalifa S, Das JR, Bhatti T, Gay M, Williams D, Taliferro-Smith L, De Marzo AM (2010) The role of PI 3-kinase p110beta in AKT signally, cell survival, and proliferation in human prostate cancer cells. Prostate 70:755–764

Hirsch E, Katanaev VL, Garlanda C, Azzolino O, Pirola L, Silengo L, Sozzani S, Mantovani A, Altruda F, Wymann MP (2000) Central role for G protein-coupled phosphoinositide 3-kinase gamma in inflammation. Science 287:1049–1053

Hirsch E, Lembo G, Montrucchio G, Rommel C, Costa C, Barberis L (2006) Signaling through PI3Kgamma: a common platform for leukocyte, platelet and cardiovascular stress sensing. Thromb Haemost 95:29–35

Hirsch E, Braccini L, Ciraolo E, Morello F, Perino A (2009) Twice upon a time: PI3K's secret double life exposed. Trends Biochem Sci 34:244–248

Imamura T, Huang J, Usui I, Satoh H, Bever J, Olefsky JM (2003) Insulin-induced GLUT4 translocation involves protein kinase C-lambda-mediated functional coupling between Rab4 and the motor protein kinesin. Mol Cell Biol 23:4892–4900

Inoue G, Cheatham B, Emkey R, Kahn CR (1998) Dynamics of insulin signaling in 3T3-L1 adipocytes. Differential compartmentalization and trafficking of insulin receptor substrate (IRS)-1 and IRS-2. J Biol Chem 273:11548–11555

Jaiswal BS, Janakiraman V, Kljavin NM, Chaudhuri S, Stern HM, Wang W, Kan Z, Dbouk HA, Peters BA, Waring P et al (2009) Somatic mutations in p85alpha promote tumorigenesis through class IA PI3K activation. Cancer Cell 16:463–474

Ji H, Rintelen F, Waltzinger C, Bertschy Meier D, Bilancio A, Pearce W, Hirsch E, Wymann MP, Ruckle T, Camps M et al (2007) Inactivation of PI3Kgamma and PI3Kdelta distorts T-cell development and causes multiple organ inflammation. Blood 110:2940–2947

Jia S, Liu Z, Zhang S, Liu P, Zhang L, Lee SH, Zhang J, Signoretti S, Loda M, Roberts TM et al (2008) Essential roles of PI(3)K-p110beta in cell growth, metabolism and tumorigenesis. Nature 454:776–779

Jou ST, Carpino N, Takahashi Y, Piekorz R, Chao JR, Carpino N, Wang D, Ihle JN (2002) Essential, nonredundant role for the phosphoinositide 3-kinase p110delta in signaling by the B-cell receptor complex. Mol Cell Biol 22:8580–8591

Kahn CR (1994) Banting lecture. Insulin action, diabetogenes, and the cause of type II diabetes. Diabetes 43:1066–1084

Kang S, Denley A, Vanhaesebroeck B, Vogt PK (2006) Oncogenic transformation induced by the p110beta, -gamma, and -delta isoforms of class I phosphoinositide 3-kinase. Proc Natl Acad Sci USA 103:1289–1294

Kerfant BG, Zhao D, Lorenzen-Schmidt I, Wilson LS, Cai S, Chen SR, Maurice DH, Backx PH (2007) PI3Kgamma is required for PDE4, not PDE3, activity in subcellular microdomains containing the sarcoplasmic reticular calcium ATPase in cardiomyocytes. Circ Res 101:400–408

Kerouz NJ, Horsch D, Pons S, Kahn CR (1997) Differential regulation of insulin receptor substrates-1 and -2 (IRS-1 and IRS-2) and phosphatidylinositol 3-kinase isoforms in liver and muscle of the obese diabetic (ob/ob) mouse. J Clin Invest 100:3164–3172

Knight ZA, Gonzalez B, Feldman ME, Zunder ER, Goldenberg DD, Williams O, Loewith R, Stokoe D, Balla A, Toth B et al (2006) A pharmacological map of the PI3-K family defines a role for p110alpha in insulin signaling. Cell 125:733–747

Knobbe CB, Reifenberger G (2003) Genetic alterations and aberrant expression of genes related to the phosphatidyl-inositol-3'-kinase/protein kinase B (Akt) signal transduction pathway in glioblastomas. Brain Pathol 13:507–518

Kotani K, Ogawa W, Hino Y, Kitamura T, Ueno H, Sano W, Sutherland C, Granner DK, Kasuga M (1999) Dominant negative forms of Akt (protein kinase B) and atypical protein kinase Clambda do not prevent insulin inhibition of phosphoenolpyruvate carboxykinase gene transcription. J Biol Chem 274:21305–21312

Kurlawalla-Martinez C, Stiles B, Wang Y, Devaskar SU, Kahn BB, Wu H (2005) Insulin hypersensitivity and resistance to streptozotocin-induced diabetes in mice lacking PTEN in adipose tissue. Mol Cell Biol 25:2498–2510

Laffargue M, Calvez R, Finan P, Trifilieff A, Barbier M, Altruda F, Hirsch E, Wymann MP (2002) Phosphoinositide 3-kinase gamma is an essential amplifier of mast cell function. Immunity 16:441–451

Lee KS, Lee HK, Hayflick JS, Lee YC, Puri KD (2006) Inhibition of phosphoinositide 3-kinase delta attenuates allergic airway inflammation and hyperresponsiveness in murine asthma model. Faseb J 20:455–465

Lee SH, Poulogiannis G, Pyne S, Jia S, Zou L, Signoretti S, Loda M, Cantley LC, Roberts TM (2010) A constitutively activated form of the p110beta isoform of PI3-kinase induces prostatic intraepithelial neoplasia in mice. Proc Natl Acad Sci USA 107:11002–11007

Lefkowitz RJ, Whalen EJ (2004) beta-arrestins: traffic cops of cell signaling. Curr Opin Cell Biol 16:162–168

Li Z, Jiang H, Xie W, Zhang Z, Smrcka AV, Wu D (2000) Roles of PLC-beta2 and -beta3 and PI3Kgamma in chemoattractant-mediated signal transduction. Science 287:1046–1049

Lim DH, Cho JY, Song DJ, Lee SY, Miller M, Broide DH (2009) PI3K gamma-deficient mice have reduced levels of allergen-induced eosinophilic inflammation and airway remodeling. Am J Physiol 296:L210–L219

Lin RC, Weeks KL, Gao XM, Williams RB, Bernardo BC, Kiriazis H, Matthews VB, Woodcock EA, Bouwman RD, Mollica JP et al (2010) PI3K(p110 alpha) protects against myocardial infarction-induced heart failure: identification of PI3K-regulated miRNA and mRNA. Arterioscler Thromb Vasc Biol 30:724–732

Link W, Rosado A, Fominaya J, Thomas JE, Carnero A (2005) Membrane localization of all class I PI 3-kinase isoforms suppresses c-Myc-induced apoptosis in Rat1 fibroblasts via Akt. J Cell Biochem 95:979–989

Liu K, Wakeland EK (2001) Delineation of the pathogenesis of systemic lupus erythematosus by using murine models. Adv Exp Med Biol 490:1–6

Liu L, Puri KD, Penninger JM, Kubes P (2007) Leukocyte PI3Kgamma and PI3Kdelta have temporally distinct roles for leukocyte recruitment in vivo. Blood 110:1191–1198

Liu D, Zhang T, Marshall AJ, Okkenhaug K, Vanhaesebroeck B, Uzonna JE (2009) The p110delta isoform of phosphatidylinositol 3-kinase controls susceptibility to Leishmania major by regulating expansion and tissue homing of regulatory T cells. J Immunol 183:1921–1933

Locke NR, Patterson SJ, Hamilton MJ, Sly LM, Krystal G, Levings MK (2009) SHIP regulates the reciprocal development of T regulatory and Th17 cells. J Immunol 183:975–983

Maffucci T, Brancaccio A, Piccolo E, Stein RC, Falasca M (2003) Insulin induces phosphatidylinositol-3-phosphate formation through TC10 activation. EMBO J 22:4178–4189

Marwick JA, Caramori G, Stevenson CS, Casolari P, Jazrawi E, Barnes PJ, Ito K, Adcock IM, Kirkham PA, Papi A (2009) Inhibition of PI3Kdelta restores glucocorticoid function in smoking-induced airway inflammation in mice. Am J Respir Crit Care Med 179:542–548

Marwick JA, Caramori G, Casolari P, Mazzoni F, Kirkham PA, Adcock IM, Chung KF, Papi A (2010) A role for phosphoinositol 3-kinase delta in the impairment of glucocorticoid responsiveness in patients with chronic obstructive pulmonary disease. J Allergy Clin Immunol 125:1146–1153

Marx SO, Reiken S, Hisamatsu Y, Jayaraman T, Burkhoff D, Rosemblit N, Marks AR (2000) PKA phosphorylation dissociates FKBP12.6 from the calcium release channel (ryanodine receptor): defective regulation in failing hearts. Cell 101:365–376

Matsui T, Li L, Wu JC, Cook SA, Nagoshi T, Picard MH, Liao R, Rosenzweig A (2002) Phenotypic spectrum caused by transgenic overexpression of activated Akt in the heart. J Biol Chem 277:22896–22901

McMullen JR, Shioi T, Zhang L, Tarnavski O, Sherwood MC, Kang PM, Izumo S (2003) Phosphoinositide 3-kinase(p110alpha) plays a critical role for the induction of physiological, but not pathological, cardiac hypertrophy. Proc Natl Acad Sci USA 100:12355–12360

McMullen JR, Amirahmadi F, Woodcock EA, Schinke-Braun M, Bouwman RD, Hewitt KA, Mollica JP, Zhang L, Zhang Y, Shioi T et al (2007) Protective effects of exercise and phosphoinositide 3-kinase(p110alpha) signaling in dilated and hypertrophic cardiomyopathy. Proc Natl Acad Sci USA 104:612–617

Mishra RK, Scaife JE, Harb Z, Gray BC, Djukanovic R, Dent G (2005) Differential dependence of eosinophil chemotactic responses on phosphoinositide 3-kinase (PI3K). Allergy 60:1204–1207

Morrow CJ, Gray A, Dive C (2005) Comparison of phosphatidylinositol-3-kinase signalling within a panel of human colorectal cancer cell lines with mutant or wild-type PIK3CA. FEBS Lett 579:5123–5128

Myers MG Jr, White MF (1993) The new elements of insulin signaling. Insulin receptor substrate-1 and proteins with SH2 domains. Diabetes 42:643–650

Myers MG Jr, Backer JM, Sun XJ, Shoelson S, Hu P, Schlessinger J, Yoakim M, Schaffhausen B, White MF (1992) IRS-1 activates phosphatidylinositol 3′-kinase by associating with src homology 2 domains of p85. Proc Natl Acad Sci USA 89:10350–10354

Naga Prasad SV, Esposito G, Mao L, Koch WJ, Rockman HA (2000) Gbetagamma-dependent phosphoinositide 3-kinase activation in hearts with in vivo pressure overload hypertrophy. J Biol Chem 275:4693–4698

Naga Prasad SV, Barak LS, Rapacciuolo A, Caron MG, Rockman HA (2001) Agonist-dependent recruitment of phosphoinositide 3-kinase to the membrane by beta-adrenergic receptor kinase 1. A role in receptor sequestration. J Biol Chem 276:18953–18959

Naga Prasad SV, Laporte SA, Chamberlain D, Caron MG, Barak L, Rockman HA (2002) Phosphoinositide 3-kinase regulates beta2-adrenergic receptor endocytosis by AP-2 recruitment to the receptor/beta-arrestin complex. J Cell Biol 158:563–575

Nakae J, Kitamura T, Silver DL, Accili D (2001) The forkhead transcription factor Foxo1 (Fkhr) confers insulin sensitivity onto glucose-6-phosphatase expression. J Clin Invest 108:1359–1367

Nienaber JJ, Tachibana H, Naga Prasad SV, Esposito G, Wu D, Mao L, Rockman HA (2003) Inhibition of receptor-localized PI3K preserves cardiac beta-adrenergic receptor function and ameliorates pressure overload heart failure. J Clin Invest 112:1067–1079

Nombela-Arrieta C, Lacalle RA, Montoya MC, Kunisaki Y, Megias D, Marques M, Carrera AC, Manes S, Fukui Y, Martinez AC et al (2004) Differential requirements for DOCK2 and phosphoinositide-3-kinase gamma during T and B lymphocyte homing. Immunity 21:429–441

Oak JS, Deane JA, Kharas MG, Luo J, Lane TE, Cantley LC, Fruman DA (2006) Sjogren's syndrome-like disease in mice with T cells lacking class 1A phosphoinositide-3-kinase. Proc Natl Acad Sci USA 103:16882–16887

Okada T, Sakuma L, Fukui Y, Hazeki O, Ui M (1994) Blockage of chemotactic peptide-induced stimulation of neutrophils by wortmannin as a result of selective inhibition as phosphatidylinositol 3-kinase. J Biol Chem 269:3563–3567

Okkenhaug K, Bilancio A, Farjot G, Priddle H, Sancho S, Peskett E, Pearce W, Meek SE, Salpekar A, Waterfield MD et al (2002) Impaired B and T cell antigen receptor signaling in p110delta PI 3-kinase mutant mice. Science 297:1031–1034

Okkenhaug K, Patton DT, Bilancio A, Garcon F, Rowan WC, Vanhaesebroeck B (2006) The p110delta isoform of phosphoinositide 3-kinase controls clonal expansion and differentiation of Th cells. J Immunol 177:5122–5128

Omori SA, Cato MH, Anzelon-Mills A, Puri KD, Shapiro-Shelef M, Calame K, Rickert RC (2006) Regulation of class-switch recombination and plasma cell differentiation by phosphatidylinositol 3-kinase signaling. Immunity 25:545–557

Palframan RT, Collins PD, Severs NJ, Rothery S, Williams TJ, Rankin SM (1998) Mechanisms of acute eosinophil mobilization from the bone marrow stimulated by interleukin 5: the role of specific adhesion molecules and phosphatidylinositol 3-kinase. J Exp Med 188:1621–1632

Patrucco E, Notte A, Barberis L, Selvetella G, Maffei A, Brancaccio M, Marengo S, Russo G, Azzolino O, Rybalkin SD et al (2004) PI3Kgamma modulates the cardiac response to chronic pressure overload by distinct kinase-dependent and -independent effects. Cell 118:375–387

Patton DT, Garden OA, Pearce WP, Clough LE, Monk CR, Leung E, Rowan WC, Sancho S, Walker LS, Vanhaesebroeck B et al (2006) Cutting edge: the phosphoinositide 3-kinase p110 delta is critical for the function of CD4+CD25+Foxp3+regulatory T cells. J Immunol 177:6598–6602

Perino A, Ghigo A, Ferrero E, Morello F, Santulli G, Baillie GS, Damilano F, Dunlop AJ, Pawson C, Walser R et al (2011) Integrating cardiac PIP3 and cAMP signaling through a PKA anchoring function of p110γ. Mol Cell 42(1):84–95

Perrino C, Naga Prasad SV, Patel M, Wolf MJ, Rockman HA (2005a) Targeted inhibition of beta-adrenergic receptor kinase-1-associated phosphoinositide-3 kinase activity preserves beta-adrenergic receptor signaling and prolongs survival in heart failure induced by calsequestrin overexpression. J Am Coll Cardiol 45:1862–1870

Perrino C, Naga Prasad SV, Schroder JN, Hata JA, Milano C, Rockman HA (2005b) Restoration of beta-adrenergic receptor signaling and contractile function in heart failure by disruption of the betaARK1/phosphoinositide 3-kinase complex. Circulation 111:2579–2587

Phu T, Haeryfar SM, Musgrave BL, Hoskin DW (2001) Phosphatidylinositol 3-kinase inhibitors prevent mouse cytotoxic T-cell development in vitro. J Leukoc Biol 69:803–814

Pinho V, Souza DG, Barsante MM, Hamer FP, De Freitas MS, Rossi AG, Teixeira MM (2005) Phosphoinositide-3 kinases critically regulate the recruitment and survival of eosinophils in vivo: importance for the resolution of allergic inflammation. J Leukoc Biol 77:800–810

Puri KD, Doggett TA, Douangpanya J, Hou Y, Tino WT, Wilson T, Graf T, Clayton E, Turner M, Hayflick JS et al (2004) Mechanisms and implications of phosphoinositide 3-kinase delta in promoting neutrophil trafficking into inflamed tissue. Blood 103:3448–3456

Puri KD, Doggett TA, Huang CY, Douangpanya J, Hayflick JS, Turner M, Penninger J, Diacovo TG (2005) The role of endothelial PI3Kgamma activity in neutrophil trafficking. Blood 106:150–157

Randis TM, Puri KD, Zhou H, Diacovo TG (2008) Role of PI3Kdelta and PI3Kgamma in inflammatory arthritis and tissue localization of neutrophils. Eur J Immunol 38:1215–1224

Reif K, Okkenhaug K, Sasaki T, Penninger JM, Vanhaesebroeck B, Cyster JG (2004) Cutting edge: differential roles for phosphoinositide 3-kinases, p110gamma and p110delta, in lymphocyte chemotaxis and homing. J Immunol 173:2236–2240

Rockman HA, Koch WJ, Lefkowitz RJ (2002) Seven-transmembrane-spanning receptors and heart function. Nature 415:206–212

Rommel C, Camps M, Ji H (2007) PI3K delta and PI3K gamma: partners in crime in inflammation in rheumatoid arthritis and beyond? Nat Rev Immunol 7:191–201

Saal LH, Holm K, Maurer M, Memeo L, Su T, Wang X, Yu JS, Malmstrom PO, Mansukhani M, Enoksson J et al (2005) PIK3CA mutations correlate with hormone receptors, node metastasis, and ERBB2, and are mutually exclusive with PTEN loss in human breast carcinoma. Cancer Res 65:2554–2559

Sadhu C, Masinovsky B, Dick K, Sowell CG, Staunton DE (2003) Essential role of phosphoinositide 3-kinase delta in neutrophil directional movement. J Immunol 170:2647–2654

Samuels Y, Wang Z, Bardelli A, Silliman N, Ptak J, Szabo S, Yan H, Gazdar A, Powell SM, Riggins GJ et al (2004) High frequency of mutations of the PIK3CA gene in human cancers. Science 304:554

Sano H, Kane S, Sano E, Miinea CP, Asara JM, Lane WS, Garner CW, Lienhard GE (2003) Insulin-stimulated phosphorylation of a Rab GTPase-activating protein regulates GLUT4 translocation. J Biol Chem 278:14599–14602

Sasaki T, Irie-Sasaki J, Jones RG, Oliveira-dos-Santos AJ, Stanford WL, Bolon B, Wakeham A, Itie A, Bouchard D, Kozieradzki I et al (2000) Function of PI3Kgamma in thymocyte development, T cell activation, and neutrophil migration. Science 287:1040–1046

Scott JD, Pawson T (2009) Cell signaling in space and time: where proteins come together and when they're apart. Science 326:1220–1224

Shepherd PR (2005) Mechanisms regulating phosphoinositide 3-kinase signalling in insulin-sensitive tissues. Acta Physiol Scand 183:3–12

Shi J, Cinek T, Truitt KE, Imboden JB (1997) Wortmannin, a phosphatidylinositol 3-kinase inhibitor, blocks antigen-mediated, but not CD3 monoclonal antibody-induced, activation of murine CD4 + T cells. J Immunol 158:4688–4695

Shioi T, Kang PM, Douglas PS, Hampe J, Yballe CM, Lawitts J, Cantley LC, Izumo S (2000) The conserved phosphoinositide 3-kinase pathway determines heart size in mice. EMBO J 19:2537–2548

Silvestris N, Tommasi S, Petriella D, Santini D, Fistola E, Russo A, Numico G, Tonini G, Maiello E, Colucci G (2009) The dark side of the moon: the PI3K/PTEN/AKT pathway in colorectal carcinoma. Oncology 77(Suppl 1):69–74

Sinclair LV, Finlay D, Feijoo C, Cornish GH, Gray A, Ager A, Okkenhaug K, Hagenbeek TJ, Spits H, Cantrell DA (2008) Phosphatidylinositol-3-OH kinase and nutrient-sensing mTOR pathways control T lymphocyte trafficking. Nat Immunol 9:513–521

Siragusa M, Katare R, Meloni M, Damilano F, Hirsch E, Emanueli C, Madeddu P (2010) Involvement of phosphoinositide 3-kinase gamma in angiogenesis and healing of experimental myocardial infarction in mice. Circ Res 106:757–768

Skeberdis VA, Gendviliene V, Zablockaite D, Treinys R, Macianskiene R, Bogdelis A, Jurevicius J, Fischmeister R (2008) beta3-adrenergic receptor activation increases human atrial tissue contractility and stimulates the L-type Ca2 + current. J Clin Invest 118:3219–3227

Smith LD, Hickman ES, Parry RV, Westwick J, Ward SG (2007) PI3Kgamma is the dominant isoform involved in migratory responses of human T lymphocytes: effects of ex vivo maintenance and limitations of non-viral delivery of siRNA. Cell Signal 19:2528–2539

Sopasakis VR, Liu P, Suzuki R, Kondo T, Winnay J, Tran TT, Asano T, Smyth G, Sajan MP, Farese RV et al (2010) Specific roles of the p110alpha isoform of phosphatidylinsositol 3-kinase in hepatic insulin signaling and metabolic regulation. Cell Metab 11:220–230

Stelzer JE, Patel JR, Walker JW, Moss RL (2007) Differential roles of cardiac myosin-binding protein C and cardiac troponin I in the myofibrillar force responses to protein kinase A phosphorylation. Circ Res 101:503–511

Stenkula KG, Lizunov VA, Cushman SW, Zimmerberg J (2010) Insulin controls the spatial distribution of GLUT4 on the cell surface through regulation of its postfusion dispersal. Cell Metab 12:250–259

Sujobert P, Bardet V, Cornillet-Lefebvre P, Hayflick JS, Prie N, Verdier F, Vanhaesebroeck B, Muller O, Pesce F, Ifrah N et al (2005) Essential role for the p110delta isoform in phosphoinositide 3-kinase activation and cell proliferation in acute myeloid leukemia. Blood 106:1063–1066

Sun XJ, Crimmins DL, Myers MG Jr, Miralpeix M, White MF (1993) Pleiotropic insulin signals are engaged by multisite phosphorylation of IRS-1. Mol Cell Biol 13:7418–7428

Szekanecz Z, Kim J, Koch AE (2003) Chemokines and chemokine receptors in rheumatoid arthritis. Sem Immunol 15:15–21

Takeda M, Ito W, Tanabe M, Ueki S, Kato H, Kihara J, Tanigai T, Chiba T, Yamaguchi K, Kayaba H et al (2009) Allergic airway hyperresponsiveness, inflammation, and remodeling do not develop in phosphoinositide 3-kinase gamma-deficient mice. J Allergy Clin Immunol 123:805–812

Taniguchi CM, Emanuelli B, Kahn CR (2006) Critical nodes in signalling pathways: insights into insulin action. Nat Rev Mol Cell Biol 7:85–96

Terauchi Y, Tsuji Y, Satoh S, Minoura H, Murakami K, Okuno A, Inukai K, Asano T, Kaburagi Y, Ueki K et al (1999) Increased insulin sensitivity and hypoglycaemia in mice lacking the p85 alpha subunit of phosphoinositide 3-kinase. Nat Genet 21:230–235

Tigani B, Hannon JP, Mazzoni L, Fozard JR (2001) Effects of wortmannin on airways inflammation induced by allergen in actively sensitised Brown Norway rats. Eur J Pharmacol 433:217–223

Tkaczyk C, Beaven MA, Brachman SM, Metcalfe DD, Gilfillan AM (2003) The phospholipase C gamma 1-dependent pathway of Fc epsilon RI-mediated mast cell activation is regulated independently of phosphatidylinositol 3-kinase. J Biol Chem 278:48474–48484

Ueki K, Yballe CM, Brachmann SM, Vicent D, Watt JM, Kahn CR, Cantley LC (2002) Increased insulin sensitivity in mice lacking p85beta subunit of phosphoinositide 3-kinase. Proc Natl Acad Sci USA 99:419–424

Vasudevan KM, Barbie DA, Davies MA, Rabinovsky R, McNear CJ, Kim JJ, Hennessy BT, Tseng H, Pochanard P, Kim SY et al (2009) AKT-independent signaling downstream of oncogenic PIK3CA mutations in human cancer. Cancer Cell 16:21–32

Velasco A, Bussaglia E, Pallares J, Dolcet X, Llobet D, Encinas M, Llecha N, Palacios J, Prat J, Matias-Guiu X (2006) PIK3CA gene mutations in endometrial carcinoma: correlation with PTEN and K-RAS alterations. Hum Pathol 37:1465–1472

Voigt P, Dorner MB, Schaefer M (2006) Characterization of p87PIKAP, a novel regulatory subunit of phosphoinositide 3-kinase gamma that is highly expressed in heart and interacts with PDE3B. J Biol Chem 281:9977–9986

Webb LM, Vigorito E, Wymann MP, Hirsch E, Turner M (2005) Cutting edge: T cell development requires the combined activities of the p110gamma and p110delta catalytic isoforms of phosphatidylinositol 3-kinase. J Immunol 175:2783–2787

Wee S, Wiederschain D, Maira SM, Loo A, Miller C, DeBeaumont R, Stegmeier F, Yao YM, Lengauer C (2008) PTEN-deficient cancers depend on PIK3CB. Proc Natl Acad Sci USA 105:13057–13062

Wong KK, Engelman JA, Cantley LC (2010) Targeting the PI3K signaling pathway in cancer. Curr Opin Genet Dev 20:87–90

Zhang TT, Okkenhaug K, Nashed BF, Puri KD, Knight ZA, Shokat KM, Vanhaesebroeck B, Marshall AJ (2008) Genetic or pharmaceutical blockade of p110delta phosphoinositide 3-kinase enhances IgE production. J Allergy Clin Immunol 122:811–819 e812

Zhao L, Vogt PK (2008) Helical domain and kinase domain mutations in p110alpha of phosphatidylinositol 3-kinase induce gain of function by different mechanisms. Proc Natl Acad Sci USA 105:2652–2657

Chapter 7
Phosphoinositide Phosphatases: Just as Important as the Kinases

Jennifer M. Dyson, Clare G. Fedele, Elizabeth M. Davies, Jelena Becanovic and Christina A. Mitchell

Abstract Phosphoinositide phosphatases comprise several large enzyme families with over 35 mammalian enzymes identified to date that degrade many phosphoinositide signals. Growth factor or insulin stimulation activates the phosphoinositide 3-kinase that phosphorylates phosphatidylinositol (4,5)-bisphosphate [PtdIns(4,5)P_2] to form phosphatidylinositol (3,4,5)-trisphosphate [PtdIns(3,4,5)P_3], which is rapidly dephosphorylated either by PTEN (phosphatase and tensin homologue deleted on chromosome 10) to PtdIns(4,5)P_2, or by the 5-phosphatases (inositol polyphosphate 5-phosphatases), generating PtdIns(3,4)P_2. 5-phosphatases also hydrolyze PtdIns(4,5)P_2 forming PtdIns(4)P. Ten mammalian 5-phosphatases have been identified, which regulate hematopoietic cell proliferation, synaptic vesicle recycling, insulin signaling, and embryonic development. Two 5-phosphatase genes, *OCRL* and *INPP5E* are mutated in Lowe and Joubert syndrome respectively. SHIP [SH2 (Src homology 2)-domain inositol phosphatase] 2, and SKIP (skeletal muscle- and kidney-enriched inositol phosphatase) negatively regulate insulin signaling and glucose homeostasis. *SHIP2* polymorphisms are associated with a predisposition to insulin resistance. SHIP1 controls hematopoietic cell proliferation and is mutated in some leukemias. The inositol polyphosphate 4-phosphatases, INPP4A and INPP4B degrade PtdIns(3,4)P_2 to PtdIns(3)P and regulate neuroexcitatory cell death, or act as a tumor suppressor in breast cancer respectively. The Sac phosphatases degrade multiple phosphoinositides, such as PtdIns(3)P, PtdIns(4)P, PtdIns(5)P and PtdIns(3,5)P_2 to form PtdIns. Mutation in the Sac phosphatase gene, *FIG4*, leads to a degenerative neuropathy. Therefore the phosphatases, like the lipid kinases, play major roles in regulating cellular functions and their mutation or altered expression leads to many human diseases.

Keywords Inositol polyphosphate 5-phosphatases · Inositol polyphosphate 4-phosphatases · Sac phosphatases · Trafficking · Hematopoietic system

---

J. M. Dyson and C. G. Fedele made equal contribution as first authors.

C. A. Mitchell (✉) · J. M. Dyson · C. G. Fedele · E. M. Davies · J. Becanovic
Department of Biochemistry and Molecular Biology,
Monash University, Wellington Rd, Clayton, 3800, Australia
e-mail: christina.mitchell@monash.edu

7.1 Introduction

Phosphoinositide phosphatases are a complex series of enzyme families that play critical roles in the regulation of insulin signaling and glucose metabolism, the progression and invasion of cancer, neurodegenerative diseases and myopathies and are implicated in the pathogenesis of many other human diseases. In mammalian cells, over 35 phosphoinositide phosphatases have been identified and some, but not all, have been extensively characterized. The results of recent mouse gene knockout studies and emerging evidence of mutations or epigenetic change in specific phosphoinositide phosphatases have revealed non-redundant roles for these phosphoinositide-metabolizing enzymes *in vivo*.

Many phosphoinositide phosphatases were originally identified and classified following their purification from tissue homogenates using enzyme assays that recognized the ability of the lipid phosphatase to degrade specific phosphoinositides and/or or inositol phosphates *in vitro*. This approach classified these phosphoinositide phosphatases on the basis of the phosphate group that was removed by the phosphatase from the inositol ring of the phosphoinositide or inositol phosphate substrate, hence 3-phosphatases, 4-phosphatases and 5-phosphatases were characterized. More recently many phosphoinositide phosphatases have been identified based on homology within specific catalytic domains and several large families of enzymes including the inositol polyphosphate 5-phosphatases (10 mammalian family members) have been identified. Other smaller families including the SAC phosphatases, and the 4-phosphatases have also been characterized. The substrates of these various lipid phosphatases are shown in Fig. 7.1.

There is some commonality of catalytic mechanism of action amongst the different lipid phosphatases. 4-phosphatases and the SAC phosphatase contain a CX_5R catalytic motif. In contrast the inositol polyphosphate 5-phosphatases are members of the apurinic/apyrimidinic (AP) endonuclease family of enzymes. Below we have described the major characteristics of the 5-, 4-phosphatases and the SAC phosphatases, concentrating on recent studies that delineate the roles these phosphatases play in human diseases. Given the breadth of enzymes and their many functions we refer throughout the text to recent excellent reviews for areas that are not extensively covered.

7.2 Inositol Polyphosphate 5-Phosphatases

The inositol polyphosphate 5-phosphatases are a relatively large family of 10 mammalian and 4 yeast enzymes and all contain a conserved 300 amino acid catalytic domain. The yeast enzymes will not be described here but are reviewed elsewhere (Strahl and Thorner 2007). These enzymes remove the 5-position phosphate from phosphorylated phosphoinositides and inositol phosphates and exhibit overlapping substrate specificities and also many share common protein-protein interaction modules. Many of the 5-phosphatases are widely expressed, in many of the

7 Phosphoinositide Phosphatases: Just as Important as the Kinases

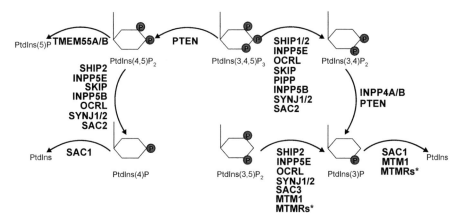

Fig. 7.1 *In vitro* and *in vivo* activities of mammalian phosphoinositide phosphatases. The reported activities of the mammalian phosphatases and their *in vitro* and known *in vivo* phosphoinositide substrates are shown. The repertoire of *bona fide in vivo* substrates of many of these enzymes remains to be delineated. *MTM and MTMR refers to myotubularin and related family members, not discussed here. Abbreviations: *PtdIns* Phosphatidylinositol, *SYNJ* synaptojanin

same cells and tissues and exhibit apparently similar enzymatic properties. Emerging evidence describing the phenotypes of 5-phosphatase knockout mice and the recent demonstration that some 5-phosphatases such as *OCRL* and *INPP5E* are mutated in human diseases indicates that although these enzymes share common features, there is little functional redundancy. It is likely specificity of function is established via distinct subcellular localization, and/or binding partners and association into specific signaling complexes. Here we describe the common features of the 5-phosphatase family and highlight some recent interesting studies that reveal the functional diversity of this complex enzyme family. The 5-phosphatases regulate glucose homeostasis, various aspects of hematopoietic function, embryonic development, neuronal and kidney development, protein trafficking, synaptic vesicle recycling, actin cytoskeleton dynamics and thereby cell migration, and cell viability to name a few.

7.2.1 5-Phosphatase Structure and Enzyme Activity

The inositol polyphosphate 5-phosphatases are Mg^{2+}-dependent phosphoesterases that hydrolyze the 5-position phosphate from the inositol ring of the water-soluble inositol phosphates $Ins(1,4,5)P_3$ and $Ins(1,3,4,5)P_4$ and the lipid-bound PtdIns-derived second messengers $PtdIns(4,5)P_2$, $PtdIns(3,4,5)P_3$, and $PtdIns(3,5)P_2$, reviewed by (Astle et al. 2007; Ooms et al. 2009). Within the 5-phosphatase family the majority of the enzymes share overlapping substrate specificities. A notable exception is the 43 kDa 5-phosphatase (5-phosphatase-1, INPP5A) which only hydrolyzes the soluble inositol phosphates, Ins(1,4,5)P$_3$ and Ins(1,3,4,5)P$_4$ forming Ins(1,4)P$_2$ and Ins(1,3,4)P$_3$ respectively. In contrast the 72 kDa 5-phosphatase (Inpp5e) only metabolizes phosphoinositides and is the most potent PtdIns(3,4,5)P$_3$ 5-phosphatase *in vitro* (Kisseleva et al. 2000). Some of the other nine family members also hydrolyze soluble inositol phosphates, such as Ins(1,4,5)P$_3$ and/or Ins(1,3,4,5)P$_4$ and potentially other inositol phosphates, but this has not been extensively characterized (Ooms et al. 2009). It has been challenging to determine what the *bona fide in vivo* substrates of the 5-phosphatases are that result in specific phenotypes following their loss of function, since these enzymes *in vitro* degrade both PtdIns(3,4,5)P$_3$, PtdIns(4,5)P$_2$ and in some cases also PtdIns(3,5)P$_2$, in addition to inositol phosphates. Few studies have attempted to correlate total cellular phosphoinositide and inositol phosphate levels with functional defects in knockout mice, siRNA-depleted cells and/or human cell lines with 5-phosphatase mutations. This is made even more challenging for the 5-phosphatases, synaptojanin-1 and 2, which contain an additional catalytic Sac domain, which contains a Cx$_5$R motif that facilitates mono- and bis-phosphorylated phosphoinositide metabolism, hydrolyzing PtdIns(3)P, PtdIns(4)P and PtdIns(3,5)P$_2$ to PtdIns (Guo et al. 1999; Nemoto et al. 2001). The Sac domain is also present in two (Inp52/3p) of the four yeast 5-phosphatases (Strahl and Thorner 2007). How the two catalytic domains function together is not clearly defined but it is interesting to speculate that given the 5-phosphatase domain of synaptojanin hydrolyzes PtdIns(4,5)P$_2$ to PtdIns(4)P, this may then allow access of the Sac domain to PtdIns(4)P, which it degrades to PtdIns. To add to this complexity, unlike PTEN which directly opposes the activity of PI3K, 5-phosphatase degradation of PtdIns(3,4,5)P$_3$ generates a new signal, PtdIns(3,4)P$_2$, which functions to activate some of the same effectors as PtdIns(3,4,5)P$_3$, including Akt (Franke et al. 1997; Scheid et al. 2002; Ma et al. 2008). In addition, 5-phosphatase degradation of PtdIns(4,5)P$_2$, generates another signaling molecule, PtdIns(4)P. Both 5-phosphatase substrates and products of their catalysis are also targets for other lipid phosphatases which degrade phosphates from the inositol ring at the 4- and 3-position (to be discussed below) thereby providing a complex metabolic cascade of signaling molecules.

Common to all 5-phosphatases is a 300 amino acid catalytic domain that folds in a manner similar to the apurinic/apyrimidinic repair endonuclease family of DNA-modifying enzymes (Whisstock et al. 2000; Tsujishita et al. 2001). Many 5-phosphatases contain additional domains that facilitate their subcellular localization and/or interaction with other proteins. These include SH2, proline-rich domains, CAAX motifs at the C-terminus, WW, SAM, ASH domains, inactive RhoGAP (RhoGTPase-activating protein) and SKICH (SKIP (skeletal muscle- and kidney-enriched inositol phosphatase)) carboxyl homology domains (see Fig. 7.2). Below we describe the characteristics of each of the 10 mammalian 5-phosphatases. These enzymes have a complex nomenclature and the reader is referred to Table 7.1 which describes their various names and also provides a summary of the reported phenotypes for mouse knockouts of each 5-phosphatase.

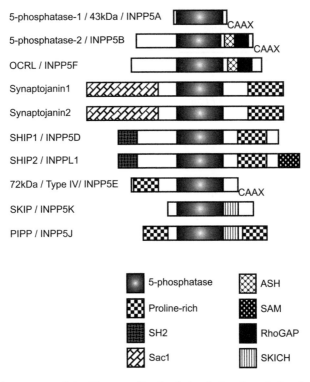

Fig. 7.2 Domain structure of the 10 mammalian inositol polyphosphate 5-phosphatases. Inositol polyphosphate 5-phosphatases contain a central 300 amino acid catalytic domain and distinct structural elements, which facilitate specific sub-cellular localization or define protein-protein or lipid-protein interactions, and may direct substrate specificity. The carboxyl-terminal CAAX motifs of INPP5A, INPP5B and INPP5E mediate plasma membrane targeting. Plasma membrane localization of SKIP and PIPP is regulated by the SKICH domain. The SH2 domains of SHIP1 and SHIP2 mediate protein interactions with tyrosine-phosphorylated receptors. The SAC domains of synaptojanin 1 and synaptojanin 2 mediate hydrolysis of PtdIns(3)P, PtdIns(4)P, or PtdIns(3,5)P_2. Proline-rich regions present in synaptojanin 1, synaptojanin 2, SHIP1, SHIP2, PIPP and INPP5E are proposed to mediate protein-protein interactions. ASH domains within INPP5B and OCRL are typically associated with ciliary proteins and may interact with microtubules

7.2.2 SHIP1

The SHIP (SH2-containing inositol phosphatase) family comprises SHIP1 (also known as INPP5D) and SHIP2 (also called INPP5L1). SHIP1 regulates many aspects of hematopoesis. Full length SHIP1 (SHIP1α) may be spliced to generate three shorter isoforms called SHIP1β, SHIP1δ and s-SHIP1 (Lucas and Rohrschneider 1999; Tu et al. 2001). SHIP1α and SHIP1δ exhibit a restricted distribution to hematopoietic and spermatogenic cells (Liu et al. 1998b). SHIP1, and the most closely related 5-phosphatase SHIP2, exhibit a similar domain structure, with an N-terminal SH2 domain, and central 5-phosphatase domain that hydrolyzes PtdIns(3,4,5)P_3 and Ins(1,3,4,5)P_4, but contain divergent C-terminal proline rich

Table 7.1 Inositol polyphosphate 5-phosphatases and phenotype of knockout mouse models

Protein name	Alias(es)	Gene name (s)	Knockout animal models	References
INPP5A	IP5-P-1 Type I IP5-P 43 kDa IP5-P	*INPP5A*	Not reported	
INPP5B	IP5-P-2 75 kDa IP5-P	*INPP5B*	Males—progressive testicular degeneration leading to sterility Double KO with *Ocrl* results in embryonic lethality	(Hellsten et al. 2001; Janne et al. 1998)
SHIP1	SHIP SHIP-1 IP5-P D	*INPP5D* *SHIP* *SHIP1*	Hematopoietic perturbations, lung pathology and shortened life span (Constitutive KO) Alterations in cytokine-mediated activation that influences Th2 response and cell cytotoxicity (T cell specific KO) Reduced B cells and an increase in basal serum IgGs. Splenic B cells exhibited increased proliferation and enhanced MAPK activation (*Ship1*$^{-/-}$/*Rag*$^{-/-}$ chimera)	(Helgason et al. 1998; Tarasenko et al. 2007; Liu et al. 1998a)
SHIP2	SHIP-2 51C protein	*INPPL1* *SHIP2*	Reduced body weight; normal insulin tolerance on standard chow diet but resistant to weight gain on high fat diet (mouse constitutive KO and rat transient SHIP2 knockdown)	(Sleeman et al. 2005; Buettner et al. 2007)
INPP5E	Pharbin Type IV IP5-P IP5-P E 72 kDa IP5-P	*INPP5E*	Embryonic or early post natal lethality. Features consistent with ciliopathy syndrome including, bilateral anophthalmos, postaxial hexadactyly, kidney cysts, anencephaly and exencephaly and ossification defects	(Jacoby et al. 2009)
OCRL	Lowe oculocerebrorenal syndrome protein IP5-P F	*OCRL* *INPP5F*	Testicular degeneration after sexual maturity	(Janne et al. 1998)

Table 7.1 (continued)

Protein name	Alias(es)	Gene name (s)	Knockout animal models	References
Synaptojanin-1	SJ1, SYNJ1	*SYNJ1* *INPP5G*	Born at mendelian frequencies however 85% of homozygotes died within 24h of birth, whilst remaining 15% failed to thrive and died within 15 days of birth. Homozygotes exhibited an accumulation of clathrin coated vesicles in nerve terminals. Brain cytosol from homozygotes had increased potency to generate clathrin coated liposomes (compared to WT)	(Cremona et al. 1999)
Synaptojanin-2	SYNJ2	*SYNJ2* *INPP5H*	Not reported	
PIPP	Phosphatidylinositol (4,5) bisphosphate 5-phosphatase A	*PIB5PA* *INPP5* *INPP*	Not reported	
SKIP	None	*SKIP* *INPP5K*	*Homozygotes:* Embryonic lethality (E10.5)—cause not reported *Heterozygotes:* Increased glucose tolerance and insulin sensitivity	(Ijuin et al. 2008)

**IP5*-P (*inositol* polyphosphate 5-*p*hosphatase)

domains. The SHIP1 C-terminus contains two NPXY motifs, that following phosphorylation, bind proteins with PTB domains including Shc, Dok 1, Dok 2; and four PxxP motifs that bind SH3-containing proteins such as Grb2, Src, Lyn, Hck, Abl, PLCg1, and PIAS1 (Liu et al. 1994), reviewed in (Hamilton et al. 2011; Sasaki et al. 2009). s-SHIP, a stem-cell specific 104 kDa isoform, lacks the SH2 domain. s-SHIP associates with receptor complexes that are important for embryonic and hematopoietic stem cell growth and survival. Recently using a transgenic mouse model, s-SHIP promoter activity was characterized in actively functioning mammary stem cells, the proposed precursor cells to basal-like human breast cancers (Bai and Rohrschneider 2010).

SHIP1 has been extensively studied within the hematopoietic system and plays a regulatory role in B and T cells, dendritic cells, macrophages, mast cells, osteoclasts, platelets and neutrophils. *Ship1$^{-/-}$* cells in general are more proliferative and exhibit greater survival as a consequence of enhanced activation of both mitogen-activated protein kinase (MAPK) and PI3K/Akt pathways, reviewed in (Hamilton et al. 2011). *Ship1*-null mice develop a chronic myelogenous leukaemia (CML)-like myeloproliferative disease, with splenomegaly, failure to thrive, elevated macrophage and granulocyte counts, associated with massive myeloid infiltration of the lungs, leading to a shortened life span (Brauweiler et al. 2000b; Helgason et al. 1998, 2000; Liu et al. 1999). Stimulation of $Ship1^{-/-}$ hematopoietic progenitors by a number of agonists including monocyte colony-stimulating factor (M-CSF), granulocyte colony-stimulating factor (G-CSF), interleukin-3, or stem cell factor (SCF) results in hyperproliferative responses and increased resistance to apoptosis, associated with amplified Akt signaling (Helgason et al. 1998). SHIP1 is also the most significant phosphoinositide phosphatase that regulates neutrophil migration. $Ship1^{-/-}$ mice exhibit significant granulocytic infiltration of many organs including the lung, in contrast to the phenotype of mice with granulocyte-specific deletion of PTEN, which do not exhibit granulocytic tissue infiltration (Nishio et al. 2007). Genetic inactivation of *Ship1* significantly impairs neutrophil polarization and motility. Studies by Nishio et al revealed SHIP1 governs the formation of the actin-rich leading edge and thereby polarization of neutrophils by regulating the spatial distribution of $PtdIns(3,4,5)P_3$ which is required for chemotaxis (Nishio et al. 2007). The early death of $Ship1^{-/-}$ mice may be in part a consequence of altered neutrophil chemotaxis, which leads to the accumulation of neutrophils in the lungs.

7.2.2.1 B and T Lymphocyte Regulation by SHIP1

SHIP1 associates with ITIM's (immunoreceptor tyrosine-based inhibitory motif) in B cells and ITAM's (immunoreceptor tyrosine-based activation motif), FcγRIIa and 2B4, in macrophages and Natural Killer (NK) cells (Bruhns et al. 2000; Kimura et al. 1997; Osborne et al. 1996). SHIP1 plays a significant role in regulating B cell numbers and function. In B lymphocytes, SHIP1 associates with the Fcγ receptor II-B complex, where the 5-phosphatase inhibits signals stimulated by immune-complexed antigen (Isnardi et al. 2006; Poe et al. 2000; Eissmann et al. 2005; Nakamura et al. 2002). In B cells and mast cells, the non receptor tyrosine kinase Lyn and SHIP1 act together to negatively regulate M-CSF-dependent Akt activation (Baran et al. 2003). SHIP1 is also a target of Lyn-dependent phosphorylation and Lyn and SHIP1 co-operate in regulating FcγRIIb-inhibitory signaling in B cells and mast cells. SHIP1 also regulates autonomous B cell receptor (BCR) signaling (Brauweiler et al. 2000a). SHIP1 is linked to the suppression of B-cell activating factor (BAFF)-induced signaling and also functions independent of the BCR to suppress signaling mediated by chemokine receptors such as CXCR4 (Crowley et al. 2009; Brauweiler et al. 2007). $Ship1^{-/-}$ mice exhibit a reduction in the size of the peripheral B cell compartment and reduced BCR-induced proliferation (Brauweiler et al. 2000a; Helgason et al. 2000; Liu et al. 1998b). In addition as $Ship1^{-/-}$ mice age, the number of B lymphocytes reduces, as a consequence of elevated IL6 secretion by macrophages (Maeda et al. 2010). SHIP1 may also regulate B cell maturation. Irradiated mouse bone marrow reconstituted with $Ship1^{-/-}$ hematopoietic cells shows a reduction in the immature and mature forms of B cells (Helgason et al. 2000; Liu et al. 1998a).

MicroRNA-155 (miR-155) is a critical regulator of immune cell development, function and disease (Baltimore et al. 2008). Recently *SHIP1* has been identified as a miR-155 target, which results in a reduction in SHIP1 protein expression in a group of diffuse large B cell lymphomas (Pedersen et al. 2009) and in a miR-155 transgenic B lymphoma mouse model (Costinean et al. 2009). MiR-155 represses SHIP1 through direct 3'UTR interactions. Retroviral delivery of a miR-155-formatted siRNA against SHIP1 results in a phenotype reminiscent of miR-155 transgenic mice or *Ship1* knockout mice, both result in a myeloproliferative-like syndrome (O'Connell et al. 2009). Moreover, in miR-155 transgenic mice, SHIP1 is gradually down-regulated in preleukemic and leukemic pre-B cells. Down regulation of SHIP1, as well as another IL-6 inhibitor, C/EBPβ (CCAAT enhancer-binding protein β), at the pre B stage (when miR-155 is maximally expressed) may block B-cell differentiation and induce a reactive proliferation of the myeloid cells (Costinean et al. 2009). Down-regulation of SHIP1 expression has also been described in macrophages responding to inflammatory stimuli (O'Connell et al. 2009).

Epigenetic down-regulation of SHIP1 expression in response to inflammation may play a role in promoting the transformation of B cells. Both PTEN and SHIP1 degrade PtdIns(3,4,5)P_3 and regulate Akt-dependent cell proliferation and survival. However, *Ship1* knockout mice do not exhibit a predisposition to lymphoma (Miletic et al. 2010). Interestingly, concomitant deletion of Pten and Ship1 (b$Pten/Ship1^{-/-}$) is associated with the development of spontaneous B cell neoplasms, consistent with marginal zone lymphoma, or with a lower frequency, follicular or centroblastic lymphoma. B cells from b$Pten/Ship1^{-/-}$ mice proliferate in response to BAFF, unlike single *Ship1* knockout cells. Therefore PTEN and SHIP1 may cooperatively suppress B cell lymphoma (Miletic et al. 2010). Ikaros are a family of transcription factors required for lymphoid development, and when functionally abnormal can contribute to lymphoid malignancy. Chip analysis reveals that the Ikaros bind to the promoter region of *INPP5D* (gene for SHIP1) and when Ikaros are not present SHIP1 expression is increased (Alinikula et al. 2010).

SHIP1 is expressed in both CD4 and CD8 lymphocytes. Peripheral T cells are constitutively active in $Ship1^{-/-}$ mice, consistent with previous *in vitro* studies suggesting SHIP1 functions downstream of T cell receptor (TCR) activation, reviewed in (Gloire et al. 2007; Parry et al. 2010). Ligation of CD3 or CD28 on T cells leads to SHIP1 phosphorylation, associated with activation of its 5-phosphatase activity (Edmunds et al. 1999). The expression of SHIP1 is reduced in a human Jurkat T-cell line. Re-expression of SHIP1 in Jurkat T-cells results in decreased levels of PtdIns(3,4,5)P_3, Akt activation and reduced cell proliferation (Horn et al. 2004; Fukuda et al. 2005). Expression of KLF2 (Krüppel-like factor 2), a negative regulator of T-cell proliferation, may be regulated by SHIP1 (Garcia-Palma et al. 2005). SHIP1 is also incorporated in multi-protein complexes which include LAT, Dok-2, Grb-2 and also interacts with and may contribute to the regulation of the Tec-kinase (Tomlinson et al. 2004; Dong et al. 2006). $Ship1^{-/-}$ mice exhibit some reduction in the total peripheral T cell numbers, and T cells show constitutive activation. Regulatory T cells

(Tregs) express CD4 and CD25 markers and limit the risk of autoimmune disease arising from TCR crossreactivity (Kashiwada et al. 2006; Stephens et al. 2005). $Ship1^{-/-}$ mice show increased numbers of Tregs that maintain their immunosuppressive capacity. However, in the constitutive $Ship1^{-/-}$ mice it was difficult to dissect the T cell specific phenotype, due to the complex effects of cytokines. In contrast T cell specific deletion of SHIP1 does not affect T cell or thymic development, T cell receptor signaling, or the number of Tregs. Rather T cell specific deletion of both SHIP1, and the smaller, s-SHIP isoform, reveals a role for SHIP1 in the generation of lymphocyte subsets and in the maintenance of inflammatory versus regulatory cells, Th1 and Th2 respectively (Tarasenko et al. 2007) favouring Th1 lymphocyte differentiation. SHIP1 thereby plays a significant role in controlling the levels of inflammatory T cells. T cell restricted *Ship1* knockout mice exhibit decreased expression of multiple cytokines including IL4, IL5 and IL13 (Th2 cytokines) (Tarasenko et al. 2007) with a diminished capacity to respond to *in vivo* challenge to this pathway.

Dendritic cells (DCs) are critical for the processing, and presentation of antigen to T cells. $Ship1^{-/-}$ mice exhibit an absence of allograft rejection (Ghansah et al. 2004; Wang et al. 2002) and SHIP1 regulates DC maturation and function, reviewed in (Hamilton et al. 2011). $Ship1^{-/-}$ splenic dendritic cells are increased in number and exhibit an altered morphology (Neill et al. 2007). SHIP1 also inhibits the generation of myeloid dendritic cells from bone marrow precursors, but promotes their maturation and function (Antignano et al. 2010).

7.2.2.2 SHIP1 Regulates Macrophage Activity and Function

Classically activated, or M1 macrophages are induced in response to the Type I cytokines, IFN-γ and TNF-α, which stimulate the production of pro-inflammatory cytokines, to phagocytose and kill intracellular microorganisms and tumor cells. Alternatively activated, or M2 macrophages are generated in response to stimulation by type II cytokines, such as IL-4, IL-10, and IL-13, anti-inflammatory cytokines that play important roles in killing extracellular microorganisms and parasites, and in promoting wound healing, reviewed in (Hamilton et al. 2011). In $Ship1^{-/-}$ mice, macrophages are M2 skewed in contrast to wildtype littermates which are M1 skewed (Rauh et al. 2005). Ship1-null peritoneal macrophages exhibit hyperactivation of components downstream of FcγR, resulting in an elevated IL-6 production, a cytokine that can exacerbate myelopoiesis (Maeda et al. 2010). SHIP1 expression can be increased in macrophages via activation of Toll-like receptors 4/9, which leads to the production of autocrine-acting TGFβ, that in turn stimulates a 10-fold increase in SHIP1 protein levels (Sly et al. 2009).

Several of the 5-phosphatases including SHIP1, SHIP2 and INPP5E inhibit macrophage phagocytosis, by degrading PtdIns(4,5)P_2 and PtdIns(3,4,5)P_3, signals that promote phagocytosis. SHIP1 and INPP5E are recruited to the macrophage phagocytic cup, the site of PtdIns(3,4,5)P_3 generation during phagocytosis. SHIP1 preferentially regulates phagocytosis mediated via the CR3 receptor, and to a lesser extent, also FcγR-mediated phagocytosis (Ai et al. 2006; Nakamura et al. 2002;

Cox et al. 2001; Horan et al. 2007). SHIP1 also regulates phagosome maturation and ROS production. The 5-phosphatase, INPP5E, hydrolyzes PtdIns(3,4,5)P_3 during phagocytosis, and like SHIP1, inhibits FcγR-mediated phagocytosis, but shows less activity for complement-mediated phagocytosis (Horan et al. 2007). The related SHIP2 also inhibits phagocytosis. SHIP2 may indirectly regulate PtdIns(4,5)P_2 levels, lipid signals that significantly regulate actin dynamics during phagocytosis via Rac-mediated activation of the PtdIns(4)P-5-kinase (Ai et al. 2006).

Some pathogens including *Francisella tularensis* (*F. tularensis*) escape phagolysosomal fusion allowing for their replication in the cytosol (Checroun et al. 2006; Chong et al. 2008). Upon host cell infection with *F. tularensis*, Fas expression increases, associated with activation of Fas-mediated apoptosis. In *Ship1*$^{-/-}$ macrophages, phagocytosis of *F. tularensis* is not affected, however, fusion events between the pathogen—containing phagocytes and lysosomes are significantly reduced, thereby promoting intra-macrophage growth of the pathogen, leading to activation of SP-1 and -3 transcription factors, induction of Fas and promotion of Fas-induced cell death (Rajaram et al. 2009). As the role of SHIP1 is only evident upon pathogenic bacterial infection, it is likely that SHIP1 may be acting indirectly. Interestingly, infection of macrophages with the less virulent form of *F. tularensis*, *F. novicida*, strongly induces miR-155 and reduces SHIP1 expression, however, infection with a virulent *F. tularensis* isolate, SCHU S4, does not alter miR-155. The differential induction of the miR-155 response in macrophages challenged with pathogenic *versus* non-pathogenic subspecies, and its subsequent effect on SHIP1 expression may contribute to the success of *F. tularensis* as an infectious agent (Cremer et al. 2009).

7.2.2.3 SHIP1 Regulation of Mast cells, Osteoclasts and Platelets

Mast cells regulate allergic reactions via interaction of surface FcεRI molecules with circulating IgE antibodies and when activated contribute to anaphylaxis and allergic asthma. Aggregation of IgE-bearing FcεRI molecules by polyvalent antigen leads to the release of proinflammatory mediators including histamine and TNFα, reviewed in (Galli and Tsai 2010). ITAMs within the β and γ subunits of the FcεRI multimer bind to the SH2 domain of SHIP1 (Kimura et al. 1997; Osborne et al. 1996). SHIP1 functions as a negative regulator of FcεRI-mediated cellular responses in mast cells by controlling the levels of PtdIns(3,4,5)P_3, signals which promote mast cell degranulation (Liu et al. 1999; Huber et al. 1998). SHIP1 is also recruited to the recently identified novel ITIM-like domain-containing receptor, Allergin 1, and facilitates Allergin-1's inhibition of IgE-stimulated mast cell degranulation and anaphylaxis in mice (Hitomi et al. 2010). Thus under normal conditions SHIP1 prevents degranulation signaling unless there is overwhelming challenge by antigens. However, SHIP1 is not required for mast cell differentiation (Rauh et al. 2003). *Ship1*$^{-/-}$ mast cells exhibit increased degranulation compared with *Ship1*$^{+/+}$ mast cells in response to IgE cross-linking (Huber et al. 2002). Naive *Ship1*$^{-/-}$ mice, under steady state conditions, show signs of allergic asthma such as airway inflammation and remodeling, with mucus hyperproduction (Roongapinun et al. 2010). *Ship1*$^{-/-}$ mice also show systemic mast cell hyperplasia, increased levels of the cytokines IL-6, TNF, and IL-5, and heightened susceptibility to anaphylaxis (Haddon et al. 2009).

Osteoclasts are bone resorbing cells of monocyte-macrophage origin. M-CSF and RANKL (receptor activator of nuclear factor-κB ligand) stimulate the differentiation of osteoclast precursors, and this is enhanced with loss of SHIP1 (Takeshita et al. 2002). In osteoclasts SHIP1 localizes to podosomes, sites of adhesion and metallomatrix proteinase secretion. *Ship1*$^{-/-}$ osteoclasts exhibit an increased ability to resorb mineralized matrix, and Ship1-null mice exhibit signs of severe osteoporosis, due to the presence of increased numbers of enlarged and hypernucleated osteoclasts (Takeshita et al. 2002). Interestingly, SHIP1, via its SH2 domain, complexes with the inhibitory adaptor and plasma membrane protein, DAP12, thereby restricting the recruitment of PI3K to DAP12 and reduces macrophage and osteoclast activation (Peng et al. 2010).

Both SHIP1 and the related SHIP2 are expressed in human platelets, essential components of the initial events in blood clot formation. SHIP1 is implicated in controlling irreversible platelet aggregation (Severin et al. 2007; Trumel et al. 1999). SHIP2 forms a complex with the adhesion receptor, glycoprotein Ib/IX, in human platelets, however the function of this complex remains to be determined (Dyson et al. 2003). In addition, studies by Giuriato et al suggest that SHIP1, rather than SHIP2, controls PtdIns(3,4,5)P$_3$ levels in response to a number of platelet agonists (Giuriato et al. 1997). Following thrombin stimulation SHIP1 is tyrosine phosphorylated, correlating with its relocation to the actin cytoskeleton and PtdIns(3,4,5)P$_3$ production (Giuriato et al. 1997). Collagen-related peptide stimulation of SHIP1-deficient platelets results in increased PtdIns(3,4,5)P$_3$ signals and activation of Bruton's tyrosine kinase (Btk), which promotes calcium entry. *Ship1*$^{-/-}$ platelets exhibit defects in platelet aggregation and abnormal platelet contractility. Ship1-null mice show prolonged bleeding and abnormal thrombus organization, suggesting SHIP1 is required for normal hemostasis (Severin et al. 2007).

7.2.2.4 SHIP1 and its Association with Human Leukemia

Emerging evidence has linked changes in SHIP1 expression and/or SHIP1 mutations in some leukemias. Recent analysis of 81 primary T-cell acute lymphatic leukemias (ALLs) has revealed inactivation of SHIP1 and PTEN, with sparing of *PIK3CA*, suggesting a role for loss of SHIP1 in disease pathogenesis (Lo et al. 2009). Mutations in the SHIP1 gene (*INPP5D*) have also been identified within the region encoding the 5-phosphatase domain that result in loss of 5-phosphatase activity in acute myeloid leukemia (Luo et al. 2003, 2004). Although there is little evidence for mutations in *SHIP1* in solid tumors, recent reports have revealed tumors grow more rapidly in *Ship1*$^{-/-}$ mice potentially resulting from SHIP1's role in regulating immune responses to tumor cells (Rauh et al. 2005) reviewed in (Hamilton et al. 2011).

In Friend murine leukemia virus (F-MuLV)-induced erythroleukemia, a proto-oncogene transcription factor, Fli-1, is activated. Fli-1 activity is downstream of PI3K in a negative feedback loop, in which SHIP1 and Fli-1 regulate each other to direct erythropoietin (Epo) signaling leading to either erythrocyte proliferation or differentiation (Lakhanpal et al. 2010). Ship1-null mice exhibit accelerated progression of F-MuLV-induced erthroleukemia and Fli-1 represses expression of Ship1 during this process.

The chronic myeloid leukemia (CML) causative fusion protein, BCR/ABL, inhibits SHIP1 expression, with elevated SHIP1 expression noted in primitive CML cells, but in terminally differentiating CML cells, SHIP1 levels are reduced at the post-transcriptional level (Jiang et al. 2003; Sattler et al. 1999). Recently it was proposed that BCR/ABL-mediated SHIP1 expression is regulated by BCR/ABL-modulated phosphorylation of SHIP1, which acts as a trigger for its ubiquitination, possibly by c-Cbl leading to its proteosomal degradation (Ruschmann et al. 2010). These findings suggest that SHIP1 may act as a tumor suppressor in CML.

7.2.3 SHIP2

The 142 kDa 5-phosphatase, SHIP2 (gene name *INPPL1*) regulates insulin signaling and metabolism. This enzyme is composed of an N-terminal SH2 domain, a central catalytic domain and a C-terminal proline-rich domain (PRD) which contains additional motifs including WW, NPXY and a sterile alpha (SAM) motif. This 5-phosphatase shares a high degree of homology and similar domain structure to the hematopoietic specific 5-phosphatase SHIP1, but differs in its PRD. SHIP2 expression overlaps in part with that of SHIP1 in hematopoietic cells, and plays a non-redundant role in macrophages (Giuriato et al. 2003; Bruyns et al. 1999; Wisniewski et al. 1999). However, SHIP2 is also widely expressed in many other cells and tissues (Muraille et al. 1999; Sleeman et al. 2005). *In vitro* studies have revealed SHIP2 utilizes a broad number of substrates including several inositol phosphates, and phosphoinositides including $PtdIns(3,4,5)P_3$, $PtdIns(3,5)P_2$ and $PtdIns(4,5)P_2$ (Taylor et al. 2000; Pesesse et al. 1997; Schmid et al. 2004). SHIP2 also hydrolyzes other inositol phosphates and phosphoinositides *in vitro* (Chi et al. 2004). However the majority of reports have concentrated on SHIP2's role in regulating $PtdIns(3,4,5)P_3$ signaling, rather than its regulation of other phosphoinositides or inositol phosphates and what role this activity plays *in vivo* is unclear.

As assessed by traditional microscopy techniques, SHIP2 is predominantly cytosolic but upon growth factor/insulin stimulation or cell-matrix contact, the 5-phosphatase translocates to the plasma membrane where it regulates the actin cytoskeleton via association with a number of actin-regulatory proteins including filamin, p130Cas, Shc, vinexin and LL5β (Takabayashi et al. 2010; Gagnon et al. 2003; Paternotte et al. 2005; Prasad et al. 2001; Dyson et al. 2001). More recently, using TIRF (total internal reflection fluorescence) microscopy, recombinant SHIP2 has been detected, like synaptojanin and OCRL, at clathrin-coated vesicles, mediated by an interaction between the SHIP2 PRD with intersectin. At this site SHIP2 acts as a negative regulator of clathrin-coated pit growth (Nakatsu et al. 2010).

In a number of insulin-sensitive cells including podocytes, SHIP2 overexpression regulates insulin-stimulated PI3K-generated PtdIns$(3,4,5)$P$_3$ degradation and thereby Akt activation resulting in reduced plasma membrane GLUT4 association and as a consequence reduced glucose uptake and glycogen synthesis (Hyvonen et al. 2010; Sasaoka et al. 2001; Wada et al. 2001). Surprisingly, however, SHIP2-depleted 3T3 L1 adipocytes do not exhibit altered insulin-mediated PI3K/Akt signaling (Tang et al. 2005). Linking the actin cytoskeletal and insulin signaling properties of SHIP2 is its interaction in the glomeruli with CD2-associated protein, which also binds with the key podocyte adhesion protein, nephrin and actin (Hyvonen et al. 2010). Overexpression of SHIP2 induces a 2-fold increase in podocyte apoptosis. Endogenous Ship2 protein levels are significantly elevated in the glomeruli of diabetic mice and rats (Hyvonen et al. 2010).

Clinical studies have identified a correlation between insulin resistance and the development of diabetic nephropathy (Parvanova et al. 2006; Orchard et al. 2002). Polymorphisms in the SHIP2 gene, *INPPL1*, have been identified in diabetic and also control subjects and are implicated in the pathogenesis of Type 2 diabetes, hypertension and the metabolic syndrome. Additionally, some reports have demonstrated altered expression of polymorphic SHIP2 *in vitro*. For example, a 16bp deletion in the 3′-untranslated region (UTR) of *INPPL1* has been detected in Type 2 diabetic patients, compared to healthy subjects, which when expressed in HEK cells, increased expression of the reporter protein. Although these studies are *in vitro*, it does suggest the 16bp deletion in the 3′UTR may enhance SHIP2 expression *in vivo* resulting in altered phosphoinositide metabolism and insulin sensitivity (Marion et al. 2002). Additionally, single nucleotide polymorphisms (SNPs) have been identified in *INPPL1*. In a British cohort study, *INPPL1* SNPs are significantly associated with diabetes and hypertension (Kaisaki et al. 2004). Surprisingly, analysis of the same SNPs in a French and British cohort did not identify association of *INPPL1* SNPs with diabetic patients *per se*, rather, an association was observed with hypertensive metabolic syndrome patients (Kaisaki et al. 2004; Marcano et al. 2007). In contrast in a Japanese cohort, *INPPL1* SNPs have been identified in control, rather than diabetic patients. Analysis of one of the several *INPPL1* SNPs identified in control subjects, that is located in the 5-phosphatase domain, results in less efficient inhibition of insulin-stimulated PtdIns$(3,4,5)$P$_3$ levels and Akt phosphorylation, consistent with reduced 5-phosphatase activity. It has been proposed that *INPPL1* SNPs in control subjects may protect subjects from Type 2 diabetes (Kagawa et al. 2005). *INPPL1* SNPs have also been identified in non-coding sequences. SNPs in the promoter and 5′UTR of the *INPPL1* gene are associated with impaired fasting glycaemia. In cell culture systems, one *INPPL1* haplotype was examined, which like the 16bp deletion in the 3′UTR, resulted in increased promoter activity, suggesting increased SHIP2 expression (Ishida et al. 2006). Collectively these studies indicate that polymorphisms in *INPPL1* are associated with hypertensive metabolic subjects, whilst at the molecular level, these polymorphisms, or at least those examined, may act to increase SHIP2 expression.

SHIP2 is also highly expressed in the brain and in NGF-stimulated PC12 cells. It localizes to lamellipodia and neurite buds, suggesting that SHIP2 may play a role in the early events of neurite budding (Aoki et al. 2007). Supporting this, siRNA-mediated knockdown of Ship2 in Pten-deficient PC12 cells, results in an elevation in the number of neurites per cell but also hyper-elongation (Aoki et al. 2007). *Ship2* knockout mice do not exhibit an overt neurological phenotype, however, detailed analysis has not been reported.

7.2.3.1 *Ship2*$^{-/-}$ Mice: Regulation of Glucose Homeostasis and Weight Gain

Two studies have reported the phenotype of *Ship2*$^{-/-}$ mice with conflicting results. In the first report Ship2-null mice were generated by targeted deletion of exons 19–29 of the *Inpp1l* gene (Clement et al. 2001). These mice die within 3 days of birth, associated with low blood glucose and insulin concentrations. However in this study the *Phox2a* gene was also inadvertently deleted, and this may contribute to the mouse phenotype. In a more recent study, *Ship2*$^{-/-}$ mice were generated by deletion of the first 18 exons of the *Inpp1l* gene, which encodes the SH2 domain and 5-phosphatase catalytic domain. *Ship2*$^{-/-}$ mice are viable but showed reduced body weight, associated with reduced serum lipids and 6-fold lower serum leptin levels on a standard chow diet (Sleeman et al. 2005). No changes in fasting serum glucose, insulin levels, or glucose/insulin tolerance were detected. On a standard chow diet, *Ship2*$^{-/-}$ mice exhibit enhanced insulin-mediated Akt and p70S6K activation in the liver and skeletal muscle. On a high fat diet, Ship2-null mice are resistant to weight gain, hyper-glycemia or insulinemia and show decreased serum lipids. This latter study has been validated by more recent studies using a variety of approaches described below.

Acute knockdown of Ship2 using antisense oligonucleotides in rats essentially recapitulates the phenotype observed in *Ship2*$^{-/-}$ mice (Buettner et al. 2007). On a standard diet, Ship2 knockdown rats show normal insulin tolerance, whilst in contrast, on a high fat diet the rats exhibit increased glucose disposal. Enhanced insulin-stimulated Akt activation is detected in the muscle of Ship2 knockdown rats, correlating with reduced Ship2 expression in this tissue. Data from two rodent models with either chronic ablation of Ship2, or acute reduction of Ship2, has revealed a role for this enzyme in regulating diet-induced insulin resistance (Buettner et al. 2007; Sleeman et al. 2005). Therapeutically, whether long-term application of SHIP2 anti-sense therapy is a viable option for the treatment of insulin resistance is unclear, however, anti-sense therapies have been successful in treating some human diseases (Jason et al. 2004).

Transgenic mice overexpressing Ship2 (Ship2-Tg) exhibit significant weight gain (5%) and elevated fasting insulin, but not glucose, leptin, adiponectin or serum lipids levels compared to wildtype controls (Kagawa et al. 2008). Additionally Ship2-Tg mice display impaired glucose tolerance and insulin sensitivity, correlating with reduced insulin-stimulated Akt activation in the liver, skeletal muscle and fat. Surprisingly, decreased phosphorylation of insulin receptor substrate 1 (IRS1) is observed in the liver of Ship2-Tg mice as well as impaired glucose homeostasis. Moreover, increased liver G6P (glucose-6-phosphatase) and PEPCK (phosphoenolpyruvate carboxykinase) mRNA coupled with reduced glycogen content and GK (glucokinase) mRNA in Ship2-Tg indicates hepatic insulin resistance. Liver-specific overexpression of Ship2 in a diabetic rodent model (db/db) has also been reported (Fukui et al. 2005). Leptin receptor-deficient db/db mice are leptin resistant and diabetic, and exhibit insulin resistance associated with decreased PI3K/Akt signaling, hyperglycemia, hyperinsulinemia, increase gluconeogenesis and other metabolic related symptoms (Kobayashi et al. 2000). Overexpression of wildtype Ship2 in db/+ m (heterozygote) mice decreases liver Akt activation, whilst overexpression of catalytically inactive Ship2 (ΔIP-Ship2) which acts as a dominant-negative, enhances liver Akt activation, but does not affect insulin signaling in peripheral insulin-sensitive tissues (Fukui et al. 2005). In contrast, overexpression of wildtype Ship2 in the liver results in reduced glucose tolerance, whilst mice expressing the catalytically inactive Ship2 display improved tolerance, suggesting Ship2 modulates liver-restricted insulin sensitivity to influence peripheral glucose and insulin tolerance. Insulin signaling is also critical for brain function (Muntzel et al. 1995). Endogenous Ship2 protein levels in db/+ m mice are elevated relative to controls and are further increased in homozygous mice (db/db). Ship2-Tg mice show impairment of brain insulin/insulin-like growth factor I (IGF-I) signaling, attenuation of the neuroprotective effects of insulin/IGF-I, and a decline in learning and memory. Inhibition of Ship2, using the specific inhibitor, AS1949490, in the db/db mice, improves synaptic plasticity and memory formation (Suwa et al. 2009). Liver-specific inhibition of Ship2 in a second diabetic mouse model, KKA^y, is associated with reduced liver G6P and PEPCK mRNA in response to pyruvate, probably as a consequence of increased gluconeogensis due to insulin resistance (Grempler et al. 2007). Collectively these studies reveal SHIP2 plays complex roles in regulating insulin signaling and metabolism in a number of insulin-sensitive tissues.

7.2.3.2 SHIP2 and Regulation of Cancer

SHIP2 has been implicated in regulating the cell cycle, although, the mechanisms are yet to be fully delineated. SHIP2 overexpression induces glioblastoma cell cycle arrest, and also suppresses platelet derived growth factor (PDGF)- and IGF-stimulated cell cycle progression in vascular smooth muscle cells (Taylor et al. 2000; Sasaoka et al. 2003). SHIP2 is predicted to be either a suppressor or enhancer of cell proliferation, depending on the cell type. In HeLa cells, SHIP2 expression regulates receptor endocytosis and ligand-induced epidermal growth factor receptor (EGFR) degradation. In breast cancer cells, SHIP2 controls the levels of this receptor, promoting cell proliferation and tumor growth and in this context SHIP2 does not act as a tumor suppressor (Prasad 2009). Under conditions of SHIP2 siRNA, EGF-stimulated Akt activation is suppressed, in contrast to findings in insulin-stimulated cells where Ship2 loss of expression enhances PI3K/Akt signaling (Prasad 2009). SHIP2 therefore plays a role in promoting EGF-stimulated Akt activation, possibly by regulating

EGFR internalization. This is consistent with recent findings that SHIP2 may regulate clathrin-coated pit formation (Nakatsu et al. 2010).

Recently SHIP2 has been implicated in the pathogenesis of squamous cell carcinoma (SSC) (Sekulic et al. 2010; Yu et al. 2008). The SHIP2 gene, *INPPL1*, is a target of small non coding microRNA (miRNA)-205, which suppresses its expression in stratified squamous epithelia. Aggressive SCCs show reduced expression of SHIP2 with increased miRNA-205 expression. Interfering with miRNA-205 by distinct mechanisms, increases SHIP2 expression and suppresses Akt activation, and increases keratinocyte cell death (Yu et al. 2008). miRNA-205 is upregulated in several carcinomas, so it will be of interest in future studies to examine its effects on SHIP2 expression and Akt regulation in these cancers (Iorio et al. 2005, 2009; Majid et al. 2010).

7.2.4 Synaptojanin 1 and 2

Synaptojanin 1 and 2 (gene names *SYNJ1* and *SYNJ2*, respectively) are two related 5-phosphatases that share amino acid sequence homology and a similar domain structure and both contain two distinct catalytic domains. Both enzymes contain an N-terminal Sac1 domain, a central 5-phosphatase domain and divergent C-terminal proline-rich domains. The 5-phosphatase domain hydrolyzes the 5-position phosphate from PtdIns(3,4,5)P_3, PtdIns(4,5)P_2, Ins(1,4,5)P_3 and Ins(1,3,4,5)P_4, whilst the Sac1 domain contains a CX_5R motif that mediates the dephosphorylation of phosphates from the inositol ring of PtdIns(3)P, PtdIns(4)P and PtdIns(3,5)P_2 to form PtdIns (Guo et al. 1999; Nemoto et al. 2001; McPherson and Marshall 1996). Synaptojanin 1 and 2 can be alternatively spliced in the C-terminal proline rich domain giving rise to 145 and 170 kDa isoforms of synaptojanin 1 and up to six isoforms of synaptojanin 2 (Nemoto et al. 1997, 2001; Ramjaun and McPherson 1996; Seet et al. 1998). The 145 kDa isoform of synaptojanin 1 is highly expressed in presynaptic nerve terminals, however, the 170 kDa synaptojanin 1 isoform is not expressed in neuronal cells (McPherson et al. 1996; Ramjaun and McPherson 1996).

Synaptojanin 1 is highly expressed in the brain and localizes to presynaptic nerve terminals where it complexes with multiple interacting proteins including Grb2 (growth-factor-receptor-bound protein 2), dynamin, syndapin, endophilin, amphiphysin I and II, Eps15, clathrin, AP-2, DAP160 (dynamin-associated protein 160)/intersectin, myosin 1E, Snx9 (sortin nexin 9) and phospholipase C (Krendel et al. 2007; Yeow-Fong et al. 2005; Haffner et al. 2000; Ahn et al. 1998; McPherson et al. 1996) reviewed by (Astle et al. 2007). In general these interactions promote synaptojanin 1 subcellular localization and/or enhance its catalytic activity (Ringstad et al. 1997; Schuske et al. 2003; Lee et al. 2004). Many of these interacting partners function to regulate endocytosis. Intersectin 1 is an endocytic scaffolding protein that may coordinate clathrin coat assembly with synaptojanin 1-mediated PtdIns(4,5)P_2 hydrolysis (Adayev et al. 2006). AP2, the adaptor protein that binds clathrin, also binds to intersectin 1, and it has been proposed this interaction may occlude the recruitment of PtdIns(4,5)P_2-bound synaptojanin 1 (Pechstein et al. 2010).

Synaptojanin 1 interacts with endophilin, which targets the 5-phosphatase to sites of endocytosis and promotes its 5-phosphatase catalytic activity in degrading PtdIns(4,5)P_2 (Ringstad et al. 1997; Schuske et al. 2003; Lee et al. 2004). The enzymatic activity of synaptojanin 1 is also regulated by its phosphorylation, including its constitutive phosphorylation in unstimulated nerve terminals by EphB2 and Cdk5 (Lee et al. 2004; Irie et al. 2005). Cdk5 (cyclin-dependent kinase 5) phosphorylation of synaptojanin 1 impairs its 5-phosphatase activity and its interaction with endophilin 1 and amphiphysin 1 (Lee et al. 2004). Following nerve depolarization, synaptojanin 1 is dephosphorylated by the serine threonine phosphatase calcineurin (McPherson et al. 1996; Bauerfeind et al. 1997). EphB2 tyrosine kinase phosphorylation of synaptojanin 1 also impairs its association with endophilin (Irie et al. 2005). In contrast, phosphorylation of synaptojanin 1 by the dual-specificity tyrosine-phosphorylated and regulated kinase 1A (MNB/DYRK1A) increases synaptojanin 1 PtdIns(4,5)P_2 5-phosphatase activity and regulates its interaction with amphiphysin 1 and intersectin 1 (Adayev et al. 2006). The subcellular targeting of synaptojanin 1 is also influenced by the fatty acid groups on PtdIns(4,5)P_2. Synaptojanin 1 exhibits a preference for PtdIns(4,5)P_2-containing long chain polyunsaturated fatty acids over synthetic PtdIns(4,5)P_2 with two saturated fatty acids (Schmid et al. 2004). In *Caenorhabditis elegans* (*C.elegans*) mutants which lack the preferred fatty acids groups on PtdIns(4,5)P_2, synaptojanin 1 is mislocalized resulting in defective synaptic vesicle recycling (Marza et al. 2008).

Deletion of synaptojanin 1 homologs in mice, *D. melanogaster* and *C.elegans* has revealed synaptojanin 1 regulates clathrin-mediated endocytosis and neuronal function. Most $Synj1^{-/-}$ mice die shortly after birth, and the 15% of knockout mice that survive up to 15 days exhibit progressive weakness, ataxia and convulsions. Neurons from mice with loss of synaptojanin show increased PtdIns(4,5)P_2, correlating with an accumulation of clathrin-coated vesicles, associated with decreased synaptic vesicles (Cremona et al. 1999). A similar phenotype is observed in synaptojanin (unc-26) deficient *C.elegans* and *D. melanogaster* (Harris et al. 2000). PtdIns(4,5)P_2 regulates the releasable pool of synaptic vesicles and secretory granules at nerve synapses (Gong et al. 2005; Milosevic et al. 2005; Di Paolo et al. 2004), and the formation of new synaptic vesicles at the plasma membrane (Cremona et al. 1999; Verstreken et al. 2003; Mani et al. 2007; Van Epps et al. 2004; Harris et al. 2000). PtdIns(4,5)P_2 may affect several steps in the synaptic vesicle cycle, a process that relies on clathrin-dependent endocytosis, and functions to supply new synaptic vesicles during nerve stimulation (Kasprowicz et al. 2008; Haucke 2003). Synaptic vesicles are released following stimulation and fuse with the presynaptic plasma membrane, and are then in turn recycled via endocytosis via clathrin-mediated reinternalization. Clathrin-coated pit formation requires the assembly of endocytic proteins at PtdIns(4,5)P_2-enriched membrane sites. By degrading PtdIns(4,5)P_2, synaptojanin 1 directs the uncoating of clathrin, and thereby the recycling of the vesicles. Hence in the absence of synaptojanin 1, there is an accumulation of clathrin-coated vesicles. Interestingly, in *D. melanogaster*, disruption of the NCA (Na^+/Ca^{2+} antiporter) ion channel, or *unc-80*, which encodes a novel protein required for ion channel subunit localization, partially suppresses the synaptojanin 1 knockout phenotype, whilst overexpression of endophilin can also partially rescue the synaptojanin-null phenotype, suggesting that aberrant NCA ion channel and endophilin function, may be required for proper synaptic vesicle recycling (Jospin et al. 2007).

Finally another function for synaptojanin 1 has been identified in *Danio rerio* (*D. rerio*). An ENU mutagenesis screen for *D. rerio* larvae with vestibular defects identified mutant *Synj1* induces abnormalities in hair cell basal blebbing and may function to regulate the number and release of synaptic vesicles at hair-cell ribbon synapses (Trapani et al. 2009).

7.2.4.1 Association of Synaptojanin 1 with Human Disease

Increased expression of synaptojanin 1 has been described in Down's syndrome by several groups using various experimental approaches. The synaptojanin gene (*SYNJ1*) is localized to chromosome 21q22.2, a region implicated in Down's syndrome, bi-polar disorder and schizophrenia (Cheon et al. 2003; Arai et al. 2002). In a Down's syndrome mouse model, Ts65Dn, and also in transgenic synaptojanin 1 mice, increased expression of synaptojanin 1 is associated with a decrease in PtdIns(4,5)P_2 levels, leading to cognitive defects (Voronov et al. 2008). The *D. melanogaster* orthologs of three genes implicated in Down's syndrome, *syn* (synaptojanin), *dap160* (intersectin) and *nla* (nebula), when overexpressed individually in *D. melanogaster* lead to abnormal synaptic morphology. However, overexpression of all three candidate genes concomitantly is required for defective endocytosis, impaired vesicle recycling and locomotor defects. Moreover, dissection of the activities of the functional complex formed between synaptojanin, intersectin and nebula, has revealed nebula regulates 5-phosphatase catalytic activity, whilst intersectin directs the subcellular localization of synaptojanin (Chang and Min 2009). Synaptojanin mutations also exacerbate polyglutamine toxicity in *C.elegans*, suggesting a potential protective role for synaptojanin, like endophilin, in Huntington's disease (Parker et al. 2007).

7.2.4.2 Synaptojanin 2

Synaptojanin 2 is much less characterized than synaptojanin 1 however, it plays a role in the regulation of clathrin-mediated endocytosis. Synaptojanin 2 knockout mice have not been described to date. Synaptojanin 2 may play a protective role as a regulator of hair cell survival and hearing. However recently, a mutant mouse called Mozart which exhibits recessively inherited non syndromic progressive hearing loss, has been shown to result from mutation in *SYNJ2*. This mutation occurs in the catalytic 5-phosphatase domain resulting in loss of PtdIns(4,5)P_2 and PtdIns(3,4,5)P_3 degrading activity. *SYNJ2* but not *SYNJ1* is expressed in the inner and outer hair cells within the cochlea. Mozart mutant mice show progressive hearing loss with hair loss (Manji et al. 2011). RNAi-mediated depletion of synaptojanin 2 in a lung carcinoma cell line impairs clathrin-mediated receptor internalization, resulting in a reduction of clathrin-coated pits and vesicles (Rusk et al. 2003). Synaptojanin 2 splice variants specifically interact with Rac1 in a GTP-dependent manner resulting in translocation of synaptojanin 2 from the cytosol to the plasma membrane, facilitating inhibition of endocytosis (Malecz et al. 2000; Nemoto et al. 2001). A role for synaptojanin 2 in Rac1-mediated cell invasion and migration has also been identified via the regulation of lamellipodia and invadipodia formation. RNAi-mediated depletion of either Rac1 or synaptojanin 2 in glioblastoma cells equally inhibits cell migration and invasion, in addition to reducing the formation of invadipodia and lamellipodia, suggesting synaptojanin 2 may act downstream of Rac1 within this signaling pathway (Chuang et al. 2004).

7.2.5 INPP5E: A Lipid Phosphatase Linked to Cilia

INPP5E (also called the Type IV 5-phosphatase, or pharbin, gene name *INPP5E*) is widely expressed in the brain, kidneys, testis and other tissues and plays a critical role in the regulation of embryonic development (Kisseleva et al. 2000; Kong et al. 2000; Asano et al. 1999). Human INPP5E and the rat enzyme, pharbin, share 74% amino acid sequence identity (Kisseleva et al. 2000). This lipid phosphatase contains the central 5-phosphatase domain, flanked by regions with proline rich motifs, with a C-terminal CAAX motif. The 5-phosphatase has been localized to the cytosol and a perinuclear/Golgi localization in proliferating cells (Kong et al. 2000). In macrophages, during phagocytosis INPP5E localizes to the phagocytic cup and regulates FcγR1-mediated phagocytosis (Horan et al. 2007). In cells which have exited the cell cycle INPP5E localizes to cilia (Jacoby et al. 2009), sensory projections that co-ordinate cell signaling pathways and play a role in regulating cell division and differentiation. Cilia are also important for many aspects of embryonic development (Tobin and Beales 2009).

INPP5E hydrolyzes the 5-position phosphate from $PtdIns(4,5)P_2$ and $PtdIns(3,4,5)P_3$, forming $PtdIns(4)P$ and $PtdIns(3,4)P_2$ respectively, and is reported to be the most potent $PtdIns(3,4,5)P_3$ 5-phosphatase *in vitro* (Kisseleva et al. 2000; Kong et al. 2000). Overexpression of mouse Inpp5e in unstimulated 3T3-L1 adipocytes also generates $PtdIns(3)P$ at the plasma membrane of transfected cells, via hydrolysis of $PtdIns(3,5)P_2$, promoting the translocation and insertion of the glucose transporter, GLUT4, into the plasma membrane, without affecting glucose uptake (Kong et al. 2006). However, these results are based on ectopic Inpp5e overexpression and this may not occur *in vivo*. Inpp5e is expressed in the hypothalamus and, following insulin stimulation, is tyrosine phosphorylated and interacts with IRS-1/IRS-2 and PI3K (Bertelli et al. 2006). Antisense-mediated reduction of Inpp5e in the hypothalamus results in reduced food intake and weight loss, accompanied by altered metabolic parameters including reduced serum insulin, leptin and glucose levels (Bertelli et al. 2006).

7.2.5.1 INPP5E is Mutated in Human Ciliopathy Syndromes

Recent exciting studies have reported the phenotype of *Inpp5e* homozygous null mice, demonstrating a significant role for this 5-phosphatase in embryonic development. In addition *INPP5E* is mutated in two human ciliopathy syndromes, genetic diseases that affect the function of cellular cilia, or anchoring structures such as the basal body (Tobin and Beales 2009). Homozygous deletion of *Inpp5e* in mice, results in late embryonic lethality or death shortly after birth, with exencephaly, polydactyly, and kidney cysts, features of a ciliopathy (Jacoby et al. 2009). An inducible global knockout of *Inpp5e* leads to polycystic kidneys by 6 months and elevated body weight, indicating the 5-phosphatase also plays a significant functional role in the adult animal (Jacoby et al. 2009). In human studies a SNP for *INPP5E* has been identified in MORM (mental retardation, obesity, congenital retinal dystrophy and micropenis in male), a rare ciliopathy syndrome that results in a truncated INPP5E protein lacking the C-terminal 18 amino acids, which encompasses the CAAX motif. This motif is required for correct 5-phosphatase cilia axoneme localization (Jacoby et al. 2009). Whether the elevated body weight in Inpp5e-null adult mice is due to a regulatory role of the 5-phosphatase in insulin signaling is unknown, however, patients with ciliopathy syndromes frequently exhibit obesity (Tobin and Beales 2009). A second study has identified SNPs in *INPP5E* in human subjects with Joubert syndrome, which is characterized by underdevelopment of the cerebral vermis, polydactyly, retinal degradation, mental retardation and other phenotypes (Bielas et al. 2009). All six SNPs localize to regions encoding the catalytic domain of the INPP5E protein. These mutants exhibit reduced activity in hydrolyzing PtdIns(3,4,5)P_3 and to a lesser extent PtdIns(4,5)P_2. Overexpression of the Joubert SNP INPP5E mutants does not suppress Akt signaling, unlike expression of wildtype INPP5E. Interestingly, a homozygous mutation in *INPP5E* has been identified in a Joubert Syndrome (ciliopathy) patient that has only mild clinical features, including minimal truncal ataxia and oculomotor apraxia and normal cognitive function. Neuroimaging of the brain revealed the classical molar tooth sign which occurs in Joubert Syndrome (Poretti et al. 2009). The molecular mechanisms by which loss of function of INPP5E leads to similar complex phenotypes in both mice and humans is yet to be fully delineated, specifically how this relates to INPP5E regulation of PtdIns(3,4,5)P_3/Akt signaling is unknown.

7.2.5.2 INPP5E and Cancer

Gene expression profiling has revealed altered expression of *INPP5E* in a variety of cancers including cervical cancer, in which *INPP5E* is one of the top five (of 74 genes) with altered expression, showing an over 50-fold change (Yoon et al. 2003). Additionally *INPP5E* is one of the top six genes upregulated in non-Hodgkin's lymphoma following treatment (Chow et al. 2006). In uterine leiomyosarcoma compared to normal myometrium *INPP5E* RNA levels are increased greater than 5-fold (Quade et al. 2004). Decreased *INPP5E* gene expression has also been identified in stomach cancer and metastatic adenocarcinoma (Ramaswamy et al. 2003; Kim et al. 2003).

Interestingly, co-overexpression of INPP5E and another 5-phosphatase, SKIP, has been detected in gemcitabine-resistant pancreatic cell lines (Akada et al. 2005). It is tempting to speculate whether the different profiles of *INPP5E* gene expression in cancer are related to the newly identified function of INPP5E at cilia, as these organelles have a recently identified role in cancer, and depending on the initiating event, may either hinder or instigate tumor growth (Toftgard 2009). Moreover, cilia assembly and disassembly is linked to the cell cycle. It is therefore of interest that overexpression of INPP5E decreases cell growth due to increased apoptosis, whilst mouse embryonic fibroblasts (MEFs) isolated from the Inpp5e-null mice exhibit cell cycle arrest, albeit at moderate levels, possibly due to altered cilia disassembly in response to platelet-derived growth factor (PDGF) stimulation (Jacoby et al. 2009).

7.2.6 OCRL

OCRL (also called Lowe protein, gene name *INPP5F*) is a 5-phosphatase that is mutated in Lowe oculocerebrorenal (OCRL) syndrome and in some cases of Dent-2 disease (Attree et al. 1992; Hoopes et al. 2005; Sekine et al. 2007; Utsch et al. 2006). Lowe syndrome is an X-linked disorder affecting $\sim$1 in 200,000 births that is characterized by growth and mental retardation, bilateral congenital cataracts and renal impairment, associated with impaired solute and protein reabsorption in the kidney proximal tubule and renal tubular acidosis (Lowe 2005). Female carriers show punctate opacities in the lens (Gardner and Brown 1976; Roschinger et al. 2000). Dent-2 disease is characterized by low molecular weight proteinuria and renal failure (Bokenkamp et al. 2009). OCRL shares 45% amino acid identity and a similar domain structure to the related 5-phosphatase, INPP5B. Both enzymes contain a recently identified NH2-terminal PH domain, a central catalytic 5-phosphatase domain, followed by an ASH domain (ASPM (abnormal spindle-like microcephaly-associated protein)/SPD2 (spindle pole body 2)/hydin) and a catalytically inactive C-terminal Rho-GAP domain (Mao et al. 2009). Mutations in some ASH-domain-containing proteins are associated with abnormalities in brain development, such as hydrocephalus (Kumar et al. 2004; Bond et al. 2003; Ponting 2006). The OCRL sequence also contains two clathrin box binding motifs, one in its N-terminal PH domain and the second within the inactive Rho-GAP domain, as well as a clathrin adaptor AP-2-binding motif that is flanked by the PH and 5-phosphatase catalytic domains (Mao et al. 2009). OCRL and INPP5B share overlapping substrate specificity, hydrolyzing the 5-position phosphate from the inositol rings of PtdIns(4,5)P_2, PtdIns(3,4,5)P_3, Ins(1,4,5)P_3 and Ins(1,3,4,5)P_4, whilst OCRL also hydrolyzes PtdIns(3,5)P_2 (Schmid et al. 2004; Zhang et al. 1995). However, several studies suggest the major substrate that OCRL regulates in intact cells is PtdIns(4,5)P_2 (Zhang et al. 1998). For example the total cellular levels of PtdIns(4,5)P_2, but not PtdIns(3,4,5)P_3, are increased in cells from Lowe syndrome patients (Zhang et al. 1995).

OCRL is expressed in a wide range of human and mouse tissues (Janne et al. 1998). Surprisingly, although mutations in human OCRL cause significant human disease, deletion of Ocrl in mice does not lead to either Lowe or Dent-2 disease-like phenotypes (Janne et al. 1998). This lack of functional redundancy in mice may be due to compensation by other 5-phosphatases, most probably Inpp5b (Janne et al. 1998). Mice which lack both Ocrl and Inpp5b are embryonic lethal (Bernard and Nussbaum 2010).

7.2.6.1 OCRL Forms Multiple Protein Complexes and Regulates Vesicular Trafficking

OCRL interacts with clathrin, AP-2, cdc42 (cell division cycle 42), Rac, APPL1 (adaptor protein containing PH domain) Ses1/2 (from the word "sesquipedalian"), GIPC (GAIP-interacting protein, C terminus) and active forms of Rab1 and Rab5/6, the latter also stimulates the PtdIns(4,5)P_2 activity of OCRL (Swan et al. 2010; Hyvola et al. 2006). Various types of mutations, including missense and truncating, have been identified in Lowe syndrome, many in the region encoding the 5-phosphatase catalytic domain. However, mutations in the ASH/RhoGAP domain have also been identified. Interestingly, APPL1 and Ses1/2 proteins bind to the same region of the ASH/RhoGAP domain of OCRL, and their association with OCRL is mutually exclusive. Furthermore, like APPL1, Ses1/2 localizes to the endocytic compartment (Swan et al. 2010). APPL1 resides on a subset of peripheral OCRL-positive endosomes that are derived from clathrin-coated pits that receive internalized receptors.

OCRL mutations occur in both Lowe and Dent-2 patients (Shrimpton et al. 2009) and although Lowe syndrome and Dent-2 disease share common kidney defects, for unknown reasons, Dent-2 disease patients do not exhibit all the abnormalities that are commonly present in Lowe syndrome (Dent and Friedman 1964; Cho et al. 2008; Hoopes et al. 2005; Utsch et al. 2006; Shrimpton et al. 2009). Disease causing mutations in OCRL have been identified throughout the coding sequence, some of which result in reduced protein expression, by generating a non-sense transcript whilst others are located in the ASH/RhoGAP domain, a region of OCRL that commonly mediates its association with other proteins (McCrea et al. 2008; Addis et al. 2004; Swan et al. 2010). Interestingly, regardless of the site of mutation in OCRL, Lowe patient fibroblasts all exhibit elevated PtdIns(4,5)P_2, suggesting OCRL localization and association with other proteins is, in addition to its 5-phosphatase activity, critical for OCRL function (Lichter-Konecki et al. 2006; Kawano et al. 1998; Lin et al. 1997). The association of OCRL with APPL is abolished by disease causing mutations in the ASH/RhoGAP domain of OCRL (Erdmann et al. 2007; McCrea et al. 2008). Significantly, 10 of the Lowe syndrome and one of the Dent-2 disease-causing mutations in OCRL inhibit its association with APPL1 or Ses1/2, suggesting that selective loss of APPL1 or Ses1/2 binding to OCRL cannot explain the clinical differences between Lowe and Dent-2 disease (Swan et al. 2010). The only disease-causing mutation in *OCRL*, A861T, that is able to associate with the endocytic proteins APPL1 and Ses1/2, is both missense and a splice-site mutation, and has been proposed to regulate OCRL protein expression by generating a non-sense transcript (Swan et al. 2010; Kawano et al. 1998). It remains to be determined whether loss of the association between OCRL and APPL1 and/or Ses1/2 is involved in Lowe and/or Dent-2 disease. It has been suggested that using its ASH/RhoGAP binding surface, OCRL may bind a progression of endocytic proteins via the same motif, as is the case for APPL and Ses1/2, and that cumulative defects in this process may contribute to altered OCRL function causing Lowe and/or Dent-2 disease (Swan et al. 2010). APPL1 binding to OCRL is abolished by disease causing mutations in the OCRL-ASH-Rho GAP domain (Swan et al. 2010). Failure of OCRL to associate with APPL1 may contribute to the neurological/cognitive defects observed in Lowe syndrome. Both APPL1 and GIPC bind the TrkA (tropomyosin receptor kinase A) NGF (nerve growth factor) receptor and together regulate TrkA endocytic trafficking (Erdmann et al. 2007; Lin et al. 2006; Varsano et al. 2006). GIPC and APPL1 also associate with megalin, a receptor expressed in the kidney proximal tubule which facilitates the uptake of low-molecular-mass proteins. Both GIPC- and megalin-knockout mice exhibit low-molecular-mass proteinuria, similar to that found in Lowe syndrome (Norden et al. 2002).

Mutations in the 5-phosphatase domain of OCRL contribute to the Lowe syndrome phenotype due to loss of PtdIns(4,5)P_2 hydrolysis, whereas mutations in the ASH and RhoGAP domains result in mislocalization of the protein (http://research.nhgri.nih.gov/lowe/) (Bond et al. 2003; Erdmann et al. 2007; McCrea et al. 2008). Some point mutations in the inactive RhoGAP domain of OCRL lead to impaired 5-phosphatase activity, possibly due to altered protein conformation (Lichter-Konecki et al. 2006). Mutations in OCRL which disrupt the interaction between the 5-phosphatase and Rab1, Rab5 or Rab6 perturb the Golgi targeting of the enzyme (Hyvola et al. 2006).

There is growing evidence OCRL regulates vesicular trafficking, however, despite many interesting studies it still remains to be determined how OCRL degradation of PtdIns(4,5)P_2 regulates many of these events. PtdIns(4)P localizes to the Golgi, but there is limited evidence for PtdIns(4,5)P_2 at this site. In non stimulated cells OCRL localizes to the TGN (*trans*-Golgi network), lysosomes and endosomes and is enriched in clathrin-coated vesicles (Zhang et al. 1998; Choudhury et al. 2005; Erdmann et al. 2007). Overexpression of OCRL (and OCRL that lacks the 5-phosphatase domain) fragments the Golgi (Choudhury et al. 2005; Hyvola et al. 2006). In addition, OCRL and mutant OCRL expression, redistributes the cation-independent mannose-6-phosphate receptor (M6PR) to enlarged endosomes, and blocks clathrin-mediated transport from early endosomes to the Golgi, and retrograde trafficking (Choudhury et al. 2005). Moreover, siRNA-mediated knockdown of OCRL leads to impaired endosome to *trans*-Golgi trafficking, as shown by an accumulation of M6PR in endosomes (Choudhury et al. 2005). Increased levels of circulating lysosomal enzymes are found in the plasma of Lowe syndrome affected individuals suggesting an endosomal trafficking/secretion defect (Ungewickell and Majerus 1999). OCRL mutant expression also affects endocytosis of the transferrin receptor, independent of its Golgi function (Choudhury et al. 2005; Hyvola et al. 2006). In addition perhaps by regulating PtdIns(4,5)P_2 levels, OCRL regulates actin cytoskeletal dynamics. Fibroblasts derived from affected individuals with Lowe syndrome exhibit abnormal cell migration, cell spreading and fluid phase uptake and an altered actin cytoskeleton, a phenotype that can be rescued by expression of INPP5B (Mao et al. 2009; Coon et al. 2009). OCRL translocates to membrane ruffles in a Rac1-dependent manner, in response to growth factor stimulation, where it co-localizes with polymerized actin and Rac1 (Faucherre et al. 2005).

The underlying molecular defects in Lowe Syndrome and Dent-2 disease are emerging, however, the kidney defects that are common to both syndromes and the significant role OCRL plays in vesicular trafficking, have lead to the prediction that the renal tubular acidosis observed in affected individuals may be the result of abnormal trafficking of receptors, that promote solute reabsorption from proximal tubules (Norden et al. 2002). Specifically, OCRL may regulate the recycling of cubulin and megalin, the latter is a proximal tubule renal receptor that facilitates the uptake of low molecular weight proteins. Significantly, megalin associates, like OCRL, with GIPC and APPL1 (Swan et al. 2010; Hyvola et al. 2006). Homozygous deletion of megalin, or GIPC in mice leads to low molecular weight proteinuria, similar to that observed in Lowe syndrome and Dent-2 disease and affected individuals exhibit reduced megalin levels in their urine. Collectively these studies suggest that OCRL may regulate the recycling of megalin to the apical surface of renal proximal tubule cells, and in its absence, this may lead to proteinuria.

7.2.7 INPP5A

INPP5A (also called the type I 5-phosphatase, 5-phosphatase-1 or 43 kDa 5-phosphatase, gene name *INPP5A*) hydrolyzes the second messenger molecules inositol 1,4,5-trisphosphate (Ins(1,4,5)P_3) and inositol 1,3,4,5-tetrakisphosphate (Ins(1,3,4,5)P_4) to form Ins(1,4)P_2 and Ins(1,3,4)P_3 (Laxminarayan et al. 1993, 1994; De Smedt et al. 1994). These inositol phosphate signaling molecules play significant roles in regulating calcium release from intracellular stores, and at the plasma membrane (reviewed by (Berridge and Irvine 1989)). There is little evidence to date that this enzyme hydrolyzes the membrane bound phosphoinositides, PtdIns(3,4,5)P_3 and/or PtdIns(4,5)P_2. INPP5A contains the 5-phosphatase domain and a C-terminal CAAX motif, that mediates its plasma membrane localization (De Smedt et al. 1996). Although INPP5A was one of the first 5-phosphatase enzymes to be identified and purified, its function *in vivo* remains relatively uncharacterized. No mouse knockout for this enzyme has been reported, although in some databases it is suggested to be embryonically lethal. INPP5A enzyme activity is regulated *in vitro* by binding the adaptor protein, 14.3.3 (Campbell et al. 1997), and also by Ca(2+)/calmodulin kinase II (De Smedt et al. 1997). Underexpression of INPP5A using an antisense strategy results in enhanced intracellular calcium oscillations and cellular transformation (Speed et al. 1996). However, there are few studies reporting a role for this enzyme in human disease. Recently expression of INPP5A was implicated in skin cancer pathogenesis (Sekulic et al. 2010). Deletion of a region on chromosome 10q, which contains the *INPP5A* gene, is detected in 24% of squamous cell carcinomas (SCC) of the skin. A decrease in INPP5A appears to be an early event in SCC development, as loss of its expression is also detected in (35%) of actinic keratoses, the earliest stage in SCC development (Sekulic et al. 2010).

7.2.8 INPP5B

INPP5B (also called 5-phosphatase-2 or 75 kDa 5-phosphatase, gene name *INPP5B*) shares 45% amino acid identity, as well the same domain structure as OCRL, but lacks the clathrin and AP-2 binding motifs (Attree et al. 1992; Erdmann et al. 2007). Although it was originally purified from human platelets as a 75 kDa 5-phosphatase, this is a cleaved fragment of a larger protein (Mitchell et al. 1989). Unlike OCRL, it contains a C-terminal CAAX motif, which facilitates its membrane attachment via isoprenylation (Matzaris et al. 1998). The N-terminal portion of INPP5B also contributes to its membrane localization. INPP5B hydrolyzes PtdIns(3,4,5)P_3, PtdIns(4,5)P_2, Ins(3,4,5)P_3 and Ins(1,3,4,5)P_4 (Matzaris et al. 1998; Jefferson and Majerus 1995; Schmid et al. 2004). Deletion of the CAAX motif reduces INPP5B catalytic activity (Erdmann et al. 2007), however deletion of both the N- and C-terminal domains in recombinant INPP5B has no effect on enzyme activity (Jefferson and Majerus 1995, Matzaris et al. 1998). INPP5B forms a direct complex with many of the endocytic Rab proteins including Rab1, Rab2 (*cis*-Golgi), Rab5 (early endosomes), Rab6 (Golgi stack) and Rab9 (late endosomes), suggesting the 5-phosphatase may regulate early secretory/trafficking events between the endoplasmic reticulum (ER) and Golgi, and/or retrograde trafficking from the Golgi to ER (Erdmann et al. 2007; Shin et al. 2005). In quiescent cells, INPP5B localizes to the Golgi, mediated by its C-terminal RhoGAP domain interaction with the Rab proteins, whilst upon growth factor stimulation it translocates to lamellipodia where it co-localizes with Rab5 and actin (Shin et al. 2005; Williams et al. 2007). INPP5B binds Rab5 via its ASH domain. Mutation of conserved key residues within the ASH domain that mediate the interaction of the 5-phosphatase with Rab5 results in mislocalization of INPP5B from the Golgi (Shin et al. 2005; Williams et al. 2007). Co-expression of an activated Rab5 mutant, together with INPP5B, recruits the 5-phosphatase to a population of enlarged early endosomes. Rab5 binding to INPP5B increases 5-phosphatase catalytic activity (Erdmann et al. 2007; Shin et al. 2005). Additionally the inactive RhoGAP domain of INPP5B interacts with the Rho family GTPases, Rac and cdc42, but not Rho. Functionally, overexpression of INPP5B regulates retrograde transport from the ER-Golgi intermediate compartment to the ER, probably dependent on its interaction with Rab proteins, but independent of its 5-phosphatase activity (Williams et al. 2007). However, in a separate study siRNA-mediated knockdown of INPP5B did not alter retrograde trafficking but rather inhibited transferrin endocytosis (Choudhury et al. 2005; Shin et al. 2005).

Interestingly, the human *INPP5B* gene is located on chromosome 1p34, and genes flanking this locus are associated with genetic disorders with features of Lowe syndrome including abnormal kidney function, lens development and mental retardation (Bisgaard et al. 2007; Konrad et al. 2006; Cormand et al. 1999; Shearman et al. 1996; Nicole et al. 1995). In contrast the murine *Inpp5b* gene is on chromosome 4, where neighbouring genes are involved in lens development (Janne et al. 1994, 1995). However, Inpp5b-null mice do not exhibit features of Lowe or Dent-2 disease, but rather show abnormalities in the testis (Janne et al. 1998). $Inpp5b^{-/-}$ testis exhibits vacuoles in seminiferous tubule epithelium and an accumulation of adherence junctions in the vacuoles of Sertoli cells. Sperm from these mice exhibit reduced motility and fertilization; abnormal processing of the sperm/egg adhesion molecule, and are unable to penetrate zona-pellucida free eggs, due to a decrease in binding and fusion with the egg plasma membrane (Marcello and Evans 2010). Reduced ADAMs processing in Inpp5b-null mice may explain the observed infertility, as ADAMs deficient mice exhibit multi-faceted infertility including abnormal sperm/egg binding (Cho et al. 2000). However, more recent studies have shown little correlation between ADAMs processing levels and infertility in Inpp5b-null mice (Marcello and Evans 2010). Conditional knockout of *Inpp5b* in spermatids does not recapitulate the fertility defects observed in the global *Inpp5b*$^{-/-}$ mice, suggesting loss of Inpp5b function in somatic cells that support sperm function may contribute to the observed infertility (Hellsten et al. 2001).

7.2.9 PIPP

The proline-rich inositol polyphosphate 5-phosphatase, PIPP (gene name *INPP5J*), is a little characterized 5-phosphatase and its function remains elusive. PIPP is expressed in brain, breast, heart, kidney, liver, stomach, and lung (Mochizuki and Takenawa 1999). This 108 kDa 5-phosphatase contains N- and C-terminal proline rich domains which may facilitate protein-protein interactions and a central 5-phosphatase catalytic domain that degrades PtdIns(3,4,5)P_3 (Mochizuki and Takenawa 1999; Ooms et al. 2006; Gurung et al. 2003). PIPP also contains a SKICH domain that mediates its constitutive association with the plasma membrane (Gurung et al. 2003). PIPP hydrolyzes PtdIns(3,4,5)P_3 decreasing Akt activation and its downstream signaling and recent reports indicate PIPP directly opposes oncogenic PI3K/Akt signaling, and its overexpression decreases colony formation in soft agar and reduces proliferation mediated by oncogenic *PIK3CA* (Ooms et al. 2006; Mochizuki and Takenawa 1999; Denley et al. 2009). In NGF-differentiated PC12 cells, PIPP localizes to the plasma membrane, the shaft of extending neurites and the growth cone where it regulates PtdIns(3,4,5)P_3 levels and the activation of Akt and GSK3β. PIPP, like SHIP2, regulates neurite outgrowth and elongation (Ooms et al. 2006). To date the phenotype of PIPP knockout mice remains unreported.

The human PIPP gene, *INPP5J*, is located on chromosome 22q12. Loss of heterozygosity (LOH) of chromosome 22q is frequently detected in breast carcinomas with the most commonly deleted region 22q13 (Castells et al. 2000; Ellsworth et al. 2003; Iida et al. 1998). Allelic loss of chromosome 22q12 has also been reported (Ellsworth et al. 2003; Iida et al. 1998; Osborne and Hamshere 2000) with one study demonstrating LOH of markers within this region in ∼ 30% of breast tumors (Iida et al. 1998). LOH of the markers D22S1150 and D22S280 that map to chromosome 22q on either side of *INPP5J* has been detected in 41% and 45% of breast carcinomas respectively (Allione et al. 1998). *INPP5J* mRNA is expressed at higher levels in estrogen receptor (ER)$^{+ve}$ tumors compared to ER^{-ve} tumors (Gruvberger et al. 2001; van't Veer et al. 2002). In a screen of ∼ 5000 genes, *INPP5J* was one of 231 genes significantly associated with disease outcome (van't Veer et al. 2002). Higher expression of *INPP5J* correlates with a better prognosis, defined as no distant metastases developing within 5 years of diagnosis (van't Veer et al. 2002). *INPP5J* has been reported to be one of the 10 highest ranked genes for predicting breast cancer patient outcome (Takahashi et al. 2004). However, the role that PIPP plays in controlling breast cancer proliferation is yet to be reported.

7.2.10 SKIP

SKIP (*s*keletal muscle and *k*idney *i*nositol *p*hosphatase, gene name *INPP5K*) is a relatively uncharacterized 51 kDa 5-phosphatase, which is most highly expressed in the heart, skeletal muscle and kidney, that regulates embryonic development, insulin signaling and glucose homeostasis (Ijuin and Takenawa 2003; Ijuin et al. 2000, 2008). SKIP contains the common catalytic 5-phosphatase domain, with a C-terminal SKICH domain (which is also found in the 5-phosphatase, PIPP), that mediates its plasma membrane association following growth factor stimulation (Ijuin and Takenawa 2003; Gurung et al. 2003). In unstimulated cells SKIP is distributed in a perinuclear distribution which may indicate its association with the ER (Gurung et al. 2003). Purified SKIP recombinant protein exhibits a preference for degrading PtdIns(4,5)P_2 over PtdIns(3,4,5)P_3, but studies in cell lines in which SKIP expression has been reduced by siRNA has revealed PtdIns(3,4,5)P_3 levels are significantly increased in response to insulin stimulation, associated with enhanced Akt activation indicating SKIP regulates PI3K/Akt signaling in intact cells (Ijuin and Takenawa 2003; Schmid et al. 2004). SKIP overexpression in L6 myotubes attenuates insulin-stimulated Akt and p70S6K phosphorylation and inhibits translocation of the glucose transporter, GLUT4, to the plasma membrane leading to decreased glucose uptake and inhibition of glycogen synthesis (Ijuin and Takenawa 2003). $Skip^{-/-}$ mice are embryonically lethal at E10.5 for unknown reasons (Ijuin et al. 2008). Interestingly $Skip^{+/-}$ mice exhibit increased glucose tolerance and insulin sensitivity, regardless of diet, however, on a high fat diet, these increases are lower compared to wild-type mice, suggesting that loss of Skip expression may provide some protection against diet-induced obesity (Ijuin et al. 2008). SKIP may play a significant role in regulating insulin signaling in skeletal muscle, the major site for post-prandial glucose uptake. However, SKIP is also highly expressed in the brain as well as skeletal muscle, therefore the observed increase in whole-body insulin sensitivity could also be mediated by SKIP regulation of brain PI3K/Akt signaling, although this has not yet been reported. The tissue-specific role of SKIP in the regulation of insulin signaling will be aided by the development of tissue-specific $Skip^{-/-}$ mouse models, which have not been described to date.

7.2.10.1 SKIP Association with Human Disease

The heterozygous deletion of eight candidate genes, including *SKIP*, is associated with the contiguous-gene syndrome Miller-Dieker syndrome (MDS), a severe form of lissencephaly (smooth brain) due to defects in neuronal cell migration, which leads to mental retardation and craniofacial/limb abnormalities (Kato and Dobyns 2003;

Cardoso et al. 2003). Recent further refinement of the critical genomic region reduced the candidate gene number to six, including *SKIP* (Bruno et al. 2010). Whether SKIP contributes to MDS is currently unknown.

The human *SKIP* gene is located on chromosome 17p13.3, a region reported to be deleted or hyper-methylated in numerous human cancers including brain, breast, and hepatocellular cancers (Cornelis et al. 1994; Saxena et al. 1992; Zhao et al. 2003a, 2003b; Biegel et al. 1992; Rood et al. 2002). Microarray analysis of gene expression profiles has revealed altered *SKIP* expression, both up and down, in a diverse range of human cancers. *SKIP* transcription is down-regulated in lung adenocarcinoma (Beer et al. 2002; Stearman et al. 2005; Su et al. 2007), prostate carcinoma (Dhanasekaran et al. 2005), chronic lymphocytic leukaemia (Haslinger et al. 2000), Burkitt lymphoma (Corcione et al. 2006) and hepatocellular carcinoma (Wurmbach et al. 2007; Ye et al. 2003b). Conversely, *SKIP* is up-regulated in bladder cancer (Sanchez-Carbayo et al. 2006), cutaneous melanoma (Talantov et al. 2005), multiple myeloma (Zhan et al. 2007) and gemcitabine-resistant pancreatic cancer cell lines (Akada et al. 2005). As yet there have been no functional studies showing the effects of altered expression of SKIP on human cancer cell proliferation and/or invasion.

SKIP also interacts with the human hepatitis B virus (HBV) core protein, leading to nuclear localization of the protein complex, resulting in the suppression of HBV gene expression and virion replication (Hung et al. 2009). The mechanism of suppression remains unknown; however, the suppressive effect is not mediated by the 5-phosphatase activity of SKIP, but by a newly identified functional domain of SKIP located within amino acids 199–226 (Hung et al. 2009). HBV infection can lead to severe liver disease including hepatocellular carcinoma (Bertoletti and Gehring 2007). The interaction of SKIP and HBV core protein is noteworthy in light of the reported LOH at 17p13.3 in hepatocellular carcinoma, and micro-array studies showing decreased *SKIP* transcription in hepatocellular carcinoma (Wurmbach et al. 2007; Ye et al. 2003a; Zhao et al. 2003a).

7.2.11 Future Studies on 5-Phosphatases

Many recent gene-linkage, *in vitro* and *in vivo* studies have demonstrated the significance of the 5-phosphatases in multiple aspects of embryonic and human development and disease. Over the last decade the 5-phosphatase enzymes have been re-classified from PI3K-terminating enzymes to PI3K-modifying enzymes, as many of the products of their catalysis exhibit signaling properties in their own right. Surprisingly, even though the 5-phosphatase family encompasses 10 mammalian enzymes there appears to be little functional redundancy.

The generation of 5-phosphatase knockout animals has provided valuable and sometimes unexpected information about these enzymes. The embryonic lethality or early death associated with some of the 5-phosphatase knockout mice indicates a significant role in development, however, the generation and characterization of inducible constitutive enzyme-null mouse models would be beneficial in determining their role in non-developmental processes. Moreover, tissue- or cell-specific knockout models and/or crossing null mice with various cancer models will also assist in determining whether the 5-phosphatases function as *bona fide* tumor suppressors, for which there is currently little evidence with the exception of SHIP1. Furthermore, the use of 5-phosphatase expression as a diagnostic and/or prognostic indicator or as a druggable target should be evaluated in the coming years.

7.3 Inositol Polyphosphate 4-Phosphatases

Inositol polyphosphate 4-phosphatases preferentially hydrolyze the D4 position phosphate of target inositol head-groups. The 4-phosphatase enzyme family comprises two inactive members, and five active members (see Table 7.2) which can be sub-classified based on substrate specificity into the PtdIns(4,5)P_2 4-phosphatases or PtdIns(3,4)P_2 4-phosphatases. The catalytic product of the PtdIns(4,5)P_2 4-phosphatases is PtdIns(5)P, while the PtdIns(3,4)P_2 4-phosphatases generate PtdIns(3)P.

7.3.1 *PtdIns(4,5)P_2 4-Phosphatases*

The PtdIns(4,5)P_2 4-phosphatases hydrolyze PtdIns(4,5)P_2 exclusively and there are two mammalian (TMEM55A and TMEM55B) and one bacterial (invasion plasmid gene D [IpgD]) isoenzymes identified to date. The mammalian TMEM55A/B enzymes share 52% amino acid identity, including a catalytic CX_5R motif, and are ubiquitously expressed in human tissues, however TMEM55A exhibits higher expression in the liver, spleen and thymus, and TMEM55B is more prevalent in regions of the brain (Ungewickell et al. 2005). In human cells, both enzymes localize to late endosomes/lysosomes. Overexpression of TMEM55A increases cellular PtdIns(5)P and is associated with increased EGFR degradation (Ungewickell et al. 2005; Zou et al. 2007). Following cell stress, TMEM55A translocates to the nucleus where it acts to promote p53 acetylation and stability, resulting in increased apoptosis through PtdIns(5)P-dependent ING2 (inhibitor of growth protein 2) activity (Zou et al. 2007). To date, no mouse knockout models or functional studies for these enzymes have been reported.

Interestingly, while the bacterial 4-phosphatase, IpgD, exhibits sequence homology with the mammalian PtdIns(3,4)P_2 4-phosphatases (Norris et al 1998), this enzyme actually functions to regulate PtdIns(4,5)P_2 levels in mammalian cells. The pathogen, *Shigella flexneri* (*S. flexneri*), responsible for causing bacillary dysentery in humans, directly injects IpgD into mammalian host cells through its type III secretion machinery during infection, resulting in PtdIns(4,5)P_2 dephosphorylation at the host cell plasma membrane, thereby facilitating actin reorganization and the formation of entry sites that promote bacterial uptake (Niebuhr et al. 2000, 2002; Allaoui et al. 1993). Interestingly, dephosphorylation of PtdIns(4,5)P_2 by IpgD also results in the accumulation of its phosphoinositide product, PtdIns(5)P, which activates the

Table 7.2 Inositol polyphosphate 4-phosphatase and murine models with gene targeted deletion of family members

Protein Name	Alias(es)	Gene	Animal models	References
Inositol polyphosphate 4-phosphatase type I	INPP4A	*INPP4A*	*Inpp4a^{wbl}* mutant mouse—early onset cerebellar ataxia, Purkinje cell loss, postnatal lethality at 2–3 weeks of age	(Nystuen et al. 2001; Sachs et al. 2009)
			Inpp4a^{-/-} constitutive KO mouse—excitotoxic neuronal death in striatum, involuntary movements, postnatal lethality at 2–3 weeks of age	(Sasaki et al. 2010)
Inositol polyphosphate 4-phosphatase type II	INPP4B	*INPP4B*	Not reported	
PtdIns(4,5)P$_2$ 4-phosphatase type I	TMEM55A	*TMEM55A*	Not reported	
PtdIns(4,5)P$_2$ 4-phosphatase type II	TMEM55B	*TMEM55B*	Not reported	
P-Rex1		*P-Rex1*	*P-Rex1^{-/-}* constitutive KO mouse—viable and healthy, decreased Rac activation, ROS and superoxide formation and chemotaxis of neutrophils and macrophages	(Welch et al. 2005; Wang et al. 2008)
			P-Rex1 GEF-dead transgenic mouse—develop normally, diminished Rac2 activation, chemotaxis, superoxide formation and actin reorganisation in neutrophils	(Dong et al. 2005)
P-Rex2	P-Rex2a, P-Rex2b	*P-Rex2*	*P-Rex2^{-/-}* constitutive KO mouse—viable, fertile and healthy, abnormal cerebellar Purkinje cells and impaired motor coordination, more severe in females and age-related	(Donald et al. 2008)
			P-Rex1^{-/-}/P-Rex2^{-/-} double KO mouse—more severe but similar phenotype as *P-Rex2^{-/-}* global KO, reduced long-term potentiation in cerebellum	(Donald et al. 2008; Jackson et al. 2010)

PI3K/Akt pathway to signal for host cell survival (Pendaries et al. 2006; Guittard et al. 2010). In addition, increased production of IpgD-dependent PtdIns(5)P may also promote tyrosine phosphorylation of Src family kinases and their substrates to regulate immunogenic T-cell responses (Guittard et al. 2009, 2010).

An IpgD-related enzyme, SopB/SigD, has also been identified in *Salmonella* species. This enzyme is integral for *Salmonella*-mediated pathogenesis in host systems (Galyov et al. 1997), through a variety of mechanisms including chloride channel and tight junction regulation, leading to cell permeability (Boyle et al. 2006; Feng et al. 2001), regulation of vesicular trafficking, and the formation of the *Salmonella*-containing vacuole (SCV) (Hernandez et al. 2004; Mallo et al. 2008; Bakowski et al. 2010) and the promotion of Akt phosphorylation and host cell survival (Knodler et al. 2005, 2009). SopB exhibits 3-, 4- and 5-phosphatase activity towards multiple soluble and insoluble inositol phosphates, and initial studies reported its ability to hydrolyze the D4 phosphate from PtdIns(3,4)P_2 and Ins(1,3,4)P_3, but not PtdIns(4,5)P_2 (Norris et al. 1998). More recent studies, however, suggest that SopB does, in fact, utilize PtdIns(4,5)P_2 as a substrate and that negative regulation of this phosphoinositide by SopB may promote actin reorganization, opening of tight junctions and inhibition of SCV-lysosome fusion in host cells (Mason et al. 2007; Bakowski et al. 2010). Given the contrasting results concerning SopB substrates, this enzyme remains to be classified as a *bona fide* PtdIns(4,5)P_2 4-phosphatase.

7.3.2 PtdIns(3,4)P_2 4-Phosphatases

There are two mammalian PtdIns(3,4)P_2 4-phosphatases; type I and type II (gene names; *INPP4A* and *INPP4B*, respectively), which share 37% amino acid identity, including a conserved amino-terminal C2 domain, responsible for their interactions with target phosphoinositides (Shearn and Norris 2007; Ivetac et al. 2005). Apart from a conserved CX_5R motif they share little resemblance to the PtdIns(4,5)P_2 4-phosphatases. INPP4A was the first mammalian 4-phosphatase identified, initially found to hydrolyze the soluble Ins(3,4)P_2 and Ins(1,3,4)P_3 (Bansal et al. 1987). It was later identified as a magnesium-, calcium- and lithium-independent phosphatase (Bansal et al. 1990) with additional activity towards insoluble PtdIns(3,4)P_2, 120-fold greater than Ins(3,4)P_2 (Norris and Majerus 1994), indicating PtdIns(3,4)P_2 is its preferred *in vivo* substrate. The catalytic activity of both enzymes is mediated by a conserved carboxy-terminal catalytic CX_5R (CKSAKDR) motif (Norris et al. 1997). INPP4A also binds the p85 subunit of PI3K, which may facilitate the recruitment of the phosphatase to substrate-enriched membranes (Munday et al. 1999). Alternatively spliced variants have been identified for both INPP4A and B, featuring a hydrophobic C-terminal transmembrane domain (Norris et al. 1997). The function of these spliced isoforms in mammalian cells remains unknown.

7.3.2.1 INNP4A Regulates Excitatory Neuronal Cell Death

INPP4A and B exhibit ubiquitous tissue distributions, with the highest expression in brain and heart (Norris et al. 1995, 1997). Evidence from recent knockout mouse studies and analysis of human tumors suggests non-redundant roles for these enzymes. Knockout of the *Inpp4a* gene in mice promotes postnatal neuronal degeneration and lethality by 2–3 weeks of age (Nystuen et al. 2001; Sachs et al. 2009; Sasaki et al. 2010). The naturally-occurring *Inpp4a* knockout mouse, *weeble* (*Inpp4a^{wbl}*), resulting from a spontaneous single nucleotide deletion in exon 10 of the *Inpp4a* gene (Δ744G), exhibits early onset cerebella ataxia, associated with Purkinje cell loss (Nystuen et al. 2001; Sachs et al. 2009). In a mouse model of targeted *Inpp4a* disruption (*Inpp4a$^{-/-}$*), neuronal degeneration is restricted to the striatum, associated with increased apoptosis of medium-sized spiny projection neurons (MSNs), resulting from glutamate receptor excitotoxicity (Sasaki et al. 2010). Both mouse models of Inpp4a loss reveal a neuroprotective role for this enzyme in postnatal animals, however, Inpp4a is dispensable during embryogenesis (Nystuen et al. 2001; Sachs et al. 2009; Sasaki et al. 2010). Although the precise molecular mechanisms by which Inpp4a fulfils its neuroprotective function during postnatal development are emerging, regulation of PtdIns(3,4)P$_2$ appears critical as *Inpp4a^{wbl}* Purkinje cells exhibit PtdIns(3,4)P$_2$ accumulation (Shin et al. 2005). Furthermore, treatment of wildtype MSNs with PtdIns(3,4)P$_2$ promotes glutamate-induced cell death (Sasaki et al. 2010).

Recent reports also indicate that the PtdIns(3,4)P$_2$ 4-phosphatases can function to regulate a wide range of additional cellular processes. For example, INPP4A acts to control normal endosome function and is recruited to endosomal membranes by binding Rab5, thereby increasing its PtdIns(3,4)P$_2$ catalytic activity and contributing to an endosomal pool of PtdIns(3)P (Ivetac et al. 2005; Shin et al. 2005). Absence or depletion of INPP4A promotes the formation of dilated endosomes and impairs transferrin internalization and its overexpression rescues PI3K inhibition-induced endosomal dilation, dependent on its catalytic activity (Ivetac et al. 2005; Shin et al. 2005). Interestingly, the endosomal function of INPP4A is not recapitulated by the INPP4B isoform, which does not localize to endosomal membranes or contribute to the production of endosomal PtdIns(3)P. Rather, INPP4B regulates plasma membrane localized PtdIns(3,4)P$_2$ downstream of PI3K activation, a role also shared by INPP4A, which translocates to the plasma membrane upon growth factor stimulation (Ivetac et al. 2005; Shin et al. 2005). The role of the 4-phosphatases at the plasma membrane is presumably to regulate signaling events downstream of PtdIns(3,4)P$_2$, including activation of the proto-oncogene, Akt. Indeed, both INPP4A and INPP4B negatively regulate Akt activation and its downstream cellular processes including cell proliferation, tumor growth, anchorage-independent colony formation and cell migration (Ivetac et al. 2009; Vyas et al. 2000; Gewinner et al. 2009; Fedele et al. 2010; Hodgson et al. 2011).

7.3.2.2 Association of INPP4A and INPP4B with Human Disease

There is some evidence of altered *INPP4A* expression in human diseases, including asthma (Sharma et al. 2010), prostate cancer (LaTulippe et al. 2002) and leukemia (Erkeland et al. 2004), but no functional studies have been reported. In contrast, a significant role for INPP4B as a putative tumor suppressor in multiple human cancers has recently been identified. For example, INPP4B expression is lost during malignant proerythroblast progression and re-introduction of INPP4B into late-stage blasts decreases Akt activation (Barnache et al. 2006). In human prostate cancer cells, *INPP4B* is an androgen receptor (AR)-responsive gene (Hodgson et al. 2011), and its transcription is diminished in late-stage, androgen-independent prostate cancer xenografts in mice (Gu et al. 2005). Significantly, INPP4B protein is frequently lost in human prostate cancers, associated with reduced recurrence-free survival (Hodgson et al. 2011) and diminished *INPP4B* gene transcription is frequently observed in metastatic prostate carcinomas (Taylor et al. 2010). Interestingly, in the human mammary gland INPP4B is expressed in a sub-population of estrogen receptor (ER)-positive cells in the terminal ductal lobuloalveolar units (TDLU), where it may act to suppress cell proliferation (Fedele et al. 2010). Indeed, in human breast cancer, *INPP4B* is frequently deleted (Naylor et al. 2005) and its transcription and protein expression positively correlates with ER and progesterone receptor (PR) expression (Yang et al. 2005; West et al. 2001; Fedele et al. 2010). LOH maps to the *INPP4B* gene in a range of human cancers, including breast and ovarian carcinomas and melanomas, associated with decreased patient survival (Gewinner et al. 2009). In human breast cancer, *INPP4B* LOH and protein loss occur most frequently in the aggressive basal-like sub-type (Gewinner et al. 2009; Fedele et al. 2010). Significantly, loss of INPP4B protein expression is frequently observed in PTEN-null breast cancers, associated with increased Akt phosphorylation, indicating co-operative promotion of tumorigenesis through the loss of multiple phosphoinositide phosphatases (Fedele et al. 2010). Overall these studies highlight the importance of PtdIns(3,4)P_2-dependent signaling and its regulation by the 4-phosphatases in maintaining normal cell proliferation and implicates this pathway in the regulation of human cancer. Collectively these studies identify INPP4B as a significant new tumor suppressor in many human cancers.

7.3.3 Inactive 4-Phosphatases: P-Rex Family Guanine Nucleotide Exchange Factors

The PtdIns(3,4,5)P_3-dependent Rac exchanger (P-REX) enzymes are related to the PtdIns(3,4)P_2 4-phosphatases, as they contain a 4-phosphatase homology domain, which shows primary amino acid sequence identity with INPP4A, with a conserved carboxy-terminal CX_5R 4-phosphatase catalytic motif (Welch et al. 2002). In contrast to the PtdIns(3,4)P_2 4-phosphatases, P-REX enzymes do not exhibit phosphoinositide phosphatase activity for unknown reasons (Welch et al. 2002). Rather, these proteins are critical for the regulation of signaling events downstream of PI3K activation in a phosphoinositide phosphatase-independent manner and are implicated in a range of human functions and diseases, including cancer and diabetes. Therefore the 4-phosphatase enzyme family is reminiscent of the myotubularins, with both active and inactive family members.

To date, three P-REX isoenzymes have been identified; P-REX1, P-REX2a and P-REX2b, all of which are multi-domain proteins, sharing 59% amino acid identity with similar protein structures, including tandem amino-terminal dbl-homology (DH) and pleckstrin-homology (PH) domains, two DEP domains and two PDZ domains (Welch et al. 2002; Donald et al. 2004; Rosenfeldt et al. 2004). The P-REX2a and P-REX2b isoforms are transcribed from the same gene, however P-REX2b lacks the carboxy-terminal 4-phosphatase domain (Rosenfeldt et al. 2004). The founding member of the P-REX family of proteins, P-REX1, was initially identified in neutrophils as the major guanine nucleotide exchange factor (GEF) required for activation of the Rho-family GTPase, Rac (Welch et al. 2002), and P-REX2a and 2b also exhibit Rac-GEF activity (Donald et al. 2004; Rosenfeldt et al. 2004; Li et al. 2005). Activation of these enzymes occurs synergistically via binding to PtdIns(3,4,5)P_3 and the G-protein, G$\beta\gamma$, at the plasma membrane in response to extracellular stimulation and receptor tyrosine kinase (RTK) and G protein-coupled receptor (GPCR) activation, dependent on their PH and DH domains, respectively (Welch et al. 2002; Hill et al. 2005; Donald et al. 2004; Barber et al. 2007). The multi-domain nature of the P-REX enzymes is critical for their activation and function. For example, the Rac-GEF activity is mediated by their DH/PH domains (Hill et al. 2005) and the PH domain may facilitate the specific recognition of certain Rac isoforms as substrates (Joseph and Norris 2005). In addition, the P-REX1 tandem DEP domains bind the mammalian target of rapamycin (mTOR) and this interaction is necessary for mTOR complex 2 (mTORC2)-dependent Rac activation (Hernandez-Negrete et al. 2007). Furthermore, the PDZ domains of P-REX1 can bind GPCR, suppressing receptor internalization and promoting downstream signaling and cell migration (Ledezma-Sanchez et al. 2010). Interestingly, the function of P-REX1 may also be self-regulated by inter-domain interactions between the DEP and PDZ domains, and also by the PH domain (Hill et al. 2005; Urano et al. 2008). Moreover, recruitment of P-REX proteins to activated RTKs at the plasma membrane may also serve to facilitate their phosphorylation and/or dephosphorylation and activation, possibly mediated by protein kinase A (PKA) (Montero et al. 2011; Mayeenuddin and Garrison 2006; Urano et al. 2008; Zhao et al. 2007). Interestingly, the 4-phosphatase domain of these enzymes appears to be dispensable for P-REX-mediated Rac activation, however it may play a role in promoting full P-REX function in an undefined manner (Hill et al. 2005; Waters et al. 2008). Once activated at the plasma membrane, P-REX enzymes catalyze the formation of functional Rac and are, therefore, critical for the regulation of a wide range of Rac-dependent cellular processes, including actin reorganization (Welch et al. 2002; Montero et al. 2011; Qin et al. 2009; Waters et al. 2008), cell migration, invasion and chemotaxis (Yoshizawa et al. 2005; Qin et al. 2009; Hernandez-Negrete et al. 2007; Ledezma-Sanchez et al. 2010; Li et al. 2005), gene transcription (Li et al. 2005) and ROS and superoxide formation (Wang et al. 2008; Welch et al. 2005; Dong et al. 2005; Nie et al. 2010).

P-REX1 is expressed most abundantly in peripheral blood leukocytes (Welch et al. 2002), however the P-REX2 isoforms are notably absent from circulating leukocytes (Donald et al. 2004; Rosenfeldt et al. 2004). Mouse models of altered P-Rex1 expression, including *P-Rex1*$^{-/-}$ mice and a transgenic mouse expressing a Rac-GEF-dead P-Rex1 mutant, while viable and healthy, exhibit impaired neutrophil and macrophage function, including decreased Rac1 and Rac2 activation, diminished ROS and superoxide formation and decreased chemotaxis and recruitment to inflammation (Welch et al. 2005; Dong et al. 2005; Wang et al. 2008). Interestingly, both P-Rex1 and P-Rex2a are highly expressed in the brain (Welch et al. 2005; Yoshizawa et al. 2005; Donald et al. 2008) and P-Rex1 negatively regulates neurite differentiation, elongation and migration of rat pheochromocytoma PC12 and primary hippocampal and cerebral cortical neurons through the regulation of actin reorganization (Waters et al. 2008; Yoshizawa et al. 2005).

P-Rex2$^{-/-}$ mice, like *P-Rex1*$^{-/-}$ mice, are viable, fertile and healthy, however they exhibit abnormal Purkinje cell development with decreased dendritic diameter and length, associated with impaired motor coordination, which is more severe in females and worsens with age (Donald et al. 2008). Interestingly, *P-Rex1*$^{-/-}$ mice do not exhibit the same differences in Purkinje cell integrity or motor skills as *P-Rex2*$^{-/-}$ mice, however, depletion of both P-Rex1 and P-Rex2 (*P-Rex1*$^{-/-}$*P-Rex2*$^{-/-}$) exacerbates the phenotype observed in single *P-Rex2*$^{-/-}$ animals (Donald et al. 2008), attributed to reduced sustained post-synaptic long term potentiation (LTP) in the cerebellum (Jackson et al. 2010).

7.3.3.1 P-REX1 Regulation of the PI3K Pathway and Disease Associations

The P-REX enzymes are also critical for the regulation of phosphoinositide signaling, even in the absence of phosphoinositide phosphatase activity, and are implicated in human diseases associated with PI3K pathway alterations, including cancer and diabetes. These enzymes achieve their phosphoinositide-regulatory function indirectly through domains distinct from the 4-phosphatase domain. For example, the DH/PH domain of P-REX2a binds the 3-phosphatase and tumor suppressor, PTEN, negatively regulating the 3-phosphatase activity of this enzyme toward its substrate, PtdIns(3,4,5)P_3 (Fine et al. 2009). Significantly, overexpression of P-REX2a in breast cancer cells decreases PTEN phosphoinositide phosphatase activity, associated with increased Akt phosphorylation and cell proliferation and rescues PTEN-induced suppression of Akt phosphorylation. Conversely, P-REX2a deficiency decreases Akt activation and cell proliferation in cells expressing PTEN. Interestingly, the inhibitory function of P-REX2a on PTEN is dependent on its binding ability, but independent of its Rac-GEF activity (Fine et al. 2009). As a negative regulator of PTEN activity, it is not surprising that P-REX2 is implicated in human cancers. The *P-REX2* gene is located at 8q13, a region frequently amplified in breast, prostate and ovarian cancers (Fejzo et al. 1998; Dimova et al. 2009; Sun et al. 2007). In breast cancer, amplified

P-REX2a expression inversely correlates with PTEN loss and is frequently observed in breast cancers expressing mutant PI3K, and overexpression of both P-REX2a and mutant PI3K drives mammary cell transformation (Fine et al. 2009).

Utilizing an alternative mechanism, the P-REX1 isoform can also function to potentiate PI3K/Akt signaling through activation of Rac1, which, in turn, promotes Akt phosphorylation resulting in reciprocal activation of Rac1, in a positive feedback loop (Nie et al. 2010). This relationship is dependent on the Rac-GEF activity of P-REX1 and the kinase activity of Akt. Significantly, amplified P-REX1 expression is associated with decreased disease-free survival in human breast cancer (Montero et al. 2011) and metastatic prostate cancer (Qin et al. 2009). Interestingly, P-REX1 may also play a role in the promotion of angiogenesis, potentially facilitating tumor growth and metastasis (Carretero-Ortega et al. 2010). Conversely, P-REX1 protein depletion in breast and prostate cancer cells decreases Rac activation, lamellipodia formation, cell motility, invasion, proliferation and xenograft tumor growth and metastasis (Montero et al. 2011; Qin et al. 2009). Recently P-REX1 was identified as an essential mediator of ErbB2 signaling in breast cancer (Sosa et al. 2010). P-REX1 is highly overexpressed in breast cancers. ErbB2 and GPCR signaling converges on P-REX1 to facilitate Rac activation. Interestingly, P-REX1 is also implicated in diabetes. The gene encoding P-REX1 is located on chromosome 20q12-13, which is linked to Type 2 diabetes, and *P-REX1* may be a type 2 diabetes-susceptibility gene (Bento et al. 2008; Lewis et al. 2010). The precise mechanisms by which the P-REX enzymes function in health and disease are, however, largely unknown and are the focus of current research.

7.4 Sac Domain Phosphoinositide Phosphatases

Sac domain phosphoinositide phosphatases are characterized by the presence of a conserved Sac phosphatase domain that was first identified in the founding member, yeast suppressor of actin (ySac1) (Novick et al. 1989; Guo et al. 1999). The Sac domain exhibits broad specificity for phosphoinositide substrates, and its intrinsic catalytic $CX_5R(S/T)$ motif can hydrolyze both the mono-phosphorylated phosphoinositides, PtdIns(3)P, PtdIns(4)P and PtdIns(5)P, in addition to dual-phosphorylated PtdIns(3,5)P_2 (Guo et al. 1999). The Sac phosphatase domain comprises ~ 400 amino acids arranged into 7 highly conserved motifs (Guo et al. 1999) and is unique in its structure and mechanism of substrate dephosphorylation, possibly involving the presence of a self-regulatory SacN domain (Manford et al. 2010). The functions of the Sac phosphatases are varied, and alterations in their expression are implicated in a range of diseases including cardiac hypertrophy and neurodegenerative disorders.

Generally, Sac domain phosphatases can be divided into two classes. The first class comprises the stand-alone Sac phosphatases that contain a conserved amino-terminal Sac domain with no other identifiable motifs, and examples include yeast ySac1 and yFig4, and human SAC1, SAC2/INPP5F and SAC3/FIG4 (Table 7.3). The second class is the Sac domain-containing inositol phosphatases (SCIPs), which comprise

Table 7.3 Mammalian Sac domain phosphatases

Protein name	Alias(es)	Gene	Animal models	References
SAC1		SAC1	Not reported	
SAC2	INPP5F	INPP5F	Inpp5f$^{-/-}$ constitutive KO mouse—normal development, increased stress-induced cardiac hypertrophy	(Zhu et al. 2009)
			Sac2 transgenic mouse—normal development, resistant to stress-induced hypertrophy	(Zhu et al. 2009)
SAC3	FIG4	FIG4	Fig4$^{-/-}$ constitutive KO mouse—"pale tremor", severe tremor, abnormal gait, neurodegeneration, juvenile lethality	(Chow et al. 2007; Zhang et al. 2008; Ferguson et al. 2009)

both a Sac phosphatase domain in addition to a central inositol polyphosphate 5-phosphatase catalytic domain, and examples include human synaptojanin 1 and 2, and the yeast phosphatases, Inp51, Inp52 and Inp53, reviewed in (Hughes et al. 2000). While in most cases the Sac domain alone is incapable of hydrolyzing inositol headgroups with adjacent phosphates, such as PtdIns(3,4)P_2 and PtdIns(4,5)P_2, the 5-phosphatase domain of SCIPs allows this class of enzyme to utilize PtdIns(4,5)P_2 as a substrate (Guo et al. 1999). Here, we have outlined the major functions of the yeast and mammalian stand-alone Sac phosphatases, with particular emphasis on the role these enzymes play in mammalian cells and in human disease.

7.4.1 SAC1

The stand alone Sac phosphatase, ySac1, was the first enzyme of this family to be identified in the yeast *Saccharomyces cerevisiae* (*S. cerevisiae*) as a regulator of actin organization and membrane trafficking events through the Golgi (Cleves et al. 1989; Novick et al. 1989; Whitters et al. 1993). In yeast, loss of ySac1 results in a range of cellular defects, including actin cytoskeletal disorganization and cold sensitivity for growth (Novick et al. 1989; Cleves et al. 1989), abnormal vacuole formation and trafficking (Foti et al. 2001; Tahirovic et al. 2005) and cell wall abnormalities (Schorr et al. 2001). Interestingly, however, ablation of ySac1 in yeast is not lethal. The function of Sac1 is evolutionarily conserved, with Sac1 orthologs identified as critical enzymes in a range of biological systems, including *D. melanogaster* (Wei et al. 2003; Yavari et al. 2010) and *Arabidopsis thaliana* (Despres et al. 2003; Zhong and Ye 2003; Zhong et al. 2005). SAC1 expression is essential for viability in mammalian systems and $Sac1^{-/-}$ mice exhibit pre-implantation lethality (Liu et al. 2008). Furthermore, SAC1 depletion in human cells results in decreased cell viability, inhibition of G_2-M cell cycle progression and suppressed cell growth rates (Liu et al. 2008; Cheong et al. 2010). Sac1 is ubiquitously expressed in adult and embryonic rat and mouse tissues (Nemoto et al. 2000; Liu et al. 2009), however, its expression is specifically up-regulated in the heart and regions of the brain during embryogenesis

(Liu et al. 2009), suggesting a critical tissue-specific role for this enzyme during development. Given the importance of SAC1 for cell viability and development, it is surprising that this enzyme has not yet been associated with any human disease.

7.4.1.1 Structure and Function

SAC1 localizes to the ER and Golgi compartments in both yeast and mammalian cells, anchored to the membranes by dual trans-membrane domains located in its carboxyl-terminus (Whitters et al. 1993). In quiescent cells, SAC1 resides predominantly in the Golgi, translocating to the ER upon growth factor stimulation (Blagoveshchenskaya et al. 2008; Cheong et al. 2010). While both yeast and human SAC1 exhibit Golgi-ER shuttling, the mechanism differs between species. In *S. cerevisiae*, retention of ySac1 in the ER is dependent on interactions between its carboxyl-terminal tail and the ER resident protein, Dpm1 (Dolichol phosphate mannosyltransferase) (Faulhammer et al. 2005), however, the mechanism for shuttling between intracellular compartments is not understood. In contrast, mammalian SAC1 forms oligomers in the ER via leucine zipper motifs located in its Sac domain, which interact with the coatomer protein complex II (COP-II) to promote its transport to the Golgi (Blagoveshchenskaya et al. 2008). Upon growth factor stimulation, activation of the mitogen activated protein kinase (MAPK) pathway promotes dissociation of SAC1 complexes to reveal a coatomer protein complex I (COP-I)-binding carboxyl-terminal "KXKXX" motif and induce retrograde trafficking back to the ER (Blagoveshchenskaya et al. 2008; Rohde et al. 2003; Liu et al. 2008). To date the precise function of SAC1 in distinct sub-cellular compartments remains poorly understood. SAC1 utilizes PtdIns(4)P as its predominant substrate, and depletion of this phosphatase in both yeast and mammalian cells results in increased cellular PtdIns(4)P levels (Guo et al. 1999; Nemoto et al. 2000; Rivas et al. 1999; Liu et al. 2008; Cheong et al. 2010). Sequestration of SAC1 in the ER following cell stimulation may allow accumulation of Golgi PtdIns(4)P, thereby promoting protein export to the plasma membrane (Blagoveshchenskaya et al. 2008). Conversely, Golgi-localized SAC1 may assist in the maintenance of normal Golgi organization through spatio-temporal regulation of PtdIns(4)P levels and its downstream effector proteins (Cheong et al. 2010). Indeed, RNAi-mediated SAC1 depletion results in disruption of Golgi integrity and distorted morphology, however, this does not affect overall rates of secretion (Liu et al. 2008; Cheong et al. 2010). Interestingly, however, expression of a Golgi-directed SAC1 mutant, resulting in PtdIns(4)P depletion, suppresses trafficking from this site (Szentpetery et al. 2010). Intriguingly, the ySac1 crystal structure suggests this phosphatase may function beyond its sub-cellular distribution, revealing the presence of a linker sequence located between the catalytic domain and the carboxyl-terminal trans-membrane domain, which may allow the Sac domain access to adjacent membranes even when embedded in the membranes of intracellular organelles (Manford et al. 2010). For example, ER-localized SAC1 could regulate phosphoinositides embedded in adjacent regions of the plasma membrane, however this function remains to be confirmed.

7.4.2 SAC2

The human SAC2 (gene name *INPP5F*) is an exceptional Sac phosphatase as it can hydrolyze the D5 position phosphate of PtdIns(4,5)P_2 and PtdIns(3,4,5)P_3, generating PtdIns(4)P and PtdIns(3,4)P_2 respectively (Minagawa et al. 2001). Although SAC2 contains a conserved catalytic phosphatase motif, it differs slightly from ySac1 in its surrounding modules, which may be responsible for the differential substrate specificities (Minagawa et al. 2001). Northern blot analysis indicates that this enzyme is ubiquitously expressed in human tissues, with highest levels detected in brain, heart, skeletal muscle, kidney and placenta (Minagawa et al. 2001). The heart-specific function of Sac2 has been characterized in mouse models with deficiency of this enzyme. Both *Inpp5f* knockout mice, and a histone deacetylase 2 (Hdac2) transgenic mice in which *Inpp5f* expression is transcriptionally repressed, exhibit stress-induced cardiac hypertrophy associated with elevated PtdIns(3,4,5)P_3 and hyper-phosphorylation of Akt and GSK3β (Trivedi et al. 2007; Zhu et al. 2009). Reciprocally, Sac2 transgenic mice are resistant to cardiac hypertrophy (Zhu et al. 2009), verifying the importance of this protein for normal cardiac myocyte function.

7.4.3 SAC3/Fig4

The related SAC3 (gene name *FIG4*) also serves a critical function in mammalian systems. Mutations in this enzyme are associated with human neuropathies. SAC3 is the human counterpart of the yFig4 protein (Rudge et al. 2004; Sbrissa et al. 2007), initially identified in *S. cerevisiae* as a pheromone-regulated or induced gene (Erdman et al. 1998). Both yFig4 and SAC3 utilize PtdIns(3,5)P_2 as their preferred *in vivo* substrate (Rudge et al. 2004; Sbrissa et al. 2007). Intriguingly, however, cellular depletion of these phosphatases results in a paradoxical suppression of PtdIns(3,5)P_2 levels, due to the loss of an activating interaction between the Sac phosphatases and the PtdIns(3,5)P_2-generating PtdIns(3)P 5-kinases, Fab1 in yeast and PIKfyve in mammalian cells (Duex et al. 2006a, 2006b; Gary et al. 2002; Rudge et al. 2004; Sbrissa et al. 2007; Chow et al. 2007; Botelho et al. 2008). This association is permitted by direct interactions between the phosphatases and the intermediate yeast Vac14 or mammalian ArPIKfyve proteins, allowing the formation of a stable ternary complex that promotes both activation of Fab1/PIKfyve, to generate PtdIns(3,5)P_2, in addition to stabilization of its agonist yFig4/SAC3 (Sbrissa et al. 2007; Ikonomov et al. 2010). This complex localizes to intracellular vesicular membranes to regulate PtdIns(3,5)P_2 production at these sites (Sbrissa et al. 2007; Rudge et al. 2004; Botelho et al. 2008).

7.4.3.1 SAC3 Function and Disease Associations

In the mouse, Sac3 protein and RNA is ubiquitously detected in all tissues, with highest expression in the brain, white adipose tissue and lung (Sbrissa et al. 2007; Chow et al. 2007). Sac3 may play a role in insulin-mediated GLUT4 translocation and glucose entry in mouse adipocytes, potentially implicating this enzyme in insulin resistance (Ikonomov et al. 2009). The rat Sac3 homolog is also important for neurite elongation in neuronal PC12 cells (Yuan et al. 2007). A naturally occurring Sac3 knockout mouse ($Fig4^{-/-}$), termed "pale tremor", exhibits severe tremor and abnormal gait, associated with neurodegeneration and juvenile lethality (Chow et al. 2007; Zhang et al. 2008; Ferguson et al. 2009). $Fig4^{-/-}$ derived fibroblasts and neurons exhibit enlarged endosomal and lysosomal compartments and impaired organelle trafficking (Chow et al. 2007; Ferguson et al. 2009) and brains from $Fig4^{-/-}$ mice exhibit inclusion bodies with markers of defective autophagy (Ferguson et al. 2009). The clinical and pathological features of the pale tremor mouse are reminiscent of human neuropathies, in particular Charcot-Marie-Tooth (CMT) disorder. Indeed, an autosomal recessive form of CMT, type 4J (CMT4J), is caused by a pathogenic mutation at amino acid 41 (Ile-to-Thr) in SAC3 (Chow et al. 2007). While the resulting mutant protein exhibits no change in its PtdIns(3,5)P_2 phosphatase activity, it is no longer stabilized by its interaction with ArPIKfyve and is thereby readily degraded (Ikonomov et al. 2010). In addition, deleterious mutations in the human *FIG4* gene are associated with the severe human neuropathy, amyotrophic lateral sclerosis (ALS) (Chow et al. 2009), further highlighting the importance of this enzyme in the regulation of neuronal function.

7.5 Concluding Remarks

The characterization of the phosphoinositide phosphatases has lagged behind the kinases but the results of the many recent studies described here reveal they play just as important a role in cellular function and human disease as the phosphoinositide kinases. There is likely to be many more exciting results that emerge in the next few years on the role the lipid phosphatases play in human diseases. Many phosphoinositide phosphatases exhibit altered expression in many different diseases as shown by gene array expression analysis but the functional consequences remain to be determined but are likely to be significant. Characterization of mouse knockout models of some of the lipid phosphatases is still emerging, which may give significant insights into phosphatase function. Furthermore most reports to date have characterized the function of individual lipid phosphatases in isolation and it will be a challenge for the future to determine how these enzymes work together to regulate the complex interactive phosphoinositide signaling pathway.

Locus	Common name (s)	SwissProt/Protein [UniProt]	Accession nr. Hs	Reference MIM	Gene map Hs/Mm
5-phosphatases					
INPP5A	IP5-P-1 Type I IP5-P 43 kDa IP5-P	*INPP5A Human*	*NM_005539.3*	MIM 600106	*10q26.3* *7 F4*
INPP5B	IP5-P-2 75 kDa IP5-P	*INPP5B Human*	*NM_005540.2*	MIM 147264	*1p34* *4 D2*
INPP5D	SHIP SHIP-1 IP5-P D	*SHIP1 Human*	*NM_001017915.1*	MIM 601582	*2q37.1* *1 C5*
INPPL1	SHIP-2 SHIP2 51C protein	*SHIP2 Human*	*NM_001567.3*	MIM 600829	*11q13* *7 F1*
INPP5E	Pharbin Type IV IP5-P IP5-P E 72 kDa IP5-P	*INPP5E Human*	*NM_019892.4*	MIM 613037	*9q34.3* *2 A3*
OCRL	Lowe oculocerebrorenal syndrome protein IP5-P F OCRL INPP5F	*OCRL-1 Human*	*NM_000276.3*	MIM 300535	*Xq25* *X A4*
INPP5G SYNJ1	Synaptojanin-1 SJ1, SYNJ1	*Synaptojanin 1 Human*	*NM 003895.3*	MIM 604297	*21q22.2* *16 C3-C4*
INPP5H SYNJ2	Synaptojanin-2 SYNJ2	*Synaptojanin 2 Human*	*NM 003898.3*	MIM 609410	*6q25.3* *17 A2-A3*
INPP5J PIPP	INPP5J PIB5PA Phosphatidylinositol (4,5) bisphosphate 5-phosphatase A	*PIPP Human*	*NM 001 002837.1*	MIM 606481	*22q12.2* *11 A1*

(continued)

Locus	Common name (s)	SwissProt/Protein [UniProt]	Accession nr. Hs	Reference MIM	Gene map Hs/Mm
INPP5K SKIP	SKIP	SKIP Human	NM 016532.3	MIM 607875	17p13.3 11 B5
4-phosphatases					
INPP4A	Inositol polyphosphate 4-phosphatase type I	Type I 4-phosphatase Human	NM 004027.2	MIM 600916	2q11.2 1 B
INPP4B	Inositol polyphosphate 4-phosphatase type II	Type II 4-phosphatase Human	NM 003866.2	MIM 607494	4q31.21 8 C2
TMEM55A	PtdIns(4,5)P₂ 4-phosphatase type I	Type I PtdIns(4,5)P2 4-phosphatase	NM 018710.2	MIM 609864	8q21.3 4 A1-A2
TMEM55B	PtdIns(4,5)P₂ 4-phosphatase type II	Type II PtdIns(4,5)P2 4-phosphatase	NM 001100814.1	MIM 609865	14q11.2 14 C1
P-Rex1	P-Rex1	P-Rex1 Human	NM 020820.3	MIM 606905	20q13.13 2 H3
P-Rex2	P-Rex2a P-Rex2b	P-Rex2 Human	NM 024870.2	MIM 612139	8q13.2 1 A3
Sac phosphatases					
SACM1L	SAC1	SAC1 Human	NM 014016	MIM 606569	3p21.3 9 F
INPP5F	SAC2	SAC2 Human	NM 014937.2	MIM 609389	10q26.11 7 F3
FIG4	SAC3	SAC3 Human	NM 014845.5	MIM 609390	6q21 10 B1

*IP5-P (inositol polyphosphate 5-phosphatase)

References

Adayev T, Chen-Hwang MC, Murakami N, Wang R, Hwang YW (2006) MNB/DYRK1A phosphorylation regulates the interactions of synaptojanin 1 with endocytic accessory proteins. Biochem Biophys Res Commun 351:1060–1065

Addis M, Loi M, Lepiani C, Cau M, Melis MA (2004) OCRL mutation analysis in Italian patients with Lowe syndrome. Hum Mutat 23:524–525

Ahn SJ, Han SJ, Mo HJ, Chung JK, Hong SH, Park TK, Kim CG (1998) Interaction of phospholipase C gamma 1 via its COOH-terminal SRC homology 2 domain with synaptojanin. Biochem Biophys Res Commun 244:62–67

Ai J, Maturu A, Johnson W, Wang Y, Marsh CB, Tridandapani S (2006) The inositol phosphatase SHIP-2 down-regulates FcgammaR-mediated phagocytosis in murine macrophages independently of SHIP-1. Blood 107:813–820

Akada M, Crnogorac-Jurcevic T, Lattimore S, Mahon P, Lopes R, Sunamura M, Matsuno S, Lemoine NR (2005) Intrinsic chemoresistance to gemcitabine is associated with decreased expression of BNIP3 in pancreatic cancer. Clin Cancer Res 11:3094–3101

Alinikula J, Kohonen P, Nera KP, Lassila O (2010) Concerted action of Helios and Ikaros controls the expression of the inositol 5-phosphatase SHIP. Eur J Immunol 40:2599–2607

Allaoui A, Menard R, Sansonetti PJ, Parsot C (1993) Characterization of the Shigella flexneri ipgD and ipgF genes, which are located in the proximal part of the mxi locus. Infect Immun 61:1707–1714

Allione F, Eisinger F, Parc P, Noguchi T, Sobol H, Birnbaum D (1998) Loss of heterozygosity at loci from chromosome arm 22Q in human sporadic breast carcinomas. Int J Cancer 75:181–186

Antignano F, Ibaraki M, Kim C, Ruschmann J, Zhang A, Helgason CD, Krystal G (2010) SHIP is required for dendritic cell maturation. J Immunol 184:2805–2813

Aoki K, Nakamura T, Inoue T, Meyer T, Matsuda M (2007) An essential role for the SHIP2-dependent negative feedback loop in neuritogenesis of nerve growth factor-stimulated PC12 cells. J Cell Biol 177:817–827

Arai Y, Ijuin T, Takenawa T, Becker LE, Takashima S (2002) Excessive expression of synaptojanin in brains with Down syndrome. Brain Dev 24:67–72

Asano T, Mochizuki Y, Matsumoto K, Takenawa T, Endo T (1999) Pharbin, a novel inositol polyphosphate 5-phosphatase, induces dendritic appearances in fibroblasts. Biochem Biophys Res Commun 261:188–195

Astle MV, Horan KA, Ooms LM, Mitchell CA (2007) The inositol polyphosphate 5-phosphatases: traffic controllers, waistline watchers and tumour suppressors? Biochem Soc Symp 74:161–181

Attree O, Olivos IM, Okabe I, Bailey LC, Nelson DL, Lewis RA, McInnes RR, Nussbaum RL (1992) The Lowe's oculocerebrorenal syndrome gene encodes a protein highly homologous to inositol polyphosphate-5-phosphatase. Nature 358:239–242

Bai L, Rohrschneider LR (2010) s-SHIP promoter expression marks activated stem cells in developing mouse mammary tissue. Genes Dev 24:1882–1892

Bakowski MA, Braun V, Lam GY, Yeung T, Heo WD, Meyer T, Finlay BB, Grinstein S, Brumell JH (2010) The phosphoinositide phosphatase SopB manipulates membrane surface charge and trafficking of the Salmonella-containing vacuole. Cell Host Microbe 7:453–462

Baltimore D, Boldin MP, O'Connell RM, Rao DS, Taganov KD (2008) MicroRNAs: new regulators of immune cell development and function. Nat Immunol 9:839–845

Bansal V, Inhorn R, Majerus P (1987) The metabolism of inositol 1,3,4-trisphosphate to inositol 1,3-bisphosphate. J Biol Chem 262:9444–9447

Bansal V, Caldwell K, Majerus P (1990) The isolation and characterization of inositol polyphosphate 4-phosphatase. J Biol Chem 265:1806–1811

Baran CP, Tridandapani S, Helgason CD, Humphries RK, Krystal G, Marsh CB (2003) The inositol 5′-phosphatase SHIP-1 and the Src kinase Lyn negatively regulate macrophage colony-stimulating factor-induced Akt activity. J Biol Chem 278:38628–38636

Barber MA, Donald S, Thelen S, Anderson KE, Thelen M, Welch HC (2007) Membrane translocation of P-Rex1 is mediated by G protein betagamma subunits and phosphoinositide 3-kinase. J Biol Chem 282:29967–29976

Barnache S, Le Scolan E, Kosmider O, Denis N, Moreau-Gachelin F (2006) Phosphatidylinositol 4-phosphatase type II is an erythropoietin-responsive gene. Oncogene 25:1420–1423

Bauerfeind R, Takei K, De Camilli P (1997) Amphiphysin I is associated with coated endocytic intermediates and undergoes stimulation-dependent dephosphorylation in nerve terminals. J Biol Chem 272:30984–30992

Beer DG, Kardia SL, Huang CC, Giordano TJ, Levin AM, Misek DE, Lin L, Chen G, Gharib TG, Thomas DG, Lizyness ML, Kuick R, Hayasaka S, Taylor JM, Iannettoni MD, Orringer MB, Hanash S (2002) Gene-expression profiles predict survival of patients with lung adenocarcinoma. Nat Med 8:816–824

Bento JL, Palmer ND, Zhong M, Roh B, Lewis JP, Wing MR, Pandya H, Freedman BI, Langefeld CD, Rich SS, Bowden DW, Mychaleckyj JC (2008) Heterogeneity in gene loci associated with type 2 diabetes on human chromosome 20q13.1. Genomics 92:226–234

Bernard DJ, Nussbaum RL (2010) X-inactivation analysis of embryonic lethality in Ocrl wt/-; Inpp5b-/- mice. Mamm Genome 21:186–194

Berridge MJ, Irvine RF (1989) Inositol phosphates and cell signalling. Nature 341:197–205

Bertelli DF, Araujo EP, Cesquini M, Stoppa GR, Gasparotto-Contessotto M, Toyama MH, Felix JV, Carvalheira JB, Michelini LC, Chiavegatto S, Boschero AC, Saad MJ, Lopes-Cendes I, Velloso LA (2006) Phosphoinositide-specific inositol polyphosphate 5-phosphatase IV inhibits inositide trisphosphate accumulation in hypothalamus and regulates food intake and body weight. Endocrinology 147:5385–5399

Bertoletti A, Gehring A (2007) Immune response and tolerance during chronic hepatitis B virus infection. Hepatol Res 37(Suppl 3):S331–S338

Biegel JA, Burk CD, Barr FG, Emanuel BS (1992) Evidence for a 17p tumor related locus distinct from p53 in pediatric primitive neuroectodermal tumors. Cancer Res 52:3391–3395

Bielas SL, Silhavy JL, Brancati F, Kisseleva MV, Al-Gazali L, Sztriha L, Bayoumi RA, Zaki MS, Abdel-Aleem A, Rosti RO, Kayserili H, Swistun D, Scott LC, Bertini E, Boltshauser E, Fazzi E, Travaglini L, Field SJ, Gayral S, Jacoby M, Schurmans S, Dallapiccola B, Majerus PW, Valente EM, Gleeson JG (2009) Mutations in INPP5E, encoding inositol polyphosphate-5-phosphatase E, link phosphatidyl inositol signaling to the ciliopathies. Nat Genet 41:1032–1036

Bisgaard AM, Kirchhoff M, Nielsen JE, Brandt C, Hove H, Jepsen B, Jensen T, Ullmann R, Skovby F (2007) Transmitted cytogenetic abnormalities in patients with mental retardation: pathogenic or normal variants? Eur J Med Genet 50:243–255

Blagoveshchenskaya A, Cheong FY, Rohde HM, Glover G, Knodler A, Nicolson T, Boehmelt G, Mayinger P (2008) Integration of Golgi trafficking and growth factor signaling by the lipid phosphatase SAC1. J Cell Biol 180:803–812

Bokenkamp A, Bockenhauer D, Cheong HI, Hoppe B, Tasic V, Unwin R, Ludwig M (2009) Dent-2 disease: a mild variant of Lowe syndrome. J Pediatr 155:94–99

Bond J, Scott S, Hampshire DJ, Springell K, Corry P, Abramowicz MJ, Mochida GH, Hennekam RC, Maher ER, Fryns JP, Alswaid A, Jafri H, Rashid Y, Mubaidin A, Walsh CA, Roberts E, Woods CG (2003) Protein-truncating mutations in ASPM cause variable reduction in brain size. Am J Hum Genet 73:1170–1177

Botelho RJ, Efe JA, Teis D, Emr SD (2008) Assembly of a Fab1 phosphoinositide kinase signaling complex requires the Fig4 phosphoinositide phosphatase. Mol Biol Cell 19:4273–4286

Boyle EC, Brown NF, Finlay BB (2006) Salmonella enterica serovar Typhimurium effectors SopB, SopE, SopE2 and SipA disrupt tight junction structure and function. Cell Microbiol 8:1946–1957

Brauweiler A, Tamir I, Dal Porto J, Benschop RJ, Helgason CD, Humphries RK, Freed JH, Cambier JC (2000a) Differential regulation of B cell development, activation, and death by the src homology 2 domain-containing 5' inositol phosphatase (SHIP). J Exp Med 191:1545–1554

Brauweiler AM, Tamir I, Cambier JC (2000b) Bilevel control of B-cell activation by the inositol 5-phosphatase SHIP. Immunol Rev 176:69–74

Brauweiler A, Merrell K, Gauld SB, Cambier JC (2007) Cutting edge: acute and chronic exposure of immature B cells to antigen leads to impaired homing and SHIP1-dependent reduction in stromal cell-derived factor-1 responsiveness. J Immunol 178:3353–3357

Bruhns P, Vely F, Malbec O, Fridman WH, Vivier E, Daeron M (2000) Molecular basis of the recruitment of the SH2 domain-containing inositol 5-phosphatases SHIP1 and SHIP2 by fcgamma RIIB. J Biol Chem 275:37357–37364

Bruno DL, Anderlid BM, Lindstrand A, Van Ravenswaaij-Arts C, Ganesamoorthy D, Lundin J, Martin CL, Douglas J, Nowak C, Adam MP, Kooy RF, Van Der AAN, Reyniers E, Vandeweyer G, Stolte-Dijkstra I, Dijkhuizen T, Yeung A, Delatycki M, Borgstrom B, Thelin L, Cardoso C, Van Bon B, Pfundt R, De Vries BB, Wallin A, Amor DJ, James PA, Slater HR, Schoumans J (2010) Further molecular and clinical delineation of co-locating 17p13.3 microdeletions and microduplications that show distinctive phenotypes. J Med Genet 47:299–311

Bruyns C, Pesesse X, Moreau C, Blero D, Erneux C (1999) The two SH2-domain-containing inositol 5-phosphatases SHIP1 and SHIP2 are coexpressed in human T lymphocytes. Biol Chem 380:969–974

Buettner R, Ottinger I, Gerhardt-Salbert C, Wrede CE, Scholmerich J, Bollheimer LC (2007) Antisense oligonucleotides against the lipid phosphatase SHIP2 improve muscle insulin sensitivity in a dietary rat model of the metabolic syndrome. Am J Physiol Endocrinol Metab 292:E1871–E1878

Campbell JK, Gurung R, Romero S, Speed CJ, Andrews RK, Berndt MC, Mitchell CA (1997) Activation of the 43 kDa inositol polyphosphate 5-phosphatase by 14-3-3zeta. Biochemistry 36:15363–15370

Cardoso C, Leventer RJ, Ward HL, Toyo-Oka K, Chung J, Gross A, Martin CL, Allanson J, Pilz DT, Olney AH, Mutchinick OM, Hirotsune S, Wynshaw-Boris A, Dobyns WB, Ledbetter DH (2003) Refinement of a 400-kb critical region allows genotypic differentiation between isolated lissencephaly, Miller-Dieker syndrome, and other phenotypes secondary to deletions of 17p13.3. Am J Hum Genet 72:918–930

Carretero-Ortega J, Walsh CT, Hernandez-Garcia R, Reyes-Cruz G, Brown JH, Vazquez-Prado J (2010) Phosphatidylinositol 3,4,5-triphosphate-dependent Rac exchanger 1 (P-Rex-1), a guanine nucleotide exchange factor for Rac, mediates angiogenic responses to stromal cell-derived factor-1/chemokine stromal cell derived factor-1 (SDF-1/CXCL-12) linked to Rac activation, endothelial cell migration, and in vitro angiogenesis. Mol Pharmacol 77:435–442

Castells A, Gusella JF, Ramesh V, Rustgi AK (2000) A region of deletion on chromosome 22q13 is common to human breast and colorectal cancers. Cancer Res 60:2836–2839

Chang KT, Min KT (2009) Upregulation of three Drosophila homologs of human chromosome 21 genes alters synaptic function: implications for Down syndrome. Proc Natl Acad Sci USA 106:17117–17122

Checroun C, Wehrly TD, Fischer ER, Hayes SF, Celli J (2006) Autophagy-mediated reentry of Francisella tularensis into the endocytic compartment after cytoplasmic replication. Proc Natl Acad Sci USA 103:14578–14583

Cheon MS, Kim SH, Ovod V, Kopitar Jerala N, Morgan JI, Hatefi Y, Ijuin T, Takenawa T, Lubec G (2003) Protein levels of genes encoded on chromosome 21 in fetal Down syndrome brain: challenging the gene dosage effect hypothesis (Part III). Amino Acids 24:127–134

Cheong FY, Sharma V, Blagoveshchenskaya A, Oorschot VM, Brankatschk B, Klumperman J, Freeze HH, Mayinger P (2010) Spatial regulation of Golgi phosphatidylinositol-4-phosphate is required for enzyme localization and glycosylation fidelity. Traffic 11:1180–1190

Chi Y, Zhou B, Wang WQ, Chung SK, Kwon YU, Ahn YH, Chang YT, Tsujishita Y, Hurley JH, Zhang ZY (2004) Comparative mechanistic and substrate specificity study of inositol polyphosphate 5-phosphatase Schizosaccharomyces pombe Synaptojanin and SHIP2. J Biol Chem 279:44987–44995

Cho C, Ge H, Branciforte D, Primakoff P, Myles DG (2000) Analysis of mouse fertilin in wild-type and fertilin beta(-/-) sperm: evidence for C-terminal modification, alpha/beta dimerization, and lack of essential role of fertilin alpha in sperm-egg fusion. Dev Biol 222:289–295

Cho HY, Lee BH, Choi HJ, Ha IS, Choi Y, Cheong HI (2008) Renal manifestations of Dent disease and Lowe syndrome. Pediatr Nephrol 23:243–249

Chong A, Wehrly TD, Nair V, Fischer ER, Barker JR, Klose KE, Celli J (2008) The early phagosomal stage of Francisella tularensis determines optimal phagosomal escape and Francisella pathogenicity island protein expression. Infect Immun 76:5488–5499

Choudhury R, Diao A, Zhang F, Eisenberg E, Saint-Pol A, Williams C, Konstantakopoulos A, Lucocq J, Johannes L, Rabouille C, Greene LE, Lowe M (2005) Lowe syndrome protein OCRL1 interacts with clathrin and regulates protein trafficking between endosomes and the trans-Golgi network. Mol Biol Cell 16:3467–3479

Chow KU, Nowak D, Kim SZ, Schneider B, Komor M, Boehrer S, Mitrou PS, Hoelzer D, Weidmann E, Hofmann WK (2006) In vivo drug-response in patients with leukemic non-Hodgkin's lymphomas is associated with in vitro chemosensitivity and gene expression profiling. Pharmacol Res 53:49–61

Chow CY, Zhang Y, Dowling JJ, Jin N, Adamska M, Shiga K, Szigeti K, Shy ME, Li J, Zhang X, Lupski JR, Weisman LS, Meisler MH (2007) Mutation of FIG4 causes neurodegeneration in the pale tremor mouse and patients with CMT4J. Nature 448:68–72

Chow CY, Landers JE, Bergren SK, Sapp PC, Grant AE, Jones JM, Everett L, Lenk GM, McKenna-Yasek DM, Weisman LS, Figlewicz D, Brown RH, Meisler MH (2009) Deleterious variants of FIG4, a phosphoinositide phosphatase, in patients with ALS. Am J Hum Genet 84:85–88

Chuang YY, Tran NL, Rusk N, Nakada M, Berens ME, Symons M (2004) Role of synaptojanin 2 in glioma cell migration and invasion. Cancer Res 64:8271–8275

Clement S, Krause U, Desmedt F, Tanti JF, Behrends J, Pesesse X, Sasaki T, Penninger J, Doherty M, Malaisse W, Dumont JE, Le Marchand-Brustel Y, Erneux C, Hue L, Schurmans S (2001) The lipid phosphatase SHIP2 controls insulin sensitivity. Nature 409:92–97

Cleves AE, Novick PJ, Bankaitis VA (1989) Mutations in the SAC1 gene suppress defects in yeast Golgi and yeast actin function. J Cell Biol 109:2939–2950

Coon BG, Mukherjee D, Hanna CB, Riese DJ 2nd, Lowe M, Aguilar RC (2009) Lowe syndrome patient fibroblasts display Ocrl1-specific cell migration defects that cannot be rescued by the homologous Inpp5b phosphatase. Hum Mol Genet 18:4478–4491

Corcione A, Arduino N, Ferretti E, Pistorio A, Spinelli M, Ottonello L, Dallegri F, Basso G, Pistoia V (2006) Chemokine receptor expression and function in childhood acute lymphoblastic leukemia of B-lineage. Leuk Res 30:365–372

Cormand B, Avela K, Pihko H, Santavuori P, Talim B, Topaloglu H, De La Chapelle A, Lehesjoki AE (1999) Assignment of the muscle-eye-brain disease gene to 1p32-p34 by linkage analysis and homozygosity mapping. Am J Hum Genet 64:126–135

Cornelis RS, Van Vliet M, Vos CB, Cleton-Jansen AM, Van de Vijver MJ, Peterse JL, Khan PM, Borresen AL, Cornelisse CJ, Devilee P (1994) Evidence for a gene on 17p13.3, distal to TP53, as a target for allele loss in breast tumors without p53 mutations. Cancer Res 54:4200–4206

Costinean S, Sandhu SK, Pedersen IM, Tili E, Trotta R, Perrotti D, Ciarlariello D, Neviani P, Harb J, Kauffman LR, Shidham A, Croce CM (2009) Src homology 2 domain-containing inositol-5-phosphatase and CCAAT enhancer-binding protein beta are targeted by miR-155 in B cells of Emicro-MiR-155 transgenic mice. Blood 114:1374–1382

Cox D, Dale BM, Kashiwada M, Helgason CD, Greenberg S (2001) A regulatory role for Src homology 2 domain-containing inositol 5'-phosphatase (SHIP) in phagocytosis mediated by Fc gamma receptors and complement receptor 3 (alpha(M)beta(2); CD11b/CD18). J Exp Med 193:61–71

Cremer TJ, Ravneberg DH, Clay CD, Piper-Hunter MG, Marsh CB, Elton TS, Gunn JS, Amer A, Kanneganti TD, Schlesinger LS, Butchar JP, Tridandapani S (2009) MiR-155 induction by F. novicida but not the virulent F. tularensis results in SHIP down-regulation and enhanced pro-inflammatory cytokine response. PLoS One 4:e8508

Cremona O, Di Paolo G, Wenk MR, Luthi A, Kim WT, Takei K, Daniell L, Nemoto Y, Shears SB, Flavell RA, McCormick DA, De Camilli P (1999) Essential role of phosphoinositide metabolism in synaptic vesicle recycling. Cell 99:179–188

Crowley JE, Stadanlick JE, Cambier JC, Cancro MP (2009) FcgammaRIIB signals inhibit BLyS signaling and BCR-mediated BLyS receptor up-regulation. Blood 113:1464–1473

De Smedt F, Verjans B, Mailleux P, Erneux C (1994) Cloning and expression of human brain type I inositol 1,4,5-trisphosphate 5-phosphatase. High levels of mRNA in cerebellar Purkinje cells. FEBS Lett 347:69–72

De Smedt F, Boom A, Pesesse X, Schiffmann SN, Erneux C (1996) Post-translational modification of human brain type I inositol-1,4,5-trisphosphate 5-phosphatase by farnesylation. J Biol Chem 271:10419–10424

De Smedt F, Missiaen L, Parys JB, Vanweyenberg V, De Smedt H, Erneux C (1997) Isoprenylated human brain type I inositol 1,4,5-trisphosphate 5-phosphatase controls Ca2+ oscillations induced by ATP in Chinese hamster ovary cells. J Biol Chem 272:17367–17375

Denley A, Gymnopoulos M, Kang S, Mitchell C, Vogt PK (2009) Requirement of phosphatidylinositol(3,4,5)trisphosphate in phosphatidylinositol 3-kinase-induced oncogenic transformation. Mol Cancer Res 7:1132–1138

Dent CE, Friedman M (1964) Hypercalcuric rickets associated with renal tubular damage. Arch Dis Child 39:240–249

Despres B, Bouissonnie F, Wu HJ, Gomord V, Guilleminot J, Grellet F, Berger F, Delseny M, Devic M (2003) Three SAC1-like genes show overlapping patterns of expression in Arabidopsis but are remarkably silent during embryo development. Plant J 34:293–306

Dhanasekaran SM, Dash A, Yu J, Maine IP, Laxman B, Tomlins SA, Creighton CJ, Menon A, Rubin MA, Chinnaiyan AM (2005) Molecular profiling of human prostate tissues: insights into gene expression patterns of prostate development during puberty. FASEB J 19:243–245

Di Paolo G, Moskowitz HS, Gipson K, Wenk MR, Voronov S, Obayashi M, Flavell R, Fitzsimonds RM, Ryan TA, De Camilli P (2004) Impaired PtdIns(4,5)P2 synthesis in nerve terminals produces defects in synaptic vesicle trafficking. Nature 431:415–422

Dimova I, Orsetti B, Negre V, Rouge C, Ursule L, Lasorsa L, Dimitrov R, Doganov N, Toncheva D, Theillet C (2009) Genomic markers for ovarian cancer at chromosomes 1, 8 and 17 revealed by array CGH analysis. Tumori 95:357–366

Donald S, Hill K, Lecureuil C, Barnouin R, Krugmann S, John Coadwell W, Andrews SR, Walker SA, Hawkins PT, Stephens LR, Welch HC (2004) P-Rex2, a new guanine-nucleotide exchange factor for Rac. FEBS Lett 572:172–176

Donald S, Humby T, Fyfe I, Segonds-Pichon A, Walker SA, Andrews SR, Coadwell WJ, Emson P, Wilkinson LS, Welch HC (2008) P-Rex2 regulates Purkinje cell dendrite morphology and motor coordination. Proc Natl Acad Sci USA 105:4483–4488

Dong X, Mo Z, Bokoch G, Guo C, Li Z, Wu D (2005) P-Rex1 is a primary Rac2 guanine nucleotide exchange factor in mouse neutrophils. Curr Biol 15:1874–1879

Dong S, Corre B, Foulon E, Dufour E, Veillette A, Acuto O, Michel F (2006) T cell receptor for antigen induces linker for activation of T cell-dependent activation of a negative signaling complex involving Dok-2, SHIP-1, and Grb-2. J Exp Med 203:2509–2518

Duex JE, Nau JJ, Kauffman EJ, Weisman LS (2006a) Phosphoinositide 5-phosphatase Fig 4p is required for both acute rise and subsequent fall in stress-induced phosphatidylinositol 3,5-bisphosphate levels. Eukaryot Cell 5:723–731

Duex JE, Tang F, Weisman LS (2006b) The Vac14p-Fig4p complex acts independently of Vac7p and couples PI3,5P2 synthesis and turnover. J Cell Biol 172:693–704

Dyson JM, O'Malley CJ, Becanovic J, Munday AD, Berndt MC, Coghill ID, Nandurkar HH, Ooms LM, Mitchell CA (2001) The SH2-containing inositol polyphosphate 5-phosphatase, SHIP-2, binds filamin and regulates submembraneous actin. J Cell Biol 155:1065–1079

Dyson JM, Munday AD, Kong AM, Huysmans RD, Matzaris M, Layton MJ, Nandurkar HH, Berndt MC, Mitchell CA (2003) SHIP-2 forms a tetrameric complex with filamin, actin, and GPIb-IX-V: localization of SHIP-2 to the activated platelet actin cytoskeleton. Blood 102:940–948

Edmunds C, Parry RV, Burgess SJ, Reaves B, Ward SG (1999) CD28 stimulates tyrosine phosphorylation, cellular redistribution and catalytic activity of the inositol lipid 5-phosphatase SHIP. Eur J Immunol 29:3507–3515

Eissmann P, Beauchamp L, Wooters J, Tilton JC, Long EO, Watzl C (2005) Molecular basis for positive and negative signaling by the natural killer cell receptor 2B4 (CD244). Blood 105:4722–4729

Ellsworth RE, Ellsworth DL, Lubert SM, Hooke J, Somiari RI, Shriver CD (2003) High-throughput loss of heterozygosity mapping in 26 commonly deleted regions in breast cancer. Cancer Epidemiol Biomarkers Prev 12:915–919

Erdman S, Lin L, Malczynski M, Snyder M (1998) Pheromone-regulated genes required for yeast mating differentiation. J Cell Biol 140:461–483

Erdmann KS, Mao Y, McCrea HJ, Zoncu R, Lee S, Paradise S, Modregger J, Biemesderfer D, Toomre D, De Camilli P (2007) A role of the Lowe syndrome protein OCRL in early steps of the endocytic pathway. Dev Cell 13:377–390

Erkeland SJ, Valkhof M, Heijmans-Antonissen C, Van Hoven-Beijen A, Delwel R, Hermans MHA, Touw IP (2004) Large-scale identification of disease genes involved in acute myeloid leukemia. J Virol 78:1971–1980

Faucherre A, Desbois P, Nagano F, Satre V, Lunardi J, Gacon G, Dorseuil O (2005) Lowe syndrome protein Ocrl1 is translocated to membrane ruffles upon Rac GTPase activation: a new perspective on Lowe syndrome pathophysiology. Hum Mol Genet 14:1441–1448

Faulhammer F, Konrad G, Brankatschk B, Tahirovic S, Knodler A, Mayinger P (2005) Cell growth-dependent coordination of lipid signaling and glycosylation is mediated by interactions between Sac1p and Dpm1p. J Cell Biol 168:185–191

Fedele CG, Ooms LM, Ho M, Vieusseux J, O'Toole SA, Millar EK, Lopez-Knowles E, Sriratana A, Gurung R, Baglietto L, Giles GG, Bailey CG, Rasko JE, Shields BJ, Price JT, Majerus PW, Sutherland RL, Tiganis T, McLean CA, Mitchell CA (2010) Inositol polyphosphate 4-phosphatase II regulates PI3K/Akt signaling and is lost in human basal-like breast cancers. Proc Natl Acad Sci USA 107:22231–22236

Fejzo MS, Godfrey T, Chen C, Waldman F, Gray JW (1998) Molecular cytogenetic analysis of consistent abnormalities at 8q12-q22 in breast cancer. Genes Chromosomes Cancer 22:105–113

Feng Y, Wente SR, Majerus PW (2001) Overexpression of the inositol phosphatase SopB in human 293 cells stimulates cellular chloride influx and inhibits nuclear mRNA export. Proc Natl Acad Sci USA 98:875–879

Ferguson CJ, Lenk GM, Meisler MH (2009) Defective autophagy in neurons and astrocytes from mice deficient in PI(3,5)P2. Hum Mol Genet 18:4868–4878

Fine B, Hodakoski C, Koujak S, Su T, Saal LH, Maurer M, Hopkins B, Keniry M, Sulis ML, Mense S, Hibshoosh H, Parsons R (2009) Activation of the PI3K pathway in cancer through inhibition of PTEN by exchange factor P-REX2a. Science 325:1261–1265

Foti M, Audhya A, Emr SD (2001) Sac1 lipid phosphatase and Stt4 phosphatidylinositol 4-kinase regulate a pool of phosphatidylinositol 4-phosphate that functions in the control of the actin cytoskeleton and vacuole morphology. Mol Biol Cell 12:2396–2411

Franke TF, Kaplan DR, Cantley LC, Toker A (1997) Direct regulation of the Akt proto-oncogene product by phosphatidylinositol-3,4-bisphosphate. Science 275:665–668

Fukuda R, Hayashi A, Utsunomiya A, Nukada Y, Fukui R, Itoh K, Tezuka K, Ohashi K, Mizuno K, Sakamoto M, Hamanoue M, Tsuji T (2005) Alteration of phosphatidylinositol 3-kinase cascade in the multilobulated nuclear formation of adult T cell leukemia/lymphoma (ATLL). Proc Natl Acad Sci USA 102:15213–15218

Fukui K, Wada T, Kagawa S, Nagira K, Ikubo M, Ishihara H, Kobayashi M, Sasaoka T (2005) Impact of the liver-specific expression of SHIP2 (SH2-containing inositol 5′-phosphatase 2) on insulin signaling and glucose metabolism in mice. Diabetes 54:1958–1967

Gagnon A, Artemenko Y, Crapper T, Sorisky A (2003) Regulation of endogenous SH2 domain-containing inositol 5-phosphatase (SHIP2) in 3T3-L1 and human preadipocytes. J Cell Physiol 197:243–250

Galli SJ, Tsai M (2010) Mast cells in allergy and infection: versatile effector and regulatory cells in innate and adaptive immunity. Eur J Immunol 40:1843–1851

Galyov EE, Wood MW, Rosqvist R, Mullan PB, Watson PR, Hedges S, Wallis TS (1997) A secreted effector protein of Salmonella dublin is translocated into eukaryotic cells and mediates inflammation and fluid secretion in infected ileal mucosa. Mol Microbiol 25: 903–912

Garcia-Palma L, Horn S, Haag F, Diessenbacher P, Streichert T, Mayr GW, Jucker M (2005) Upregulation of the T cell quiescence factor KLF2 in a leukaemic T-cell line after expression of the inositol 5′-phosphatase SHIP-1. Br J Haematol 131:628–631

Gardner RJ, Brown N (1976) Lowe's syndrome: identification of carriers by lens examination. J Med Genet 13:449–454

Gary JD, Sato TK, Stefan CJ, Bonangelino CJ, Weisman LS, Emr SD (2002) Regulation of Fab1 phosphatidylinositol 3-phosphate 5-kinase pathway by Vac7 protein and Fig4, a polyphosphoinositide phosphatase family member. Mol Biol Cell 13:1238–1251

Gewinner C, Wang ZC, Richardson A, Teruya-Feldstein J, Etemadmoghadam D, Bowtell D, Barretina J, Lin WM, Rameh L, Salmena L, Pandolfi PP, Cantley LC (2009) Evidence that inositol polyphosphate 4-phosphatase type II is a tumor suppressor that inhibits PI3K signaling. Cancer Cell 16:115–125

Ghansah T, Paraiso KH, Highfill S, Desponts C, May S, McIntosh JK, Wang JW, Ninos J, Brayer J, Cheng F, Sotomayor E, Kerr WG (2004) Expansion of myeloid suppressor cells in SHIP-deficient mice represses allogeneic T cell responses. J Immunol 173:7324–7330

Giuriato S, Payrastre B, Drayer AL, Plantavid M, Woscholski R, Parker P, Erneux C, Chap H (1997) Tyrosine phosphorylation and relocation of SHIP are integrin-mediated in thrombin-stimulated human blood platelets. J Biol Chem 272:26857–26863

Giuriato S, Pesesse X, Bodin S, Sasaki T, Viala C, Marion E, Penninger J, Schurmans S, Erneux C, Payrastre B (2003) SH2-containing inositol 5-phosphatases 1 and 2 in blood platelets: their interactions and roles in the control of phosphatidylinositol 3,4,5-trisphosphate levels. Biochem J 376:199–207

Gloire G, Erneux C, Piette J (2007) The role of SHIP1 in T-lymphocyte life and death. Biochem Soc Trans 35:277–280

Gong XQ, Frandsen A, Lu WY, Wan Y, Zabek RL, Pickering DS, Bai D (2005) D-aspartate and NMDA, but not L-aspartate, block AMPA receptors in rat hippocampal neurons. Br J Pharmacol 145:449–459

Grempler R, Zibrova D, Schoelch C, Van Marle A, Rippmann JF, Redemann N (2007) Normalization of prandial blood glucose and improvement of glucose tolerance by liver-specific inhibition of SH2 domain containing inositol phosphatase 2 (SHIP2) in diabetic KKAy mice: SHIP2 inhibition causes insulin-mimetic effects on glycogen metabolism, gluconeogenesis, and glycolysis. Diabetes 56:2235–2241

Gruvberger S, Ringner M, Chen Y, Panavally S, Saal LH, Borg A, Ferno M, Peterson C, Meltzer PS (2001) Estrogen receptor status in breast cancer is associated with remarkably distinct gene expression patterns. Cancer Res 61:5979–5984

Gu Z, Rubin MA, Yang Y, Deprimo SE, Zhao H, Horvath S, Brooks JD, Loda M, Reiter RE (2005) Reg IV: a promising marker of hormone refractory metastatic prostate cancer. Clin Cancer Res 11:2237–2243

Guittard G, Gerard A, Dupuis-Coronas S, Tronchere H, Mortier E, Favre C, Olive D, Zimmermann P, Payrastre B, Nunes JA (2009) Cutting edge: Dok-1 and Dok-2 adaptor molecules are regulated by phosphatidylinositol 5-phosphate production in T cells. J Immunol 182:3974–3978

Guittard G, Mortier E, Tronchere H, Firaguay G, Gerard A, Zimmermann P, Payrastre B, Nunes JA (2010) Evidence for a positive role of PtdIns5P in T-cell signal transduction pathways. FEBS Lett 584:2455–2460

Guo S, Stolz LE, Lemrow SM, York JD (1999) SAC1-like domains of yeast SAC1, INP52, and INP53 and of human synaptojanin encode polyphosphoinositide phosphatases. J Biol Chem 274:12990–12995

Gurung R, Tan A, Ooms LM, McGrath MJ, Huysmans RD, Munday AD, Prescott M, Whisstock JC, Mitchell CA (2003) Identification of a novel domain in two mammalian inositol-polyphosphate 5-phosphatases that mediates membrane ruffle localization. The inositol 5-phosphatase skip localizes to the endoplasmic reticulum and translocates to membrane ruffles following epidermal growth factor stimulation. J Biol Chem 278:11376–11385

Haddon DJ, Antignano F, Hughes MR, Blanchet MR, Zbytnuik L, Krystal G, McNagny KM (2009) SHIP1 is a repressor of mast cell hyperplasia, cytokine production, and allergic inflammation in vivo. J Immunol 183:228–236

Haffner C, Di Paolo G, Rosenthal JA, De Camilli P (2000) Direct interaction of the 170 kDa isoform of synaptojanin 1 with clathrin and with the clathrin adaptor AP-2. Curr Biol 10:471–474

Hamilton MJ, Ho VW, Kuroda E, Ruschmann J, Antignano F, Lam V, Krystal G (2011) Role of SHIP in cancer. Exp Hematol 39:2–13

Harris TW, Hartwieg E, Horvitz HR, Jorgensen EM (2000) Mutations in synaptojanin disrupt synaptic vesicle recycling. J Cell Biol 150:589–600

Haslinger B, Mandl-Weber S, Sitter T (2000) Thrombin suppresses matrix metalloproteinase 2 activity and increases tissue inhibitor of metalloproteinase 1 synthesis in cultured human peritoneal mesothelial cells. Perit Dial Int 20:778–783

Haucke V (2003) Where proteins and lipids meet: membrane trafficking on the move. Dev Cell 4:153–157

Helgason CD, Damen JE, Rosten P, Grewal R, Sorensen P, Chappel SM, Borowski A, Jirik F, Krystal G, Humphries RK (1998) Targeted disruption of SHIP leads to hemopoietic perturbations, lung pathology, and a shortened life span. Genes Dev 12:1610–1620

Helgason CD, Kalberer CP, Damen JE, Chappel SM, Pineault N, Krystal G, Humphries RK (2000) A dual role for Src homology 2 domain-containing inositol-5-phosphatase (SHIP) in immunity: aberrant development and enhanced function of b lymphocytes in ship -/- mice. J Exp Med 191:781–794

Hellsten E, Evans JP, Bernard DJ, Janne PA, Nussbaum RL (2001) Disrupted sperm function and fertilin beta processing in mice deficient in the inositol polyphosphate 5-phosphatase Inpp5b. Dev Biol 240:641–653

Hernandez LD, Hueffer K, Wenk MR, Galan JE (2004) Salmonella modulates vesicular traffic by altering phosphoinositide metabolism. Science 304:1805–1807

Hernandez-Negrete I, Carretero-Ortega J, Rosenfeldt H, Hernandez-Garcia R, Calderon-Salinas JV, Reyes-Cruz G, Gutkind JS, Vazquez-Prado J (2007) P-Rex1 links mammalian target of rapamycin signaling to Rac activation and cell migration. J Biol Chem 282:23708–23715

Hill K, Krugmann S, Andrews SR, Coadwell WJ, Finan P, Welch HC, Hawkins PT, Stephens LR (2005) Regulation of P-Rex1 by phosphatidylinositol (3,4,5)-trisphosphate and Gbetagamma subunits. J Biol Chem 280:4166–4173

Hitomi K, Tahara-Hanaoka S, Someya S, Fujiki A, Tada H, Sugiyama T, Shibayama S, Shibuya K, Shibuya A (2010) An immunoglobulin-like receptor, Allergin-1, inhibits immunoglobulin E-mediated immediate hypersensitivity reactions. Nat Immunol 11:601–607

Hodgson MC, Shao LJ, Frolov A, Li R, Peterson LE, Ayala G, Ittmann MM, Weigel NL, Agoulnik IU (2011) Decreased expression and androgen regulation of the tumor suppressor gene INPP4B in prostate cancer. Cancer Res 71:572–582

Hoopes RR Jr, Shrimpton AE, Knohl SJ, Hueber P, Hoppe B, Matyus J, Simckes A, Tasic V, Toenshoff B, Suchy SF, Nussbaum RL, Scheinman SJ (2005) Dent Disease with mutations in OCRL1. Am J Hum Genet 76:260–267

Horan KA, Watanabe K, Kong AM, Bailey CG, Rasko JE, Sasaki T, Mitchell CA (2007) Regulation of FcgammaR-stimulated phagocytosis by the 72-kDa inositol polyphosphate 5-phosphatase: SHIP1, but not the 72-kDa 5-phosphatase, regulates complement receptor 3 mediated phagocytosis by differential recruitment of these 5-phosphatases to the phagocytic cup. Blood 110:4480–4491

Horn S, Endl E, Fehse B, Weck MM, Mayr GW, Jucker M (2004) Restoration of SHIP activity in a human leukemia cell line downregulates constitutively activated phosphatidylinositol 3-kinase/Akt/GSK-3beta signaling and leads to an increased transit time through the G1 phase of the cell cycle. Leukemia 18:1839–1849

Huber M, Helgason CD, Damen JE, Liu L, Humphries RK, Krystal G (1998) The src homology 2-containing inositol phosphatase (SHIP) is the gatekeeper of mast cell degranulation. Proc Natl Acad Sci USA 95:11330–11335

Huber M, Kalesnikoff J, Reth M, Krystal G (2002) The role of SHIP in mast cell degranulation and IgE-induced mast cell survival. Immunol Lett 82:17–21

Hughes WE, Cooke FT, Parker PJ (2000) Sac phosphatase domain proteins. Biochem J 350(Pt 2):337–352

Hung CS, Lin YL, Wu CI, Huang CJ, Ting LP (2009) Suppression of hepatitis B viral gene expression by phosphoinositide 5-phosphatase SKIP. Cell Microbiol 11:37–50

Hyvola N, Diao A, McKenzie E, Skippen A, Cockcroft S, Lowe M (2006) Membrane targeting and activation of the Lowe syndrome protein OCRL1 by rab GTPases. EMBO J 25:3750–3761

Hyvonen ME, Saurus P, Wasik A, Heikkila E, Havana M, Trokovic R, Saleem M, Holthofer H, Lehtonen S (2010) Lipid phosphatase SHIP2 downregulates insulin signalling in podocytes. Mol Cell Endocrinol 328:70–79

Iida A, Kurose K, Isobe R, Akiyama F, Sakamoto G, Yoshimoto M, Kasumi F, Nakamura Y, Emi M (1998) Mapping of a new target region of allelic loss to a 2-cM interval at 22q13.1 in primary breast cancer. Genes Chromosomes Cancer 21:108–112

Ijuin T, Takenawa T (2003) SKIP negatively regulates insulin-induced GLUT4 translocation and membrane ruffle formation. Mol Cell Biol 23:1209–1220

Ijuin T, Mochizuki Y, Fukami K, Funaki M, Asano T, Takenawa T (2000) Identification and characterization of a novel inositol polyphosphate 5-phosphatase. J Biol Chem 275:10870–10875

Ijuin T, Yu YE, Mizutani K, Pao A, Tateya S, Tamori Y, Bradley A, Takenawa T (2008) Increased insulin action in SKIP heterozygous knockout mice. Mol Cell Biol 28:5184–5195

Ikonomov OC, Sbrissa D, Ijuin T, Takenawa T, Shisheva A (2009) Sac3 is an insulin-regulated phosphatidylinositol 3,5-bisphosphate phosphatase: gain in insulin responsiveness through Sac3 down-regulation in adipocytes. J Biol Chem 284:23961–23971

Ikonomov OC, Sbrissa D, Fligger J, Delvecchio K, Shisheva A (2010) ArPIKfyve regulates Sac3 protein abundance and turnover: disruption of the mechanism by Sac3I41T mutation causing Charcot-Marie-Tooth 4J disorder. J Biol Chem 285:26760–26764

Iorio MV, Ferracin M, Liu CG, Veronese A, Spizzo R, Sabbioni S, Magri E, Pedriali M, Fabbri M, Campiglio M, Menard S, Palazzo JP, Rosenberg A, Musiani P, Volinia S, Nenci I, Calin GA, Querzoli P, Negrini M, Croce CM (2005) MicroRNA gene expression deregulation in human breast cancer. Cancer Res 65:7065–7070

Iorio MV, Casalini P, Piovan C, Di Leva G, Merlo A, Triulzi T, Menard S, Croce CM, Tagliabue E (2009) microRNA-205 regulates HER3 in human breast cancer. Cancer Res 69:2195–2200

Irie F, Okuno M, Pasquale EB, Yamaguchi Y (2005) EphrinB-EphB signalling regulates clathrin-mediated endocytosis through tyrosine phosphorylation of synaptojanin 1. Nat Cell Biol 7:501–509

Ishida S, Funakoshi A, Miyasaka K, Shimokata H, Ando F, Takiguchi S (2006) Association of SH-2 containing inositol 5′-phosphatase 2 gene polymorphisms and hyperglycemia. Pancreas 33:63–67

Isnardi I, Bruhns P, Bismuth G, Fridman WH, Daeron M (2006) The SH2 domain-containing inositol 5-phosphatase SHIP1 is recruited to the intracytoplasmic domain of human FcgammaRIIB and is mandatory for negative regulation of B cell activation. Immunol Lett 104:156–165

Ivetac I, Munday AD, Kisseleva MV, Zhang X-M, Luff S, Tiganis T, Whisstock JC, Rowe T, Majerus PW, Mitchell CA (2005) The type I{alpha} inositol polyphosphate 4-phosphatase generates and terminates phosphoinositide 3-kinase signals on endosomes and the plasma membrane. Mol Biol Cell 16:2218–2233

Ivetac I, Gurung R, Hakim S, Horan KA, Sheffield DA, Binge LC, Majerus PW, Tiganis T, Mitchell CA (2009) Regulation of PI(3)K/Akt signalling and cellular transformation by inositol polyphosphate 4-phosphatase-1. EMBO Rep 10:487–493

Jackson C, Welch HC, Bellamy TC (2010) Control of cerebellar long-term potentiation by P-Rex-family guanine-nucleotide exchange factors and phosphoinositide 3-kinase. PLoS One 5:e11962

Jacoby M, Cox JJ, Gayral S, Hampshire DJ, Ayub M, Blockmans M, Pernot E, Kisseleva MV, Compere P, Schiffmann SN, Gergely F, Riley JH, Perez-Morga D, Woods CG, Schurmans S (2009)

INPP5E mutations cause primary cilium signaling defects, ciliary instability and ciliopathies in human and mouse. Nat Genet 41:1027–1031

Janne PA, Dutra AS, Dracopoli NC, Charnas LR, Puck JM, Nussbaum RL (1994) Localization of the 75-kDa inositol polyphosphate-5-phosphatase (INPP5B) to human chromosome band 1p34. Cytogenet Cell Genet 66:164–166

Janne PA, Rochelle JM, Martin-Deleon PA, Stambolian D, Seldin MF, Nussbaum RL (1995) Mapping of the 75-kDa inositol polyphosphate-5-phosphatase (Inpp5b) to distal mouse chromosome 4 and its exclusion as a candidate gene for dysgenetic lens. Genomics 28:280–285

Janne PA, Suchy SF, Bernard D, Macdonald M, Crawley J, Grinberg A, Wynshaw-Boris A, Westphal H, Nussbaum RL (1998) Functional overlap between murine Inpp5b and Ocrl1 may explain why deficiency of the murine ortholog for OCRL1 does not cause Lowe syndrome in mice. J Clin Invest 101:2042–2053

Jason TL, Koropatnick J, Berg RW (2004) Toxicology of antisense therapeutics. Toxicol Appl Pharmacol 201:66–83

Jefferson AB, Majerus PW (1995) Properties of type II inositol polyphosphate 5-phosphatase. J Biol Chem 270:9370–9377

Jiang X, Stuible M, Chalandon Y, Li A, Chan WY, Eisterer W, Krystal G, Eaves A, Eaves C (2003) Evidence for a positive role of SHIP in the BCR-ABL-mediated transformation of primitive murine hematopoietic cells and in human chronic myeloid leukemia. Blood 102:2976–2984

Joseph RE, Norris FA (2005) Substrate specificity and recognition is conferred by the pleckstrin homology domain of the Dbl family guanine nucleotide exchange factor P-Rex2. J Biol Chem 280:27508–27512

Jospin M, Watanabe S, Joshi D, Young S, Hamming K, Thacker C, Snutch TP, Jorgensen EM, Schuske K (2007) UNC-80 and the NCA ion channels contribute to endocytosis defects in synaptojanin mutants. Curr Biol 17:1595–1600

Kagawa S, Sasaoka T, Yaguchi S, Ishihara H, Tsuneki H, Murakami S, Fukui K, Wada T, Kobayashi S, Kimura I, Kobayashi M (2005) Impact of SRC homology 2-containing inositol 5′-phosphatase 2 gene polymorphisms detected in a Japanese population on insulin signaling. J Clin Endocrinol Metab 90:2911–2919

Kagawa S, Soeda Y, Ishihara H, Oya T, Sasahara M, Yaguchi S, Oshita R, Wada T, Tsuneki H, Sasaoka T (2008) Impact of transgenic overexpression of SH2-containing inositol 5′-phosphatase 2 on glucose metabolism and insulin signaling in mice. Endocrinology 149:642–650

Kaisaki PJ, Delepine M, Woon PY, Sebag-Montefiore L, Wilder SP, Menzel S, Vionnet N, Marion E, Riveline JP, Charpentier G, Schurmans S, Levy JC, Lathrop M, Farrall M, Gauguier D (2004) Polymorphisms in type II SH2 domain-containing inositol 5-phosphatase (INPPL1, SHIP2) are associated with physiological abnormalities of the metabolic syndrome. Diabetes 53:1900–1904

Kashiwada M, Cattoretti G, McKeag L, Rouse T, Showalter BM, Al-Alem U, Niki M, Pandolfi PP, Field EH, Rothman PB (2006) Downstream of tyrosine kinases-1 and Src homology 2-containing inositol 5′-phosphatase are required for regulation of CD4+CD25+ T cell development. J Immunol 176:3958–3965

Kasprowicz J, Kuenen S, Miskiewicz K, Habets RL, Smitz L, Verstreken P (2008) Inactivation of clathrin heavy chain inhibits synaptic recycling but allows bulk membrane uptake. J Cell Biol 182:1007–1016

Kato M, Dobyns WB (2003) Lissencephaly and the molecular basis of neuronal migration. Hum Mol Genet 12(Spec No 1):R89–R96

Kawano T, Indo Y, Nakazato H, Shimadzu M, Matsuda I (1998) Oculocerebrorenal syndrome of Lowe: three mutations in the OCRL1 gene derived from three patients with different phenotypes. Am J Med Genet 77:348–355

Kim B, Bang S, Lee S, Kim S, Jung Y, Lee C, Choi K, Lee SG, Lee K, Lee Y, Kim SS, Yeom YI, Kim YS, Yoo HS, Song K, Lee I (2003) Expression profiling and subtype-specific expression of stomach cancer. Cancer Res 63:8248–8255

Kimura T, Sakamoto H, Appella E, Siraganian RP (1997) The negative signaling molecule SH2 domain-containing inositol-polyphosphate 5-phosphatase (SHIP) binds to the tyrosine-phosphorylated beta subunit of the high affinity IgE receptor. J Biol Chem 272:13991–13996

Kisseleva MV, Wilson MP, Majerus PW (2000) The isolation and characterization of a cDNA encoding phospholipid-specific inositol polyphosphate 5-phosphatase. J Biol Chem 275:20110–20116

Knodler LA, Finlay BB, Steele-Mortimer O (2005) The Salmonella effector protein SopB protects epithelial cells from apoptosis by sustained activation of Akt. J Biol Chem 280:9058–9064

Knodler LA, Winfree S, Drecktrah D, Ireland R, Steele-Mortimer O (2009) Ubiquitination of the bacterial inositol phosphatase, SopB, regulates its biological activity at the plasma membrane. Cell Microbiol 11:1652–1670

Kobayashi K, Amemiya S, Higashida K, Ishihara T, Sawanobori E, Mochizuki M, Kikuchi N, Tokuyama K, Nakazawa S (2000) Pathogenic factors of glucose intolerance in obese Japanese adolescents with type 2 diabetes. Metabolism 49:186–191

Kong AM, Speed CJ, O'Malley CJ, Layton MJ, Meehan T, Loveland KL, Cheema S, Ooms LM, Mitchell CA (2000) Cloning and characterization of a 72-kDa inositol-polyphosphate 5-phosphatase localized to the Golgi network. J Biol Chem 275:24052–24064

Kong AM, Horan KA, Sriratana A, Bailey CG, Collyer LJ, Nandurkar HH, Shisheva A, Layton MJ, Rasko JE, Rowe T, Mitchell CA (2006) Phosphatidylinositol 3-phosphate [PtdIns3P] is generated at the plasma membrane by an inositol polyphosphate 5-phosphatase: endogenous PtdIns3P can promote GLUT4 translocation to the plasma membrane. Mol Cell Biol 26:6065–6081

Konrad M, Schaller A, Seelow D, Pandey AV, Waldegger S, Lesslauer A, Vitzthum H, Suzuki Y, Luk JM, Becker C, Schlingmann KP, Schmid M, Rodriguez-Soriano J, Ariceta G, Cano F, Enriquez R, Juppner H, Bakkaloglu SA, Hediger MA, Gallati S, Neuhauss SC, Nurnberg P, Weber S (2006) Mutations in the tight-junction gene claudin 19 (CLDN19) are associated with renal magnesium wasting, renal failure, and severe ocular involvement. Am J Hum Genet 79:949–957

Krendel M, Osterweil EK, Mooseker MS (2007) Myosin 1E interacts with synaptojanin-1 and dynamin and is involved in endocytosis. FEBS Lett 581:644–650

Kumar A, Blanton SH, Babu M, Markandaya M, Girimaji SC (2004) Genetic analysis of primary microcephaly in Indian families: novel ASPM mutations. Clin Genet 66:341–348

Lakhanpal GK, Vecchiarelli-Federico LM, Li YJ, Cui JW, Bailey ML, Spaner DE, Dumont DJ, Barber DL, Ben-David Y (2010) The inositol phosphatase SHIP-1 is negatively regulated by Fli-1 and its loss accelerates leukemogenesis. Blood 116:428–436

Latulippe E, Satagopan J, Smith A, Scher H, Scardino P, Reuter V, Gerald WL (2002) Comprehensive gene expression analysis of prostate cancer reveals distinct transcriptional programs associated with metastatic disease. Cancer Res 62:4499–4506

Laxminarayan KM, Matzaris M, Speed CJ, Mitchell CA (1993) Purification and characterization of a 43-kDa membrane-associated inositol polyphosphate 5-phosphatase from human placenta. J Biol Chem 268:4968–4974

Laxminarayan KM, Chan BK, Tetaz T, Bird PI, Mitchell CA (1994) Characterization of a cDNA encoding the 43-kDa membrane-associated inositol-polyphosphate 5-phosphatase. J Biol Chem 269:17305–17310

Ledezma-Sanchez BA, Garcia-Regalado A, Guzman-Hernandez ML, Vazquez-Prado J (2010) Sphingosine-1-phosphate receptor S1P1 is regulated by direct interactions with P-Rex1, a Rac guanine nucleotide exchange factor. Biochem Biophys Res Commun 391:1647–1652

Lee SY, Wenk MR, Kim Y, Nairn AC, De Camilli P (2004) Regulation of synaptojanin 1 by cyclin-dependent kinase 5 at synapses. Proc Natl Acad Sci USA 101:546–551

Lewis JP, Palmer ND, Ellington JB, Divers J, Ng MC, Lu L, Langefeld CD, Freedman BI, Bowden DW (2010) Analysis of candidate genes on chromosome 20q12-13.1 reveals evidence for BMI mediated association of PREX1 with type 2 diabetes in European Americans. Genomics 96:211–219

Li Z, Paik JH, Wang Z, Hla T, Wu D (2005) Role of guanine nucleotide exchange factor P-Rex-2b in sphingosine 1-phosphate-induced Rac1 activation and cell migration in endothelial cells. Prostaglandins Other Lipid Mediat 76:95–104

Lichter-Konecki U, Farber LW, Cronin JS, Suchy SF, Nussbaum RL (2006) The effect of missense mutations in the RhoGAP-homology domain on ocrl1 function. Mol Genet Metab 89:121–128

Lin T, Orrison BM, Leahey AM, Suchy SF, Bernard DJ, Lewis RA, Nussbaum RL (1997) Spectrum of mutations in the OCRL1 gene in the Lowe oculocerebrorenal syndrome. Am J Hum Genet 60:1384–1388

Lin DC, Quevedo C, Brewer NE, Bell A, Testa JR, Grimes ML, Miller FD, Kaplan DR (2006) APPL1 associates with TrkA and GIPC1 and is required for nerve growth factor-mediated signal transduction. Mol Cell Biol 26:8928–8941

Liu L, Damen JE, Cutler RL, Krystal G (1994) Multiple cytokines stimulate the binding of a common 145-kilodalton protein to Shc at the Grb2 recognition site of Shc. Mol Cell Biol 14:6926–6935

Liu Q, Oliveira-Dos-Santos AJ, Mariathasan S, Bouchard D, Jones J, Sarao R, Kozieradzki I, Ohashi PS, Penninger JM, Dumont DJ (1998a) The inositol polyphosphate 5-phosphatase ship is a crucial negative regulator of B cell antigen receptor signaling. J Exp Med 188:1333–1342

Liu Q, Shalaby F, Jones J, Bouchard D, Dumont DJ (1998b) The SH2-containing inositol polyphosphate 5-phosphatase, ship, is expressed during hematopoiesis and spermatogenesis. Blood 91:2753–2759

Liu Q, Sasaki T, Kozieradzki I, Wakeham A, Itie A, Dumont DJ, Penninger JM (1999) SHIP is a negative regulator of growth factor receptor-mediated PKB/Akt activation and myeloid cell survival. Genes Dev 13:786–791

Liu Y, Boukhelifa M, Tribble E, Morin-Kensicki E, Uetrecht A, Bear JE, Bankaitis VA (2008) The Sac1 phosphoinositide phosphatase regulates Golgi membrane morphology and mitotic spindle organization in mammals. Mol Biol Cell 19:3080–3096

Liu Y, Boukhelifa M, Tribble E, Bankaitis VA (2009) Functional studies of the mammalian Sac1 phosphoinositide phosphatase. Adv Enzyme Regul 49:75–86

Lo TC, Barnhill LM, Kim Y, Nakae EA, Yu AL, Diccianni MB (2009) Inactivation of SHIP1 in T-cell acute lymphoblastic leukemia due to mutation and extensive alternative splicing. Leuk Res 33:1562–1566

Lowe M (2005) Structure and function of the Lowe syndrome protein OCRL1. Traffic 6:711–719

Lucas DM, Rohrschneider LR (1999) A novel spliced form of SH2-containing inositol phosphatase is expressed during myeloid development. Blood 93:1922–1933

Luo JM, Yoshida H, Komura S, Ohishi N, Pan L, Shigeno K, Hanamura I, Miura K, Iida S, Ueda R, Naoe T, Akao Y, Ohno R, Ohnishi K (2003) Possible dominant-negative mutation of the SHIP gene in acute myeloid leukemia. Leukemia 17:1–8

Luo JM, Liu ZL, Hao HL, Wang FX, Dong ZR, Ohno R (2004) Mutation analysis of SHIP gene in acute leukemia. Zhongguo Shi Yan Xue Ye Xue Za Zhi 12:420–426

Ma K, Cheung SM, Marshall AJ, Duronio V (2008) PI(3,4,5)P3 and PI(3,4)P2 levels correlate with PKB/akt phosphorylation at Thr308 and Ser473, respectively; PI(3,4)P2 levels determine PKB activity. Cell Signal 20:684–694

Maeda K, Mehta H, Drevets DA, Coggeshall KM (2010) IL-6 increases B-cell IgG production in a feed-forward proinflammatory mechanism to skew hematopoiesis and elevate myeloid production. Blood 115:4699–4706

Majid S, Dar AA, Saini S, Yamamura S, Hirata H, Tanaka Y, Deng G, Dahiya R (2010) MicroRNA-205-directed transcriptional activation of tumor suppressor genes in prostate cancer. Cancer 116:5637–5649

Malecz N, McCabe PC, Spaargaren C, Qiu R, Chuang Y, Symons M (2000) Synaptojanin 2, a novel Rac1 effector that regulates clathrin-mediated endocytosis. Curr Biol 10:1383–1386

Mallo GV, Espina M, Smith AC, Terebiznik MR, Aleman A, Finlay BB, Rameh LE, Grinstein S, Brumell JH (2008) SopB promotes phosphatidylinositol 3-phosphate formation on Salmonella vacuoles by recruiting Rab5 and Vps34. J Cell Biol 182:741–752

Manford A, Xia T, Saxena AK, Stefan C, Hu F, Emr SD, Mao Y (2010) Crystal structure of the yeast Sac1: implications for its phosphoinositide phosphatase function. EMBO J 29:1489–1498

Mani M, Lee SY, Lucast L, Cremona O, Di Paolo G, De Camilli P, Ryan TA (2007) The dual phosphatase activity of synaptojanin1 is required for both efficient synaptic vesicle endocytosis and reavailability at nerve terminals. Neuron 56:1004–1018

Manji SS M., Williams LH, Miller KA, Ooms LM, Bahlo M, Mitchell CA, Dahl HM (2011) A mutation in Synaptojanin 2 causes progressive hearing loss in the ENU-mutagenised mouse strain Mozart. PLoS One 6(3):e17607

Mao Y, Balkin DM, Zoncu R, Erdmann KS, Tomasini L, Hu F, Jin MM, Hodsdon ME, De Camilli P (2009) A PH domain within OCRL bridges clathrin-mediated membrane trafficking to phosphoinositide metabolism. EMBO J 28:1831–1842

Marcano AC, Burke B, Gungadoo J, Wallace C, Kaisaki PJ, Woon PY, Farrall M, Clayton D, Brown M, Dominiczak A, Connell JM, Webster J, Lathrop M, Caulfield M, Samani N, Gauguier D, Munroe PB (2007) Genetic association analysis of inositol polyphosphate phosphatase-like 1 (INPPL1, SHIP2) variants with essential hypertension. J Med Genet 44:603–605

Marcello MR, Evans JP (2010) Multivariate analysis of male reproductive function in Inpp5b-/- mice reveals heterogeneity in defects in fertility, sperm-egg membrane interaction and proteolytic cleavage of sperm ADAMs. Mol Hum Reprod 16:492–505

Marion E, Kaisaki PJ, Pouillon V, Gueydan C, Levy JC, Bodson A, Krzentowski G, Daubresse JC, Mockel J, Behrends J, Servais G, Szpirer C, Kruys V, Gauguier D, Schurmans S (2002) The gene INPPL1, encoding the lipid phosphatase SHIP2, is a candidate for type 2 diabetes in rat and man. Diabetes 51:2012–2017

Marza E, Long T, Saiardi A, Sumakovic M, Eimer S, Hall DH, Lesa GM (2008) Polyunsaturated fatty acids influence synaptojanin localization to regulate synaptic vesicle recycling. Mol Biol Cell 19:833–842

Mason D, Mallo GV, Terebiznik MR, Payrastre B, Finlay BB, Brumell JH, Rameh L, Grinstein S (2007) Alteration of epithelial structure and function associated with PtdIns(4,5)P2 degradation by a bacterial phosphatase. J Gen Physiol 129:267–283

Matzaris M, O'Malley CJ, Badger A, Speed CJ, Bird PI, Mitchell CA (1998) Distinct membrane and cytosolic forms of inositol polyphosphate 5-phosphatase II. Efficient membrane localization requires two discrete domains. J Biol Chem 273:8256–8267

Mayeenuddin LH, Garrison JC (2006) Phosphorylation of P-Rex1 by the cyclic AMP-dependent protein kinase inhibits the phosphatidylinositiol (3,4,5)-trisphosphate and Gbetagamma-mediated regulation of its activity. J Biol Chem 281:1921–1928

McCrea HJ, Paradise S, Tomasini L, Addis M, Melis MA, De Matteis MA, De Camilli P (2008) All known patient mutations in the ASH-RhoGAP domains of OCRL affect targeting and APPL1 binding. Biochem Biophys Res Commun 369:493–499

McPherson PS, Garcia EP, Slepnev VI, David C, Zhang X, Grabs D, Sossin WS, Bauerfeind R, Nemoto Y, De Camilli P (1996) A presynaptic inositol-5-phosphatase. Nature 379:353–357

McPherson RJ, Marshall JF (1996) Intrastriatal AP5 differentially affects behaviors induced by local infusions of D1 vs. D2 dopamine agonists. Brain Res 739:19–25

Miletic AV, Anzelon-Mills AN, Mills DM, Omori SA, Pedersen IM, Shin DM, Ravetch JV, Bolland S, Morse HC 3rd, Rickert RC (2010) Coordinate suppression of B cell lymphoma by PTEN and SHIP phosphatases. J Exp Med 207:2407–2420

Milosevic I, Sorensen JB, Lang T, Krauss M, Nagy G, Haucke V, Jahn R, Neher E (2005) Plasmalemmal phosphatidylinositol-4,5-bisphosphate level regulates the releasable vesicle pool size in chromaffin cells. J Neurosci 25:2557–2565

Minagawa T, Ijuin T, Mochizuki Y, Takenawa T (2001) Identification and characterization of a sac domain-containing phosphoinositide 5-phosphatase. J Biol Chem 276:22011–22015

Mitchell CA, Connolly TM, Majerus PW (1989) Identification and isolation of a 75-kDa inositol polyphosphate-5-phosphatase from human platelets. J Biol Chem 264:8873–8877

Mochizuki Y, Takenawa T (1999) Novel inositol polyphosphate 5-phosphatase localizes at membrane ruffles. J Biol Chem 274:36790–36795

Montero JC, Seoane S, Ocaña A, Pandiella A (2011) P-Rex1 participates in Neuregulin-ErbB signal transduction and its expression correlates with patient outcome in breast cancer. Oncogene 30:1059–1071

Munday AD, Norris FA, Caldwell KK, Brown S, Majerus PW, Mitchell CA (1999) The inositol polyphosphate 4-phosphatase forms a complex with phosphatidylinositol 3-kinase in human platelet cytosol. Proc Natil Acad Sci USA 96:3640–3645

Muntzel MS, Anderson EA, Johnson AK, Mark AL (1995) Mechanisms of insulin action on sympathetic nerve activity. Clin Exp Hypertens 17:39–50

Muraille E, Pesesse X, Kuntz C, Erneux C (1999) Distribution of the src-homology-2-domain-containing inositol 5-phosphatase SHIP-2 in both non-haemopoietic and haemopoietic cells and possible involvement of SHIP-2 in negative signalling of B-cells. Biochem J 342(Pt 3):697–705

Nakamura K, Malykhin A, Coggeshall KM (2002) The Src homology 2 domain-containing inositol 5-phosphatase negatively regulates Fcgamma receptor-mediated phagocytosis through immunoreceptor tyrosine-based activation motif-bearing phagocytic receptors. Blood 100:3374–3382

Nakatsu F, Perera RM, Lucast L, Zoncu R, Domin J, Gertler FB, Toomre D, De Camilli P (2010) The inositol 5-phosphatase SHIP2 regulates endocytic clathrin-coated pit dynamics. J Cell Biol 190:307–315

Naylor T, Greshock J, Wang Y, Colligon T, Yu Q, Clemmer V, Zaks T, Weber B (2005) High resolution genomic analysis of sporadic breast cancer using array-based comparative genomic hybridization. Breast Cancer Res 7:R1186–R1198

Neill L, Tien AH, Rey-Ladino J, Helgason CD (2007) SHIP-deficient mice provide insights into the regulation of dendritic cell development and function. Exp Hematol 35:627–639

Nemoto Y, Arribas M, Haffner C, De Camilli P (1997) Synaptojanin 2, a novel synaptojanin isoform with a distinct targeting domain and expression pattern. J Biol Chem 272:30817–30821

Nemoto Y, Kearns BG, Wenk MR, Chen H, Mori K, Alb JG Jr, De Camilli P, Bankaitis VA (2000) Functional characterization of a mammalian Sac1 and mutants exhibiting substrate-specific defects in phosphoinositide phosphatase activity. J Biol Chem 275:34293–34305

Nemoto Y, Wenk MR, Watanabe M, Daniell L, Murakami T, Ringstad N, Yamada H, Takei K, De Camilli P (2001) Identification and characterization of a synaptojanin 2 splice isoform predominantly expressed in nerve terminals. J Biol Chem 276:41133–41142

Nicole S, Ben Hamida C, Beighton P, Bakouri S, Belal S, Romero N, Viljoen D, Ponsot G, Sammoud A, Weissenbach J et al (1995) Localization of the Schwartz-Jampel syndrome (SJS) locus to chromosome 1p34-p36.1 by homozygosity mapping. Hum Mol Genet 4:1633–1636

Nie B, Cheng N, Dinauer MC, Ye RD (2010) Characterization of P-Rex1 for its role in fMet-Leu-Phe-induced superoxide production in reconstituted COS(phox) cells. Cell Signal 22:770–782

Niebuhr K, Jouihri N, Allaoui A, Gounon P, Sansonetti PJ, Parsot C (2000) IpgD, a protein secreted by the type III secretion machinery of Shigella flexneri, is chaperoned by IpgE and implicated in entry focus formation. Mol Microbiol 38:8–19

Niebuhr K, Giuriato S, Pedron T, Philpott DJ, Gaits F, Sable J, Sheetz MP, Parsot C, Sansonetti PJ, Payrastre B (2002) Conversion of PtdIns(4,5)P(2) into PtdIns(5)P by the S.flexneri effector IpgD reorganizes host cell morphology. EMBO J 21:5069–5078

Nishio M, Watanabe K, Sasaki J, Taya C, Takasuga S, Iizuka R, Balla T, Yamazaki M, Watanabe H, Itoh R, Kuroda S, Horie Y, Forster I, Mak TW, Yonekawa H, Penninger JM, Kanaho Y, Suzuki A, Sasaki T (2007) Control of cell polarity and motility by the PtdIns(3,4,5)P3 phosphatase SHIP1. Nat Cell Biol 9:36–44

Norden AG, Lapsley M, Igarashi T, Kelleher CL, Lee PJ, Matsuyama T, Scheinman SJ, Shiraga H, Sundin DP, Thakker RV, Unwin RJ, Verroust P, Moestrup SK (2002) Urinary megalin deficiency implicates abnormal tubular endocytic function in Fanconi syndrome. J Am Soc Nephrol 13:125–133

Norris F, Majerus P (1994) Hydrolysis of phosphatidylinositol 3,4-bisphosphate by inositol polyphosphate 4-phosphatase isolated by affinity elution chromatography. J Biol Chem 269:8716–8720

Norris FA, Auethavekiat V, Majerus PW (1995) The isolation and characterization of cDNA encoding human and rat brain inositol polyphosphate 4-phosphatase. J Biol Chem 270:16128–16133

Norris FA, Atkins RC, Majerus PW (1997) The cDNA cloning and characterization of inositol polyphosphate 4-phosphatase type II. Evidence for conserved alternative splicing in the 4-phosphatase family. J Biol Chem 272:23859–23864

Norris FA, Wilson MP, Wallis TS, Galyov EE, Majerus PW (1998) SopB, a protein required for virulence of Salmonella dublin, is an inositol phosphate phosphatase. Proc Natl Acad Sci USA 95:14057–14059

Novick P, Osmond BC, Botstein D (1989) Suppressors of yeast actin mutations. Genetics 121:659–674

Nystuen A, Legare ME, Shultz LD, Frankel WN (2001) A null mutation in inositol polyphosphate 4-phosphatase type I causes selective neuronal loss in weeble mutant mice. Neuron 32:203–212

O'Connell RM, Chaudhuri AA, Rao DS, Baltimore D (2009) Inositol phosphatase SHIP1 is a primary target of miR-155. Proc Natl Acad Sci USA 106:7113–7118

Ooms LM, Fedele CG, Astle MV, Ivetac I, Cheung V, Pearson RB, Layton MJ, Forrai A, Nandurkar HH, Mitchell CA (2006) The inositol polyphosphate 5-phosphatase, PIPP, Is a novel regulator of phosphoinositide 3-kinase-dependent neurite elongation. Mol Biol Cell 17:607–622

Ooms LM, Horan KA, Rahman P, Seaton G, Gurung R, Kethesparan DS, Mitchell CA (2009) The role of the inositol polyphosphate 5-phosphatases in cellular function and human disease. Biochem J 419:29–49

Orchard TJ, Chang YF, Ferrell RE, Petro N, Ellis DE (2002) Nephropathy in type 1 diabetes: a manifestation of insulin resistance and multiple genetic susceptibilities? Further evidence from the Pittsburgh Epidemiology of Diabetes Complication Study. Kidney Int 62:963–970

Osborne RJ, Hamshere MG (2000) A genome-wide map showing common regions of loss of heterozygosity/allelic imbalance in breast cancer. Cancer Res 60:3706–3712

Osborne MA, Zenner G, Lubinus M, Zhang X, Songyang Z, Cantley LC, Majerus P, Burn P, Kochan JP (1996) The inositol 5′-phosphatase SHIP binds to immunoreceptor signaling motifs and responds to high affinity IgE receptor aggregation. J Biol Chem 271:29271–29278

Parker JA, Metzler M, Georgiou J, Mage M, Roder JC, Rose AM, Hayden MR, Neri C (2007) Huntingtin-interacting protein 1 influences worm and mouse presynaptic function and protects Caenorhabditis elegans neurons against mutant polyglutamine toxicity. J Neurosci 27:11056–11064

Parry RV, Harris SJ, Ward SG (2010) Fine tuning T lymphocytes: a role for the lipid phosphatase SHIP-1. Biochim Biophys Acta 1804:592–597

Parvanova AI, Trevisan R, Iliev IP, Dimitrov BD, Vedovato M, Tiengo A, Remuzzi G, Ruggenenti P (2006) Insulin resistance and microalbuminuria: a cross-sectional, case-control study of 158 patients with type 2 diabetes and different degrees of urinary albumin excretion. Diabetes 55:1456–1462

Paternotte N, Zhang J, Vandenbroere I, Backers K, Blero D, Kioka N, Van der Winden JM, Pirson I, Erneux C (2005) SHIP2 interaction with the cytoskeletal protein Vinexin. FEBS J 272:6052–6066

Pechstein A, Bacetic J, Vahedi-Faridi A, Gromova K, Sundborger A, Tomlin N, Krainer G, Vorontsova O, Schafer JG, Owe SG, Cousin MA, Saenger W, Shupliakov O, Haucke V (2010) Regulation of synaptic vesicle recycling by complex formation between intersectin 1 and the clathrin adaptor complex AP2. Proc Natl Acad Sci USA 107:4206–4211

Pedersen IM, Otero D, Kao E, Miletic AV, Hother C, Ralfkiaer E, Rickert RC, Gronbaek K, David M (2009) Onco-miR-155 targets SHIP1 to promote TNFalpha-dependent growth of B cell lymphomas. EMBO Mol Med 1:288–295

Pendaries C, Tronchere H, Arbibe L, Mounier J, Gozani O, Cantley L, Fry MJ, Gaits-Iacovoni F, Sansonetti PJ, Payrastre B (2006) PtdIns5P activates the host cell PI3-kinase/Akt pathway during Shigella flexneri infection. EMBO J 25:1024–1034

Peng Q, Malhotra S, Torchia JA, Kerr WG, Coggeshall KM, Humphrey MB (2010) TREM2- and DAP12-dependent activation of PI3K requires DAP10 and is inhibited by SHIP1. Sci Signal 3:ra38

Pesesse X, Deleu S, De Smedt F, Drayer L, Erneux C (1997) Identification of a second SH2-domain-containing protein closely related to the phosphatidylinositol polyphosphate 5-phosphatase SHIP. Biochem Biophys Res Commun 239:697–700

Poe JC, Fujimoto M, Jansen PJ, Miller AS, Tedder TF (2000) CD22 forms a quaternary complex with SHIP, Grb2, and Shc. A pathway for regulation of B lymphocyte antigen receptor-induced calcium flux. J Biol Chem 275:17420–17427

Ponting CP (2006) A novel domain suggests a ciliary function for ASPM, a brain size determining gene. Bioinformatics 22:1031–1035

Poretti A, Dietrich Alber F, Brancati F, Dallapiccola B, Valente EM, Boltshauser E (2009) Normal cognitive functions in joubert syndrome. Neuropediatrics 40:287–290

Prasad N, Topping RS, Decker SJ (2001) SH2-containing inositol 5′-phosphatase SHIP2 associates with the p130(Cas) adapter protein and regulates cellular adhesion and spreading. Mol Cell Biol 21:1416–1428

Prasad NK (2009) SHIP2 phosphoinositol phosphatase positively regulates EGFR-Akt pathway, CXCR4 expression, and cell migration in MDA-MB-231 breast cancer cells. Int J Oncol 34:97–105

Qin J, Xie Y, Wang B, Hoshino M, Wolff DW, Zhao J, Scofield MA, Dowd FJ, Lin MF, Tu Y (2009) Upregulation of PIP3-dependent Rac exchanger 1 (P-Rex1) promotes prostate cancer metastasis. Oncogene 28:1853–1863

Quade BJ, Wang TY, Sornberger K, Dal CIN P, Mutter GL, Morton CC (2004) Molecular pathogenesis of uterine smooth muscle tumors from transcriptional profiling. Genes Chromosomes Cancer 40:97–108

Rajaram MV, Butchar JP, Parsa KV, Cremer TJ, Amer A, Schlesinger LS, Tridandapani S (2009) Akt and SHIP modulate Francisella escape from the phagosome and induction of the Fas-mediated death pathway. PLoS One 4:e7919

Ramaswamy S, Ross KN, Lander ES, Golub TR (2003) A molecular signature of metastasis in primary solid tumors. Nat Genet 33:49–54

Ramjaun AR, McPherson PS (1996) Tissue-specific alternative splicing generates two synaptojanin isoforms with differential membrane binding properties. J Biol Chem 271:24856–24861

Rauh MJ, Kalesnikoff J, Hughes M, Sly L, Lam V, Krystal G (2003) Role of Src homology 2-containing-inositol 5′-phosphatase (SHIP) in mast cells and macrophages. Biochem Soc Trans 31:286–291

Rauh MJ, Ho V, Pereira C, Sham A, Sly LM, Lam V, Huxham L, Minchinton AI, Mui A, Krystal G (2005) SHIP represses the generation of alternatively activated macrophages. Immunity 23:361–374

Ringstad N, Nemoto Y, De Camilli P (1997) The SH3p4/Sh3p8/SH3p13 protein family: binding partners for synaptojanin and dynamin via a Grb2-like Src homology 3 domain. Proc Natl Acad Sci USA 94:8569–8574

Rivas MP, Kearns BG, Xie Z, Guo S, Sekar MC, Hosaka K, Kagiwada S, York JD, Bankaitis VA (1999) Pleiotropic alterations in lipid metabolism in yeast sac1 mutants: relationship to "bypass Sec14p" and inositol auxotrophy. Mol Biol Cell 10:2235–2250

Rohde HM, Cheong FY, Konrad G, Paiha K, Mayinger P, Boehmelt G (2003) The human phosphatidylinositol phosphatase SAC1 interacts with the coatomer I complex. J Biol Chem 278:52689–52699

Rood BR, Zhang H, Weitman DM, Cogen PH (2002) Hypermethylation of HIC-1 and 17p allelic loss in medulloblastoma. Cancer Res 62:3794–3797

Roongapinun S, Oh SY, Wu F, Panthong A, Zheng T, Zhu Z (2010) Role of SHIP-1 in the adaptive immune responses to aeroallergen in the airway. PLoS One 5:e14174

Roschinger W, Muntau AC, Rudolph G, Roscher AA, Kammerer S (2000) Carrier assessment in families with lowe oculocerebrorenal syndrome: novel mutations in the OCRL1 gene and correlation of direct DNA diagnosis with ocular examination. Mol Genet Metab 69:213–222

Rosenfeldt H, Vazquez-Prado J, Gutkind JS (2004) P-REX2, a novel PI-3-kinase sensitive Rac exchange factor. FEBS Lett 572:167–171

Rudge SA, Anderson DM, Emr SD (2004) Vacuole size control: regulation of PtdIns(3,5)P2 levels by the vacuole-associated Vac14-Fig4 complex, a PtdIns(3,5)P2-specific phosphatase. Mol Biol Cell 15:24–36

Ruschmann J, Ho V, Antignano F, Kuroda E, Lam V, Ibaraki M, Snyder K, Kim C, Flavell RA, Kawakami T, Sly L, Turhan AG, Krystal G (2010) Tyrosine phosphorylation of SHIP promotes its proteasomal degradation. Exp Hematol 38:392–402, 402 e1

Rusk N, Le PU, Mariggio S, Guay G, Lurisci C, Nabi IR, Corda D, Symons M (2003) Synaptojanin 2 functions at an early step of clathrin-mediated endocytosis. Curr Biol 13:659–663

Sachs AJ, David SA, Haider NB, Nystuen AM (2009) Patterned neuroprotection in the Inpp4a(wbl) mutant mouse cerebellum correlates with the expression of Eaat4. PLoS One 4:e8270

Sanchez-Carbayo M, Socci ND, Lozano J, Saint F, Cordon-Cardo C (2006) Defining molecular profiles of poor outcome in patients with invasive bladder cancer using oligonucleotide microarrays. J Clin Oncol 24:778–789

Sasaki T, Takasuga S, Sasaki J, Kofuji S, Eguchi S, Yamazaki M, Suzuki A (2009) Mammalian phosphoinositide kinases and phosphatases. Prog Lipid Res 48:307–343

Sasaki J, Kofuji S, Itoh R, Momiyama T, Takayama K, Murakami H, Chida S, Tsuya Y, Takasuga S, Eguchi S, Asanuma K, Horie Y, Miura K, Davies EM, Mitchell C, Yamazaki M, Hirai H, Takenawa T, Suzuki A, Sasaki T (2010) The PtdIns(3,4)P(2) phosphatase INPP4A is a suppressor of excitotoxic neuronal death. Nature 465:497–501

Sasaoka T, Hori H, Wada T, Ishiki M, Haruta T, Ishihara H, Kobayashi M (2001) SH2-containing inositol phosphatase 2 negatively regulates insulin-induced glycogen synthesis in L6 myotubes. Diabetologia 44:1258–1267

Sasaoka T, Kikuchi K, Wada T, Sato A, Hori H, Murakami S, Fukui K, Ishihara H, Aota R, Kimura I, Kobayashi M (2003) Dual role of SRC homology domain 2-containing inositol phosphatase 2 in the regulation of platelet-derived growth factor and insulin-like growth factor I signaling in rat vascular smooth muscle cells. Endocrinology 144:4204–4214

Sattler M, Verma S, Byrne CH, Shrikhande G, Winkler T, Algate PA, Rohrschneider LR, Griffin JD (1999) BCR/ABL directly inhibits expression of SHIP, an SH2-containing polyinositol-5-phosphatase involved in the regulation of hematopoiesis. Mol Cell Biol 19:7473–7480

Saxena A, Clark WC, Robertson JT, Ikejiri B, Oldfield EH, Ali IU (1992) Evidence for the involvement of a potential second tumor suppressor gene on chromosome 17 distinct from p53 in malignant astrocytomas. Cancer Res 52:6716–6721

Sbrissa D, Ikonomov OC, Fu Z, Ijuin T, Gruenberg J, Takenawa T, Shisheva A (2007) Core protein machinery for mammalian phosphatidylinositol 3,5-bisphosphate synthesis and turnover that regulates the progression of endosomal transport. Novel Sac phosphatase joins the ArPIKfyve-PIKfyve complex. J Biol Chem 282:23878–23891

Scheid MP, Huber M, Damen JE, Hughes M, Kang V, Neilsen P, Prestwich GD, Krystal G, Duronio V (2002) Phosphatidylinositol (3,4,5)P3 is essential but not sufficient for protein kinase B (PKB) activation; phosphatidylinositol (3,4)P2 is required for PKB phosphorylation at Ser-473: studies using cells from SH2-containing inositol-5-phosphatase knockout mice. J Biol Chem 277:9027–9035

Schmid AC, Wise HM, Mitchell CA, Nussbaum R, Woscholski R (2004) Type II phosphoinositide 5-phosphatases have unique sensitivities towards fatty acid composition and head group phosphorylation. FEBS Lett 576:9–13

Schorr M, Then A, Tahirovic S, Hug N, Mayinger P (2001) The phosphoinositide phosphatase Sac1p controls trafficking of the yeast Chs3p chitin synthase. Curr Biol 11:1421–1426

Schuske KR, Richmond JE, Matthies DS, Davis WS, Runz S, Rube DA, Van der Bliek AM, Jorgensen EM (2003) Endophilin is required for synaptic vesicle endocytosis by localizing synaptojanin. Neuron 40:749–762

Seet LF, Cho S, Hessel A, Dumont DJ (1998) Molecular cloning of multiple isoforms of synaptojanin 2 and assignment of the gene to mouse chromosome 17A2-3.1. Biochem Biophys Res Commun 247:116–122

Sekine T, Nozu K, Iyengar R, Fu XJ, Matsuo M, Tanaka R, Iijima K, Matsui E, Harita Y, Inatomi J, Igarashi T (2007) OCRL1 mutations in patients with Dent disease phenotype in Japan. Pediatr Nephrol 22:975–980

Sekulic A, Kim SY, Hostetter G, Savage S, Einspahr JG, Prasad A, Sagerman P, Curiel-Lewandrowski C, Krouse R, Bowden GT, Warneke J, Alberts DS, Pittelkow MR, Dicaudo D, Nickoloff BJ, Trent JM, Bittner M (2010) Loss of inositol polyphosphate 5-phosphatase is an early event in development of cutaneous squamous cell carcinoma. Cancer Prev Res (Phila) 3:1277–1283

Severin S, Gratacap MP, Lenain N, Alvarez L, Hollande E, Penninger JM, Gachet C, Plantavid M, Payrastre B (2007) Deficiency of Src homology 2 domain-containing inositol 5-phosphatase 1 affects platelet responses and thrombus growth. J Clin Invest 117:944–952

Sharma M, Batra J, Mabalirajan U, Sharma S, Nagarkatti R, Aich J, Sharma SK, Niphadkar PV, Ghosh B (2008) A genetic variation in inositol polyphosphate 4 phosphatase A enhances susceptibility to asthma. Am J Respir Crit Care Med 177:712–719

Shearman AM, Hudson TJ, Andresen JM, Wu X, Sohn RL, Haluska F, Housman DE, Weiss JS (1996) The gene for schnyder's crystalline corneal dystrophy maps to human chromosome 1p34.1-p36. Hum Mol Genet 5:1667–1672

Shearn CT, Norris FA (2007) Biochemical characterization of the type I inositol polyphosphate 4-phosphatase C2 domain. Biochem Biophys Res Commun 356:255–259

Shin HW, Hayashi M, Christoforidis S, Lacas-Gervais S, Hoepfner S, Wenk MR, Modregger J, Uttenweiler-Joseph S, Wilm M, Nystuen A, Frankel WN, Solimena M, De Camilli P, Zerial M (2005) An enzymatic cascade of Rab5 effectors regulates phosphoinositide turnover in the endocytic pathway. J Cell Biol 170:607–618

Shrimpton AE, Hoopes RR Jr, Knohl SJ, Hueber P, Reed AA, Christie PT, Igarashi T, Lee P, Lehman A, White C, Milford DV, Sanchez MR, Unwin R, Wrong OM, Thakker RV, Scheinman SJ (2009) OCRL1 mutations in Dent 2 patients suggest a mechanism for phenotypic variability. Nephron Physiol 112:p27–p36

Sleeman MW, Wortley KE, Lai KM, Gowen LC, Kintner J, Kline WO, Garcia K, Stitt TN, Yancopoulos GD, Wiegand SJ, Glass DJ (2005) Absence of the lipid phosphatase SHIP2 confers resistance to dietary obesity. Nat Med 11:199–205

Sly LM, Hamilton MJ, Kuroda E, Ho VW, Antignano FL, Omeis SL, Van Netten-Thomas CJ, Wong D, Brugger HK, Williams O, Feldman ME, Houseman BT, Fiedler D, Shokat KM, Krystal G (2009) SHIP prevents lipopolysaccharide from triggering an antiviral response in mice. Blood 113:2945–2954

Sosa MS, Lopez-Haber C, Yang C, Wang H, Lemmon MA, Busillo JM, Luo J, Benovic JL, Klein-Szanto A, Yagi H, Gutkind JS, Parsons RE, Kazanietz MG (2010) Identification of the Rac-GEF P-Rex1 as an essential mediator of ErbB signaling in breast cancer. Mol Cell 40:877–892

Speed CJ, Little PJ, Hayman JA, Mitchell CA (1996) Underexpression of the 43 kDa inositol polyphosphate 5-phosphatase is associated with cellular transformation. EMBO J 15:4852–4861

Stearman RS, Dwyer-Nield L, Zerbe L, Blaine SA, Chan Z, Bunn PA Jr, Johnson GL, Hirsch FR, Merrick DT, Franklin WA, Baron AE, Keith RL, Nemenoff RA, Malkinson AM, Geraci MW (2005) Analysis of orthologous gene expression between human pulmonary adenocarcinoma and a carcinogen-induced murine model. Am J Pathol 167:1763–1775

Stephens LA, Gray D, Anderton SM (2005) CD4+CD25+ regulatory T cells limit the risk of autoimmune disease arising from T cell receptor crossreactivity. Proc Natl Acad Sci USA 102:17418–17423

Strahl T, Thorner J (2007) Synthesis and function of membrane phosphoinositides in budding yeast, Saccharomyces cerevisiae. Biochim Biophys Acta 1771:353–404

Su D, Ma S, Liu P, Jiang Z, Lv W, Zhang Y, Deng Q, Smith S, Yu H (2007) Genetic polymorphisms and treatment response in advanced non-small cell lung cancer. Lung Cancer 56:281–288

Sun J, Liu W, Adams TS, Li X, Turner AR, Chang B, Kim JW, Zheng SL, Isaacs WB, Xu J (2007) DNA copy number alterations in prostate cancers: a combined analysis of published CGH studies. Prostate 67:692–700

Suwa A, Yamamoto T, Sawada A, Minoura K, Hosogai N, Tahara A, Kurama T, Shimokawa T, Aramori I (2009) Discovery and functional characterization of a novel small molecule inhibitor of the intracellular phosphatase, SHIP2. Br J Pharmacol 158:879–887

Swan LE, Tomasini L, Pirruccello M, Lunardi J, De Camilli P (2010) Two closely related endocytic proteins that share a common OCRL-binding motif with APPL1. Proc Natl Acad Sci USA 107:3511–3516

Szentpetery Z, Varnai P, Balla T (2010) Acute manipulation of Golgi phosphoinositides to assess their importance in cellular trafficking and signaling. Proc Natl Acad Sci USA 107:8225–8230

Tahirovic S, Schorr M, Mayinger P (2005) Regulation of intracellular phosphatidylinositol-4-phosphate by the Sac1 lipid phosphatase. Traffic 6:116–130

Takabayashi T, Xie MJ, Takeuchi S, Kawasaki M, Yagi H, Okamoto M, Tariqur RM, Malik F, Kuroda K, Kubota C, Fujieda S, Nagano T, Sato M (2010) LL5beta directs the translocation of filamin A and SHIP2 to sites of phosphatidylinositol 3,4,5-triphosphate (PtdIns(3,4,5)P3) accumulation, and PtdIns(3,4,5)P3 localization is mutually modified by co-recruited SHIP2. J Biol Chem 285:16155–16165

Takahashi H, Masuda K, Ando T, Kobayashi T, Honda H (2004) Prognostic predictor with multiple fuzzy neural models using expression profiles from DNA microarray for metastases of breast cancer. J Biosci Bioeng 98:193–199

Takeshita S, Namba N, Zhao JJ, Jiang Y, Genant HK, Silva MJ, Brodt MD, Helgason CD, Kalesnikoff J, Rauh MJ, Humphries RK, Krystal G, Teitelbaum SL, Ross FP (2002) SHIP-deficient mice are severely osteoporotic due to increased numbers of hyper-resorptive osteoclasts. Nat Med 8:943–949

Talantov D, Mazumder A, Yu JX, Briggs T, Jiang Y, Backus J, Atkins D, Wang Y (2005) Novel genes associated with malignant melanoma but not benign melanocytic lesions. Clin Cancer Res 11:7234–7242

Tang J, Qi X, Mercola D, Han J, Chen G (2005) Essential role of p38gamma in K-Ras transformation independent of phosphorylation. Journal of Biological Chemistry 280:23910–23917

Tarasenko T, Kole HK, Chi AW, Mentink-Kane MM, Wynn TA, Bolland S (2007) T cell-specific deletion of the inositol phosphatase SHIP reveals its role in regulating Th1/Th2 and cytotoxic responses. Proc Natl Acad Sci USA 104:11382–11387

Taylor V, Wong M, Brandts C, Reilly L, Dean NM, Cowsert LM, Moodie S, Stokoe D (2000) 5′ phospholipid phosphatase SHIP-2 causes protein kinase B inactivation and cell cycle arrest in glioblastoma cells. Mol Cell Biol 20:6860–6871

Taylor BS, Schultz N, Hieronymus H, Gopalan A, Xiao Y, Carver BS, Arora VK, Kaushik P, Cerami E, Reva B, Antipin Y, Mitsiades N, Landers T, Dolgalev I, Major JE, Wilson M, Socci ND, Lash AE, Heguy A, Eastham JA, Scher HI, Reuter VE, Scardino PT, Sander C, Sawyers CL, Gerald WL (2010) Integrative genomic profiling of human prostate cancer. Cancer Cell 18:11–22

Tobin JL, Beales PL (2009) The nonmotile ciliopathies. Genet Med 11:386–402

Toftgard R (2009) Two sides to cilia in cancer. Nat Med 15:994–996

Tomlinson MG, Heath VL, Turck CW, Watson SP, Weiss A (2004) SHIP family inositol phosphatases interact with and negatively regulate the Tec tyrosine kinase. J Biol Chem 279:55089–55096

Trapani JG, Obholzer N, Mo W, Brockerhoff SE, Nicolson T (2009) Synaptojanin1 is required for temporal fidelity of synaptic transmission in hair cells. PLoS Genet 5:e1000480

Trivedi CM, Luo Y, Yin Z, Zhang M, Zhu W, Wang T, Floss T, Goettlicher M, Noppinger PR, Wurst W, Ferrari VA, Abrams CS, Gruber PJ, Epstein JA (2007) Hdac2 regulates the cardiac hypertrophic response by modulating Gsk3 beta activity. Nat Med 13:324–331

Trumel C, Payrastre B, Plantavid M, Hechler B, Viala C, Presek P, Martinson EA, Cazenave JP, Chap H, Gachet C (1999) A key role of adenosine diphosphate in the irreversible platelet aggregation induced by the PAR1-activating peptide through the late activation of phosphoinositide 3-kinase. Blood 94:4156–4165

Tsujishita Y, Guo S, Stolz LE, York JD, Hurley JH (2001) Specificity determinants in phosphoinositide dephosphorylation: crystal structure of an archetypal inositol polyphosphate 5-phosphatase. Cell 105:379–389

Tu Z, Ninos JM, Ma Z, Wang JW, Lemos MP, Desponts C, Ghansah T, Howson JM, Kerr WG (2001) Embryonic and hematopoietic stem cells express a novel SH2-containing inositol 5′-phosphatase isoform that partners with the Grb2 adapter protein. Blood 98:2028–2038

Ungewickell AJ, Majerus PW (1999) Increased levels of plasma lysosomal enzymes in patients with Lowe syndrome. Proc Natl Acad Sci USA 96:13342–13344

Ungewickell A, Hugge C, Kisseleva M, Chang SC, Zou J, Feng Y, Galyov EE, Wilson M, Majerus PW (2005) The identification and characterization of two phosphatidylinositol-4,5-bisphosphate 4-phosphatases. Proc Natl Acad Sci USA 102:18854–18859

Urano D, Nakata A, Mizuno N, Tago K, Itoh H (2008) Domain-domain interaction of P-Rex1 is essential for the activation and inhibition by G protein betagamma subunits and PKA. Cell Signal 20:1545–1554

Utsch B, Bokenkamp A, Benz MR, Besbas N, Dotsch J, Franke I, Frund S, Gok F, Hoppe B, Karle S, Kuwertz-Broking E, Laube G, Neb M, Nuutinen M, Ozaltin F, Rascher W, Ring T, Tasic V, Van Wijk JA, Ludwig M (2006) Novel OCRL1 mutations in patients with the phenotype of Dent disease. Am J Kidney Dis 48:942 e1–e14

Van Epps HA, Hayashi M, Lucast L, Stearns GW, Hurley JB, De Camilli P, Brockerhoff SE (2004) The zebrafish nrc mutant reveals a role for the polyphosphoinositide phosphatase synaptojanin 1 in cone photoreceptor ribbon anchoring. J Neurosci 24:8641–8650

Van'T Veer LJ, Dai H, Van de Vijver MJ, He YD, Hart AA, Mao M, Peterse HL, Van der Kooy K, Marton MJ, Witteveen AT, Schreiber GJ, Kerkhoven RM, Roberts C, Linsley PS, Bernards R, Friend SH (2002) Gene expression profiling predicts clinical outcome of breast cancer. Nature 415:530–536

Varsano T, Dong MQ, Niesman I, Gacula H, Lou X, Ma T, Testa JR, Yates JR 3rd, Farquhar MG (2006) GIPC is recruited by APPL to peripheral TrkA endosomes and regulates TrkA trafficking and signaling. Mol Cell Biol 26:8942–8952

Verstreken P, Koh TW, Schulze KL, Zhai RG, Hiesinger PR, Zhou Y, Mehta SQ, Cao Y, Roos J, Bellen HJ (2003) Synaptojanin is recruited by endophilin to promote synaptic vesicle uncoating. Neuron 40:733–748

Voronov SV, Frere SG, Giovedi S, Pollina EA, Borel C, Zhang H, Schmidt C, Akeson EC, Wenk MR, Cimasoni L, Arancio O, Davisson MT, Antonarakis SE, Gardiner K, De Camilli P, Di Paolo G (2008) Synaptojanin 1-linked phosphoinositide dyshomeostasis and cognitive deficits in mouse models of Down's syndrome. Proc Natl Acad Sci USA 105:9415–9420

Vyas P, Norris FA, Joseph R, Majerus PW, Orkin SH (2000) Inositol polyphosphate 4-phosphatase type I regulates cell growth downstream of transcription factor GATA-1. Proc Natl Acad Sci USA 97:13696–13701

Wada T, Sasaoka T, Funaki M, Hori H, Murakami S, Ishiki M, Haruta T, Asano T, Ogawa W, Ishihara H, Kobayashi M (2001) Overexpression of SH2-containing inositol phosphatase 2 results in negative regulation of insulin-induced metabolic actions in 3T3-L1 adipocytes via its 5′-phosphatase catalytic activity. Mol Cell Biol 21:1633–1646

Wang JW, Howson JM, Ghansah T, Desponts C, Ninos JM, May SL, Nguyen KH, Toyama-Sorimachi N, Kerr WG (2002) Influence of SHIP on the NK repertoire and allogeneic bone marrow transplantation. Science 295:2094–2097

Wang Z, Dong X, Li Z, Smith JD, Wu D (2008) Lack of a significant role of P-Rex1, a major regulator of macrophage Rac1 activation and chemotaxis, in atherogenesis. Prostaglandins Other Lipid Mediat 87:9–13

Waters JE, Astle MV, Ooms LM, Balamatsias D, Gurung R, Mitchell CA (2008) P-Rex1—a multidomain protein that regulates neurite differentiation. J Cell Sci 121:2892–2903

Wei HC, Sanny J, Shu H, Baillie DL, Brill JA, Price JV, Harden N (2003) The Sac1 lipid phosphatase regulates cell shape change and the JNK cascade during dorsal closure in Drosophila. Curr Biol 13:1882–1887

Welch HC, Coadwell WJ, Ellson CD, Ferguson GJ, Andrews SR, Erdjument-Bromage H, Tempst P, Hawkins PT, Stephens LR (2002) P-Rex1, a PtdIns(3,4,5)P3- and Gbetagamma-regulated guanine-nucleotide exchange factor for Rac. Cell 108:809–821

Welch HC, Condliffe AM, Milne LJ, Ferguson GJ, Hill K, Webb LM, Okkenhaug K, Coadwell WJ, Andrews SR, Thelen M, Jones GE, Hawkins PT, Stephens LR (2005) P-Rex1 regulates neutrophil function. Curr Biol 15:1867–1873

West M, Blanchette C, Dressman H, Huang E, Ishida S, Spang R, Zuzan H, Olson JA, Marks JR, Nevins JR (2001) Predicting the clinical status of human breast cancer by using gene expression profiles. Proc Natl Acad Sci USA 98:11462–11467

Whisstock JC, Romero S, Gurung R, Nandurkar H, Ooms LM, Bottomley SP, Mitchell CA (2000) The inositol polyphosphate 5-phosphatases and the apurinic/apyrimidinic base excision repair endonucleases share a common mechanism for catalysis. J Biol Chem 275:37055–37061

Whitters EA, Cleves AE, McGee TP, Skinner HB, Bankaitis VA (1993) SAC1p is an integral membrane protein that influences the cellular requirement for phospholipid transfer protein function and inositol in yeast. J Cell Biol 122:79–94

Williams C, Choudhury R, McKenzie E, Lowe M (2007) Targeting of the type II inositol polyphosphate 5-phosphatase INPP5B to the early secretory pathway. J Cell Sci 120:3941–3951

Wisniewski D, Strife A, Swendeman S, Erdjument-Bromage H, Geromanos S, Kavanaugh WM, Tempst P, Clarkson B (1999) A novel SH2-containing phosphatidylinositol 3,4,5-trisphosphate 5-phosphatase (SHIP2) is constitutively tyrosine phosphorylated and associated with src homologous and collagen gene (SHC) in chronic myelogenous leukemia progenitor cells. Blood 93:2707–2720

Wurmbach E, Chen YB, Khitrov G, Zhang W, Roayaie S, Schwartz M, Fiel I, Thung S, Mazzaferro V, Bruix J, Bottinger E, Friedman S, Waxman S, Llovet JM (2007) Genome-wide molecular profiles of HCV-induced dysplasia and hepatocellular carcinoma. Hepatology 45:938–947

Yang F, Foekens JA, Yu J, Sieuwerts AM, Timmermans M, Klijn JG M., Atkins D, Wang Y, Jiang Y (2005) Laser microdissection and microarray analysis of breast tumors reveal ER-[alpha] related genes and pathways. Oncogene 25:1413–1419

Yavari A, Nagaraj R, Owusu-Ansah E, Folick A, Ngo K, Hillman T, Call G, Rohatgi R, Scott MP, Banerjee U (2010) Role of lipid metabolism in smoothened derepression in hedgehog signaling. Dev Cell 19:54–65

Ye C, Xi PC, Hu XG (2003a) Clinical analysis of uncinate process carcinoma of the pancreas. Hepatobiliary Pancreat Dis Int 2:605–608

Ye QH, Qin LX, Forgues M, He P, Kim JW, Peng AC, Simon R, Li Y, Robles AI, Chen Y, Ma ZC, Wu ZQ, Ye SL, Liu YK, Tang ZY, Wang XW (2003b) Predicting hepatitis B virus-positive metastatic hepatocellular carcinomas using gene expression profiling and supervised machine learning. Nat Med 9:416–423

Yeow-Fong L, Lim L, Manser E (2005) SNX9 as an adaptor for linking synaptojanin-1 to the Cdc42 effector ACK1. FEBS Lett 579:5040–5048

Yoon JH, Lee JM, Namkoong SE, Bae SM, Kim YW, Han SJ, Cho YL, Nam GH, Kim CK, Seo JS, Ahn WS (2003) cDNA microarray analysis of gene expression profiles associated with cervical cancer. Cancer Res Treat 35:451–459

Yoshizawa M, Kawauchi T, Sone M, Nishimura YV, Terao M, Chihama K, Nabeshima Y, Hoshino M (2005) Involvement of a Rac activator, P-Rex1, in neurotrophin-derived signaling and neuronal migration. J Neurosci 25:4406–4419

Yu J, Ryan DG, Getsios S, Oliveira-Fernandes M, Fatima A, Lavker RM (2008) MicroRNA-184 antagonizes microRNA-205 to maintain SHIP2 levels in epithelia. Proc Natl Acad Sci USA 105:19300–19305

Yuan Y, Gao X, Guo N, Zhang H, Xie Z, Jin M, Li B, Yu L, Jing N (2007) rSac3, a novel Sac domain phosphoinositide phosphatase, promotes neurite outgrowth in PC12 cells. Cell Res 17:919–932

Zhan FH, Barlogie B, John DS Jr (2007) Gene expression profiling defines a high-risk entity of multiple myeloma. Zhong Nan Da Xue Xue Bao Yi Xue Ban 32:191–203

Zhang X, Jefferson AB, Auethavekiat V, Majerus PW (1995) The protein deficient in Lowe syndrome is a phosphatidylinositol-4,5-bisphosphate 5-phosphatase. Proc Natl Acad Sci USA 92:4853–4856

Zhang X, Hartz PA, Philip E, Racusen LC, Majerus PW (1998) Cell lines from kidney proximal tubules of a patient with Lowe syndrome lack OCRL inositol polyphosphate 5-phosphatase and accumulate phosphatidylinositol 4,5-bisphosphate. J Biol Chem 273:1574–1582

Zhang X, Chow CY, Sahenk Z, Shy ME, Meisler MH, Li J (2008) Mutation of FIG4 causes a rapidly progressive, asymmetric neuronal degeneration. Brain 131:1990–2001

Zhao W, Li ZG, Wu MY, Geng LZ, Shi HW, Zhang YH, Wu RH (2003a) [Analysis of 21 children with acute non-lymphoid leukemia carrying AML1/ETO fusion gene]. Zhonghua Er Ke Za Zhi 41:325–328

Zhao X, He M, Wan D, Ye Y, He Y, Han L, Guo M, Huang Y, Qin W, Wang MW, Chong W, Chen J, Zhang L, Yang N, Xu B, Wu M, Zuo L, Gu J (2003b) The minimum LOH region defined on chromosome 17p13.3 in human hepatocellular carcinoma with gene content analysis. Cancer Lett 190:221–232

Zhao T, Nalbant P, Hoshino M, Dong X, Wu D, Bokoch GM (2007) Signaling requirements for translocation of P-Rex1, a key Rac2 exchange factor involved in chemoattractant-stimulated human neutrophil function. J Leukoc Biol 81:1127–1136

Zhong R, Ye ZH (2003) The SAC domain-containing protein gene family in Arabidopsis. Plant Physiol 132:544–555

Zhong R, Burk DH, Nairn CJ, Wood-Jones A, Morrison WH 3RD, Ye ZH (2005) Mutation of SAC1, an Arabidopsis SAC domain phosphoinositide phosphatase, causes alterations in cell morphogenesis, cell wall synthesis, and actin organization. Plant Cell 17:1449–1466

Zhu W, Trivedi CM, Zhou D, Yuan L, Lu MM, Epstein JA (2009) Inpp5f is a polyphosphoinositide phosphatase that regulates cardiac hypertrophic responsiveness. Circ Res 105:1240–1247

Zou J, Marjanovic J, Kisseleva MV, Wilson M, Majerus PW (2007) Type I phosphatidylinositol-4,5-bisphosphate 4-phosphatase regulates stress-induced apoptosis. Proc Natl Acad Sci USA 104:16834–16839

Chapter 8
The PTEN and Myotubularin Phosphoinositide 3-Phosphatases: Linking Lipid Signalling to Human Disease

Elizabeth M. Davies, David A. Sheffield, Priyanka Tibarewal, Clare G. Fedele, Christina A. Mitchell and Nicholas R. Leslie

Abstract Two classes of lipid phosphatases selectively dephosphorylate the 3 position of the inositol ring of phosphoinositide signaling molecules: the PTEN and the Myotubularin families. PTEN dephosphorylates PtdIns(3,4,5)P_3, acting in direct opposition to the Class I PI3K enzymes in the regulation of cell growth, proliferation and polarity and is an important tumor suppressor. Although there are several PTEN-related proteins encoded by the human genome, none of these appear to fulfill the same functions. In contrast, the Myotubularins dephosphorylate both PtdIns(3)P and PtdIns(3,5)P_2, making them antagonists of the Class II and Class III PI 3-kinases and regulators of membrane traffic. Both phosphatase groups were originally identified through their causal mutation in human disease. Mutations in specific myotubularins result in myotubular myopathy and Charcot-Marie-Tooth peripheral neuropathy; and loss of PTEN function through mutation and other mechanisms is evident in as many as a third of all human tumors. This chapter will discuss these two classes of phosphatases, covering what is known about their biochemistry, their functions at the cellular and whole body level and their influence on human health.

Keywords PTEN · Myotubularin · Phosphoinositide · Phosphatase · PI 3-kinase

---

Elizabeth M. Davies and David A. Sheffield made equal contribution as first authors.

N. R. Leslie (✉) · P. Tibarewal
Division of Cell Signalling and Immunology, Wellcome Trust Biocentre,
College of Life Sciences, University of Dundee,
Dundee, Scotland, United Kingdom
e-mail: n.r.leslie@dundee.ac.uk

E. M. Davies · D. A. Sheffield · C. G. Fedele · C. A. Mitchell
Department of Biochemistry and Molecular Biology,
Monash University, Clayton, Australia

8.1 Introduction

Phosphoinositides are second messengers that relay extracellular signals to initiate cellular signaling cascades. They are derived from the precursor phosphatidylinositol (PtdIns), which can be transiently phosphorylated at the D3, D4 or D5 position of the inositol head group. Seven phosphoinositide species have been currently identified, each with unique subcellular localization patterns and distinct roles in cellular signaling pathways. Generation of phosphoinositides phosphorylated at the D3 position of the inositol head group is a critical component in phosphoinositide metabolism, and in the coordination of cellular responses required for appropriate physiological development. Phosphatidylinositol 3,4,5-trisphosphate (PtdIns(3,4,5)P$_3$) levels are low in quiescent cells but increase transiently in response to agonist stimulation. Agonist-induced activation of the class I phosphatidylinositol 3-kinase (PI3K) results in the generation of PtdIns(3,4,5)P$_3$ at the plasma membrane through the phosphorylation of phosphatidylinositol 4,5-bisphosphate (PtdIns(4,5)P$_2$) at the D3 position of its inositol head group. PtdIns(3,4,5)P$_3$ directs numerous cellular processes, including cell proliferation, growth, survival, cell polarity and migration. Phosphatidylinositol 3-phosphate (PtdIns(3)P) is constitutively generated at the site of early endosomes by the class III PI3K (Vps34) or by the class II PI3K. PtdIns(3)P regulates endosomal fusion and motility; receptor sorting and recycling; and vesicle trafficking. Phosphatidylinositol 3,5-bisphosphate (PtdIns(3,5)P$_2$) is generated by phosphorylation of PtdIns(3)P by the PIKfyve kinase on early and/or late endosomes and regulates endosomal sorting and endomembrane homeostasis. The downstream cellular effects of 3-phosphorylated phosphoinositides are transmitted via the recruitment of specific phosphoinositide-binding proteins. This occurs either through interaction of the phosphoinositide inositol head group with basic amino acid residues or alternatively via interaction with discrete phosphoinositide-binding domains, for example pleckstrin homology (PH) domains. Dysregulation of 3-phosphoinositide metabolism leads to the disruption of cellular function and the development of disease. Therefore, their levels are tightly regulated by the activity of the phosphoinositide kinases, which generate them; and the activity of phosphoinositide 3-phosphatases, which selectively remove the phosphate group at the D3 position of their inositol head-groups. The phosphoinositide 3-phosphatases include PTEN and its related homologs; and the multiple members of the myotubularin (MTM) family. These enzymes share a highly conserved phosphatase domain, containing a CX$_5$R catalytic motif, and both will be described within this chapter. While *in vitro* kinetic analyses have demonstrated enzyme activity of PTEN and MTMs toward membrane-bound phosphoinositides and soluble inositol phosphates, membrane-bound phosphoinositides are recognized as their preferred physiological substrates. Therefore, at the functional level, the 3-phosphatases act preferentially at membrane microdomains at the plasma membrane or on intracellular organelles such as early endosomes.

PTEN is a recognized tumor suppressor gene, which is frequently mutated at the 10q23 chromosomal locus in both spontaneous cancers and hereditary cancer predisposition syndromes. PtdIns(3,4,5)P$_3$ is the recognized physiological substrate of PTEN, generating PtdIns(4,5)P_2. Therefore, PTEN directly antagonizes the Class I PI3K at the plasma membrane, and thereby regulates cell proliferation, survival, cell cycle progression, cell polarity, migration, invasion, embryonic development, immune function, insulin signaling and glucose metabolism. The myotubularins are a large family consisting of 9 catalytically active and 7 catalytically inactive family members. Heterodimeric interaction between active and inactive members of the myotubularin family regulates catalytic activity and/or sub-cellular protein localization of the active family members. Myotubularins hydrolyze PtdIns(3)P and PtdIns(3,5)P_2, to generate PtdIns and PtdIns(5)P respectively. Therefore, MTMs antagonize Class II and III PI3Ks, thereby regulating phosphoinositide-dependent endosomal membrane homeostasis. Mutations in various members of the myotubularin family are associated with human disease, including the peripheral neuropathy Charcot-Marie Tooth disease; and mytobular or centronuclear myopathies. Dephosphorylation of phosphoinositides at the D3 position of the inositol head-group is also a function of Sac1, which has been extensively described in Chap. 7, and will not be further described here. The following chapter will discuss the prominent features of these two families of 3-phosphoinositide phosphatase enzymes, focusing on current studies that enhance our knowledge of how the loss of function of these proteins contributes to human diseases.

8.2 PTEN

PTEN/MMAC/TEP1 (phosphatase and tensin homolog deleted on chromosome ten/mutated in multiple advanced cancers/TGFβ-regulated and epithelial cell-enriched phosphatase) is a tumor suppressor that is frequently mutated in sporadic human cancers and also in the inherited autosomal dominant cancer predisposition syndromes, Cowden disease, Lhermitte-Duclos disease, Bannayan-Zonana syndrome, and Proteus and Proteus-like syndromes (Yin and Shen 2008). These syndromes are characterized by developmental disorders, including neurological abnormalities, multiple hamartomas, and an associated increased risk of cancer development in later life, including breast, thyroid, and endometrial cancers (Liaw et al. 1997; Marsh et al. 1997; Tsuchiya et al. 1998). Other PTEN-like phosphatases have been identified in humans including the Trans-membrane Phosphatase with Tensin homology (TPTE), and the TPTE and PTEN homologous inositol lipid phosphatase (TPIP/TPTE2). However, these enzymes appear to be expressed predominantly in the testis, and although their functions are poorly defined, they seem quite distinct from those of PTEN, reviewed in (Sasaki et al. 2009).

PTEN shares sequence homology with the protein tyrosine phosphatase family and initial reports identified PTEN as a dual-specificity protein phosphatase (Li and Sun 1997; Myers et al. 1997). Subsequent studies using recombinant PTEN identified its 3-phosphatase activity toward PtdIns(3,4,5)P_3, PtdIns(3,4)P_2, PtdIns(3)P, and Ins(1,3,4,5)P_4 (Maehama and Dixon 1998). Whilst constitutive elevation of both PtdIns(3,4)P_2 and PtdIns(3,4,5)P_3 have been identified in PTEN-null cells (Haas-Kogan et al. 1998; Taylor et al. 2000b), there is evidence to indicate that

PtdIns(3,4,5)P$_3$ is the major physiological target of PTEN. Firstly, a H93A PTEN mutant selectively reduces PTEN activity toward PtdIns(3,4,5)P$_3$, but not PtdIns(3,4)P$_2$ (Lee et al. 1999). Furthermore, the catalytic efficiency of PTEN for PtdIns(3,4,5)P$_3$ as a substrate is 200-fold greater than that for PtdIns(3,4)P$_2$ (McConnachie et al. 2003). Therefore, in the physiological context, it is likely PTEN is a phosphoinositide phosphatase that preferentially hydrolyzes the D3-position phosphate from PtdIns(3,4,5)P$_3$, to generate PtdIns(4,5)P$_2$, and thereby directly antagonizes phosphoinositide 3-kinase (PI3K) signaling and attenuates Akt activation to regulate cell survival and proliferation (Salmena et al. 2008).

8.2.1 PTEN Structure

Several domains and motifs have been identified in PTEN that contribute to its activity, stability or localization (Fig. 8.1). PTEN contains two major domains, which associate across an extensive interface through hydrogen bonding (Lee et al. 1999; Li et al. 1997; Steck et al. 1997). Within the amino-terminal domain is the catalytic phosphatase domain, that contains a conserved CX$_5$R catalytic motif; and also an extreme amino-terminal PtdIns(4,5)P$_2$—binding motif (Walker et al. 2004; Campbell et al. 2003; Iijima et al. 2004). The carboxyl-terminal domain contains a calcium-independent phospholipid binding C2 domain that regulates its plasma membrane localization; two PEST (proline, glutamic acid, serine, threonine) sequences, and a PDZ (Post synaptic density protein, Drosophila disc large tumor suppressor, zonula occludens-1 protein)-binding domain, that mediates the interaction with several binding partners and can affect PTEN protein stability (Salmena et al. 2008). Examination of PTEN's crystal structure reveals the presence of a more enlarged catalytic pocket, in comparison to protein tyrosine phosphatases, which also have a CX$_5$R catalytic motif, facilitating the association of PtdIns(3,4,5)P$_3$ with particular basic catalytic residues (Lee et al. 1999). The amino-terminal PtdIns(4,5)P$_2$—binding motif, in addition to the C2 domain, regulates the transient association of PTEN from the cytosol to the plasma membrane, positioning the phosphatase for maximal access to its membrane-bound phosphoinositide substrate.

8.2.2 Regulation of PTEN

PTEN can be regulated both at the level of transcription, or by post-translational modification. Indeed, regulation of PTEN at the transcriptional level plays a prominent role in those cancers or cases of Cowden disease in which mutation of *PTEN* is absent, but PTEN expression is lost. Naturally occurring alternative splice variants of *PTEN* have been identified in both normal and cancerous tissue (Sharrard and Maitland 2000). However, in identified *PTEN* mutation-negative cases of Cowden disease or sporadic breast cancers, alternative splice variants of *PTEN* are found to associate with decreased transcription of full-length *PTEN* (Sarquis et al. 2006; Agrawal and Eng 2006). Epigenetic silencing of the *PTEN* promoter through methylation and the

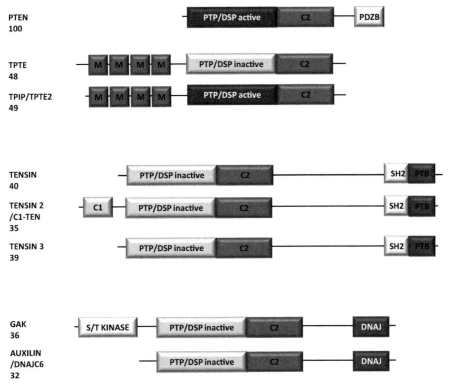

Fig. 8.1 Schematic representation of the major domains of the human PTEN family. The domain structure of human PTEN is shown, along with its closest 7 relatives within the human genome. More distantly related phosphatases display both lower sequence identity through the phosphatase domain and lack an adjacent recognizable C2 domain. Numbers below each protein name show the sequence identity with PTEN through the phosphatase domain. Domains are those identified in the NCBI/CDD database. Abbreviations: *GAK* Cyclin G-associated Kinase, *PTP/DSP* Protein tyrosine phosphatase/dual specific phosphatase, M Transmembrane domain, *SH2* Src homology 2, *PTB* Phosphotyrosine binding, S/T Kinase Serine/Threonine Protein Kinase, PDZB binding motif for PDZ domains (Post synaptic density protein, Drosophila disc large tumor suppressor, zonula occludens-1 protein)

actions of several oncogenic microRNAs also decrease PTEN protein expression in sporadic cancers (Khan et al. 2004; Whang et al. 1998; Mirmohammadsadegh et al. 2006; Salvesen et al. 2004; Poliseno et al. 2010a).

Post-translational modifications of PTEN may regulate protein stability, expression, catalytic activity or sub-cellular localization. PTEN interacts with PCAF (p300/CBP-associated factor), a histone acetyltransferase that regulates gene transcription (Okumura et al. 2006; Yao and Nyomba 2008). Interaction of PTEN and PCAF results in increased acetylation of Lys^{125} and Lys^{128}, within the catalytic cleft of PTEN, inhibiting PTEN's phosphatase activity. Several studies have shown that phosphorylation of specific residues within the carboxyl-terminal of PTEN by a number of kinases also regulates protein stability and turnover. Phosphorylation of PTEN at a cluster of phosphorylation sites (Ser^{380}, Thr^{382}, Thr^{383} and Ser^{385}) in the carboxyl-terminal tail region by casein kinase 2 promotes a conformational change in PTEN, indirectly increasing PTEN's resistance to proteosome-mediated degradation, but decreasing its membrane association and cellular activity (Vazquez et al. 2000, 2001, 2006; Torres and Pulido 2001; Maccario et al. 2010). Conversely, glycogen synthase kinase 3β (GSK3β)-mediated phosphorylation of PTEN at Thr^{366} in its carboxyl-terminal tail results in protein destabilization (Maccario et al. 2007). Phosphorylation of PTEN by the Src family of protein-tyrosine kinases, probably within the C2 domain, may also regulate protein stability or its sub-cellular localization (Lu et al. 2003). Additionally, the candidate tumor suppressor, PICT-1 (also known as GLTSCR2) promotes PTEN phosphorylation and stability, although the precise mechanism is unclear (Okahara et al. 2004). PEST sequences are commonly found in proteins that are targeted for degradation within the ubiquitin pathway. Phosphorylation-dependent polyubiquitination has been proposed as a potential molecular mechanism targeting PTEN for proteosomal degradation (Tolkacheva et al. 2001), however, the identification of the physiological ubiquitin ligase remains to be confirmed. Studies in which NEDD4-1 levels were manipulated through ectopic expression or RNA interference, identified this protein as the E3 ubiquitin ligase that polyubiquinates PTEN (Wang et al. 2007, 2008). However, studies in Nedd4-1 knockout mice, showed that Nedd4-1 was dispensable for the regulation of PTEN stability, activity and/or localization (Fouladkou et al. 2008). The reasons for these apparent contradictory findings are yet to be resolved.

Recently PTEN was demonstrated to interact with P-Rex2, a multi-domain protein that contains a Rac GEF domain and a domain with homology to the inositol polyphosphate 4-phosphatases. Within this latter domain P-Rex2 contains a CX$_5$R motif, but there is no evidence that P-Rex2 or the related P-Rex1 are catalytically active phosphoinositide phosphatases (see Chap. 7). The interaction between P-Rex2 and PTEN inhibits PTEN catalytic activity, and as a consequence cell proliferation and survival is enhanced (Fine et al. 2009).

8.2.3 Functional Roles of the Protein Versus Lipid Phosphatase Activity of PTEN

Lipid phosphatase-independent roles for PTEN are currently emerging; however, the identification of the G129E missense mutation in Cowden disease kindred, which selectively eliminates the lipid phosphatase activity of PTEN, while retaining its protein phosphatase activity, demonstrates that the lipid phosphatase activity of PTEN is essential for tumor suppression (Furnari et al. 1998; Myers et al. 1998). However, while PTEN is the central regulator of the PI3K signaling pathway, reports from many studies suggest a role for the protein phosphatase activity of PTEN, particularly in adhesion and cell migration, and the functional role of PTEN may indeed require both its lipid and protein phosphatase activities. Potential PTEN protein substrates include FAK, Shc and platelet-derived growth factor receptor (PDGFR) (Gu et al. 1999; Tamura et al. 1998; Mahimainathan and Choudhury 2004); however, whether these are *bona fide* physiological targets of PTEN remains unresolved (Davidson et al. 2010).

In *Dictyostelium discoideum* (*D. discoideum*), PTEN sub-cellular localization is restricted to the rear and lateral aspects of the cell, ensuring PtdIns(3,4,5)P$_3$ is localized to the leading edge of chemotaxing cells (Funamoto et al. 2002; Iijima and Devreotes 2002). The reconstitution of wild-type PTEN into PTEN-null mouse fibroblasts inhibits cell migration, and decreases the activation of the small GTPases Rac1 and Cdc42, dependent on the lipid phosphatase activity of PTEN (Liliental et al. 2000). However, studies using the G129E mutant (Furnari et al. 1998; Myers et al. 1998) have shown that PTEN can inhibit mammalian cell migration through a mechanism that is dependent on PTEN's protein phosphatase activity (Tamura et al. 1998, 1999a; Dey et al. 2008; Leslie et al. 2007; Gildea et al. 2004; Gu et al. 1999). The most compelling function of the protein phosphatase activity of PTEN in the regulation of mammalian cell migration is its proposed role in the auto-dephosphorylation of its carboxyl-terminal tail to reveal its lipid-binding C2 domain. PTEN inhibits cell migration in glioblastoma cells, independent of its lipid phosphatase activity, but reliant on its protein phosphatase activity (Raftopoulou et al. 2004). The C2 domain alone can also inhibit cellular migration in microinjection experiments, suggesting that this activity of the C2 domain may be regulated by the full-length protein. The dephosphorylation of PTEN is essential to C2 domain activation and is dependent solely on the protein phosphatase activity of PTEN. Raftopoulou et al. identified the specific dephosphorylation of residue Thr^{383} as important in this process; however analysis in many cell types by multiple groups using phospho-specific antibodies has failed to delineate the significance of this site as compared to the other phosphorylation cluster sites Ser^{380}, Thr^{382} and Ser^{385} in the regulation of PTEN's activity and function (Odriozola et al. 2007; Leslie et al. 2007; Rahdar et al. 2009). PTEN has previously been shown to exhibit preferential protein substrate specificity toward highly acidic proteins and peptides (Myers et al. 1997). The PTEN carboxyl-terminal tail is predominantly acidic, which may implicate the protein phosphatase activity of PTEN in its autodephosphorylation, leading to the activation of the C2 domain and inhibition of cellular migration. In support of this, a 71 amino acid region within the carboxyl-terminal tail of PTEN has been identified as an auto-inhibitory domain that regulates membrane localization and catalytic activity through an intramolecular association with the CBRIII motif of the C2 domain (Odriozola et al. 2007). One report suggests that PTEN is involved in the regulation of two distinct processes that require the co-operation of both its lipid and protein phosphatase activities to mediate cell migration during embryonic development (Leslie et al. 2007). The protein phosphatase activity of PTEN is required for the control of epithelial-to-mesenchymal transition (EMT) via the autodephosphorylation of its carboxyl-terminal domain, while the lipid phosphatase activity of PTEN regulates PtdIns(3,4,5)P$_3$-dependent cell polarization and directionality of mesodermal cell migration. A recent study has also contributed to the contention that both the lipid and phosphatase activities of PTEN are required to act in co-operation to regulate physiological processes. Davidson et al. generated a novel PTEN mutant, Y138L, which retains lipid phosphatase activity, but lacks phosphatase activity toward protein substrates (Davidson et al.

2010). Using this mutant alongside the well-described G129E mutant, the role of the lipid and protein phosphatase activities of PTEN in physiological processes was further delineated. The lipid phosphatase activity of PTEN regulated cell proliferation in soft agar and cell spreading. Adherent cell migration was regulated by either the protein or lipid phosphatase activities of PTEN; whereas cellular invasion required the coordinated actions of both activities. Therefore, the lipid and protein phosphatase activities of PTEN may be required for the regulation of cellular processes important in development and disease prevention.

8.2.4 A Nuclear Function for PTEN

Apart from its function as a negative regulator of PI3K-mediated signaling pathways at the plasma membrane, a role for PTEN within the cell nucleus is currently emerging. The localization of PTEN within the nucleus has been described in a range of both normal and tumor cells, with nuclear exclusion of PTEN associated with increased cancer progression (Zhou et al. 2002; Perren et al. 2000; Fridberg et al. 2007). At the functional level, targeted expression of PTEN within the nucleus does not affect catalytic activity *in vitro*, but leads to loss of PTEN function in cellular assays of proliferation, promotes cell cycle arrest and inhibits anchorage-independent cell growth (Ginn-Pease and Eng 2003; Liu et al. 2005b; Chung and Eng 2005; Denning et al. 2007). Therefore, control of PTEN localization may become a future therapeutic tool.

While a traditional NLS (nuclear localization signal) has not been identified in PTEN to date, a number of mechanisms regulating PTEN nuclear localization have been proposed. PTEN has been shown to enter the nucleus through passive diffusion in a RAN (Ras-related nuclear protein)-independent manner (Liu et al. 2005a). Putative NLS-like sequences have additionally been identified in PTEN that are required for nuclear import, mediated through interaction with MVP (major vault protein) (Chung et al. 2005). A further mechanism has identified both putative NLS and nuclear exclusion motifs as necessary for nuclear localization of PTEN, mediated through currently unidentified importin proteins and RAN (Gil et al. 2006). Finally, mono-ubiquitination is emerging as a critical means of regulating PTEN localization. NEDD4-1-mediated mono-ubiquitination of Lys^{289} or Lys^{13} residues within PTEN has been identified as a molecular mechanism that regulates nuclear import of the protein (Trotman et al. 2007). Although the nuclear function of PTEN remains to be fully characterized, the K289E mutation in the carboxyl-terminal tail of PTEN is associated with Cowden disease. This point mutation does not affect catalytic activity or plasma membrane localization (Georgescu et al. 2000), but prevents mono-ubiquitination at this site (Trotman et al. 2007). Alternative regulation of mono-ubiquitination of PTEN occurs through the opposing actions of HAUSP (herpesvirus-associated ubiquitin-specific protease) and PML (promyelocytic leukemia protein) via the adapter protein DAAX (death domain-associated protein) (Song et al. 2008). PTEN localization is abnormal in acute promyelocytic leukemia where PML function is impaired; and HAUSP is over-expressed in human prostate cancer and is associated with nuclear exclusion of PTEN.

While a nuclear pool of PtdIns(3,4,5)P$_3$ has been identified, this may be insensitive to PTEN expression (Lindsay et al. 2006), suggesting a phosphoinositide phosphatase-independent role for nuclear PTEN. In support of this, several groups have described phosphatase-independent functions of PTEN within in the nucleus, which may promote chromosome stability. These functions are predominantly associated with the regulation of protein interactions within the nucleus. Phosphatase-independent protein interactions between PTEN and p300 in the nucleus induce hyper-acetylation of p53, inducing cell cycle arrest in response to DNA damage (Liu et al. 2006). An association between loss of PTEN and chromosomal fragmentation has recently been described, suggesting a possible role for PTEN in DNA repair mechanisms. Endogenous PTEN was identified at the centromere, where it associated with the core centromeric protein Cenp-C, a protein required for kinetochore assembly and also during mitosis for metaphase to anaphase transition (Shen et al. 2007). This association was mediated via the carboxyl-terminus of PTEN, independently of catalytic activity.

8.2.5 *PTEN Function as Revealed by Mouse Knock-out Studies*

To dissect PTEN function, both global and tissue-specific deletion of PTEN in mice have been undertaken over the last 10 years. These studies have revealed roles for PTEN in autoimmune disease, non-alcoholic steatohepatitis, insulin hypersensitivity, heart failure, angiogenesis via regulation of endothelial cell function, macroencephaly, bone density, respiratory distress syndrome, immunoglobulin class switching, and resistance to hair graying to name a few (Knobbe et al. 2008). Homozygosity for a null mutation of *Pten* in mice results in early embryonic lethality (Di Cristofano et al. 1998, 1999; Podsypanina et al. 1999; Stambolic et al. 1998; Suzuki et al. 1998). Many different tissue-specific mouse *Pten* knockouts have been generated and their phenotypes are summarized in Table 8.1. For more detailed descriptions of conditional Pten mutant mice the reader is referred to recent reviews (Suzuki et al. 2008; Knobbe et al. 2008).

8.2.6 *Disruption of PTEN Correlates with Tumorigenesis and Cancer Progression*

After p53, *PTEN* is the second most frequently mutated tumor suppressor gene in human cancer. It was identified as the tumor suppressor gene at the 10q23 human chromosomal locus, a region frequently mutated in a vast range of sporadic cancers (Li and Sun 1997; Steck et al. 1997). The classification of PTEN as a tumor suppressor is further sustained through the identification of germline mutations of *PTEN* in the autosomal dominant cancer predisposition syndromes, Cowden disease, Lhermitte-Duclos disease and Bannayan-Zonana syndrome. While targeted disruption of *Pten* in mice results in early embryonic death between embryonic day 6.5 and 9.5, the phenotype exhibited in heterozygotes varies, possibly as a result of variations in targeting constructs; however, increased susceptibility to tumor formation is

Table 8.1 Genetically modified PTEN murine models

PTEN knockout mouse model	Phenotype	Reference
Constitutive Pten knockout mice		
Global	Homozygotes: Embryonic death at E6.5-E9.5. Heterozygotes: Increased susceptibility to tumor development in multiple tissues. Increased autoimmune responses	(Di Cristofano et al. 1998, 1999; Podsypanina et al. 1999; Stambolic et al. 1998; Suzuki et al. 1998)
Conditional Pten knockout mice (single mutants)		
Adipocyte-specific (*aP2Cre*)	Improved systemic glucose tolerance and insulin sensitivity. Increased resistance to diabetes	(Kurlawalla-Martinez et al. 2005)
B-cell-specific (*CD19Cre*)	Impaired immunoglobulin class switching and defective B-cell homeostasis. Hyper-proliferation, resistance to apoptosis and enhanced migration of splenic B-cells. Abrogation of BCR-mediated apoptosis and restoration of BCR-induced cell cycle progression via PtdIns(3,4,5)P$_3$-dependent signaling pathways in immature B cells	(Suzuki et al. 2003b; Anzelon et al. 2003; Cheng et al. 2009)
Cardiomyocyte-specific (*MckCre*)	Cardiac hypertrophy from 10 weeks of age and decreased cardiac contractility	(Crackower et al. 2002)
Cerebellum-specific (*En2Cre-neuronal and glial cells of the vermis of the cerebellum*) (*L7Cre-Purkinje cells*)	Reduced proliferation and progressive loss of Purkinje cells, beginning in early postnatal development, characterized by increasing vacuolation of the cells and the accumulation of fibrillary inclusions. Increased cerebellar size, neurons with larger soma size and thickened dendrites, dysplastic astrocytes and abnormally localized oligodendrocytes	(Marino et al. 2002)
Chrondrocyte-specific (*Col2a1Cre*)	Contrasting phenotype reported in two independent studies. Ford-Hutchinson et al. showed increased skeletal size, increased vertebrae size, and primary spongiosa development. They described disorganization of long bone growth plates, matrix overproduction and accelerated hypertrophic differentiation. No evidence of hamartoma, benign bone lesions, or chondrosarcoma was reported, however, 2/12 mice followed for a period of 12 months, developed metastatic osteosarcoma. However, Yang et al. described a phenotype with chondrocyte-specific deletion of *Pten* resulting in dyschondroplasia, as a result of delayed chondrocyte differentiation and decreased proliferation. Pathological cartilaginous neoplasms were evident from birth resembling human enchondroma	(Ford-Hutchinson et al. 2007; Yang et al. 2008)

Table 8.1 (continued)

PTEN knockout mouse model	Phenotype	Reference
Endothelial cell-specific (*Tie2Cre*)	*Homozygotes:* Embryonic death prior to E11.5, associated with bleeding and cardiac failure as a result of impaired recruitment of pericytes and vascular smooth muscle cells to blood vessels, and cardiomyocytes to the endocardium. *Heterozygotes:* Enhanced tumorigenesis due to increased angiogenesis, associated with altered expression of endothelial cell receptor proteins, vascular adhesion molecules and vascular growth factors	(Hamada et al. 2005)
Hematopoietic stem cells (*pIpc-inducible Mx1Cre*)	Rapid development of myeloproliferative disorders within 4–6 weeks of age, which progressed to acute myeloid leukemia or acute lymphoblastic leukemia	(Yilmaz et al. 2006; Zhang et al. 2006)
Hepatocyte-specific (*AlbCre*)	Increased hepatomegaly, steatohepatitis and an accumulation of triglycerides similar to that in human non-alcoholic steatohepatitis (NASH). Increased levels of C16:1 and C18:1 acids within the liver. Increased induction of adipocyte-specific genes (adipsin, adiponectin, and aP2) and lipogenic genes. Increased development of liver adenoma and hepatocellular carcinoma within 78 weeks of age. Decreased serum glucose levels due to insulin hypersensitivity, and reduced serum insulin. Hepatic steatosis, inflammation, and carcinogenesis in Pten-deficient mice were attenuated in females compared to males. Decreased hepatic protein levels of apoB100 and microsomal triglyceride transfer protein	(Horie et al. 2004; Anezaki et al. 2009; Stiles et al. 2004; Qiu et al. 2008)
Hypothalamic POMC-specific (*PomcCre*)	Development of hyperphagia and sexually-differential diet-induced obesity. In male mice, increased body weight was associated with increased consumption of a normal diet. Females maintained normal body weight on a normal chow diet, but became obese on a high-fat diet. Pomc-expressing neurons were larger in Pten-null cells and had more efferent fibers than wild-type neurons	(Plum et al. 2006)
Intestinal epithelial cell-specific (*Tg(Cyp1a1-cre)1Dwi* or *Tg(Vil-cre/ESR1)23Syr*)	No effect on the normal architecture or homeostasis of the epithelium within adult or embryonic epithelial cells	(Marsh et al. 2008)
Intestinal stem cell-specific (*Mx1Cre*)	Deletion of PTEN in the epithelial and stromal cells of the small intestine results in increased proliferative intestinal stem cells that initiate intestinal polyp formation, resembling intestinal polyposis, within 1 month	(He et al. 2007)

Table 8.1 (continued)

PTEN knockout mouse model	Phenotype	Reference
Keratinocyte-specific (*k5Cre, MMTVCre*)	Epidermal hyperplasia, hyperkeratosis, and shaggy hair. Decreased body weight, with approximately 90% of mutants dying from malnutrition within 3 weeks of birth, as a result of esophageal hyperkeratosis. Mice that survive beyond 2 months of age show increased susceptibility to the development of squamous papilloma squamous cell carcinoma, sebaceous carcinoma and adenocarcinoma of the sweat gland within 9 months	(Suzuki et al. 2003a; Yang et al. 2005; Backman et al. 2004)
Lung epithelial cell-specific (*Doxycycline-inducible SP-C-rtTA/(tetO)$_7$-Cre*)	Ninety percent of mutant mice that receive doxycycline in utero (E10-16) die of hypoxia within 2 h of birth. Hyperplasia of bronchioalveolar epithelial cells and myofibroblast precursors. Enlarged undifferentiated alveolar epithelial cells, and impaired production of surfactant proteins. Increased numbers of bronchioalveolar stem cells, which are putative initiators of lung adenocarcinoma. Increased susceptibility of the surviving mutants, and mice receiving doxycycline postnatally (P21-27), to spontaneous or induced lung adenocarcinoma	(Yanagi et al. 2007)
Macrophage-specific (*LysMCre*)	Increased susceptibility to infection and reduced clearance of infection, resulting from decreased secretion of tumor necrosis factor, correlating with reduced expression of inducible nitric oxide synthase and reduction in nitric oxide production	(Kuroda et al. 2008)
Mammary-specific (*MMTVCre*)	Enhanced lobulo-alveolar development, excessive ductal branching, delayed involution and severely reduced apoptosis in mutant mammary tissue. Increased development of mammary tumors in mutant females within 2 months of age	(Li et al. 2002a)
Melanocyte-specific (*DctCre*)	Increased melanocytes in the dermis of perinatal mice. Protection against hair graying. No change in spontaneous tumor development, but increased susceptibility to the development of large nevi and melanoma after carcinogen exposure	(Inoue-Narita et al. 2008)

Table 8.1 (continued)

PTEN knockout mouse model	Phenotype	Reference
Neuronal cell-specific (*GfapCre*)	Increased neurological defects including seizures and ataxia within 6–9 weeks of birth, death within 29–48 weeks of age. Macroencephaly, hydrocephaly and dysplasia of several neural cell populations, including enlarged soma in Pten-null neurons, accompanied by enlarged caliber of neuronal projections and increased dendritic spine density. Abnormal synaptic structures and severe myelination defects, with weakened synaptic transmission and plasticity at excitatory synapses. Disorganized architecture as a result of neuronal migratory defects that lead to abnormal accumulation of granule cells in the external granule cell layer	(Backman et al. 2001; Kwon et al. 2001; Yue et al. 2005; Fraser et al. 2008)
Neuronal (differentiated cells of cortical layer III-IV *NseCre*)	Increased soma hypertrophy, macroencephaly and premature death in the forebrain and hippocampus. Increased axonal outgrowth and altered spine morphology. Abnormal social interaction, increased anxiety, hyperactivity and increased sensory sensitivity, reminiscent of autistic spectrum disorders in humans. Increased sporadic spontaneous seizures	(Kwon et al. 2006; Ogawa et al. 2007)
Neuronal progenitor cell-specific (*NestinCre*)	Early perinatal death. Enlarged, histoarchitecturally abnormal brains, as a result of increased cell proliferation, decreased cell death and enlarged cell size	(Groszer et al. 2001)
Neutrophil-specific (*LysMCre*)	Increased superoxide production. Enhanced actin polymerization, membrane ruffling, and pseudopod formation, which resulted in increased chemotaxis and migratory speed, but loss of directionality	(Zhu et al. 2006a)
NKT cell-specific (*LckCre*)	Reduced levels of serum γ-interferon, in response to NKT activation. Impaired NKT cell development, with increased numbers of immature NKT cells and decreased numbers of mature NKT cells, with impaired functionality. Increased tumorigenesis, resulting from decreased immune surveillance	(Kishimoto et al. 2007)
Oocyte-specific (*GDF9Cre*)	Premature activation of the complete primordial follicle pool, resulting in ovarian failure	(Reddy et al. 2008)
Osteoblast-specific (*OcCre*)	Progressively increasing bone mineral density throughout life, as a result of increased differentiation and reduced apoptosis	(Liu et al. 2007)

Table 8.1 (continued)

PTEN knockout mouse model	Phenotype	Reference
Ovarian granulosa cell-specific (*Cyp19Cre*)	Increased ovary volume, increased production of oocytes during ovulation and increased number of pups	(Fan et al. 2008)
Pancreatic-specific (*Pdx1Cre, RipCre*)	Increased islet cell numbers and total islet mass evident at P15 and persisting through adulthood in RipCrePten$^{flox/flox}$ mice, in which Pten is deleted in pancreatic β-cells only. Hypoglycemic and diabetic resistance in adult mice, with smaller body mass and reduced lifespan, but no evidence of pancreatic tumor development. However, Pdx1CrePten$^{flox/flox}$ mice, which lack Pten expression in all pancreatic cells, show progressive replacement of the acinar pancreas with highly proliferative ductal structures, containing mucins and expressing markers of pancreatic progenitor cells, with increased development of ductal malignancy. Mice exhibited delayed onset of streptozotocin-induced diabetes and sex-biased resistance to high-fat-diet -induced diabetes	(Stanger et al. 2005; Stiles et al. 2006; Nguyen et al. 2006; Tong et al. 2009)
Primordial germ cell-specific (*TNAPCre*)	Bilateral testicular teratoma development in all new-born males, as a result of impaired mitotic arrest and outgrowth of cells with immature characters. Increased pluripotent embryonic germ cell production in both sexes	(Kimura et al. 2003)
Prostate-specific (*PbCre, PbCre4, MMTVCre PSACre, PSACreER(T2)*)	Development of prostatic hypoplasia within the early postnatal period, rapidly progressing to high-grade prostatic intraepithelial neoplasia, then to invasive adenocarcinoma, and metastatic carcinoma	(Ma et al. 2005; Backman et al. 2004; Trotman et al. 2003; Wang et al. 2003; Ratnacaram et al. 2008)
Retinal ganglionic cell-specific (*AAVCre surgical delivery*)	Increased RGC survival, protein synthesis and axon regeneration following optic nerve injury	(Park et al. 2008)
Retinal pigment epithelium-specific (*TRP1Cre*)	Progressive degeneration of both RPE cells and their photoreceptors due to an inability of RPE cells to maintain basolateral adhesions, the development of an epithelial-to-mesenchymal transition (EMT), and subsequent cellular migration out of the retina	(Kim et al. 2008)

Table 8.1 (continued)

PTEN knockout mouse model	Phenotype	Reference
Smooth muscle cell-specific (*TaglnCre, Sm22αCre*)	Elevated incidence of smooth muscle cell hyperplasia and abdominal leiomyosarcomas, within 2 months of age. Early perinatal lethality, increased development of medial and intimal smooth muscle cell hyperplasia, and vascular recruitment of progenitor/proinflammatory cells	(Hernando et al. 2007; Nemenoff et al. 2008)
Skeletal muscle-specific (*MckCre*)	Enhanced protection from insulin resistance and diabetes on a high-fat diet. Reversal of high-fat diet-induced impairment of muscle regeneration	(Wijesekara et al. 2005)
T-cell-specific (*LckCre, CD4Cre*)	Defective thymic negative selection, increased autoimmune responses. Lymphadenopathy, splenomegaly, and enlarged thymus within 6–8 weeks. Post-natal death prior to 20 weeks of age as a result of malignant T-cell lymphoma	(Suzuki et al. 2001; Hagenbeek and Spits 2008; Hagenbeek et al. 2004; Xue et al. 2008)
Thyroid follicular cell-specific (*TpoCre*)	Increased induction of thyroid hyperplasia and diffuse colloid goiter, caused by an increased thyroid mitotic index. Increased neoplastic transformation within 10 months of age	(Yeager et al. 2007)
Ureteric bud epithelial cell-specific (*HoxB7Cre*)	Defective branching morphogenesis in developing mouse kidneys, mislocalization of glomeruli and post-natal lethality before P26	(Kim and Dressler 2007)
Urothelial cell-specific (*FabpCre*)	Increased urothelial hyperplasia with complete penetrance at 6–8 weeks after birth. Increased spontaneous pedicellate papillary transitional cell carcinoma and increased susceptibility to chemically-induced carcinogenesis	(Tsuruta et al. 2006; Yoo et al. 2006)
Conditional Pten knockout mice (double mutants)		
B-cell-specific (*CD19Cre*) Pten & Ship1	Development of spontaneous and lethal B cell neoplasms consistent with marginal zone lymphoma or follicular or centroblastic lymphoma	(Miletic et al. 2010)
Central nervous system-specific (*hGFAPCre*) Pten & p53	Complete CNS deletion of *Pten* resulted in lethal hydrocephalus in early postnatal life. hGFAP-Cre$^+$;p53$^{lox/lox}$;Pten$^{lox/+}$ mice presented with acute-onset neurological symptoms, including seizure, ataxia and/or paralysis. Mice developed penetrant acute-onset high-grade malignant glioma with clinical, pathological and molecular resemblance to primary glioblastoma multiforme in humans	(Zheng et al. 2008)

Table 8.1 (continued)

PTEN knockout mouse model	Phenotype	Reference
Endothelial cell-specific (*Tie2Cre*) *Pten* & *Pdk1*	Slight delay in embryonic lethality, resulting from defective vascular remodeling and cardiac development	(Feng et al. 2010)
Intestinal epithelial cell-specific (*Tg(Cyp1a1-cre)1Dwi* or *Tg(Vil-cre/ESR1)23Syr*) *Pten* & *Apc*	Accelerated tumorigenesis through increased activation of Akt, resulting in the rapid development of adenocarcinoma	(Marsh et al. 2008)
Keratinocyte-specific (*k5Cre*) *Pten* & *Smad4*	Development of early onset hyperplasia and dysplasia in the esophageal and forestomach epithelia and accelerated tumor formation in the forestomach. Squamous cell carcinomas developed at 1 month of age, which progressed to invasive SCC with 100% penetrance by 2 months. Mice exhibited progressive growth retardation and post-natal death between P10 and P100	(Teng et al. 2006)
Prostate-specific (*PBCre4*) *Pten* & *Trp53*	Development of invasive prostate cancer by 2 weeks post-puberty, with disease progression and death prior to 7 months of age	(Chen et al. 2005)
Renal tubular cell-specific (*NseCre*) *Pten* & *TSC1*	Development of severe polycystic kidney disease and increased post-natal lethality	(Zhou et al. 2009)
T-cell-specific (*LckCre*) *Pten* & *Mnk1/Mnk2*	Suppression of malignant T-cell lymphoma associated with T-cell-specific deletion of Pten in mice	(Ueda et al. 2010)
Urothelial cell-specific (*AdenoCre—surgical delivery*)*Pten* & *p53*	Development of bladder tumors with 100% penetrance by 6 months of age, displaying the histological features of carcinoma in situ, as well as high-grade invasive carcinoma	(Puzio-Kuter et al. 2009)

observed in multiple tissues, signifying haploinsufficiency of *Pten* in mouse models (Di Cristofano et al. 1998; Podsypanina et al. 1999; Suzuki et al. 1998).

Indeed, PTEN is frequently mutated or its expression lost in many human cancers including glioblastomas, breast, kidney and uterine endometrioid carcinomas, lung cancer, colon cancer, and melanoma (Jiang and Liu 2008; Salmena et al. 2008; Steck et al. 1997). The incidence of somatic mutation or deletion of *PTEN* is high in high-grade glioblastoma (estimated prevalence 30–40%), breast (10%), melanoma (7–20%), prostate cancer (15%), and endometrial cancer (50%). *PTEN* mutations have been reported at lower rates in bladder, lung, ovary, colon cancers, and in lymphoma (Cairns et al. 1997; Gronbaek et al. 1998; Kim et al. 1998; Kohno et al. 1998;

Sansal and Sellers 2004). *PTEN* inactivation results in multiple abnormal processes leading to abnormal cell polarity and invasion, cell proliferation and survival, cell architecture, chromosomal integrity, cell cycle progression, and stem cell self-renewal. In the early stages of some cancers, including prostate, breast, colon, or lung cancers, monoallelic *PTEN* mutation or deletion is detected, however, the second *PTEN* allele remains active (Salmena et al. 2008), and it is only in late stage or metastatic cancers that biallelic loss of *PTEN* is observed. Most *PTEN* mutations lead to loss of phosphatase activity both to phosphoinositide and protein substrates.

Examination of tissue-specific *Pten* deletion in mice further supports a role for Pten in the prevention of tumorigenesis, with hyper-proliferation and neoplastic change observed in tissues where *Pten* is selectively inactivated (Backman et al. 2004; Horie et al. 2004; Yanagi et al. 2007; Yeager et al. 2007). Furthermore, injection of *Pten*-deficient ES cells, PTEN-null tumor cells or catalytically inactive PTEN mutants in nude or syngeneic mice results in enhanced tumor generation, due to increased anchorage-independent cell growth and abnormal differentiation (Di Cristofano et al. 1998; Li and Sun 1998). Studies in cell culture models have revealed loss of PTEN function results in increased proliferation, growth and survival, correlating with increased basal PtdIns(3,4,5)P_3 levels and enhanced Akt activation (Subramanian et al. 2007; Li and Sun 1998; Furnari et al. 1998; Myers et al. 1998; Stambolic et al. 1998; Lee et al. 1999; Davies et al. 1998; Haas-Kogan et al. 1998). Conversely, expression of PTEN induces apoptosis and promotes cell cycle arrest (Furnari et al. 1998; Li and Sun 1998; Maehama and Dixon 1998; Davies et al. 1998; Koul et al. 2001; Tamura et al. 1999a; Stambolic et al. 1998).

PTEN's haploinsufficiency is sufficient to promote tumor formation. Analysis of a series of hypomorphic mouse mutants, developed in order to assess a correlation between PTEN levels and cancer progression, reveals that heterozygous *Pten* loss in a mouse model of prostate cancer leads to prostate epithelial hyperplasia and low-grade lesions with incomplete penetrance (Trotman et al. 2003). Further reduction in *Pten* is associated with massive prostate hyperplasia in all mice, with accelerated tumor progression, and complete loss of *Pten* leads to highly invasive and aggressive cancer. These studies suggest the reduction in PTEN expression below the heterozygous loss of function level may lead to more aggressive cancer. Recently it was reported even a subtle reduction (20%) in *Pten* expression may promote cancer susceptibility in mice. *Pten* hypermorphic mice (*Pten*$^{hy/+}$), which express 80% of normal levels of Pten, develop a variety of tumors, with the most common being breast cancer (Alimonti et al. 2010). These mice show reduced survival, develop autoimmune disease, with lymphadenopathy and splenomegaly, but significantly, females show an increased susceptibility to epithelial cancers including breast cancer (∼75%) and uterine cancer (67%). Even this small reduction in PTEN by 20% is sufficient to promote the activation of a pro-proliferative gene expression signature.

Interestingly, in some tissues complete loss of PTEN from untransformed cells may trigger cellular senescence, a process which protects against tumor initiation (Alimonti et al. 2010). Cell senescence is a very stable form of cell cycle arrest, which is activated in response to stress, including oncogenic signaling and telomere shortening (Collado and Serrano 2010). Recent studies have revealed constitutive activation of Akt promotes senescence via inhibition of the transcription factors FOXO1/O3 (Nogueira et al. 2008).

Altered PTEN expression in cancer may be due to inherited germ line mutations, sporadic mutations or chromosomal alterations, transcriptional repression, epigenetic silencing, post-transcriptional gene regulation, post-translational modification, and aberrant PTEN localization, which can in turn regulate the initiation, progression and long term survival from cancer (Sasaki et al. 2009). Recent studies have revealed a significant role for post-transcriptional silencing of PTEN by multiple microRNAs (miRNAs) from precursors with a single hairpin structure (*miR-21*, *miR-22*, *miR-214* and *miR-205*) or from a polycistronic structure (*mir-17-92*, *mir-106b*, *mir-367-302b* and *mir-221-22*) that recognizes target sequences in PTEN and thereby regulates its expression, reviewed in (He 2010). Interestingly, the related *PTENP1* may act as a decoy for the same miRNA species to rescue *PTEN* loss of expression (He 2010). Consistent with this contention, *PTENP1* chromosome deletion has been reported in colon and breast cancer associated with decreased PTEN expression (Poliseno et al. 2010b).

PTEN localizes to the cytosol, plasma membrane and nucleus, and its function as a tumor suppressor appears to be dependent on its appropriate localization within the cell. Numerous inactivating mutations of *PTEN* have been identified in both hereditary and spontaneous human cancers that map to the phosphatase domain of *PTEN*, however, many mutations map to areas outside of this region (Marsh et al. 1998). These residues may be crucial for the regulation of protein stability, or localization patterns. For example, mutations that map to the C2 domain, or that disrupt the phosphatase/C2 domain interface, have been identified, which prevent the correct positioning of PTEN at the plasma membrane and thereby access to PtdIns(3,4,5)P_3 (Lee et al. 1999). Loss-of-function hereditary mutations have also been described within the promoter region of *PTEN*, leading to increased Akt signaling (Zhou et al. 2003). As described earlier, the nuclear import-defective PTEN mutants, K289E and K13E, are associated with hereditary and spontaneous cancer respectively (Trotman et al. 2007; Duerr et al. 1998). PTEN mutations within the conserved polybasic amino-terminal motif required for PtdIns(4,5)P_2 binding have also been identified in sporadic cancer. While retaining catalytic activity, mutation of PTEN within this region disrupts PtdIns(4,5)P_2 binding, thus preventing the correct orientation of PTEN at the plasma membrane and access to PtdIns(3,4,5)P_3 (Han et al. 2000; Walker et al. 2004).

Loss of PTEN is associated with highly invasive cancer, implicating this enzyme in cell motility and invasion. Indeed, disruption of *Pten* in *D. discoideum* impairs cell polarization, actin polymerization and both the speed and directionality of cell migration (Iijima and Devreotes 2002; Funamoto et al. 2002). In contrast, disruption of PTEN in numerous mammalian cells increases cell migration (Gao et al. 2005; Gu et al. 1999; Liliental et al. 2000; Suzuki et al. 2003b; Tamura et al. 1998, 1999b). Chemoattractant-induced Transwell migration of Pten-deficient murine neutrophils is increased, however, defects in directionality are evident in single-cell chemotaxis assays (Subramanian et al. 2007) and in the prioritization of chemoattractant cues (Heit et al. 2008). On the other hand, these results contrast with studies in human cell lines that show that while loss of PTEN increases the rate of chemokine-induced migration, it does not affect directionality (Lacalle et al. 2004). Also in primary neutrophils a more dominant role for the PtdIns(3,4,5)P$_3$ 5-phosphatase, SHIP1, than PTEN has been shown (Nishio et al. 2007). Interestingly, subventricular zone precursor cells (Li et al. 2002a) and isolated primary B cells (Fox et al. 2002) from *Pten*$^{+/-}$ mice also exhibit greater motility than wild-type cells, providing further evidence that in mice, *Pten* is haploinsufficient.

PTEN may exert a tumor suppressive effect via regulation of surrounding fibroblasts. Recently PTEN loss of function has been linked to the regulation of stroma and tumor cell signaling (Trimboli et al. 2009). Genetic inactivation of *Pten* in stromal fibroblasts of mouse mammary glands accelerates the initiation and malignant transformation of mammary epithelial tumors. Notably there was significant remodeling of the extracellular matrix with increased angiogenesis. Global gene expression profiling of PTEN-depleted mammary stromal cells reveals activation of the Ets2-specific transcription program (Trimboli et al. 2009). *Ets2* inactivation in *Pten* stroma-deleted tumors decreases tumor growth and progression, revealing the Pten–Ets2 axis as a stroma-specific signaling pathway that suppresses mammary epithelial tumors.

8.2.7 The Tumor Context of PTEN Loss of Function

There are several points which will be considered here regarding the occurrence of changes in other oncogenes and tumor suppressors in tumors, in addition to PTEN. Firstly, does additional activation of other components of the broad PI3K/growth factor receptor signaling network occur in PTEN null tumors? Secondly, are there other independent pathways that are favorably mutated in parallel with PTEN in certain tumor types? And thirdly, are there oncogenes and tumor suppressors that affect malignant transformation indirectly through their influence on PTEN function, in the way that murine double minute 2 (MDM2) amplification drives tumorigenesis through suppression of p53 function?

In other major functional pathways implicated in cancer, such as the Rb and MDM2/p53 pathways, mutation of more than one component of the same definable functional pathway is generally very rare (Cancer Genome Atlas Research Network 2008). However, in the PI3K pathway, mutation of multiple components, such as both PTEN and p110α PI3K is not uncommon (Yuan and Cantley 2008; Cancer Genome Atlas Research Network 2008). Because many factors independently influence cellular PtdIns(3,4,5)P$_3$ levels and localization, and because there are many PtdIns(3,4,5)P$_3$-binding proteins and Akt substrates, that regulate diverse biological processes with different dose responses to changes in PtdIns(3,4,5)P$_3$ levels, it seems unsurprising that multiple mutations within the pathway are able to drive further selective advantages to a tumor.

Extensive genomic and expression analyses of large numbers of tumors has confirmed their classification into identifiable groups that share patterns of genomic and expression changes. For example, 94% of the "classical" sub-type of glioblastomas share high level amplification of EGFR and homozygous deletion of CDKN2A (Verhaak et al. 2010). However, in many tumor types, PTEN loss is common in many independently identifiable sub-groups of tumors. For example, in a categorization of glioblastomas, loss of one allele of PTEN occurs at very high levels in all four categories of glioblastoma, with lowest frequency being in "proneural" tumors, but still as high as 67% (Verhaak et al. 2010). However, there are cases with notable association of PTEN mutation with other events. For example, several studies have identified a strong association between TMPRSS2-ERG translocations and loss of PTEN in a large fraction of human prostate cancers (Taylor et al. 2010) and have demonstrated the functional significance of this genetic interaction in transgenic mice (Carver et al. 2009). Similarly, co-operating pathways have been identified in experiments in mice, with for example, experimental deletion of Pten from T-cells leading to lymphomas that are also found to display myc translocations (Liu et al. 2010).

Finally, data has emerged over the last few years showing that several regulators of PTEN function are mutated or over-expressed in tumors and that this may represent a mechanism by which they affect tumor development indirectly through the regulation of PTEN. These include P-REX2 and SIPL1 that appear to directly inhibit PTEN activity, RAK and PICT1/GLTSCR2 that stabilize the PTEN protein and the E3 ubiquitin ligase, NEDD4 (Fine et al. 2009; He et al. 2010; Okahara et al. 2006; Wang et al. 2007; Yim et al. 2009). These findings have been recently reviewed in depth (Leslie et al. 2010).

8.2.8 PTEN and Cellular Senescence

Loss of PTEN function appears to be a common early event in endometrial cancer (Mutter et al. 2000). However, this appears not to be the case in most other tumor types, and PTEN loss has been described at highest frequencies in late stage tumors in many tissues (Salmena et al. 2008). In particular, complete *PTEN* loss is infrequent in early stages of prostate cancer (Taylor et al. 2010). Interestingly, acute complete loss of PTEN may induce cell senescence, a cellular program in mitotic cells that induces irreversible growth arrest as an anti-tumor mechanism. This program may be initiated by tumor suppressor genes in response to DNA damage and/or oncogene activation (Campisi and d'Adda di Fagagna 2007). Complete acute loss of *Pten* from untransformed cells, rather than promoting enhanced proliferation, induces a significant senescence response that opposes tumor progression associated with enhanced p19Arf–p53 signaling (Chen et al. 2005). Therefore, complete loss of PTEN can oppose tumorigenesis by triggering a p53-dependent cellular senescence response. Inactivation of PTEN concomitantly with *p53* allows escape from senescence, promoting invasive cancer. PTEN-induced cell senescence has also been shown in primary human epithelial cells (Kim et al. 2007). PTEN-loss—induced cellular senescence occurs rapidly after Pten inactivation, in the absence of cellular proliferation and DNA damage checkpoint responses, and is associated with enhanced p53 translation. Pharmacological inhibition of PTEN or p53-stabilizing drugs potentiates senescence and its tumor suppressive potential (Alimonti et al. 2010).

Pten-loss—induced cellular senescence exerts its effects using molecular pathways distinct from oncogenic-induced cell senescence. Skp2 protein is an E3-ubiquitin ligase that mediates degradation of a number of proteins, including p21 and p27. Mice lacking Skp2 are viable, but *Skp2* inactivation restricts tumorigenesis by promoting cellular senescence during oncogenic conditions, such as following expression of the Ras oncoprotein. In the absence of Skp2, cells lacking Pten become more sensitized to senescence (Lin et al. 2010). Mice with loss of one copy of *Pten* and deficient in Skp2 ($Pten^{+/-}$; $Skp2^{-/-}$) are protected from cancer. Recent studies have revealed PTEN also acts as a critical determinant of cell fate between senescence and apoptosis in several glioma cell lines in response to ionizing radiation. Depletion of Akt or scavenging of reactive oxygen species (ROS) prevents radiation-induced senescence in PTEN-deficient glioma (Lee et al. 2011).

8.2.9 PTEN and the Brain

As previously described, germline mutations in *PTEN* are associated with Cowden disease, Lhermitte-Duclos disease and Bannayan-Zonana syndrome, which are collectively described as the PTEN hamartoma tumor syndromes (PHTS) (Salmena et al. 2008). Whether these are indeed separate syndromes, or only one syndrome with a broad clinical spectrum remains to be determined, however, neurological abnormalities are a prominent feature of these disorders. PTEN is widely expressed in the mammalian brain and is enriched in large pyramidal, Purkinje, olfactory and mitral neurons. Conditional deletion of *Pten* specifically in the mouse brain results in macrocephaly and increased neurological defects including seizures and ataxia, commencing within 6–9 weeks of birth (Groszer et al. 2001; Backman et al. 2001; Kwon et al. 2001). The macrocephaly is associated with abnormal brain patterning, as a consequence of increased neuronal cell number, enhanced cell survival, and enlarged cell size, as a result of enhanced PI3K/Akt signaling, subsequent activation of the mTOR complex and the promotion of protein translation. $Pten^{-/-}$ mice additionally exhibit dysplasia of neuronal cell populations, abnormal synaptic structures and severe myelination defects. The number of neuronal stem cells in the fetal brain is also regulated by PTEN, with loss of PTEN in neuronal stem cells shown to stimulate stem cell proliferation and enhance self-renewal capacity (Groszer et al. 2006). The observed phenotypes in $Pten^{-/-}$ mice recapitulate many of those described in PHTS, in particular Lhermitte-Duclos disease (Waite and Eng 2002). In addition, PTEN function is implicated in depression and anxiety (Bandaru et al. 2009), and also in response to drug addiction (Ji et al. 2006). Pharmacological inhibition of PI3K by wortmannin treatment leads to increased depression and anxiety (Bandaru et al. 2009) and impaired fear memory (Lin et al. 2001). Conditional knockout of

Pten in specific differentiated neurons results in increased anxiety and decreased social interaction suggesting that either up- or down-regulation of PI3K signaling, and thus too much or too little PtdIns(3,4,5)P$_3$, is critical to these processes (Kwon et al. 2006).

PTEN not only controls neuronal cell number and size but is also implicated in regulating responses to neuronal injury, including brain ischemia (Chang et al. 2007). Increased expression of PTEN in hippocampal neurons promotes neuronal cell death following exposure to the excitatory amino acid glutamate via regulation of the Akt signaling pathway, while decreased expression of endogenous Pten increases neuron resistance to seizure-induced cell death (Gary and Mattson 2002). In a rat model of transient cerebral ischemia, PTEN phosphorylation is enhanced in the ischemic core (Omori et al. 2002), which is proposed to decrease PTEN phosphatase activity (Torres and Pulido 2001) and thereby increase Akt signaling. In a rat model of chronic exposure to ethanol, increased PTEN expression is associated with cerebellar hypoplasia and increased neuronal cell death (reviewed in (Chang et al. 2007)). Therefore, downregulating the activity and/or expression of PTEN may be a novel therapeutic treatment for brain injury.

The precise molecular mechanisms surrounding the role of PTEN in ischemic neurodegeneration remain to be determined, and PTEN may co-operatively regulate numerous down-stream effectors. A direct role for PTEN has been described in the regulation of neurodegeneration during oxidative stress via mitochondria-dependent apoptosis. Neuronal precursor cells, that lack one copy of *PTEN*, exhibit increased resistance to oxidative stress-induced apoptosis (Li et al. 2002b). In many cells in response to growth factor stimulation, PTEN translocates to the plasma membrane to degrade PtdIns(3,4,5)P$_3$ and inhibit Akt signaling; however in staurosporine-treated hippocampal neurons, mitochondrial accumulation of PTEN is observed, consistent with a role for PTEN in the regulation of mitochondria-dependent apoptotic pathways. Knockdown of PTEN protects hippocampal cells from apoptotic damage by inhibiting staurosporine-induced release of cytochrome c and caspase-3 activity (Zhu et al. 2006b). Down-regulation of PTEN decreases the activity of apoptosis signal-regulating kinase 1 (ASK1), an upstream component of the mitogen-activated protein kinases (Wu et al. 2006). The activity of the pro-apoptotic c Jun N-terminal kinases 1/2 is reduced as a result of decreased PTEN activity and increased Akt activation following ischemic injury (Zhang et al. 2007). PTEN regulates the generation of intracellular ROS in response to oxygen–glucose deprivation (OGD) and neurotoxin 1-methyl-4-phenylpyridinium iodide (Zhu et al. 2007). Suppression of PTEN activity also complexes with the *N*-methyl-d-aspartate (NMDA) receptor (NMDAR), a subtype of excitatory glutamate receptor that promotes excitotoxicity-induced neuronal death, to inhibit extrasynaptic NMDAR activity, leading to protection against ischemic neuronal death *in vitro* and *in vivo* (Ning et al. 2004).

PTEN may also play a role in regulating neurodegeneration associated with Alzheimer disease (AD) and Parkinson disease. Decreased PTEN expression has been reported in the hippocampus of AD brains (Griffin et al. 2005). Three mitochondrial associated genes *parkin*, *DJ-1* and *PINK1* (PTEN-induced putative kinase 1), have recently been shown to be associated with early-onset recessive parkinsonism (reviewed in (van der Vegt et al. 2009)). PINK1 is a serine/threonine kinase, which is transcriptionally activated by PTEN (Valente et al. 2004), and phosphorylates TRAP1, preventing oxidative stress-induced mitochondrial cytochrome *c* release (Pridgeon et al. 2007). Parkin is an E3 ubiquitin ligase and DJ-1 functions in parallel to the PINK1/parkin pathway to maintain mitochondrial function in the presence of oxidative stress (Inzelberg and Jankovic 2007). Mutations in *PINK1* occur in familial Parkinson disease and loss of PINK1 function increases sensitivity to oxidative stress, followed by neuronal death (Gispert et al. 2009). DJ-1 inhibits PTEN's negative regulation of the PI3K/Akt/mTOR pathway, which exerts a pro-survival effect during oxidative stress (Delgado-Esteban et al. 2007). siRNA knock-down of DJ-1 enhances cellular sensitivity to oxidative stress in Drosophila and mice brains (Kim et al. 2005; Shendelman et al. 2004). DJ-1 and PINK1, with PTEN, may contribute to the regulation of neuroprotection and neurodegeneration. Thus when PTEN levels are high, DJ-1 signaling will decrease and the PI3K/Akt/mTOR pathway will be suppressed, enhancing sensitivity of neuronal cells to oxidative stress. When PTEN levels are low, PINK1 levels will also be low, and oxidative stress-induced neuronal mitochondrial death pathways may be enhanced (Bonifati et al. 2003).

8.2.10 The PTEN and p53 Association

Over several years, many studies have implied a specific functional connection between PTEN and p53, the stress-induced transcription factor that is itself the most frequently mutated tumor suppressor in human cancers (Liu et al. 2008). p53 may influence PTEN expression through the presence of potential p53-binding sites in the *PTEN* promoter. This mediates induction of *PTEN* transcription, as part of a p53 transcriptional program in response to stresses, such as gamma irradiation (Stambolic et al. 2001). Conversely, PTEN has been reported to affect p53 in several ways. p53 expression is normally maintained at low levels by the action of a proto-oncogenic E3 ubiquitin ligase, MDM2. Two reports simultaneously described the phosphorylation of MDM2 by the PTEN-regulated kinase, AKT, and showed that MDM2 ligase activity is increased by phosphorylation on these sites (Mayo and Donner 2001; Zhou et al. 2001). PTEN activity, or inhibition of PI3K, suppresses MDM2 transcription by negatively regulating its P1 promoter (Chang et al. 2004), thereby enhancing p53 levels. A direct physical interaction between PTEN and p53 has also been reported, and PTEN-null ES cells and immortalized mouse embryonic fibroblasts (MEFs) exhibit reduced p53 levels (Freeman et al. 2003). Together these studies provide good evidence that loss of PTEN can lead to reduced p53 function in several settings. However, more recent work has added a complication to this picture, showing that genetic deletion of both *Pten* copies from primary MEFs, or in the murine prostate, leads to an induction of p53 expression and cellular senescence as described here earlier (Chen et al. 2005). A resolution to these somewhat contradictory findings and a better understanding of the complex relationship(s) between PTEN and p53 will require further study, and may be critical given the frequency with which both tumor suppressors are deregulated in cancer.

8.2.11 PTEN Function and Stem Cell Fate

Much of our knowledge of PTEN function comes from tissue-specific genetic deletion of *Pten* in mouse models (see Table 8.1). Several such mouse models have revealed the specific effects of Pten in the regulation of stem cell fate in different lineages. Deletion of Pten early in development from the mouse brain leads to increased numbers of neural stem cells and increases in differentiated cell numbers and size (Groszer et al. 2001). These increases in neural stem cell number are caused by promotion of the G_0-G_1 cell cycle progression and a gene expression signature associated with rapid proliferation and a rapid cell cycle (Groszer et al. 2006). In contrast, studies of mice lacking Pten in hematopoietic stem cells (HSCs) have indicated that Pten loss promotes a short term proliferation of these cells, yet leads to a long term depletion of the self-renewing stem cell compartment (Yilmaz et al. 2006). Crucially, Pten loss in HSCs leads to the rapid development of leukemia in these mice, yet this disease is characterized by the presence of Pten-null self-renewing leukemic stem cells (LSCs). These apparently distinct characteristics of normal HSCs versus LSCs suggests that their self-renewal is regulated by different mechanisms and provides hope for therapies that target LSCs, without killing normal HSCs. The short term proliferation of HSCs driven by Pten loss is associated with enhanced G_0-G_1 cell cycle transition, as observed in neuronal stem cells. However, why in HSCs, this leads to a transient, rather than long term expansion in stem cell numbers, as seen in the brain, is unclear. Studies of other lineages imply that other stem cell populations can be similarly classified, with Pten loss in melanocyte stem cells also leading to long term stem cell expansion (Inoue-Narita et al. 2008). This proposed role for PTEN in the maintenance of the normal stem cell compartment is supported by the identification of PTEN as the most significantly reduced transcript in microarray analyses of gene expression differences between normal fetal neuronal stem cells and glioblastoma stem cells isolated from brain tumor patients (Pollard et al. 2009). By regulating PTEN expression this may allow in future therapeutic approaches that target cancer stem cells (or cancer-initiating cells) without affecting normal stem cell compartments (Rossi and Weissman 2006).

8.2.12 PTEN and the Immune System

The role of the PI3K signaling network in the immune system has been studied extensively. This research activity has been driven by the recognition that PI3K signaling plays key roles in the survival and proliferation of several normal and transformed immune cell populations and also in other immune cell functions such as migration, phagocytosis and reactive oxygen species generation (Fruman and Bismuth 2009). Of particular significance was the initial discovery that mice selectively lacking the gamma catalytic isoform of PI3K (p110γ) were viable and fertile, yet their T-cells and neutrophils displayed reduced chemotaxis *in vitro* and recruitment in *vivo* (Hirsch et al. 2000; Sasaki et al. 2000; Li et al. 2000). This has lead to intense drug discovery activity attempting to target p110γ, and also p110δ, for inflammatory diseases, that have been sustained by subsequent discoveries (Ruckle et al. 2006).

The tumor suppressor activity of PTEN is important in the avoidance of both T cell and B cell malignancies, due to its roles in promoting non-proliferative states and apoptosis. Studies of mice, in which Pten has been specifically deleted in immune cell populations have also identified roles for the phosphatase in the suppression of B-cell and neutrophil migration *in vivo* (Suzuki et al. 2003b; Subramanian et al. 2007; Heit et al. 2008; Li et al. 2009). Furthermore, Pten-null macrophages exhibit elevated phagocytosis and Fcγ-mediated signaling (Cao et al. 2004). Given these specific actions of PTEN to limit the selection and proliferation of immune cell populations and also in suppressing immune cell activities, it is not surprising that autoimmunity is a striking phenotype of Pten loss in mice. As discussed, tight regulation of PI3K and thus PtdIns(3,4,5)P$_3$ levels is important for normal development and responsiveness of the immune system and also to prevent immunopathology. Some studies have shown that diminished PI3K activity can also lead to similar autoimmune conditions (Oak et al. 2006), while hyperactivation of this pathway in T-cells also leads to lymphoproliferation and systemic immunity (Suzuki et al. 2001). Consistent with this, Pten heterozygous (Pten$^{+/-}$) mice develop lymphoproliferative syndrome with autoimmune features (Di Cristofano et al. 1999; Suzuki et al. 2001). Several mechanisms have been proposed for PTEN's role in suppressing autoimmunity. PTEN can regulate Fas-mediated apoptosis (an autoimmunity repression mechanism) and also IL-4 production and thus regulate CD4(+) T-cell function (Di Cristofano et al. 1999; Liu et al. 2010). Mice with higher expression of microRNA miR-17-92 have been shown to have similar autoimmune disorders and this is believed to be due to miR-17-92-mediated suppression of PTEN expression (Xiao et al. 2008).

8.2.13 PTEN Function in Epithelial Biology

The frequent loss of PTEN function observed in epithelial-derived solid tumors (carcinomas) has stimulated investigations into the functions of PTEN in normal and transformed epithelial biology. Significantly, several pieces of evidence point towards distinct roles for PTEN and PI3K signaling in epithelia, regulating both cell growth, proliferation and also cell polarity/architecture (Liu et al. 2004; Leslie et al. 2008). In this context, it is notable that PtdIns(3,4,5)P$_3$ is found to be enriched in the basolateral membrane of several epithelial cell types (Liu et al. 2004; Watton and Downward 1999) and PTEN appears to be strongly enriched in the apical domains of some but not all, epithelial cell types, such as the *Drosophila melanogaster* (*D. melanogaster*) photoreceptors and embryonic epithelia (von Stein et al. 2005; Pinal et al. 2006), chick epiblast (Leslie et al. 2007) and murine retinal pigment epithelium (Kim et al. 2008). These localization data and the identified interactions of PTEN with proteins such as PAR3 and MAGI1 indicate that PTEN is enriched in apical cell-cell junctional complexes, almost certainly including adherens junctions (Pinal et al. 2006; von Stein et al. 2005; Kotelevets et al. 2005). Functional studies in cultured epithelial cells suggest that PTEN acts to suppress PI3K-dependent epithelial cell-cell junctional destabilization (Kotelevets et al. 2001; Martin-Belmonte et al. 2007; Liu et al. 2004). In particular, in epithelial cell lines cultured in 3D matrices, the knockdown of PTEN expression leads to a dramatic loss of cell polarity and tissue organization (Martin-Belmonte et al. 2007). Knockdown of PTEN in normal human mammary epithelial cells causes the formation of disorganized hyperplastic lesions when introduced into humanized murine mammary fat pads (Korkaya et al. 2009).

The phenotypes of mice with different epithelial tissue-specific deletion of Pten, reveals a range of related effects but dramatic differences in the severity of the phenotype. Inducible deletion of Pten from the intestinal epithelium has no overt effects on the morphology of the tissue in the absence of further alterations (Marsh et al. 2008). Deletion of PTEN from the kidney leads to defects in branching morphogenesis within the organ, which although not affecting the integrity of the epithelium still appear responsible for the death of these animals within a month of birth (Kim and Dressler 2007). Similarly, MMTV-Cre driven mammary-specific deletion of PTEN in mice leads to hyperproliferative development of the gland with dysregulated ductal branching and mammary tumors (Li et al. 2002a). In addition, deletion of Pten in the lung leads to increased epithelial cell size, abnormal lung morphogenesis and adenocarcinoma in mice that escape the hypoxic postnatal lethality (Yanagi et al. 2007). Perhaps most remarkably, deletion of Pten from the retinal pigment epithelium, that, significantly, displays a strongly polarized localization of the Pten protein, causes a full EMT and complete disruption of the tissue (Kim et al. 2008). This supports a model in which PTEN function plays a role in establishing, although perhaps less significantly maintaining, epithelial cell-cell junctions, with loss of PTEN driving a shift to a more motile, mesenchymal phenotype (Leslie et al. 2007; Song et al. 2009; Kim et al. 2008).

8.3 Myotubularins

The myotubularins are a large family of conserved proteins in eukaryotes. They dephosphorylate PtdIns(3)P and PtdIns(3,5)P_2, to form PtdIns and PtdIns(5)P respectively. Despite the large number of myotubularins (MTMs) and shared specificity for their substrates, mutations in individual myotubularins can cause different human diseases. Myotubularins exhibit varied sub-cellular localization, and integrate into protein/signaling complexes. These lipid phosphatases are required for the cellular control of diverse processes, including ion channel-stimulated excitation-contraction coupling and signaling, endocytic trafficking, autophagy and cell proliferation.

The myotubularins are defined by conserved myotubularin-related and dual specific protein tyrosine phosphatase CX_5R-containing domains. 16 myotubularins have been described in mammalian cells, the first was designated myotubularin (MTM1), and the other family members "myotubularin-related phosphatases" (MTMR1-15) (Table 8.2). Seven of the myotubularins possess inactivating mutations within the catalytic motif, and are phosphatase dead, however, a number of these proteins still play an active role in cellular functions and when mutated can cause disease (Begley et al. 2006). Myotubularins possess a number of additional domains, conserved across most family members; including PH-GRAM (Pleckstrin-Homology-Glucosyltransferase, Rab-like GTPase Activator), and coiled-coil domains (Laporte et al. 2003) (Fig. 8.2). However, the two most recently described myotubularins, MTMR14 and MTMR15, exhibit very little homology at the PH-GRAM or coiled-coil regions. Myotubularins also contain conserved SID (Set Interacting Domain), and RID (Rac Induced Recruitment Domain) domains (Begley et al. 2003, 2006; Laporte et al. 2003). Individual myotubularins also contain a number of additional protein or lipid interacting modules, described in more detail below.

A phylogenetic survey of 31 different species across broad taxonomic groups counted varied complements of different myotubularins, ranging from one in fungi such as *Saccharomyces cerevisiae* and *Aspergillus niger*, to 19 in *Entamoeba histolytica* (Kerk and Moorhead 2010). Additionally, the inactive myotubularin homologs appear to have evolved on three separate occasions in different eukaryotic lineages, highlighting their presence as more than simply an interesting observation. Phylogenetic analysis also permits the grouping of the myotubularins into similarity clusters—three for active myotubularins, and three for inactive (See Fig. 8.2) (Laporte et al. 2003; Kerk and Moorhead 2010). Some organisms with lower complements of myotubularins have representatives of these groups (Robinson and Dixon 2006; Laporte et al. 2003). For example, for *Caenorhabditis elegans* (*C. elegans*) these are: M1-mtm-1 (CeY110A7A.5); R3-mtm-3 (CeT24A11.1); R6-mtm-6 (CeF53A2.8); R5-mtm-5 (CeH28G03.6) and mtmr-9 (CeY39H10A) (Xue et al. 2003). Recently, MTMR14 (also known as hJumpy or MIP) has been characterized, and it contains an active CX_5R phosphatase motif and homology to the myotubularin-related domain, but no PH-GRAM domain (Alonso et al. 2004; Tosch et al. 2006; Vergne et al. 2009). A sixteenth myotubularin has also been identified, MTMR15, which shares little homology with other myotubularins, and has an inactive phosphatase site (Alonso et al. 2004). This protein functions as a DNA repair enzyme and as such, its characterization as a myotubularin awaits further clarification. Many myotubularins (MTM1, MTMR1-6 and MTMR12) are expressed in a broad range of tissues and cell types (Laporte et al. 1998; Nandurkar et al. 2001; Zhao et al. 2001). In contrast, MTMR5 is concentrated in the testis, and MTMR7 shows a brain-specific expression (Laporte et al. 1998; Firestein et al. 2002). Non-redundancy may be a result of differences in the sub-cellular distribution of various myotubularins, and also by different protein and lipid interactions. This in turn may regulate myotubularin access to specific pools of PtdIns(3)P and/or PtdIns(3,5)P_2 in a directed manner, with different functional consequences (Kim et al. 2002; Lorenzo et al. 2006), as described below.

Table 8.2 Genetically modified myotubularin animal models

Protein name	Alias(es)	Activity	Knockout animal models	References
MTM1 (myotubularin)	CNM, MTMX, XLMTM	Active	Mouse constitutive KO—some embryonic lethality, growth retardation, muscle weakness, premature death, ultrastructural changes in muscle triads, decreased calcium release from SR Labrador retriever model—muscle weakness, ultrastructural changes in triads	(Buj-Bello et al. 2002b; Beggs et al. 2010)
MTMR1		Active	Not reported	
MTMR2	CMT4B, CMT4B1, KIAA1073	Active	Mouse constitutive KO report (1)—some embryonic lethality, growth retardation, minimal motor change, abnormal neurophysiology and histopathology, azospermia	(Bonneick et al. 2005; Bolino et al. 2004)
			Mouse constitutive KO report (2)—minimal motor change, abnormal neurophysiology and histopathology	
MTMR3	FYVE-DSP1, KIAA0371	Active	Not reported	
MTMR4	FYVE-DSP2, KIAA0647	Active	Not reported	
MTMR5	sbf1	Inactive	Mouse constitutive KO—smaller testes and infertile, azospermia	(Firestein et al. 2002)
MTMR6		Active	Not reported	
MTMR7		Active	Not reported	
MTMR8	FLJ20126, FLJ60798	Active	Not reported	
MTMR9	LIP-STYX	Inactive	Not reported	
MTMR10	FLJ20313	Inactive	Not reported	
MTMR11	CRA	Inactive	Not reported	
MTMR12	3-PAP	Inactive	Not reported	
MTMR13	sbf2, CMT4B2, FLJ22918, FLJ41627, KIAA1766	Inactive	Mouse constitutive KO (two reports)—mild motor phenotype, abnormal neurophysiology and nerve histopathology	(Tersar et al. 2007; Robinson et al. 2008)
MTMR14	hJumpy, MIP, FLJ22405	Active	Mouse constitutive KO—decreased performance on strenuous exercise, abnormalities of excitation contraction coupling, calcium leak from SR, decreased stimulated calcium release from SR	(Shen et al. 2009)
MTMR15	KIA1018	Inactive	Not reported	

XLMTM X-linked myotubular myopathy, *CMT* Charcot-Marie-Tooth, *SR* sarcoplasmic reticulum

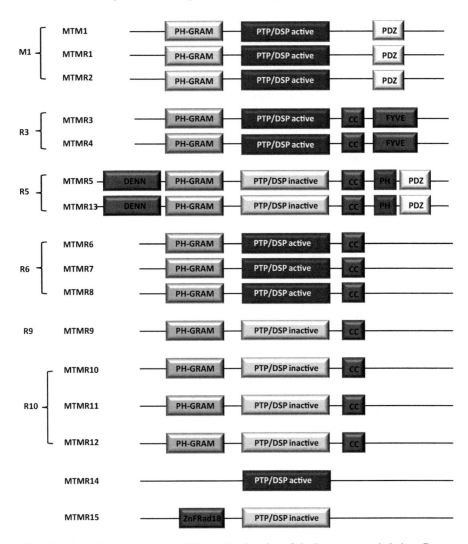

Fig. 8.2 Schematic representation of the major domains of the human myotubularins. Groups are named as reported by (Laporte et al. 2003; Kerk and Moorhead 2010). Domain associations were obtained from searches on conserved domains using NCBI/CDD and SMART databases, as well as those reported by (Laporte et al. 2003). Abbreviations: *PH-GRAM* Pleckstrin-Homology-Glucosyltransferase, Rab-like GTPase Activator; *PTP/DSP* Protein tyrosine phosphatase/dual specific phosphatase; *CC* Coiled-coil; *PDZ* Post synaptic density protein, Drosophila disc large tumor suppressor, zonula occludens-1 protein; FYVE, Fab1, Yotb, Vac1p and EEA1; *DENN* Differentially Expressed in Neoplastic versus Normal cells; ZnFRad18, Rad18-like CCHC Zinc finger

8.3.1 Myotubularin Structure and Substrate Specificity

The crystal structure of MTMR2 has been solved, and shows a larger active site pocket than most other protein tyrosine phosphatases, which is postulated to be required to accommodate the large inositol head group of phosphoinositides (Begley et al. 2003). The "WPD" loop also differs from most protein tyrosine phosphatases, lacking an aspartic acid, prompting Begley et al. to speculate a unique phosphatase action for the myotubularins (Begley et al. 2003). The tertiary structure also reveals that the then-named "GRAM domain" (Doerks et al. 2000) is similar to the PH domain, a phosphoinositide, phosphotyrosine and protein-protein interacting region, and that the SID domain is part of the protein phosphatase domain (Begley et al. 2003). The PH-GRAM domain may interact with PtdIns(5)P (Lorenzo et al. 2005), PtdIns(3,5)P_2, PtdIns(4)P, and PtdIns(3,4,5)P_3 (Berger et al. 2003), although this predicted role in lipid binding is not fully established. MTMR2 binds PtdIns(3)P and PtdIns(3,5)P_2 in its active site, however, no lipid binding was detected at the PH-GRAM domain (Begley et al. 2006).

Despite showing the properties consistent with the tyrosine phosphatase superfamily, the myotubularins exhibit predominant activity against 3-phosphorylated phosphoinositides (Taylor et al. 2000a; Zhao et al. 2001). PtdIns(3)P and PtdIns(3,5)P_2 are hydrolyzed at their 3-position phosphate by the MTMs, *in vitro* forming PtdIns and PtdIns(5)P respectively (Velichkova et al. 2010; Naughtin et al. 2010; Tosch et al. 2006; Berger et al. 2002, 2006; Tronchere et al. 2004; Schaletzky et al. 2003; Begley et al. 2003: Zhao et al. 2001; Walker et al. 2001; Taylor et al. 2000a; Blondeau et al. 2000). Myotubularins exhibit significantly less catalytic activity against PtdIns(3,4,5)P_3, PtdIns(3,4)P_2, PtdIns(4,5)P_2, PtdIns(4)P or PtdIns(5)P (Blondeau et al. 2000; Walker et al. 2001; Zhao et al. 2001; Berger et al. 2002; Schaletzky et al. 2003; Tronchere et al. 2004; Tosch et al. 2006). One exception is MTMR14, which exhibits phosphatase activity against a larger number of phosphoinositides (PtdIns(3,5)P_2 > PtdIns(3,4)P_2 > PtdIns(4,5)$_2$ > PtdIns(3)P) (Shen et al. 2009). However, in a separate study using immunoprecipitates from HEK293 cells, MTMR14 selectively degraded only PtdIns(3)P and PtdIns(3,5)P_2 (Tosch et al. 2006).

The cellular localization of PtdIns(3)P and PtdIns(3,5)P_2 is predominantly endosomal, and some of the myotubularins are predicted to control endosomal PtdIns(3)P levels, although there are conflicting reports. The different effects of MTM over-expression or decreased expression on endosomal PtdIns(3)P may relate to different experimental approaches. Many microscopy-based studies have detected endosomal PtdIns(3)P, by expression of GFP-tagged constructs containing two FYVE domains in tandem, with co-localization with endosomal markers. Another approach has utilized purified GST-2xFYVE as a probe, detected by GST antibodies after cellular fixation and permeabilization (Gillooly et al. 2000). Over-expression of MTM1, MTMR2-4, and MTMR14 decreases endosomal PtdIns(3)P in both cell lines as well as in primary cells (Kim et al. 2002; Kelley and Schorey 2004; Lorenzo et al. 2006; Tosch et al. 2006; Cao et al. 2007; Naughtin et al. 2010; Taguchi-Atarashi et al. 2010). Expression of the *D. melanogaster* mtm decreases endosomal PtdIns(3)P in hemocytes (Velichkova et al. 2010). However, other studies have shown that expression of MTM1 (Laporte et al. 2002), or MTMR2 (Kim et al. 2002) does not influence endosomal PtdIns(3)P levels, and in fibroblasts from human patients with different MTM1 mutations, no alteration in endosomal PtdIns(3)P probe fluorescence was detected (Tronchere et al. 2004). Fili and colleagues used a chimeric MTM1 protein designed to be recruited to Rab5 positive membranes only upon addition of rapalogue (Fili et al. 2006). Expression of the MTM1-FKB chimera in HeLa cells did not decrease GST-2xFYVE staining until rapalogue was added, which induced MTM1 targeting to early endosomes (Fili et al. 2006). In contrast, siRNA knockdown of different myotubularins increases endosomal PtdIns(3)P in some studies (Cao et al. 2008; Naughtin et al. 2010; Taguchi-Atarashi et al. 2010; Velichkova et al. 2010). Some of the differences in the effects of myotubularin expression on PtdIns(3)P might be accounted for by the frequent reliance of over-expression of myotubularins and/or GST-2xFYVE probe over-expression and the associated difficulties in comparing cells with different expression levels. Expression of a GFP-2xFYVE probe *per se*, as opposed to application of a recombinant GST-2xFYVE probe after cellular fixation, itself interferes with PtdIns(3)P-dependent events (Vieira et al. 2004; Gillooly et al. 2000). In addition, most studies have presented qualitative representative images of cells with little quantification. Of special note, Cao et al. used siRNA to MTM1 or MTMR2, and with quantitation demonstrated increased PtdIns(3)P in cells by analysis of GFP-2xFYVE fluorescence (Cao et al. 2008). Interestingly, MTM1 knockdown did not alter total GFP-2xFYVE fluorescence, but increased the fraction of GFP-2xFYVE which co-localized with the early endosomal marker, EEA1. In contrast MTMR2 siRNA increased late endosomal GFP-2xFYVE (Cao et al. 2008). Additionally, GST-2xFYVE staining was quantified in HeLa cells treated with MTMR4 siRNA (Naughtin et al. 2010). This study reported that in knockdown cells, the increased GST-2xFYVE signal did not reflect an increase in the fluorescence associated with individual endosomes, but rather was a result of a large increase in the number of GST-2xFYVE-positive endosomes per cell. This suggests a function of MTMR4 is to restrict PtdIns(3)P from certain endosomal populations.

Despite being characterized as a dual specific phosphatase, the ability of a myotubularin to utilize phosphoprotein(s) as substrates has only been identified in one study. Recently, negative regulation of TGFβ signaling by MTMR4 was demonstrated, via binding to and dephosphorylation of the R-SMADS, SMAD2 and SMAD3 (Yu et al. 2010). Expression of MTMR4 rendered cells resistant to the anti-proliferative effects induced by TGFβ treatment (Yu et al. 2010). This raises the possibility that MTMR4 and other myotubularins may exhibit biological activity against additional phosphoprotein substrates.

To determine the sub-cellular localization of specific MTMs, many studies have relied on ectopic over-expression of myotubularin in cells (Blondeau et al. 2000; Taylor et al. 2000a; Firestein et al. 2002; Walker et al. 2001; Kim et al. 2002; Berger et al. 2003, 2006; Chaussade et al. 2003; Kim et al. 2003; Tsujita et al. 2004; Robinson and Dixon 2005; Lorenzo et al. 2006; Tosch et al. 2006; Vergne et al. 2009; Taguchi-Atarashi et al. 2010; Plant et al. 2009). These various studies have reported discordant results for the sub-cellular localization of some of the myotubularins, perhaps as a result of different experimental techniques or differences in cell types and/or the level of over-expression achieved. Multiple reports have localized recombinant MTM1 to the cytosol, at the plasma membrane and on membrane ruffles (Blondeau et al. 2000; Fili et al. 2006; Kim et al. 2002; Laporte et al. 2002; Lorenzo et al. 2006; Nandurkar et al. 2003; Taylor et al. 2000a; Tsujita et al. 2004). In contrast, endogenous myotubularin is reported to localize to early endosomes, where it regulates a pool of PtdIns(3)P (Cao et al. 2007, 2008). The only myotubularin which has been consistently reported to localize to endosomes is MTMR4, which contains a FYVE domain. When expressed in cells recombinant MTMR4 exhibits an endosomal distribution, co-localizing with early, late and recycling endosomal markers (Lorenzo et al. 2006; Naughtin et al. 2010; Plant et al. 2009; Yu et al. 2010).

Several reports have localized recombinant MTMR2 to the cytosol, with some perinuclear enrichment (Kim et al. 2002, 2007; Robinson and Dixon 2005) as well as the nucleus (Lorenzo et al. 2006), whilst endogenous MTMR2 is detected on late endosomes (Cao et al. 2008). Some of the inconsistencies in these reports on the different sub-cellular localizations of myotubularin and the MTMs may be a consequence of different protein:protein interactions between myotubularins, and/or the localization may change with agonist specific-cell stimulation, or may be cell type-specific. For example stimulation of COS7 or FlpIn293 cells with EGF results in a MTM1 or MTMR2 cytosolic redistribution to punctate and vesicular structures (Tsujita et al. 2004; Berger et al. 2011).

Several other myotubularins exhibit a cytosolic localization, with non-specific patterns of punctate or reticular localization (Lorenzo et al. 2006). Over-expressed wild-type MTMR3 localizes to a cytosolic and reticular distribution, with some overlap with the endoplasmic reticulum (Walker et al. 2001), and despite the presence of a FYVE domain it shows little endosomal localization. In addition, mutations in MTMR3 have been associated with significant alterations in its sub-cellular distribution (Walker et al. 2001; Lorenzo et al. 2005). A single mutation at the cysteine in the MTMR3 catalytic site (C413S) results in its punctate cellular distribution, co-localizing with structures which by ultrastructural analysis have the appearance of autophagosomes, and by light microscopy co-localize with autophagosome markers such as WIPI-1α and DFCP1 (Walker et al. 2001; Taguchi-Atarashi et al. 2010). Double mutation of MTMR3 C413S, together with deletion of the PH domain, surprisingly induces Golgi localization of MTMR3 (Lorenzo et al. 2005).

8.3.2 Myotubularin Association with Human Disease

Mutations in either active or inactive myotubularins lead to human disease. MTM1 is mutated in X-linked centronuclear (myotubular) myopathy (Laporte et al. 1996). This congenital disease is characterized by severe muscle weakness in affected males, which often results in death in infancy from respiratory failure (Laporte et al. 1996).

Pathological analysis of muscles from affected patients reveals small rounded muscle cells with centrally placed nuclei and a surrounding halo devoid of contractile elements (Laporte et al. 1996). $Mtm1^{-/-}$ mice exhibit growth retardation with a shortened life span, associated with a progressive muscle phenotype that resembles the human disease, attributed to a defect in muscle maintenance, but not myogenesis (Biancalana et al. 2003). MTMR2 mutation is associated with the peripheral neuropathy Charcot-Marie-Tooth disease type 4B1, and mutation in its inactive binding partner MTMR13 leads to a similar clinical syndrome, Charcot-Marie-Tooth disease type 4B2 (Bolino et al. 2000; Berger et al. 2002; Azzedine et al. 2003; Senderek et al. 2003). Pathological analysis of affected patients reveals a demyelinating neuropathy with focally folded myelin sheaths. Mice have also been generated with a deletion of *Mtmr2*, or *Mtmr13* genes (Bolino et al. 2004; Tersar et al. 2007; Robinson and Dixon 2005). $Mtmr2^{-/-}$ mice with a complete deletion are viable, but underrepresented according to Mendelian predictions, and exhibit growth retardation, but no apparent weakness (Bolino et al. 2004). These mice show a peripheral neuropathy on nerve conduction and histopathological analyses, and also exhibit azospermia (Bolino et al. 2004). Conditional loss of *Mtmr2* in Schwann cells, but not neurons, reproduces the neuropathy phenotype (Bolis et al. 2005). In a separate study, mice expressing a truncated transcript lacking the phosphatase domain of *Mtmr2* were born within the expected Mendelian frequency, and showed normal growth and physical appearance (Bonneick et al. 2005). Histopathological analysis, however, revealed the characteristic features observed in humans with Charcot-Marie-Tooth 4B1 (Bonneick et al. 2005). $Mtmr13^{-/-}$ mice are viable and are born at the predicted Mendelian frequency (Tersar et al. 2007; Robinson et al. 2008). These mice do not exhibit an obvious disease phenotype, however, defects were observed on detailed motor testing at 12 months age, with electrophysiological and histopathological evidence of neuropathy (Tersar et al. 2007). A mouse knock-out of *Mtmr5* exhibits azospermia, perhaps related to abnormalities in Sertoli cell function (Firestein et al. 2002). Mutations of *MTMR14* are implicated in some cases of human autosomal recessive centronuclear myopathy (Tosch et al. 2006). $Mtmr14^{-/-}$ mice have a normal appearance, but exhibit abnormal motor testing during strenuous exercise, and abnormalities of excitation-contraction coupling (Shen et al. 2009).

8.3.3 Inactive Myotubularins and Protein Complex Formation

Inactive myotubularins have been speculated to regulate the catalytic activity of the active myotubularins, or alternatively to interact with active myotubularins and direct their sub-cellular localization (Berger et al. 2003, 2006; Kim et al. 2003; Nandurkar et al. 2003). For example, the inactive MTMR13 increases the activity of MTMR2 toward PtdIns(3)P and PtdIns(3,5)P_2, by 10- and 25-fold amounts respectively (Berger et al. 2006). This might provide a molecular basis for the observation that mutations in the inactive MTMR13 result in Charcot Marie Tooth type 4B2 neuropathy in humans and in mouse models. Both endogenous and recombinant MTMR2 and MTMR13 form homodimers in cells via interaction of their coiled-coil domains (Berger et al. 2006). These myotubularins also interact with each other in a tetrameric protein complex (Berger et al. 2006). The sub-cellular distribution of MTMR13 overlaps with MTMR2 under resting conditions, but diverges under conditions of hypo-osmotic stress, suggesting that in addition to regulating enzyme activity of an active phosphatase, the inactive phosphatase regulates the sub-cellular localization of the active phosphatase (Berger et al. 2006). Similarly, MTMR12 forms a complex with MTM1, which requires the SID domain. This interaction directs the localization of MTM1 away from the plasma membrane with reversal of MTM1-induced changes in filopodial formation in COS7 cells (Nandurkar et al. 2003). Interactions have also been observed for many other recombinant myotubularins, but whether these interactions occur *in vivo* is unclear due to a general difficulty in this field in making specific high affinity antibodies to any of the myotubularins: MTM1 interacts with MTMR10 (Lorenzo et al. 2006); MTMR2 interacts with MTMR5, MTMR10, MTMR12 and MTMR13, and the interaction between MTMR5 and MTMR13 requires the coiled-coil domain and increases phosphatase activity (Nandurkar et al. 2003; Kim et al. 2003; Robinson and Dixon 2005; Lorenzo et al. 2006; Berger et al. 2006). MTMR3 interacts with MTMR4 and co-expression of MTMR3 and MTMR4 alters the sub-cellular localization of each phosphatase (Lorenzo et al. 2006); MTMR6, MTMR7 and MTMR8 interact with MTMR9; the MTMR7/MTMR9 interaction requires the coiled-coil domain and results in increased MTMR7 phosphatase activity (Mochizuki and Majerus 2003; Lorenzo et al. 2006). Protein complex formation is not restricted only to other myotubularins. MTM1 and MTMR2 interact in a multi-protein complex with the Class III PI3K (hVps34) and hVps15 (Cao et al. 2007, 2008). MTMR2 also interacts with PSD95, an excitatory postsynaptic scaffolding protein, as well as with disc-large1 (Dlg1) (Bolis et al. 2009; Lee et al. 2010). MTRM4 interacts with R-SMADs, and the ubiquitin ligase, NEDD4 (Plant et al. 2009; Yu et al. 2010).

8.3.4 Myotubularins and Cellular Function

The functional role that individual myotubularins play in the cellular control of phosphoinositide signaling is currently a target of much work. PtdIns(3)P localizes to early and late endosomes, the plasma membrane and the forming autophagosome (Gillooly et al. 2000, 2003; Tooze and Yoshimori 2010). The localization of PtdIns(3,5)P_2 is within endosomes/lysosomes (Dove et al. 2009). A number of the myotubularins may regulate endosomal PtdIns(3)P. This suggests potential roles for the myotubularins in regulating endocytic trafficking, including endocytosis, degradative and recycling pathways and autophagy. In *C. elegans*, mutations in mtm-6 or mtmr-9 impair endocytosis by coelomocytes (Fares and Greenwald 2001; Xue et al. 2003). siRNA depletion of MTMR4 decreases endocytosis of transferrin by HeLa cells (Naughtin et al. 2010). Conversely, siRNA depletion of MTMR2 in neuronal cells enhances endocytosis and impairs synaptic maintenance (Lee et al. 2010). Sorting of EGFR for degradation is an important PI3K-dependent pathway. Expression of MTM1 or

MTMR2 inhibits EGFR degradation (Tsujita et al. 2004; Berger et al. 2011). In contrast, Lorenzo and colleagues found that over-expression of MTMR4, but not MTM1, MTMR2 or MTMR3, impaired EGFR degradation in some cells (Lorenzo et al. 2006). Expression of a catalytically inactive MTMR4, or treatment with MTM1 or MTMR2 siRNA also results in impairment of EGFR degradation, possibly suggesting that tight control of phosphoinositide levels is necessary for normal degradative receptor sorting (Lorenzo et al. 2006; Cao et al. 2008). Recycling of transferrin can occur by both PI3K-dependent and independent pathways, and over-expression of MTMR4 impairs the sorting of transferrin from early endosomes (Naughtin et al. 2010). In contrast, expression of a phosphatase inactive MTMR4 mutant increases transferrin recycling (Naughtin et al. 2010). Recycling of MIG-14/Wls is necessary for Wnt secretion, a signaling process important in tissue development and disease, and it has recently been reported that this is disrupted in *C. elegans* mtm-6 and mtmr-9 mutants (Silhankova et al. 2010). The authors also reported that mtm-6 and mtmr-9 regulate the localization of sorting nexin 3 to PtdIns(3)P-positive endosomes and maintain normal trafficking of MIG-14/Wls to control Wnt secretion (Silhankova et al. 2010).

Autophagy is a cellular process whereby portions of the cellular contents are internalized in double membrane autophagosomes and degraded. PtdIns(3)P plays a significant role in autophagosome initiation (Tooze and Yoshimori 2010). MTMR3 and MTMR14 localize to the site of autophagosome formation and negatively regulate autophagy (Vergne et al. 2009; Dowling et al. 2010; Taguchi-Atarashi et al. 2010). Additionally MTMR14 *Danio rerio* (*D. rerio*) morphants show elevated autophagic markers, and a double morphant embryo of MTM1 and MTMR14 exhibits significantly greater autophagic markers and autophagic structures by ultrastructural analysis (Dowling et al. 2010). Taken together, this data suggests that at least three of the myotubularins may inhibit autophagy.

A number of reports have recently linked the muscle defects observed in myotubularin mutations with alterations in skeletal muscle triad structure and function. Ultrastructural defects in T-tubules and the sarcoplasmic reticulum have been described in animal models of myotubular myopathy, including $Mtm1^{-/-}$ mice, knockdown of mtm1 in *D. melanogaster* and *D. rerio* and a spontaneous $Mtm1$ mutation in Labrador retrievers, as well as in the tissues of patients with myotubular myopathy (Al-Qusairi et al. 2009; Dowling et al. 2009, 2010; Beggs et al. 2010; Toussaint et al. 2010). In skeletal muscle, MTM1 localizes to the region of the T-tubule (Buj-Bello et al. 2008; Dowling et al. 2009). $Mtm1^{-/-}$ mice show depressed calcium release from the sarcoplasmic reticulum with lower protein levels of the type 1 ryanodine receptor (RyR1). MTM1 *D. rerio* morphants show abnormal excitation-contraction coupling (Al-Qusairi et al. 2009; Dowling et al. 2009). MTMR14 *D. rerio* morphants show normal muscle triad ultrastructure, but defective excitation-contraction coupling and a motor phenotype (Dowling et al. 2010). A mouse $Mtmr14^{-/-}$ model exhibits prolonged muscle relaxation and fatigability, a result of spontaneous calcium leakage from the sarcoplasmic reticulum (Shen et al. 2009). Intriguingly, the $Mtmr14^{-/-}$ mouse shows increased PtdIns(3,5)P$_2$ (Shen et al. 2009). PtdIns(3,5)P$_2$ and PtdIns(3,4)P$_2$ directly activate the RyR1 calcium channel, linking the observed muscle weakness with an accumulation of these phosphoinositides (Shen et al. 2009). In addition to the effect of MTM1 and MTMR14 on muscle calcium channel signaling, a third myotubularin has also been linked to ion channel signaling. MTMR6 negatively regulates the calcium-dependent activated potassium channel, KCa3.1 (Srivastava et al. 2005; Choudhury et al. 2006). MTMR6 functions as a negative regulator of calcium influx and proliferation of reactivated CD4 + T cells (Srivastava et al. 2006). Very recently myotubularin has been shown to bind desmin, and regulate desmin function (Hnia et al. 2010). Desmin is a major intermediate filament protein in skeletal muscle, and mutations in desmin are associated with cardiomyopathy and myopathy (Omary et al. 2004). Loss of MTM1 results in abnormal desmin intermediate filament assembly and mitochondrial positioning, and this is independent of the phosphatase activity of MTM1 (Hnia et al. 2010).

Several MTMs have also been associated with abnormal muscle maturation and atrophy. Abnormal regulation of the MTMR1 gene has been described in myotonic dystrophies types 1 and 2 (Buj-Bello et al. 2002a; Santoro et al. 2010). MTMR4 interacts with the ubiquitin ligase NEDD4, and in atrophying muscle increased levels of NEDD4 correlate with decreased levels of MTMR4 (Plant et al. 2009). Additionally, aged mice show progressive loss of Mtmr14, together with altered motor function and calcium homeostasis, suggesting a role for the loss of MTMR14 in sarcopenia (Romero-Suarez et al. 2010).

The molecular basis of myotubularin mutations and resultant neurological manifestations are also slowly emerging. MTMR2 forms a protein complex, interacting with the scaffolding protein Dlg1, the plus-end kinesin motor protein, Kif13b, and the exocyst component Sec8 (Bolis et al. 2009). Dlg1 and Kif13b transport MTMR2 to sites of membrane remodeling, where MTMR2 restricts and Sec8 promotes membrane addition to regulate Schwann cell myelination (Bolis et al. 2009). Additionally, expression of recombinant MTM1 or MTMR2 mutants, but not wild type MTM1, results in aggregation of neurofilaments in an adrenal carcinoma cell line (Goryunov et al. 2008). MTMR2 interacts with the scaffolding protein, PSD95, and knockdown of MTMR2 results in reduced excitatory synapse number and synapse transmission (Lee et al. 2010).

Two of the myotubularins have also been linked in gene association studies to metabolic defects. Single nucleotide polymorphisms (SNPs) of *MTMR4* are associated with elevated plasma total cholesterol values, and SNPs in *MTMR9* associate with obesity and hypertension (Dolley et al. 2009; Yanagiya et al. 2007). However the molecular basis of these interesting observations remains currently unknown.

A number of the myotubularins may also positively regulate cell proliferation and/or inhibit apoptosis. MTMR6 expression promotes resistance to etoposide-induced apoptosis, an effect enhanced by MTMR9 (Zou et al. 2009a). Silencing of MTMR2 in cultured Schwann cells results in decreased proliferation and enhanced caspase-dependent cell death (Chojnowski et al. 2007). MTMR2 and MTMR13 positively regulate Akt signaling and prevent the degradation of EGFR (Berger et al. 2011). MTMR4 expression suppresses growth inhibition induced by TGFβ, through its action in dephosphorylating R-SMADS (Yu et al. 2010). Taken together this data suggests that at least five of the myotubularins are pro-proliferative, by yet to be

Table 8.3 PTEN and Myotubularin family members

Locus	Common name	Swiss prot protein	Accession	Reference MIM	Gene map Hs/Mm
PTEN	PTEN	PTEN_HUMAN	NM_000314	MIM: 601728	10q23.3
MTM1	Myotubularin	MTM1_HUMAN	NM_000252	MIM: 300415	Xq28
MTMR1	Myotubularin related protein 1	MTMR1_HUMAN	NM_003828	MIM: 300171	Xq28
MTMR2	Myotubularin related protein 2	MTMR2_HUMAN	NM_016156	MIM: 603557	11q22
MTMR3	Myotubularin related protein 3	MTMR3_HUMAN	NM_153051	MIM: 603558	22q12.2
MTMR4	Myotubularin related protein 4	MTMR4_HUMAN	NM_004687	MIM: 603559	17q22-q23
SBF1	SET binding factor 1 Myotubularin related protein 5	MTMR5_HUMAN	NM_002972	MIM: 603560	22q13.33
MTMR6	Myotubularin related protein 6	MTMR6_HUMAN	NM_004685	MIM: 603561	13q12
MTMR7	Myotubularin related protein 7	MTMR7_HUMAN	NM_004686	MIM: 603562	8p22
MTMR8	Myotubularin related protein 8	MTMR8_HUMAN	NM_017677		Xq11.2
MTMR9	Myotubularin related protein 9	MTMR9_HUMAN	NM_015458	MIM: 606260	8p23-p22
MTMR10	Myotubularin related protein 10	MTMRA_HUMAN	NM_017762		15q13.3
MTMR11	Myotubularin related protein 11	MTMRB_HUMAN	NM_181873		1q12-q21
MTMR12	Myotubularin related protein 12	MTMRC_HUMAN	NM_001040446	MIM: 606501	5p13.3
SBF2	SET binding factor 2 Myotubularin related protein 13	MTMRD_HUMAN	NM_030962	MIM: 607697	11p15.4
MTMR14	Myotubularin related protein 14 Jumpy	MTMRE_HUMAN	NM_001077526	MIM: 611089	3p26
FAN1	Myotubularin related protein 15 Fanconi associated nuclease 1	FAN1_HUMAN	NM_014967	MIM: 613534	15q13.2-q13.3

defined molecular mechanisms. RNAi-mediated knockdown of the *D. melanogaster* homologs of human MTM1, MTMR5/13 or MTMR 6/7/8 result in mitotic defects, suggesting a role for myotubularins in the regulation of cell division (Chen et al. 2007). These findings are of interest given recent evidence that shows PtdIns(3)P localizes to the midbody during division (Sagona et al. 2010).

Additional diverse functions of myotubularins have been reported. MTMR8 may be involved in angiogenesis as Mtmr8 knockdown in *D. rerio* results in abnormal vascular development (Mei et al. 2010). Mtm1 knockdown in *D. melanogaster* produces defects in cell cytoskeletal responses to hormone stimulation, and alters the wound recruitment of hemocytes (Velichkova et al. 2010). Mtm1 also functions in *C. elegans* as a negative regulator of corpse engulfment during development (Zou et al. 2009b).

8.4 Concluding Remarks

We have described a wealth of research conducted over 15 years or so, that has shown us how these two contrasting groups of PI 3-phosphatases, which play roles in the fundamental cellular processes of signal transduction and of membrane traffic, have important influences on human health and disease. The ability to use biochemistry and genetics to link strongly the related catalytic activities of these phosphatases to human diseases provides an excellent framework for future studies, which are both biological and translational. Ongoing work should provide a deeper understanding of the complex biology by which aberrant PI 3-kinase functions cause these pathologies and it is hoped that drug discovery efforts in these areas, many targeting the lipid kinases, may lead to successful treatments Table 8.3.

Additional information regarding sequences and online database entries is shown on Table 8.3.

References

Agrawal S, Eng C (2006) Differential expression of novel naturally occurring splice variants of PTEN and their functional consequences in Cowden syndrome and sporadic breast cancer. Hum Mol Genet 15:777–787

Al-Qusairi L, Weiss N, Toussaint A, Berbey C, Messaddeq N, Kretz C, Sanoudou D, Beggs AH, Allard B, Mandel JL, Laporte J, Jacquemond V, Buj-Bello A (2009) T-tubule disorganization and defective excitation-contraction coupling in muscle fibers lacking myotubularin lipid phosphatase. Proc Natl Acad Sci USA 106:18763–18768

Alimonti A, Nardella C, Chen Z, Clohessy JG, Carracedo A, Trotman LC, Cheng K, Varmeh S, Kozma SC, Thomas G, Rosivatz E, Woscholski R, Cognetti F, Scher HI, Pandolfi PP (2010) A novel type of cellular senescence that can be enhanced in mouse models and human tumor xenografts to suppress prostate tumorigenesis. J Clin Invest 120:681–693

Alonso A, Sasin J, Bottini N, Friedberg I, Osterman A, Godzik A, Hunter T, Dixon J, Mustelin T (2004) Protein tyrosine phosphatases in the human genome. Cell 117:699–711

Anezaki Y, Ohshima S, Ishii H, Kinoshita N, Dohmen T, Kataoka E, Sato W, Iizuka M, Goto T, Sasaki J, Sasaki T, Suzuki A, Ohnishi H, Horie Y (2009) Sex difference in the liver of hepatocyte-specific Pten-deficient mice: a model of nonalcoholic steatohepatitis. Hepatol Res 39:609–618

Anzelon AN, Wu H, Rickert RC (2003) Pten inactivation alters peripheral B lymphocyte fate and reconstitutes CD19 function. Nat Immunol 4:287–294

Azzedine H, Bolino A, Taieb T, Birouk N, Di DUCA M, Bouhouche A, Benamou S, Mrabet A, Hammadouche T, Chkili T, Gouider R, Ravazzolo R, Brice A, Laporte J, Leguern E (2003) Mutations in MTMR13, a new pseudophosphatase homologue of MTMR2 and Sbf1, in two families with an autosomal recessive demyelinating form of Charcot-Marie-Tooth disease associated with early-onset glaucoma. Am J Hum Genet 72:1141–1153

Backman SA, Stambolic V, Suzuki A, Haight J, Elia A, Pretorius J, Tsao MS, Shannon P, Bolon B, Ivy GO, Mak TW (2001) Deletion of Pten in mouse brain causes seizures, ataxia and defects in soma size resembling Lhermitte-Duclos disease. Nat Genet 29:396–403

Backman SA, Ghazarian D, So K, Sanchez O, Wagner KU, Hennighausen L, Suzuki A, Tsao MS, Chapman WB, Stambolic V, Mak TW (2004) Early onset of neoplasia in the prostate and skin of mice with tissue-specific deletion of Pten. Proc Natl Acad Sci USA 101:1725–1730

Bandaru SS, Lin K, Roming SL, Vellipuram R, Harney JP (2009) Effects of PI3K inhibition and low docosahexaenoic acid on cognition and behavior. Physiol Behav 100:239–244

Beggs AH, Bohm J, Snead E, Kozlowski M, Maurer M, Minor K, Childers MK, Taylor SM, Hitte C, Mickelson JR, Guo LT, Mizisin AP, Buj-Bello A, Tiret L, Laporte J, Shelton GD (2010) MTM1 mutation associated with X-linked myotubular myopathy in Labrador Retrievers. Proc Natl Acad Sci USA 107:14697–14702

Begley MJ, Taylor GS, Kim S-A, Veine DM, Dixon JE, Stuckey JA (2003) Crystal structure of a phosphoinositide phosphatase, MTMR2: insights into myotubular myopathy and Charcot-Marie-Tooth syndrome. Mol Cell 12:1391

Begley MJ, Taylor GS, Brock MA, Ghosh P, Woods VL, Dixon JE (2006) Molecular basis for substrate recognition by MTMR2, a myotubularin family phosphoinositide phosphatase. Proc Natl Acad Sci USA 103:927–932

Berger P, Bonneick S, Willi S, Wymann M, Suter U (2002) Loss of phosphatase activity in myotubularin-related protein 2 is associated with Charcot-Marie-Tooth disease type 4B1. Hum Mol Genet 11:1569–1579

Berger P, Schaffitzel C, Berger I, Ban N, Suter U (2003) Membrane association of myotubularin-related protein 2 is mediated by a pleckstrin homology-GRAM domain and a coiled-coil dimerization module. Proc Natl Acad Sci USA 100:12177–12182

Berger P, Berger I, Schaffitzel C, Tersar K, Volkmer B, Suter U (2006) Multi-level regulation of myotubularin-related protein-2 (mtmr2) phosphatase activity by myotubularin-related protein-13/set-binding factor-2 (MTMR13/SBF2). Hum Mol Genet 15:569–579

Berger P, Tersar K, Ballmer-Hofer K, Suter U (2011) The CMT4B disease-causing proteins MTMR2 and MTMR13/SBF2 regulate AKT signalling. J Cell Mol Med 15:307–315

Biancalana V, Caron O, Gallati S, Baas F, Kress W, Novelli G, D'Apice MR, Lagier-Tourenne C, Buj-Bello A, Romero NB, Mandel JL (2003) Characterisation of mutations in 77 patients with X-linked myotubular myopathy, including a family with a very mild phenotype. Hum Genet 112:135–142

Blondeau F, Laporte J, Bodin S, Superti-Furga G, Payrastre B, Mandel J-L (2000) Myotubularin, a phosphatase deficient in myotubular myopathy, acts on phosphatidylinositol 3-kinase and phosphatidylinositol 3-phosphate pathway. Hum Mol Genet 9:2223–2229

Bolino A, Muglia M, Conforti FL, Leguern E, Salih MA, Georgiou DM, Christodoulou K, Hausmanowa-Petrusewicz I, Mandich P, Schenone A, Gambardella A, Bono F, Quattrone A, Devoto M, Monaco AP (2000) Charcot-Marie-Tooth type 4B is caused by mutations in the gene encoding myotubularin-related protein-2. Nat Genet 25:17–19

Bolino A, Bolis A, Previtali SC, Dina G, Bussini S, Dati G, Amadio S, Del Carro U, Mruk DD, Feltri ML, Cheng CY, Quattrini A, Wrabetz L (2004) Disruption of Mtmr2 produces CMT4B1-like neuropathy with myelin outfolding and impaired spermatogenesis. J Cell Biol 167:711–721

Bolis A, Coviello S, Bussini S, Dina G, Pardini C, Previtali SC, Malaguti M, Morana P, Del Carro U, Feltri ML, Quattrini A, Wrabetz L, Bolino A (2005) Loss of Mtmr2 phosphatase in Schwann cells but not in motor neurons causes Charcot-Marie-Tooth type 4B1 neuropathy with myelin outfoldings. J Neurosci 25:8567–8577

Bolis A, Coviello S, Visigalli I, Taveggia C, Bachi A, Chishti AH, Hanada T, Quattrini A, Previtali SC, Biffi A, Bolino A (2009) Dlg1, Sec8, and Mtmr2 regulate membrane homeostasis in Schwann cell myelination. J Neurosci 29:8858–8870

Bonifati V, Rizzu P, Van BAREN MJ, Schaap O, Breedveld GJ, Krieger E, Dekker MC, Squitieri F, Ibanez P, Joosse M, Van Dongen JW, Vanacore N, Van Swieten JC, Brice A, Meco G, Van Duijn CM, Oostra BA, Heutink P (2003) Mutations in the DJ-1 gene associated with autosomal recessive early-onset parkinsonism. Science 299:256–259

Bonneick S, Boentert M, Berger P, Atanasoski S, Mantei N, Wessig C, Toyka KV, Young P, Suter U (2005) An animal model for Charcot-Marie-Tooth disease type 4B1. Hum Mol Genet 14:3685–3695

Buj-Bello A, Furling D, Tronchere H, Laporte J, Lerouge T, Butler-Browne GS, Mandel JL (2002a) Muscle-specific alternative splicing of myotubularin-related 1 gene is impaired in DM1 muscle cells. Hum Mol Genet 11:2297–2307

Buj-Bello A, Laugel V, Messaddeq N, Zahreddine H, Laporte J, Pellissier J-F, Mandel J-L (2002b) The lipid phosphatase myotubularin is essential for skeletal muscle maintenance but not for myogenesis in mice. Proc Natl Acad Sci USA 99:15060–15065

Buj-Bello A, Fougerousse F, Schwab Y, Messaddeq N, Spehner D, Pierson CR, Durand M, Kretz C, Danos O, Douar AM, Beggs AH, Schultz P, Montus M, Denefle P, Mandel JL (2008) AAV-mediated intramuscular delivery of myotubularin corrects the myotubular myopathy phenotype in targeted murine muscle and suggests a function in plasma membrane homeostasis. Hum Mol Genet 17:2132–2143

Cairns P, Okami K, Halachmi S, Halachmi N, Esteller M, Herman JG, Jen J, Isaacs WB, Bova GS, Sidransky D (1997) Frequent inactivation of PTEN/MMAC1 in primary prostate cancer. Cancer Res 57:4997–5000

Campbell RB, Liu F, Ross AH (2003) Allosteric activation of PTEN phosphatase by phosphatidylinositol 4,5-bisphosphate. J Biol Chem 278:33617–33620

Campisi J, D'Adda Di Fagagna F (2007) Cellular senescence: when bad things happen to good cells. Nat Rev Mol Cell Biol 8:729–740

Cancer Genomeatlas Research Network (2008) Comprehensive genomic characterization defines human glioblastoma genes and core pathways. Nature 455:1061–1068

Cao X, Wei G, Fang H, Guo J, Weinstein M, Marsh CB, Ostrowski MC, Tridandapani S (2004) The inositol 3-phosphatase PTEN negatively regulates Fc gamma receptor signaling, but supports Toll-like receptor 4 signaling in murine peritoneal macrophages. J Immunol 172:4851–4857

Cao C, Laporte J, Backer JM, Wandinger-Ness A, Stein M-P (2007) Myotubularin lipid phosphatase binds the hVPS15/hVPS34 lipid kinase complex on endosomes. Traffic 8:1052–1067

Cao C, Backer JM, Laporte J, Bedrick EJ, Wandinger-Ness A (2008) Sequential actions of myotubularin lipid phosphatases regulate endosomal PI(3)P and growth factor receptor trafficking. Mol Biol Cell 19:3334–3346

Carver BS, Tran J, Gopalan A, Chen Z, Shaikh S, Carracedo A, Alimonti A, Nardella C, Varmeh S, Scardino PT, Cordon-Cardo C, Gerald W, Pandolfi PP (2009) Aberrant ERG expression cooperates with loss of PTEN to promote cancer progression in the prostate. Nat Genet 41:619–624

Chang CJ, Freeman DJ, Wu H (2004) PTEN regulates Mdm2 expression through the P1 promoter. J Biol Chem 279:29841–29848

Chang N, El-Hayek YH, Gomez E, Wan Q (2007) Phosphatase PTEN in neuronal injury and brain disorders. Trends Neurosci 30:581–586

Chaussade C, Pirola L, Bonnafous S, Blondeau F, Brenz-Verca S, Tronchere H, Portis F, Rusconi S, Payrastre B, Laporte J, Van Obberghen E (2003) Expression of myotubularin by an adenoviral vector demonstrates its function as a phosphatidylinositol 3-phosphate [PtdIns(3)P] phosphatase in muscle cell lines: involvement of PtdIns(3)P in insulin-stimulated glucose transport. Mol Endocrinol 17:2448–2460

Chen Z, Trotman LC, Shaffer D, Lin HK, Dotan ZA, Niki M, Koutcher JA, Scher HI, Ludwig T, Gerald W, Cordon-Cardo C, Pandolfi PP (2005) Crucial role of p53-dependent cellular senescence in suppression of Pten-deficient tumorigenesis. Nature 436:725–730

Chen F, Archambault V, Kar A, Lio P, D'Avino PP, Sinka R, Lilley K, Laue ED, Deak P, Capalbo L, Glover DM (2007) Multiple protein phosphatases are required for mitosis in Drosophila. Curr Biol 17:293–303

Cheng S, Hsia CY, Feng B, Liou ML, Fang X, Pandolfi PP, Liou HC (2009) BCR-mediated apoptosis associated with negative selection of immature B cells is selectively dependent on Pten. Cell Res 19:196–207

Chojnowski A, Ravise N, Bachelin C, Depienne C, Ruberg M, Brugg B, Laporte J, Baron-Van Evercooren A, Leguern E (2007) Silencing of the Charcot-Marie-Tooth associated MTMR2 gene decreases proliferation and enhances cell death in primary cultures of Schwann cells. Neurobiol Dis 26:323–331

Choudhury P, Srivastava S, Li Z, Ko K, Albaqumi M, Narayan K, Coetzee WA, Lemmon MA, Skolnik EY (2006) Specificity of the myotubularin family of phosphatidylinositol-3-phosphatase is determined by the PH/GRAM domain. J Biol Chem 281:31762–31769

Chung JH, Eng C (2005) Nuclear-cytoplasmic partitioning of phosphatase and tensin homologue deleted on chromosome 10 (PTEN) differentially regulates the cell cycle and apoptosis. Cancer Res 65:8096–8100

Chung JH, Ginn-Pease ME, Eng C (2005) Phosphatase and tensin homologue deleted on chromosome 10 (PTEN) has nuclear localization signal-like sequences for nuclear import mediated by major vault protein. Cancer Res 65:4108–4116

Collado M, Serrano M (2010) Senescence in tumours: evidence from mice and humans. Nat Rev Cancer 10:51–57

Crackower MA, Oudit GY, Kozieradzki I, Sarao R, Sun H, Sasaki T, Hirsch E, Suzuki A, Shioi T, Irie-Sasaki J, Sah R, Cheng HY, Rybin VO, Lembo G, Fratta L, Oliveira-Dos-Santos AJ, Benovic JL, Kahn CR, Izumo S, Steinberg SF, Wymann MP, Backx PH, Penninger JM (2002) Regulation of myocardial contractility and cell size by distinct PI3K-PTEN signaling pathways. Cell 110:737–749

Davidson L, Maccario H, Perera NM, Yang X, Spinelli L, Tibarewal P, Glancy B, Gray A, Weijer CJ, Downes CP, Leslie NR (2010) Suppression of cellular proliferation and invasion by the concerted lipid and protein phosphatase activities of PTEN. Oncogene 29:687–697

Davies MA, Lu Y, Sano T, Fang X, Tang P, Lapushin R, Koul D, Bookstein R, Stokoe D, Yung WK, Mills GB, Steck PA (1998) Adenoviral transgene expression of MMAC/PTEN in human glioma cells inhibits Akt activation and induces anoikis. Cancer Res 58:5285–5290

Delgado-Esteban M, Martin-Zanca D, Andres-Martin L, Almeida A, Bolanos JP (2007) Inhibition of PTEN by peroxynitrite activates the phosphoinositide-3-kinase/Akt neuroprotective signaling pathway. J Neurochem 102:194–205

Denning G, Jean-Joseph B, Prince C, Durden DL, Vogt PK (2007) A short N-terminal sequence of PTEN controls cytoplasmic localization and is required for suppression of cell growth. Oncogene 26:3930–3940

Dey N, Crosswell HE, De P, Parsons R, Peng Q, Su JD, Durden DL (2008) The protein phosphatase activity of PTEN regulates SRC family kinases and controls glioma migration. Cancer Res 68:1862–1871

Di Cristofano A, Pesce B, Cordon-Cardo C, Pandolfi PP (1998) Pten is essential for embryonic development and tumour suppression. Nat Genet 19:348–355

Di Cristofano A, Kotsi P, Peng YF, Cordon-Cardo C, Elkon KB, Pandolfi PP (1999) Impaired Fas response and autoimmunity in Pten+/- mice. Science 285:2122–2125

Doerks T, Strauss M, Brendel M, Bork P (2000) GRAM, a novel domain in glucosyltransferases, myotubularins and other putative membrane-associated proteins. Trends Biochem Sci 25:483–485

Dolley G, Lamarche B, Despres JP, Bouchard C, Perusse L, Vohl MC (2009) Phosphoinositide cycle gene polymorphisms affect the plasma lipid profile in the Quebec Family Study. Mol Genet Metab 97:149–154

Dove SK, Dong K, Kobayashi T, Williams FK, Michell RH (2009) Phosphatidylinositol 3,5-bisphosphate and Fab1p/PIKfyve underPPIn endo-lysosome function. Biochem J 419:1–13

Dowling JJ, Vreede AP, Low SE, Gibbs EM, Kuwada JY, Bonnemann CG, Feldman EL (2009) Loss of myotubularin function results in T-tubule disorganization in zebrafish and human myotubular myopathy. PLoS Genet 5:e1000372

Dowling JJ, Low SE, Busta AS, Feldman EL (2010) Zebrafish MTMR14 is required for excitation-contraction coupling, developmental motor function and the regulation of autophagy. Hum Mol Genet 19:2668–2681

Duerr EM, Rollbrocker B, Hayashi Y, Peters N, Meyer-Puttlitz B, Louis DN, Schramm J, Wiestler OD, Parsons R, Eng C, Von Deimling A (1998) PTEN mutations in gliomas and glioneuronal tumors. Oncogene 16:2259–2264

Fan HY, Liu Z, Cahill N, Richards JS (2008) Targeted disruption of Pten in ovarian granulosa cells enhances ovulation and extends the life span of luteal cells. Mol Endocrinol 22:2128–2140

Fares H, Greenwald I (2001) Genetic analysis of endocytosis in Caenorhabditis elegans: coelomocyte uptake defective mutants. Genetics 159:133–145

Feng Q, Di R, Tao F, Chang Z, Lu S, Fan W, Shan C, Li X, Yang Z (2010) PDK1 regulates vascular remodeling and promotes epithelial-mesenchymal transition in cardiac development. Mol Cell Biol 30:3711–3721

Fili N, Calleja V, Woscholski R, Parker PJ, Larijani B (2006) Compartmental signal modulation: endosomal phosphatidylinositol 3-phosphate controls endosome morphology and selective cargo sorting. Proc Natl Acad Sci USA 103:15473–15478

Fine B, Hodakoski C, Koujak S, Su T, Saal LH, Maurer M, Hopkins B, Keniry M, Sulis ML, Mense S, Hibshoosh H, Parsons R (2009) Activation of the PI3K pathway in cancer through inhibition of PTEN by exchange factor P-REX2a. Science 325:1261–1265

Firestein R, Nagy PL, Daly M, Huie P, Conti M, Cleary ML (2002) Male infertility, impaired spermatogenesis, and azoospermia in mice deficient for the pseudophosphatase Sbf1. J Clin Invest 109:1165–1172

Ford-Hutchinson AF, Ali Z, Lines SE, Hallgrimsson B, Boyd SK, Jirik FR (2007) Inactivation of Pten in osteo-chondroprogenitor cells leads to epiphyseal growth plate abnormalities and skeletal overgrowth. J Bone Miner Res 22:1245–1259

Fouladkou F, Landry T, Kawabe H, Neeb A, Lu C, Brose N, Stambolic V, Rotin D (2008) The ubiquitin ligase Nedd4-1 is dispensable for the regulation of PTEN stability and localization. Proc Natl Acad Sci USA 105:8585–8590

Fox JA, Ung K, Tanlimco SG, Jirik FR (2002) Disruption of a single Pten allele augments the chemotactic response of B lymphocytes to stromal cell-derived factor-1. J Immunol 169:49–54

Fraser MM, Bayazitov IT, Zakharenko SS, Baker SJ (2008) Phosphatase and tensin homolog, deleted on chromosome 10 deficiency in brain causes defects in synaptic structure, transmission and plasticity, and myelination abnormalities. Neuroscience 151:476–488

Freeman DJ, Li AG, Wei G, Li HH, Kertesz N, Lesche R, Whale AD, Martinez-Diaz H, Rozengurt N, Cardiff RD, Liu X, Wu H (2003) PTEN tumor suppressor regulates p53 protein levels and activity through phosphatase-dependent and -independent mechanisms. Cancer Cell 3:117–130

Fridberg M, Servin A, Anagnostaki L, Linderoth J, Berglund M, Soderberg O, Enblad G, Rosen A, Mustelin T, Jerkeman M, Persson JL, Wingren AG (2007) Protein expression and cellular localization in two prognostic subgroups of diffuse large B-cell lymphoma: higher expression of ZAP70 and PKC-beta II in the non-germinal center group and poor survival in patients deficient in nuclear PTEN. Leuk Lymphoma 48:2221–2232

Fruman DA, Bismuth G (2009) Fine tuning the immune response with PI3K. Immunol Rev 228:253–272

Funamoto S, Meili R, Lee S, Parry L, Firtel RA (2002) Spatial and temporal regulation of 3-phosphoinositides by PI 3-kinase and PTEN mediates chemotaxis. Cell 109:611–623

Furnari FB, Huang HJ, Cavenee WK (1998) The phosphoinositol phosphatase activity of PTEN mediates a serum-sensitive G1 growth arrest in glioma cells. Cancer Res 58:5002–5008

Gao P, Wange RL, Zhang N, Oppenheim JJ, Howard OM (2005) Negative regulation of CXCR4-mediated chemotaxis by the lipid phosphatase activity of tumor suppressor PTEN. Blood 106:2619–2626

Gary DS, Mattson MP (2002) PTEN regulates Akt kinase activity in hippocampal neurons and increases their sensitivity to glutamate and apoptosis. Neuromolecular Med 2:261–269

Georgescu MM, Kirsch KH, Kaloudis P, Yang H, Pavletich NP, Hanafusa H (2000) Stabilization and productive positioning roles of the C2 domain of PTEN tumor suppressor. Cancer Res 60:7033–7038

Gil A, Andres-Pons A, Fernandez E, Valiente M, Torres J, Cervera J, Pulido R (2006) Nuclear localization of PTEN by a Ran-dependent mechanism enhances apoptosis: involvement of an N-terminal nuclear localization domain and multiple nuclear exclusion motifs. Mol Biol Cell 17:4002–4013

Gildea JJ, Herlevsen M, Harding MA, Gulding KM, Moskaluk CA, Frierson HF, Theodorescu D (2004) PTEN can inhibit in vitro organotypic and in vivo orthotopic invasion of human bladder cancer cells even in the absence of its lipid phosphatase activity. Oncogene 23:6788–6797

Gillooly DJ, Morrow IC, Lindsay M, Gould R, Bryant NJ, Gaullier J-M, Parton RG, Stenmark H (2000) Localization of phosphatidylinositol 3-phosphate in yeast and mammalian cells. EMBO J 19:4577–4588

Gillooly DJ, Raiborg C, Stenmark H (2003) Phosphatidylinositol 3-phosphate is found in microdomains of early endosomes. Histochem Cell Biol 120:445–453

Ginn-Pease ME, Eng C (2003) Increased nuclear phosphatase and tensin homologue deleted on chromosome 10 is associated with G0-G1 in MCF-7 cells. Cancer Res 63:282–286

Gispert S, Ricciardi F, Kurz A, Azizov M, Hoepken HH, Becker D, Voos W, Leuner K, Muller WE, Kudin AP, Kunz WS, Zimmermann A, Roeper J, Wenzel D, Jendrach M, Garcia-Arencibia M, Fernandez-Ruiz J, Huber L, Rohrer H, Barrera M, Reichert AS, Rub U, Chen A, Nussbaum RL, Auburger G (2009) Parkinson phenotype in aged PINK1-deficient mice is accompanied by progressive mitochondrial dysfunction in absence of neurodegeneration. PLoS One 4:e5777

Goryunov D, Nightingale A, Bornfleth L, Leung C, Liem RK (2008) Multiple disease-linked myotubularin mutations cause NFL assembly defects in cultured cells and disrupt myotubularin dimerization. J Neurochem 104:1536–1552

Griffin RJ, Moloney A, Kelliher M, Johnston JA, Ravid R, Dockery P, O'Connor R, O'Neill C (2005) Activation of Akt/PKB, increased phosphorylation of Akt substrates and loss and altered distribution of Akt and PTEN are features of Alzheimer's disease pathology. J Neurochem 93:105–117

Gronbaek K, Zeuthen J, Guldberg P, Ralfkiaer E, Hou-Jensen K (1998) Alterations of the MMAC1/PTEN gene in lymphoid malignancies. Blood 91:4388–4390

Groszer M, Erickson R, Scripture-Adams DD, Lesche R, Trumpp A, Zack JA, Kornblum HI, Liu X, Wu H (2001) Negative regulation of neural stem/progenitor cell proliferation by the Pten tumor suppressor gene in vivo. Science 294:2186–2189

Groszer M, Erickson R, Scripture-ADAMS DD, Dougherty JD, Le Belle J, Zack JA, Geschwind DH, Liu X, Kornblum HI, Wu H (2006) PTEN negatively regulates neural stem cell self-renewal by modulating G0-G1 cell cycle entry. Proc Natl Acad Sci USA 103:111–116

Gu J, Tamura M, Pankov R, Danen EH, Takino T, Matsumoto K, Yamada KM (1999) Shc and FAK differentially regulate cell motility and directionality modulated by PTEN. J Cell Biol 146:389–403

Haas-Kogan D, Shalev N, Wong M, Mills G, Yount G, Stokoe D (1998) Protein kinase B (PKB/Akt) activity is elevated in glioblastoma cells due to mutation of the tumor suppressor PTEN/MMAC. Curr Biol 8:1195–1198

Hagenbeek TJ, Spits H (2008) T-cell lymphomas in T-cell-specific Pten-deficient mice originate in the thymus. Leukemia 22:608–619

Hagenbeek TJ, Naspetti M, Malergue F, Garcon F, Nunes JA, Cleutjens KB, Trapman J, Krimpenfort P, Spits H (2004) The loss of PTEN allows TCR alphabeta lineage thymocytes to bypass IL-7 and Pre-TCR-mediated signaling. J Exp Med 200:883–894

Hamada K, Sasaki T, Koni PA, Natsui M, Kishimoto H, Sasaki J, Yajima N, Horie Y, Hasegawa G, Naito M, Miyazaki J, Suda T, Itoh H, Nakao K, Mak TW, Nakano T, Suzuki A (2005) The PTEN/PI3K pathway governs normal vascular development and tumor angiogenesis. Genes Dev 19:2054–2065

Han SY, Kato H, Kato S, Suzuki T, Shibata H, Ishii S, Shiiba K, Matsuno S, Kanamaru R, Ishioka C (2000) Functional evaluation of PTEN missense mutations using in vitro phosphoinositide phosphatase assay. Cancer Res 60:3147–3151

He L (2010) Posttranscriptional regulation of PTEN dosage by noncoding RNAs. Sci Signal 3:pe39

He XC, Yin T, Grindley JC, Tian Q, Sato T, Tao WA, Dirisina R, Porter-Westpfahl KS, Hembree M, Johnson T, Wiedemann LM, Barrett TA, Hood L, Wu H, Li L (2007) PTEN-deficient intestinal stem cells initiate intestinal polyposis. Nat Genet 39:189–198

He L, Ingram A, Rybak AP, Tang D (2010) Shank-interacting protein-like 1 promotes tumorigenesis via PTEN inhibition in human tumor cells. J Clin Invest 120:2094–2108

Heit B, Robbins SM, Downey CM, Guan Z, Colarusso P, Miller BJ, Jirik FR, Kubes P (2008) PTEN functions to 'prioritize' chemotactic cues and prevent 'distraction' in migrating neutrophils. Nat Immunol 9:743–752

Hernando E, Charytonowicz E, Dudas ME, Menendez S, Matushansky I, Mills J, Socci ND, Behrendt N, Ma L, Maki RG, Pandolfi PP, Cordon-Cardo C (2007) The AKT-mTOR pathway plays a critical role in the development of leiomyosarcomas. Nat Med 13:748–753

Hirsch E, Katanaev VL, Garlanda C, Azzolino O, Pirola L, Silengo L, Sozzani S, Mantovani A, Altruda F, Wymann MP (2000) Central role for G protein-coupled phosphoinositide 3-kinase gamma in inflammation. Science 287:1049–1053

Hnia K, Tronchere H, Tomczak KK, Amoasii L, Schultz P, Beggs AH, Payrastre B, Mandel JL, Laporte J (2010) Myotubularin controls desmin intermediate filament architecture and mitochondrial dynamics in human and mouse skeletal muscle. J Clin Invest 121:70–85

Horie Y, Suzuki A, Kataoka E, Sasaki T, Hamada K, Sasaki J, Mizuno K, Hasegawa G, Kishimoto H, Iizuka M, Naito M, Enomoto K, Watanabe S, Mak TW, Nakano T (2004) Hepatocyte-specific Pten deficiency results in steatohepatitis and hepatocellular carcinomas. J Clin Invest 113:1774–1783

Iijima M, Devreotes P (2002) Tumor suppressor PTEN mediates sensing of chemoattractant gradients. Cell 109:599–610

Iijima M, Huang YE, Luo HR, Vazquez F, Devreotes PN (2004) Novel mechanism of PTEN regulation by its phosphatidylinositol 4,5-bisphosphate binding motif is critical for chemotaxis. J Biol Chem 279:16606–16613

Inoue-Narita T, Hamada K, Sasaki T, Hatakeyama S, Fujita S, Kawahara K, Sasaki M, Kishimoto H, Eguchi S, Kojima I, Beermann F, Kimura T, Osawa M, Itami S, Mak TW, Nakano T, Manabe M, Suzuki A (2008) Pten deficiency in melanocytes results in resistance to hair graying and susceptibility to carcinogen-induced melanomagenesis. Cancer Res 68:5760–5768

Inzelberg R, Jankovic J (2007) Are Parkinson disease patients protected from some but not all cancers? Neurology 69:1542–1550

Ji SP, Zhang Y, Van Cleemput J, Jiang W, Liao M, Li L, Wan Q, Backstrom JR, Zhang X (2006) Disruption of PTEN coupling with 5-HT2 C receptors suppresses behavioral responses induced by drugs of abuse. Nat Med 12:324–329

Jiang BH, Liu LZ (2008) PI3K/PTEN signaling in tumorigenesis and angiogenesis. Biochim Biophys Acta 1784:150–158

Kelley VA, Schorey JS (2004) Modulation of cellular phosphatidylinositol 3-phosphate levels in primary macrophages affects heat-killed but not viable Mycobacterium avium's transport through the phagosome maturation process. Cell Microbiol 6:973–985

Kerk D, Moorhead GB (2010) A phylogenetic survey of myotubularin genes of eukaryotes: distribution, protein structure, evolution, and gene expression. BMC Evol Biol 10:196

Khan S, Kumagai T, Vora J, Bose N, Sehgal I, Koeffler PH, Bose S (2004) PTEN promoter is methylated in a proportion of invasive breast cancers. Int J Cancer 112:407–410

Kim D, Dressler GR (2007) PTEN modulates GDNF/RET mediated chemotaxis and branching morphogenesis in the developing kidney. Dev Biol 307:290–299

Kim SK, Su LK, Oh Y, Kemp BL, Hong WK, Mao L (1998) Alterations of PTEN/MMAC1, a candidate tumor suppressor gene, and its homologue, PTH2, in small cell lung cancer cell lines. Oncogene 16:89–93

Kim S-A, Taylor GS, Torgersen KM, Dixon JE (2002) Myotubularin and MTMR2, Phosphatidylinositol 3-phosphatases mutated in myotubular myopathy and type 4B Charcot-Marie-Tooth disease. J Biol Chem 277:4526–4531

Kim S-A, Vacratsis PO, Firestein R, Cleary ML, Dixon JE (2003) Regulation of myotubularin-related (MTMR)2 phosphatidylinositol phosphatase by MTMR5, a catalytically inactive phosphatase. Proc Natl Acad Sci USA 100:4492–4497

Kim RH, Smith PD, Aleyasin H, Hayley S, Mount MP, Pownall S, Wakeham A, You-Ten AJ, Kalia SK, Horne P, Westaway D, Lozano AM, Anisman H, Park DS, Mak TW (2005) Hypersensitivity of DJ-1-deficient mice to 1-methyl-4-phenyl-1,2,3,6-tetrahydropyrindine (MPTP) and oxidative stress. Proc Natl Acad Sci USA 102:5215–5220

Kim JS, Lee C, Bonifant CL, Ressom H, Waldman T (2007) Activation of p53-dependent growth suppression in human cells by mutations in PTEN or PIK3CA. Mol Cell Biol 27:662–677

Kim JW, Kang KH, Burrola P, Mak TW, Lemke G (2008) Retinal degeneration triggered by inactivation of PTEN in the retinal pigment epithelium. Genes Dev 22:3147–3157

Kimura T, Suzuki A, Fujita Y, Yomogida K, Lomeli H, Asada N, Ikeuchi M, Nagy A, Mak TW, Nakano T (2003) Conditional loss of PTEN leads to testicular teratoma and enhances embryonic germ cell production. Development 130:1691–1700

Kishimoto H, Ohteki T, Yajima N, Kawahara K, Natsui M, Kawarasaki S, Hamada K, Horie Y, Kubo Y, Arase S, Taniguchi M, Vanhaesebroeck B, Mak TW, Nakano T, Koyasu S, Sasaki T, Suzuki A (2007) The Pten/PI3K pathway governs the homeostasis of Valpha14iNKT cells. Blood 109:3316–3324

Knobbe CB, Lapin V, Suzuki A, Mak TW (2008) The roles of PTEN in development, physiology and tumorigenesis in mouse models: a tissue-by-tissue survey. Oncogene 27:5398–5415

Kohno T, Takahashi M, Manda R, Yokota J (1998) Inactivation of the PTEN/MMAC1/TEP1 gene in human lung cancers. Genes Chromosomes Cancer 22:152–156

Korkaya H, Paulson A, Charafe-Jauffret E, Ginestier C, Brown M, Dutcher J, Clouthier SG, Wicha MS (2009) Regulation of mammary stem/progenitor cells by PTEN/Akt/beta-catenin signaling. PLoS Biol 7:e1000121

Kotelevets L, Van Hengel J, Bruyneel E, Mareel M, Van Roy F, Chastre E (2001) The lipid phosphatase activity of PTEN is critical for stabilizing intercellular junctions and reverting invasiveness. J Cell Biol 155:1129–1135

Kotelevets L, Van Hengel J, Bruyneel E, Mareel M, Van Roy F, Chastre E (2005) Implication of the MAGI-1b/PTEN signalosome in stabilization of adherens junctions and suppression of invasiveness. Faseb J 19:115–117

Koul D, Parthasarathy R, Shen R, Davies MA, Jasser SA, Chintala SK, Rao JS, Sun Y, Benvenisite EN, Liu TJ, Yung WK (2001) Suppression of matrix metalloproteinase-2 gene expression and invasion in human glioma cells by MMAC/PTEN. Oncogene 20:6669–6678

Kurlawalla-Martinez C, Stiles B, Wang Y, Devaskar SU, Kahn BB, Wu H (2005) Insulin hypersensitivity and resistance to streptozotocin-induced diabetes in mice lacking PTEN in adipose tissue. Mol Cell Biol 25:2498–2510

Kuroda S, Nishio M, Sasaki T, Horie Y, Kawahara K, Sasaki M, Natsui M, Matozaki T, Tezuka H, Ohteki T, Forster I, Mak TW, Nakano T, Suzuki A (2008) Effective clearance of intracellular Leishmania major in vivo requires Pten in macrophages. Eur J Immunol 38:1331–1340

Kwon CH, Zhu X, Zhang J, Knoop LL, Tharp R, Smeyne RJ, Eberhart CG, Burger PC, Baker SJ (2001) Pten regulates neuronal soma size: a mouse model of Lhermitte-Duclos disease. Nat Genet 29:404–411

Kwon CH, Luikart BW, Powell CM, Zhou J, Matheny SA, Zhang W, Li Y, Baker SJ, Parada LF (2006) Pten regulates neuronal arborization and social interaction in mice. Neuron 50:377–388

Lacalle RA, Gomez-Mouton C, Barber DF, Jimenez-Baranda S, Mira E, Martinez AC, Carrera AC, Manes S (2004) PTEN regulates motility but not directionality during leukocyte chemotaxis. J Cell Sci 117:6207–6215

Laporte J, Hu LJ, Kretz C, Mandel, J.-L., Kioschis P, Coy JF, Klauck SM, Poustka A, Dahl N (1996) A gene mutated in X-linked myotubular myopathy defines a new putative tyrosine phosphatase family conserved in yeast. Nat Genet 13:175–182

Laporte J, Blondeau F, Buj-Bello A, Tentler D, Kretz C, Dahl N, Mandel J (1998) Characterization of the myotubularin dual specificity phosphatase gene family from yeast to human. Hum Mol Genet 7:1703–1712

Laporte J, Blondeau F, Gansmuller A, Lutz Y, Vonesch J-L, Mandel J-L (2002) The PtdIns3P phosphatase myotubularin is a cytoplasmic protein that also localizes to Rac1-inducible plasma membrane ruffles. J Cell Sci 115:3105–3117

Laporte J, Bedez F, Bolino A, Mandel J-L (2003) Myotubularins, a large disease-associated family of cooperating catalytically active and inactive phosphoinositides phosphatases. Hum Mol Genet 12:R285–R292

Lee JO, Yang H, Georgescu MM, Di Cristofano A, Maehama T, Shi Y, Dixon JE, Pandolfi P, Pavletich NP (1999) Crystal structure of the PTEN tumor suppressor: implications for its phosphoinositide phosphatase activity and membrane association. Cell 99:323–334

Lee HW, Kim Y, Han K, Kim H, Kim E (2010) The phosphoinositide 3-phosphatase MTMR2 interacts with PSD-95 and maintains excitatory synapses by modulating endosomal traffic. J Neurosci 30:5508–5518

Lee JJ, Kim BC, Park MJ, Lee YS, Kim YN, Lee BL, Lee JS (2011) PTEN status switches cell fate between premature senescence and apoptosis in glioma exposed to ionizing radiation. Cell Death Differ 18:666–677

Leslie NR, Yang X, Downes CP, Weijer CJ (2007) PtdIns(3,4,5)P3-dependent and -independent roles for PTEN in the control of cell migration. Curr Biol 17:115–125

Leslie NR, Batty IH, Maccario H, Davidson L, Downes CP (2008) Understanding PTEN regulation: PIP2, polarity and protein stability. Oncogene 27:5464–5476

Leslie NR, Spinelli L, Tibarewal P, Zilidis G, Weerasinghe N, Lim JC, Maccario H, Downes CP (2010) Indirect mechanisms of carcinogenesis via downregulation of PTEN function. Adv Enzyme Regul 50:112–118

Li DM, Sun H (1997) TEP1, encoded by a candidate tumor suppressor locus, is a novel protein tyrosine phosphatase regulated by transforming growth factor beta. Cancer Res 57:2124–2129

Li DM, Sun H (1998) PTEN/MMAC1/TEP1 suppresses the tumorigenicity and induces G1 cell cycle arrest in human glioblastoma cells. Proc Natl Acad Sci USA 95:15406–15411

Li J, Yen C, Liaw D, Podsypanina K, Bose S, Wang SI, Puc J, Miliaresis C, Rodgers L, McCombie R, Bigner SH, Giovanella BC, Ittmann M, Tycko B, Hibshoosh H, Wigler MH, Parsons R (1997) PTEN, a putative protein tyrosine phosphatase gene mutated in human brain, breast, and prostate cancer. Science 275:1943–1947

Li Z, Jiang H, Xie W, Zhang Z, Smrcka AV, Wu D (2000) Roles of PLC-beta2 and -beta3 and PI3Kgamma in chemoattractant-mediated signal transduction. Science 287:1046–1049

Li G, Robinson GW, Lesche R, Martinez-DIAZ H, Jiang Z, Rozengurt N, Wagner KU, Wu DC, Lane TF, Liu X, Hennighausen L, Wu H (2002a) Conditional loss of PTEN leads to precocious development and neoplasia in the mammary gland. Development 129:4159–4170

Li L, Liu F, Salmonsen RA, Turner TK, Litofsky NS, Di Cristofano A, Pandolfi PP, Jones SN, Recht LD, Ross AH (2002b) PTEN in neural precursor cells: regulation of migration, apoptosis, and proliferation. Mol Cell Neurosci 20:21–29

Li Y, Jia Y, Pichavant M, Loison F, Sarraj B, Kasorn A, You J, Robson BE, Umetsu DT, Mizgerd JP, Ye K, Luo HR (2009) Targeted deletion of tumor suppressor PTEN augments neutrophil function and enhances host defense in neutropenia-associated pneumonia. Blood 113:4930–4941

Liaw D, Marsh DJ, Li J, Dahia PL, Wang SI, Zheng Z, Bose S, Call KM, Tsou HC, Peacocke M, Eng C, Parsons R (1997) Germline mutations of the PTEN gene in Cowden disease, an inherited breast and thyroid cancer syndrome. Nat Genet 16:64–67

Liliental J, Moon SY, Lesche R, Mamillapalli R, Li D, Zheng Y, Sun H, Wu H (2000) Genetic deletion of the Pten tumor suppressor gene promotes cell motility by activation of Rac1 and Cdc42 GTPases. Curr Biol 10:401–404

Lin CH, Yeh SH, Lu KT, Leu TH, Chang WC, Gean PW (2001) A role for the PI-3 kinase signaling pathway in fear conditioning and synaptic plasticity in the amygdala. Neuron 31:841–851

Lin HK, Chen Z, Wang G, Nardella C, Lee SW, Chan CH, Yang WL, Wang J, Egia A, Nakayama KI, Cordon-Cardo C, Teruya-Feldstein J, Pandolfi PP (2010) Skp2 targeting suppresses tumorigenesis by Arf-p53-independent cellular senescence. Nature 464:374–379

Lindsay Y, McCoull D, Davidson L, Leslie NR, Fairservice A, Gray A, Lucocq J, Downes CP (2006) Localization of agonist-sensitive PtdIns(3,4,5)P3 reveals a nuclear pool that is insensitive to PTEN expression. J Cell Sci 119:5160–5168

Liu H, Radisky DC, Wang F, Bissell MJ (2004) Polarity and proliferation are controlled by distinct signaling pathways downstream of PI3-kinase in breast epithelial tumor cells. J Cell Biol 164:603–612

Liu F, Wagner S, Campbell RB, Nickerson JA, Schiffer CA, Ross AH (2005a) PTEN enters the nucleus by diffusion. J Cell Biochem 96:221–234

Liu JL, Sheng X, Hortobagyi ZK, Mao Z, Gallick GE, Yung WK (2005b) Nuclear PTEN-mediated growth suppression is independent of Akt down-regulation. Mol Cell Biol 25:6211–6224

Liu X, Shi Y, Giranda VL, Luo Y (2006) Inhibition of the phosphatidylinositol 3-kinase/Akt pathway sensitizes MDA-MB468 human breast cancer cells to cerulenin-induced apoptosis. Mol Cancer Ther 5:494–501

Liu X, Bruxvoort KJ, Zylstra CR, Liu J, Cichowski R, Faugere MC, Bouxsein ML, Wan C, Williams BO, Clemens TL (2007) Lifelong accumulation of bone in mice lacking Pten in osteoblasts. Proc Natl Acad Sci USA 104:2259–2264

Liu W, Zhou Y, Reske SN, Shen C (2008) PTEN mutation: many birds with one stone in tumorigenesis. Anticancer Res 28:3613–3619

Liu X, Karnell JL, Yin B, Zhang R, Zhang J, Li P, Choi Y, Maltzman JS, Pear WS, Bassing CH, Turka LA (2010) Distinct roles for PTEN in prevention of T cell lymphoma and autoimmunity in mice. J Clin Invest 120:2497–2507

Lorenzo O, Urbe S, Clague MJ (2005) Analysis of phosphoinositide binding domain properties within the myotubularin-related protein MTMR3. J Cell Sci 118:2005–2012

Lorenzo O, Urbe S, Clague MJ (2006) Systematic analysis of myotubularins: heteromeric interactions, subcellular localisation and endosome related functions. J Cell Sci 119:2953–2959

Lu Y, Yu Q, Liu JH, Zhang J, Wang H, Koul D, McMurray JS, Fang X, Yung WK, Siminovitch KA, Mills GB (2003) Src family protein-tyrosine kinases alter the function of PTEN to regulate phosphatidylinositol 3-kinase/AKT cascades. J Biol Chem 278:40057–40066

Ma X, Ziel-Van der Made AC, Autar B, Van der Korput HA, Vermeij M, Van Duijn P, Cleutjens KB, De Krijger R, Krimpenfort P, Berns A, Van der Kwast TH, Trapman J (2005) Targeted biallelic inactivation of Pten in the mouse prostate leads to prostate cancer accompanied by increased epithelial cell proliferation but not by reduced apoptosis. Cancer Res 65:5730–5739

Maccario H, Perera NM, Davidson L, Downes CP, Leslie NR (2007) PTEN is destabilized by phosphorylation on Thr366. Biochem J 405:439–444

Maccario H, Perera NM, Gray A, Downes CP, Leslie NR (2010) Ubiquitination of PTEN (phosphatase and tensin homolog) inhibits phosphatase activity and is enhanced by membrane targeting and hyperosmotic stress. J Biol Chem 285:12620–12628

Maehama T, Dixon JE (1998) The Tumor Suppressor, PTEN/MMAC1, dephosphorylates the lipid second messenger, phosphatidylinositol 3,4,5-trisphosphate. J Biol Chem 273:13375–13378

Mahimainathan L, Choudhury GG (2004) Inactivation of platelet-derived growth factor receptor by the tumor suppressor PTEN provides a novel mechanism of action of the phosphatase. J Biol Chem 279:15258–15268

Marino S, Krimpenfort P, Leung C, Van der Korput HA, Trapman J, Camenisch I, Berns A, Brandner S (2002) PTEN is essential for cell migration but not for fate determination and tumourigenesis in the cerebellum. Development 129:3513–3522

Marsh DJ, Dahia PL, Zheng Z, Liaw D, Parsons R, Gorlin RJ, Eng C (1997) Germline mutations in PTEN are present in Bannayan-Zonana syndrome. Nat Genet 16:333–334

Marsh DJ, Coulon V, Lunetta KL, Rocca-SERRA P, Dahia PL, Zheng Z, Liaw D, Caron S, Duboue B, Lin AY, Richardson AL, Bonnetblanc JM, Bressieux JM, Cabarrot-Moreau A, Chompret A, Demange L, Eeles RA, Yahanda AM, Fearon ER, Fricker JP, Gorlin RJ, Hodgson SV, Huson S, Lacombe D, Eng C et al (1998) Mutation spectrum and genotype-phenotype analyses in Cowden disease and Bannayan-Zonana syndrome, two hamartoma syndromes with germline PTEN mutation. Hum Mol Genet 7:507–515

Marsh V, Winton DJ, Williams GT, Dubois N, Trumpp A, Sansom OJ, Clarke AR (2008) Epithelial Pten is dispensable for intestinal homeostasis but suppresses adenoma development and progression after Apc mutation. Nat Genet 40:1436–1444

Martin-Belmonte F, Gassama A, Datta A, Yu W, Rescher U, Gerke V, Mostov K (2007) PTEN-mediated apical segregation of phosphoinositides controls epithelial morphogenesis through Cdc42. Cell 128:383–397

Mayo LD, Donner DB (2001) A phosphatidylinositol 3-kinase/Akt pathway promotes translocation of Mdm2 from the cytoplasm to the nucleus. Proc Natl Acad Sci USA 98:11598–11603

McConnachie G, Pass I, Walker SM, Downes CP (2003) Interfacial kinetic analysis of the tumour suppressor phosphatase, PTEN: evidence for activation by anionic phospholipids. Biochem J 371:947–955

Mei J, Liu S, Li Z, Gui JF (2010) Mtmr8 is essential for vasculature development in zebrafish embryos. BMC Dev Biol 10:96

Miletic AV, Anzelon-Mills AN, Mills DM, Omori SA, Pedersen IM, Shin DM, Ravetch JV, Bolland S, Morse HC 3RD, Rickert RC (2010) Coordinate suppression of B cell lymphoma by PTEN and SHIP phosphatases. J Exp Med 207:2407–2420

Mirmohammadsadegh A, Marini A, Nambiar S, Hassan M, Tannapfel A, Ruzicka T, Hengge UR (2006) Epigenetic silencing of the PTEN gene in melanoma. Cancer Res 66:6546–6552

Mochizuki Y, Majerus PW (2003) Characterization of myotubularin-related protein 7 and its binding partner, myotubularin-related protein 9. Proc Natl Acad Sci USA 100:9768–9773

Mutter GL, Lin MC, Fitzgerald JT, Kum JB, Baak JP, Lees JA, Weng LP, Eng C (2000) Altered PTEN expression as a diagnostic marker for the earliest endometrial precancers. J Natl Cancer Inst 92:924–930

Myers MP, Stolarov JP, Eng C, Li J, Wang SI, Wigler MH, Parsons R, Tonks NK (1997) P-TEN, the tumor suppressor from human chromosome 10q23, is a dual-specificity phosphatase. Proc Natl Acad Sci USA 94:9052–9057

Myers MP, Pass I, Batty IH, Van der Kaay J, Stolarov JP, Hemmings BA, Wigler MH, Downes CP, Tonks NK (1998) The lipid phosphatase activity of PTEN is critical for its tumor supressor function. Proc Natl Acad Sci USA 95:13513–13518

Nandurkar HH, Caldwell KK, Whisstock JC, Layton MJ, Gaudet EA, Norris FA, Majerus PW, Mitchell CA (2001) Characterization of an adapter subunit to a phosphatidylinositol (3)P 3-phosphatase: Identification of a myotubularin-related protein lacking catalytic activity. Proc Natl Acad Sci USA 98:9499–9504

Nandurkar HH, Layton M, Laporte J, Selan C, Corcoran L, Caldwell KK, Mochizuki Y, Majerus PW, Mitchell CA (2003) Identification of myotubularin as the lipid phosphatase catalytic subunit associated with the 3-phosphatase adapter protein, 3-PAP. Proc Natl Acad Sci USA 100:8660–8665

Naughtin MJ, Sheffield DA, Rahman P, Hughes WE, Gurung R, Stow JL, Nandurkar HH, Dyson JM, Mitchell CA (2010) The myotubularin phosphatase MTMR4 regulates sorting from early endosomes. J Cell Sci 123:3071–3083

Nemenoff RA, Simpson PA, Furgeson SB, Kaplan-Albuquerque N, Crossno J, Garl PJ, Cooper J, Weiser-EVANS MC (2008) Targeted deletion of PTEN in smooth muscle cells results in vascular remodeling and recruitment of progenitor cells through induction of stromal cell-derived factor-1alpha. Circ Res 102:1036–1045

Nguyen KT, Tajmir P, Lin CH, Liadis N, Zhu XD, Eweida M, Tolasa-Karaman G, Cai F, Wang R, Kitamura T, Belsham DD, Wheeler MB, Suzuki A, Mak TW, Woo M (2006) Essential role of Pten in body size determination and pancreatic beta-cell homeostasis in vivo. Mol Cell Biol 26:4511–4518

Ning K, Pei L, Liao M, Liu B, Zhang Y, Jiang W, Mielke JG, Li L, Chen Y, El-Hayek YH, Fehlings MG, Zhang X, Liu F, Eubanks J, Wan Q (2004) Dual neuroprotective signaling mediated by downregulating two distinct phosphatase activities of PTEN. J Neurosci 24:4052–4060

Nishio M, Watanabe K, Sasaki J, Taya C, Takasuga S, Iizuka R, Balla T, Yamazaki M, Watanabe H, Itoh R, Kuroda S, Horie Y, Forster I, Mak TW, Yonekawa H, Penninger JM, Kanaho Y, Suzuki A, Sasaki T (2007) Control of cell polarity and motility by the PtdIns(3,4,5)P3 phosphatase SHIP1. Nat Cell Biol 9:36–44

Nogueira V, Park Y, Chen CC, Xu PZ, Chen ML, Tonic I, Unterman T, Hay N (2008) Akt determines replicative senescence and oxidative or oncogenic premature senescence and sensitizes cells to oxidative apoptosis. Cancer Cell 14:458–470

Oak JS, Deane JA, Kharas MG, Luo J, Lane TE, Cantley LC, Fruman DA (2006) Sjogren's syndrome-like disease in mice with T cells lacking class 1A phosphoinositide-3-kinase. Proc Natl Acad Sci USA 103:16882–16887

Odriozola L, Singh G, Hoang T, Chan AM (2007) Regulation of PTEN activity by its carboxyl-terminal autoinhibitory domain. J Biol Chem 282:23306–23315

Ogawa S, Kwon CH, Zhou J, Koovakkattu D, Parada LF, Sinton CM (2007) A seizure-prone phenotype is associated with altered free-running rhythm in Pten mutant mice. Brain Res 1168:112–123

Okahara F, Ikawa H, Kanaho Y, Maehama T (2004) Regulation of PTEN phosphorylation and stability by a tumor suppressor candidate protein. J Biol Chem 279:45300–45303

Okahara F, Itoh K, Nakagawara A, Murakami M, Kanaho Y, Maehama T (2006) Critical role of PICT-1, a tumor suppressor candidate, in phosphatidylinositol 3,4,5-trisphosphate signals and tumorigenic transformation. Mol Biol Cell 17:4888–4895

Okumura K, Mendoza M, Bachoo RM, Depinho RA, Cavenee WK, Furnari FB (2006) PCAF modulates PTEN activity. J Biol Chem 281:26562–26568

Omary MB, Coulombe PA, McLean WH (2004) Intermediate filament proteins and their associated diseases. N Engl J Med 351:2087–2100

Omori N, Jin G, Li F, Zhang WR, Wang SJ, Hamakawa Y, Nagano I, Manabe Y, Shoji M, Abe K (2002) Enhanced phosphorylation of PTEN in rat brain after transient middle cerebral artery occlusion. Brain Res 954:317–322

Park KK, Liu K, Hu Y, Smith PD, Wang C, Cai B, Xu B, Connolly L, Kramvis I, Sahin M, He Z (2008) Promoting axon regeneration in the adult CNS by modulation of the PTEN/mTOR pathway. Science 322:963–966

Perren A, Komminoth P, Saremaslani P, Matter C, Feurer S, Lees JA, Heitz PU, Eng C (2000) Mutation and expression analyses reveal differential subcellular compartmentalization of PTEN in endocrine pancreatic tumors compared to normal islet cells. Am J Pathol 157:1097–1103

Pinal N, Goberdhan DC, Collinson L, Fujita Y, Cox IM, Wilson C, Pichaud F (2006) Regulated and polarized PtdIns(3,4,5)P3 accumulation is essential for apical membrane morphogenesis in photoreceptor epithelial cells. Curr Biol 16:140–149

Plant PJ, Correa J, Goldenberg N, Bain J, Batt J (2009) The inositol phosphatase MTMR4 is a novel target of the ubiquitin ligase Nedd4. Biochem J 419:57–63

Plum L, Ma X, Hampel B, Balthasar N, Coppari R, Munzberg H, Shanabrough M, Burdakov D, Rother E, Janoschek R, Alber J, Belgardt BF, Koch L, Seibler J, Schwenk F, Fekete C, Suzuki A, Mak TW, Krone W, Horvath TL, Ashcroft FM, Bruning JC (2006) Enhanced PIP3 signaling in POMC neurons causes KATP channel activation and leads to diet-sensitive obesity. J Clin Invest 116:1886–1901

Podsypanina K, Ellenson LH, Nemes A, Gu J, Tamura M, Yamada KM, Cordon-Cardo C, Catoretti G, Fisher PE, Parsons R (1999) Mutation of Pten/Mmac1 in mice causes neoplasia in multiple organ systems. Proc Natl Acad Sci USA 96:1563–1568

Poliseno L, Salmena L, Riccardi L, Fornari A, Song MS, Hobbs RM, Sportoletti P, Varmeh S, Egia A, Fedele G, Rameh L, Loda M, Pandolfi PP (2010a) Identification of the miR-106b~25 microRNA cluster as a proto-oncogenic PTEN-targeting intron that cooperates with its host gene MCM7 in transformation. Sci Signal 3:ra29

Poliseno L, Salmena L, Zhang J, Carver B, Haveman WJ, Pandolfi PP (2010b) A coding-independent function of gene and pseudogene mRNAs regulates tumour biology. Nature 465:1033–1038

Pollard SM, Yoshikawa K, Clarke ID, Danovi D, Stricker S, Russell R, Bayani J, Head R, Lee M, Bernstein M, Squire JA, Smith A, Dirks P (2009) Glioma stem cell lines expanded in adherent culture have tumor-specific phenotypes and are suitable for chemical and genetic screens. Cell Stem Cell 4:568–580

Pridgeon JW, Olzmann JA, Chin LS, Li L (2007) PINK1 protects against oxidative stress by phosphorylating mitochondrial chaperone TRAP1. PLoS Biol 5:e172

Puzio-Kuter AM, Castillo-Martin M, Kinkade CW, Wang X, Shen TH, Matos T, Shen MM, Cordon-Cardo C, Abate-Shen C (2009) Inactivation of p53 and Pten promotes invasive bladder cancer. Genes Dev 23:675–680

Qiu W, Federico L, Naples M, Avramoglu RK, Meshkani R, Zhang J, Tsai J, Hussain M, Dai K, Iqbal J, Kontos CD, Horie Y, Suzuki A, Adeli K (2008) Phosphatase and tensin homolog (PTEN) regulates hepatic lipogenesis, microsomal triglyceride transfer protein, and the secretion of apolipoprotein B-containing lipoproteins. Hepatology 48:1799–1809

Raftopoulou M, Etienne-Manneville S, Self A, Nicholls S, Hall A (2004) Regulation of cell migration by the C2 domain of the tumor suppressor PTEN. Science 303:1179–1181

Rahdar M, Inoue T, Meyer T, Zhang J, Vazquez F, Devreotes PN (2009) A phosphorylation-dependent intramolecular interaction regulates the membrane association and activity of the tumor suppressor PTEN. Proc Natl Acad Sci USA 106:480–485

Ratnacaram CK, Teletin M, Jiang M, Meng X, Chambon P, Metzger D (2008) Temporally controlled ablation of PTEN in adult mouse prostate epithelium generates a model of invasive prostatic adenocarcinoma. Proc Natl Acad Sci USA 105:2521–2526

Reddy P, Liu L, Adhikari D, Jagarlamudi K, Rajareddy S, Shen Y, Du C, Tang W, Hamalainen T, Peng SL, Lan ZJ, Cooney AJ, Huhtaniemi I, Liu K (2008) Oocyte-specific deletion of Pten causes premature activation of the primordial follicle pool. Science 319:611–613

Robinson FL, Dixon JE (2005) The Phosphoinositide-3-phosphatase MTMR2 Associates with MTMR13, a membrane-associated pseudophosphatase also mutated in type 4B Charcot-Marie-Tooth disease. J Biol Chem 280:31699–31707

Robinson FL, Dixon JE (2006) Myotubularin phosphatases: policing 3-phosphoinositides. Trends Cell Biol 16:403

Robinson FL, Niesman IR, Beiswenger KK, Dixon JE (2008) Loss of the inactive myotubularin-related phosphatase Mtmr13 leads to a Charcot-Marie-Tooth 4B2-like peripheral neuropathy in mice. Proc Natl Acad Sci USA 105:4916–4921

Romero-Suarez S, Shen J, Brotto L, Hall T, Mo C, Valdivia HH, Andresen J, Wacker M, Nosek TM, Qu CK, Brotto M (2010) Muscle-specific inositide phosphatase (MIP/MTMR14) is reduced with age and its loss accelerates skeletal muscle aging process by altering calcium homeostasis. Aging (Albany NY) 2:504–513

Rossi DJ, Weissman IL (2006) Pten, tumorigenesis, and stem cell self-renewal. Cell 125:229–231

Ruckle T, Schwarz MK, Rommel C (2006) PI3Kgamma inhibition: towards an 'aspirin of the 21st century'? Nat Rev Drug Discov 5:903–918

Sagona AP, Nezis IP, Pedersen NM, Liestol K, Poulton J, Rusten TE, Skotheim RI, Raiborg C, Stenmark H (2010) PtdIns(3)P controls cytokinesis through KIF13A-mediated recruitment of FYVE-CENT to the midbody. Nat Cell Biol 12:362–371

Salmena L, Carracedo A, Pandolfi PP (2008) Tenets of PTEN tumor suppression. Cell 133:403–414

Salvesen HB, Stefansson I, Kretzschmar EI, Gruber P, Macdonald ND, Ryan A, Jacobs IJ, Akslen LA, Das S (2004) Significance of PTEN alterations in endometrial carcinoma: a population-based study of mutations, promoter methylation and PTEN protein expression. Int J Oncol 25:1615–1623

Sansal I, Sellers WR (2004) The biology and clinical relevance of the PTEN tumor suppressor pathway. J Clin Oncol 22:2954–2963

Santoro M, Modoni A, Masciullo M, Gidaro T, Broccolini A, Ricci E, Tonali PA, Silvestri G (2010) Analysis of MTMR1 expression and correlation with muscle pathological features in juvenile/adult onset myotonic dystrophy type 1 (DM1) and in myotonic dystrophy type 2 (DM2). Exp Mol Pathol 89:158–168

Sarquis MS, Agrawal S, Shen L, Pilarski R, Zhou XP, Eng C (2006) Distinct expression profiles for PTEN transcript and its splice variants in Cowden syndrome and Bannayan-Riley-Ruvalcaba syndrome. Am J Hum Genet 79:23–30

Sasaki T, Irie-SASAKI J, Jones RG, Oliveira-Dos-Santos AJ, Stanford WL, Bolon B, Wakeham A, Itie A, Bouchard D, Kozieradzki I, Joza N, Mak TW, Ohashi PS, Suzuki A, Penninger JM (2000) Function of PI3Kgamma in thymocyte development, T cell activation, and neutrophil migration. Science 287:1040–1046

Sasaki T, Takasuga S, Sasaki J, Kofuji S, Eguchi S, Yamazaki M, Suzuki A (2009) Mammalian phosphoinositide kinases and phosphatases. Prog Lipid Res 48:307–343

Schaletzky J, Dove SK, Short B, Lorenzo O, Clague MJ, Barr FA (2003) Phosphatidylinositol-5-phosphate activation and conserved substrate specificity of the myotubularin phosphatidylinositol 3-phosphatases. Curr Biol 13:504–509

Senderek J, Bergmann C, Weber S, Ketelsen U-P, Schorle H, Rudnik-Schoneborn S, Buttner R, Buchheim E, Zerres K (2003) Mutation of the SBF2 gene, encoding a novel member of the myotubularin family, in Charcot-Marie-Tooth neuropathy type 4B2/11p15. Hum Mol Genet 12:349–356

Sharrard RM, Maitland NJ (2000) Alternative splicing of the human PTEN/MMAC1/TEP1 gene. Biochim Biophys Acta 1494:282–285

Shen WH, Balajee AS, Wang J, Wu H, Eng C, Pandolfi PP, Yin Y (2007) Essential role for nuclear PTEN in maintaining chromosomal integrity. Cell 128:157–170

Shen J, Yu WM, Brotto M, Scherman JA, Guo C, Stoddard C, Nosek TM, Valdivia HH, Qu CK (2009) Deficiency of MIP/MTMR14 phosphatase induces a muscle disorder by disrupting Ca(2+) homeostasis. Nat Cell Biol 11:769–776

Shendelman S, Jonason A, Martinat C, Leete T, Abeliovich A (2004) DJ-1 is a redox-dependent molecular chaperone that inhibits alpha-synuclein aggregate formation. PLoS Biol 2:e362

Silhankova M, Port F, Harterink M, Basler K, Korswagen HC (2010) Wnt signalling requires MTM-6 and MTM-9 myotubularin lipid-phosphatase function in Wnt-producing cells. EMBO J 29:4094–4105

Song MS, Salmena L, Carracedo A, Egia A, Lo-Coco F, Teruya-Feldstein J, Pandolfi PP (2008) The deubiquitinylation and localization of PTEN are regulated by a HAUSP-PML network. Nature 455:813–817

Song LB, Li J, Liao WT, Feng Y, Yu CP, Hu LJ, Kong QL, Xu LH, Zhang X, Liu WL, Li MZ, Zhang L, Kang TB, Fu LW, Huang WL, Xia YF, Tsao SW, Li M, Band V, Band H, Shi QH, Zeng YX, Zeng MS (2009) The polycomb group protein Bmi-1 represses the tumor suppressor PTEN and induces epithelial-mesenchymal transition in human nasopharyngeal epithelial cells. J Clin Invest 119:3626–3636

Srivastava S, Li Z, Lin L, Liu G, Ko K, Coetzee WA, Skolnik EY (2005) The phosphatidylinositol 3-phosphate phosphatase myotubularin-related protein 6 (MTMR6) is a negative regulator of the Ca2+-Activated K+Channel KCa3.1. Mol Cell Biol 25:3630–3638

Srivastava S, Ko K, Choudhury P, Li Z, Johnson AK, Nadkarni V, Unutmaz D, Coetzee WA, Skolnik EY (2006) Phosphatidylinositol-3 phosphatase myotubularin-related protein 6 negatively regulates CD4T cells. Mol Cell Biol 26:5595–5602

Stambolic V, Suzuki A, De La Pompa JL, Brothers GM, Mirtsos C, Sasaki T, Ruland J, Penninger JM, Siderovski DP, Mak TW (1998) Negative regulation of PKB/Akt-dependent cell survival by the tumor suppressor PTEN. Cell 95:29–39

Stambolic V, Macpherson D, Sas D, Lin Y, Snow B, Jang Y, Benchimol S, Mak TW (2001) Regulation of PTEN transcription by p53. Mol Cell 8:317–325

Stanger BZ, Stiles B, Lauwers GY, Bardeesy N, Mendoza M, Wang Y, Greenwood A, Cheng KH, McLaughlin M, Brown D, Depinho RA, Wu H, Melton DA, Dor Y (2005) Pten constrains centroacinar cell expansion and malignant transformation in the pancreas. Cancer Cell 8:185–195

Steck PA, Pershouse MA, Jasser SA, Yung WK, Lin H, Ligon AH, Langford LA, Baumgard ML, Hattier T, Davis T, Frye C, Hu R, Swedlund B, Teng DH, Tavtigian SV (1997) Identification of a candidate tumour suppressor gene, MMAC1, at chromosome 10q23.3 that is mutated in multiple advanced cancers. Nat Genet 15:356–362

Stiles B, Wang Y, Stahl A, Bassilian S, Lee WP, Kim YJ, Sherwin R, Devaskar S, Lesche R, Magnuson MA, Wu H (2004) Liver-specific deletion of negative regulator Pten results in fatty liver and insulin hypersensitivity [corrected]. Proc Natl Acad Sci USA 101:2082–2087

Stiles BL, Kuralwalla-Martinez C, Guo W, Gregorian C, Wang Y, Tian J, Magnuson MA, Wu H (2006) Selective deletion of Pten in pancreatic beta cells leads to increased islet mass and resistance to STZ-induced diabetes. Mol Cell Biol 26:2772–2781

Subramanian KK, Jia Y, Zhu D, Simms BT, Jo H, Hattori H, You J, Mizgerd JP, Luo HR (2007) Tumor suppressor PTEN is a physiologic suppressor of chemoattractant-mediated neutrophil functions. Blood 109:4028–4037

Suzuki A, De La Pompa JL, Stambolic V, Elia AJ, Sasaki T, Del Barco Barrantes I, Ho A, Wakeham A, Itie A, Khoo W, Fukumoto M, Mak TW (1998) High cancer susceptibility and embryonic lethality associated with mutation of the PTEN tumor suppressor gene in mice. Curr Biol 8:1169–1178

Suzuki A, Yamaguchi MT, Ohteki T, Sasaki T, Kaisho T, Kimura Y, Yoshida R, Wakeham A, Higuchi T, Fukumoto M, Tsubata T, Ohashi PS, Koyasu S, Penninger JM, Nakano T, Mak TW (2001) T cell-specific loss of Pten leads to defects in central and peripheral tolerance. Immunity 14:523–534

Suzuki A, Itami S, Ohishi M, Hamada K, Inoue T, Komazawa N, Senoo H, Sasaki T, Takeda J, Manabe M, Mak TW, Nakano T (2003a) Keratinocyte-specific Pten deficiency results in epidermal hyperplasia, accelerated hair follicle morphogenesis and tumor formation. Cancer Res 63:674–681

Suzuki A, Kaisho T, Ohishi M, Tsukio-Yamaguchi M, Tsubata T, Koni PA, Sasaki T, Mak TW, Nakano T (2003b) Critical roles of Pten in B cell homeostasis and immunoglobulin class switch recombination. J Exp Med 197:657–667

Suzuki A, Nakano T, Mak TW, Sasaki T (2008) Portrait of PTEN: messages from mutant mice. Cancer Sci 99:209–213

Taguchi-Atarashi N, Hamasaki M, Matsunaga K, Omori H, Ktistakis NT, Yoshimori T, Noda T (2010) Modulation of local Ptdins3P levels by the PI phosphatase MTMR3 regulates constitutive autophagy. Traffic 11:468–478

Tamura M, Gu J, Matsumoto K, Aota S, Parsons R, Yamada KM (1998) Inhibition of cell migration, spreading, and focal adhesions by tumor suppressor PTEN. Science 280:1614–1617

Tamura M, Gu J, Danen EH, Takino T, Miyamoto S, Yamada KM (1999a) PTEN interactions with focal adhesion kinase and suppression of the extracellular matrix-dependent phosphatidylinositol 3-kinase/Akt cell survival pathway. J Biol Chem 274:20693–20703

Tamura M, Gu J, Takino T, Yamada KM (1999b) Tumor suppressor PTEN inhibition of cell invasion, migration, and growth: differential involvement of focal adhesion kinase and p130Cas. Cancer Res 59:442–449

Taylor GS, Maehama T, Dixon JE (2000a) Inaugural article: myotubularin, a protein tyrosine phosphatase mutated in myotubular myopathy, dephosphorylates the lipid second messenger, phosphatidylinositol 3-phosphate. Proc Natl Acad Sci USA 97:8910–8915

Taylor V, Wong M, Brandts C, Reilly L, Dean NM, Cowsert LM, Moodie S, Stokoe D (2000b) 5′ phospholipid phosphatase SHIP-2 causes protein kinase B inactivation and cell cycle arrest in glioblastoma cells. Mol Cell Biol 20:6860–6871

Taylor BS, Schultz N, Hieronymus H, Gopalan A, Xiao Y, Carver BS, Arora VK, Kaushik P, Cerami E, Reva B, Antipin Y, Mitsiades N, Landers T, Dolgalev I, Major JE, Wilson M, Socci ND, Lash AE, Heguy A, Eastham JA, Scher HI, Reuter VE, Scardino PT, Sander C, Sawyers CL, Gerald WL (2010) Integrative genomic profiling of human prostate cancer. Cancer Cell 18:11–22

Teng Y, Sun AN, Pan XC, Yang G, Yang LL, Wang MR, Yang X (2006) Synergistic function of Smad4 and PTEN in suppressing forestomach squamous cell carcinoma in the mouse. Cancer Res 66:6972–6981

Tersar K, Boentert M, Berger P, Bonneick S, Wessig C, Toyka KV, Young P, Suter U (2007) Mtmr13/Sbf2-deficient mice: an animal model for CMT4B2. Hum Mol Genet 16:2991–3001

Tolkacheva T, Boddapati M, Sanfiz A, Tsuchida K, Kimmelman AC, Chan AM (2001) Regulation of PTEN binding to MAGI-2 by two putative phosphorylation sites at threonine 382 and 383. Cancer Res 61:4985–4989

Tong Z, Fan Y, Zhang W, Xu J, Cheng J, Ding M, Deng H (2009) Pancreas-specific Pten deficiency causes partial resistance to diabetes and elevated hepatic AKT signaling. Cell Res 19:710–719

Tooze SA, Yoshimori T (2010) The origin of the autophagosomal membrane. Nat Cell Biol 12:831–835

Torres J, Pulido R (2001) The tumor suppressor PTEN is phosphorylated by the protein kinase CK2 at its C terminus. Implications for PTEN stability to proteasome-mediated degradation. J Biol Chem 276:993–998

Tosch V, Rohde HM, Tronchere H, Zanoteli E, Monroy N, Kretz C, Dondaine N, Payrastre B, Mandel JL, Laporte J (2006) A novel PtdIns3P and PtdIns(3,5)P2 phosphatase with an inactivating variant in centronuclear myopathy. Hum Mol Genet 15:3098–3106

Toussaint A, Cowling BS, Hnia K, Mohr M, Oldfors A, Schwab Y, Yis U, Maisonobe T, Stojkovic T, Wallgren-Pettersson C, Laugel V, Echaniz-Laguna A, Mandel JL, Nishino I, Laporte J (2010) Defects in amphiphysin 2 (BIN1) and triads in several forms of centronuclear myopathies. Acta Neuropathol 121:253–266

Trimboli AJ, Cantemir-Stone CZ, Li F, Wallace JA, Merchant A, Creasap N, Thompson JC, Caserta E, Wang H, Chong JL, Naidu S, Wei G, Sharma SM, Stephens JA, Fernandez SA, Gurcan MN, Weinstein MB, Barsky SH, Yee L, Rosol TJ, Stromberg PC, Robinson ML, Pepin F, Hallett M, Park M, Ostrowski MC, Leone G (2009) Pten in stromal fibroblasts suppresses mammary epithelial tumours. Nature 461:1084–1091

Tronchere H, Laporte J, Pendaries C, Chaussade C, Liaubet L, Pirola L, Mandel JL, Payrastre B (2004) Production of phosphatidylinositol 5-phosphate by the phosphoinositide 3-phosphatase myotubularin in mammalian cells. J Biol Chem 279:7304–7312

Trotman LC, Niki M, Dotan ZA, Koutcher JA, Di Cristofano A, Xiao A, Khoo AS, Roy-Burman P, Greenberg NM, Van Dyke T, Cordon-Cardo C, Pandolfi PP (2003) Pten dose dictates cancer progression in the prostate. Plos Biol 1:E59

Trotman LC, Wang X, Alimonti A, Chen Z, Teruya-Feldstein J, Yang H, Pavletich NP, Carver BS, Cordon-Cardo C, Erdjument-Bromage H, Tempst P, Chi SG, Kim HJ, Misteli T, Jiang X, Pandolfi PP (2007) Ubiquitination regulates PTEN nuclear import and tumor suppression. Cell 128:141–156

Tsuchiya KD, Wiesner G, Cassidy SB, Limwongse C, Boyle JT, Schwartz S (1998) Deletion 10q23.2-q23.33 in a patient with gastrointestinal juvenile polyposis and other features of a Cowden-like syndrome. Genes Chromosomes Cancer 21:113–118

Tsujita K, Itoh T, Ijuin T, Yamamoto A, Shisheva A, Laporte J, Takenawa T (2004) Myotubularin regulates the function of the late endosome through the GRAM domain-phosphatidylinositol 3,5-bisphosphate interaction. J Biol Chem 279:13817–13824

Tsuruta H, Kishimoto H, Sasaki T, Horie Y, Natsui M, Shibata Y, Hamada K, Yajima N, Kawahara K, Sasaki M, Tsuchiya N, Enomoto K, Mak TW, Nakano T, Habuchi T, Suzuki A (2006) Hyperplasia and carcinomas in Pten-deficient mice and reduced PTEN protein in human bladder cancer patients. Cancer Res 66:8389–8396

Ueda T, Sasaki M, Elia AJ, Chio II, Hamada K, Fukunaga R, Mak TW (2010) Combined deficiency for MAP kinase-interacting kinase 1 and 2 (Mnk1 and Mnk2) delays tumor development. Proc Natl Acad Sci USA 107:13984–13990

Valente EM, Salvi S, Ialongo T, Marongiu R, Elia AE, Caputo V, Romito L, Albanese A, Dallapiccola B, Bentivoglio AR (2004) PINK1 mutations are associated with sporadic early-onset parkinsonism. Ann Neurol 56:336–341

Van der Vegt JP, Van Nuenen BF, Bloem BR, Klein C, Siebner HR (2009) Imaging the impact of genes on Parkinson's disease. Neuroscience 164:191–204

Vazquez F, Ramaswamy S, Nakamura N, Sellers WR (2000) Phosphorylation of the PTEN tail regulates protein stability and function. Mol Cell Biol 20:5010–5018

Vazquez F, Grossman SR, Takahashi Y, Rokas MV, Nakamura N, Sellers WR (2001) Phosphorylation of the PTEN tail acts as an inhibitory switch by preventing its recruitment into a protein complex. J Biol Chem 276:48627–48630

Vazquez F, Matsuoka S, Sellers WR, Yanagida T, Ueda M, Devreotes PN (2006) Tumor suppressor PTEN acts through dynamic interaction with the plasma membrane. Proc Natl Acad Sci USA 103:3633–3638

Velichkova M, Juan J, Kadandale P, Jean S, Ribeiro I, Raman V, Stefan C, Kiger AA (2010) Drosophila Mtm and class II PI3K coregulate a PI(3)P pool with cortical and endolysosomal functions. J Cell Biol 190:407–425

Vergne I, Roberts E, Elmaoued RA, Tosch V, Delgado MA, Proikas-Cezanne T, Laporte J, Deretic V (2009) Control of autophagy initiation by phosphoinositide 3-phosphatase Jumpy. EMBO J 28:2244–2258

Verhaak RG, Hoadley KA, Purdom E, Wang V, Qi Y, Wilkerson MD, Miller CR, Ding L, Golub T, Mesirov JP, Alexe G, Lawrence M, O'Kelly M, Tamayo P, Weir BA, Gabriel S, Winckler W, Gupta S, Jakkula L, Feiler HS, Hodgson JG, James CD, Sarkaria JN, Brennan C, Kahn A, Spellman PT, Wilson RK, Speed TP, Gray JW, Meyerson M, Getz G, Perou CM, Hayes DN (2010) Integrated genomic analysis identifies clinically relevant subtypes of glioblastoma characterized by abnormalities in PDGFRA, IDH1, EGFR, and NF1. Cancer Cell 17:98–110

Vieira OV, Harrison RE, Scott CC, Stenmark H, Alexander D, Liu J, Gruenberg J, Schreiber AD, Grinstein S (2004) Acquisition of Hrs, an essential component of phagosomal maturation, is impaired by mycobacteria. Mol Cell Biol 24:4593–4604

Von Stein W, Ramrath A, Grimm A, Muller-Borg M, Wodarz A (2005) Direct association of Bazooka/PAR-3 with the lipid phosphatase PTEN reveals a link between the PAR/aPKC complex and phosphoinositide signaling. Development 132:1675–1686

Waite KA, Eng C (2002) Protean PTEN: form and function. Am J Hum Genet 70:829–844

Walker DM, Urbe S, Dove SK, Tenza D, Raposo G, Clague MJ (2001) Characterization of MTMR3, an inositol lipid 3-phosphatase with novel substrate specificity. Curr Biol 11:1600–1605

Walker SM, Leslie NR, Perera NM, Batty IH, Downes CP (2004) The tumour-suppressor function of PTEN requires an N-terminal lipid-binding motif. Biochem J 379:301–307

Wang S, Gao J, Lei Q, Rozengurt N, Pritchard C, Jiao J, Thomas GV, Li G, Roy-Burman P, Nelson PS, Liu X, Wu H (2003) Prostate-specific deletion of the murine Pten tumor suppressor gene leads to metastatic prostate cancer. Cancer Cell 4:209–221

Wang X, Trotman LC, Koppie T, Alimonti A, Chen Z, Gao Z, Wang J, Erdjument-Bromage H, Tempst P, Cordon-Cardo C, Pandolfi PP, Jiang X (2007) NEDD4-1 is a proto-oncogenic ubiquitin ligase for PTEN. Cell 128:129–139

Wang X, Shi Y, Wang J, Huang G, Jiang X (2008) Crucial role of the C-terminus of PTEN in antagonizing NEDD4-1-mediated PTEN ubiquitination and degradation. Biochem J 414:221–229

Watton SJ, Downward J (1999) Akt/PKB localisation and 3' phosphoinositide generation at sites of epithelial cell-matrix and cell-cell interaction. Curr Biol 9:433–436

Whang YE, Wu X, Suzuki H, Reiter RE, Tran C, Vessella RL, Said JW, Isaacs WB, Sawyers CL (1998) Inactivation of the tumor suppressor PTEN/MMAC1 in advanced human prostate cancer through loss of expression. Proc Natl Acad Sci USA 95:5246–5250

Wijesekara N, Konrad D, Eweida M, Jefferies C, Liadis N, Giacca A, Crackower M, Suzuki A, Mak TW, Kahn CR, Klip A, Woo M (2005) Muscle-specific Pten deletion protects against insulin resistance and diabetes. Mol Cell Biol 25:1135–1145

Wu DN, Pei DS, Wang Q, Zhang GY (2006) Down-regulation of PTEN by sodium orthovanadate inhibits ASK1 activation via PI3-K/Akt during cerebral ischemia in rat hippocampus. Neurosci Lett 404:98–102

Xiao C, Srinivasan L, Calado DP, Patterson HC, Zhang B, Wang J, Henderson JM, Kutok JL, Rajewsky K (2008) Lymphoproliferative disease and autoimmunity in mice with increased miR-17-92 expression in lymphocytes. Nat Immunol 9:405–414

Xue Y, Fares H, Grant B, Li Z, Rose AM, Clark SG, Skolnik EY (2003) Genetic analysis of the myotubularin family of phosphatases in Caenorhabditis elegans. J Biol Chem 278:34380–34386

Xue L, Nolla H, Suzuki A, Mak TW, Winoto A (2008) Normal development is an integral part of tumorigenesis in T cell-specific PTEN-deficient mice. Proc Natl Acad Sci USA 105:2022–2027

Yanagi S, Kishimoto H, Kawahara K, Sasaki T, Sasaki M, Nishio M, Yajima N, Hamada K, Horie Y, Kubo H, Whitsett JA, Mak TW, Nakano T, Nakazato M, Suzuki A (2007) Pten controls lung morphogenesis, bronchoalveolar stem cells, and onset of lung adenocarcinomas in mice. J Clin Invest 117:2929–2940

Yanagiya T, Tanabe A, Iida A, Saito S, Sekine A, Takahashi A, Tsunoda T, Kamohara S, Nakata Y, Kotani K, Komatsu R, Itoh N, Mineo I, Wada J, Masuzaki H, Yoneda M, Nakajima A, Miyazaki S, Tokunaga K, Kawamoto M, Funahashi T, Hamaguchi K, Tanaka K, Yamada K, Hanafusa T, Oikawa S, Yoshimatsu H, Nakao K, Sakata T, Matsuzawa Y, Kamatani N, Nakamura Y, Hotta K (2007) Association of single-nucleotide polymorphisms in MTMR9 gene with obesity. Hum Mol Genet 16:3017–3026

Yang L, Mao C, Teng Y, Li W, Zhang J, Cheng X, Li X, Han X, Xia Z, Deng H, Yang X (2005) Targeted disruption of Smad4 in mouse epidermis results in failure of hair follicle cycling and formation of skin tumors. Cancer Res 65:8671–8678

Yang G, Sun Q, Teng Y, Li F, Weng T, Yang X (2008) PTEN deficiency causes dyschondroplasia in mice by enhanced hypoxia-inducible factor 1alpha signaling and endoplasmic reticulum stress. Development 135:3587–3597

Yao XH, Nyomba BL (2008) Hepatic insulin resistance induced by prenatal alcohol exposure is associated with reduced PTEN and TRB3 acetylation in adult rat offspring. Am J Physiol Regul Integr Comp Physiol 294:R1797–R1806

Yeager N, Klein-Szanto A, Kimura S, Di Cristofano A (2007) Pten loss in the mouse thyroid causes goiter and follicular adenomas: insights into thyroid function and Cowden disease pathogenesis. Cancer Res 67:959–966

Yilmaz OH, Valdez R, Theisen BK, Guo W, Ferguson DO, Wu H, Morrison SJ (2006) Pten dependence distinguishes haematopoietic stem cells from leukaemia-initiating cells. Nature 441:475–482

Yim EK, Peng G, Dai H, Hu R, Li K, Lu Y, Mills GB, Meric-Bernstam F, Hennessy BT, Craven RJ, Lin SY (2009) Rak functions as a tumor suppressor by regulating PTEN protein stability and function. Cancer Cell 15:304–314

Yin Y, Shen WH (2008) PTEN: a new guardian of the genome. Oncogene 27:5443–5453

Yoo LI, Liu DW, Le VU S, Bronson RT, Wu H, Yuan J (2006) Pten deficiency activates distinct downstream signaling pathways in a tissue-specific manner. Cancer Res 66:1929–1939

Yu J, Pan L, Qin X, Chen H, Xu Y, Chen Y, Tang H (2010) MTMR4 attenuates transforming growth factor beta (TGFbeta) signaling by dephosphorylating R-Smads in endosomes. J Biol Chem 285:8454–8462

Yuan TL, Cantley LC (2008) PI3K pathway alterations in cancer: variations on a theme. Oncogene 27:5497–5510

Yue Q, Groszer M, Gil JS, Berk AJ, Messing A, Wu H, Liu X (2005) PTEN deletion in Bergmann glia leads to premature differentiation and affects laminar organization. Development 132:3281–3291

Zhang J, Grindley JC, Yin T, Jayasinghe S, He XC, Ross JT, Haug JS, Rupp D, Porter-Westpfahl KS, Wiedemann LM, Wu H, Li L (2006) PTEN maintains haematopoietic stem cells and acts in lineage choice and leukaemia prevention. Nature 441:518–522

Zhang QG, Wu DN, Han D, Zhang GY (2007) Critical role of PTEN in the coupling between PI3K/Akt and JNK1/2 signaling in ischemic brain injury. FEBS Lett 581:495–505

Zhao R, Qi Y, Chen J, Zhao ZJ (2001) FYVE-DSP2, a FYVE domain-containing dual specificity protein phosphatase that dephosphorylates phosphotidylinositol 3-phosphate. Exp Cell Res 265:329–338

Zheng H, Ying H, Yan H, Kimmelman AC, Hiller DJ, Chen AJ, Perry SR, Tonon G, Chu GC, Ding Z, Stommel JM, Dunn KL, Wiedemeyer R, You MJ, Brennan C, Wang YA, Ligon KL, Wong WH, Chin L, Depinho RA (2008) p53 and Pten control neural and glioma stem/progenitor cell renewal and differentiation. Nature 455:1129–1133

Zhou BP, Liao Y, Xia W, Zou Y, Spohn B, Hung MC (2001) HER-2/neu induces p53 ubiquitination via Akt-mediated MDM2 phosphorylation. Nat Cell Biol 3:973–982

Zhou XP, Loukola A, Salovaara R, Nystrom-Lahti M, Peltomaki P, De La Chapelle A, Aaltonen LA, Eng C (2002) PTEN mutational spectra, expression levels, and subcellular localization in microsatellite stable and unstable colorectal cancers. Am J Pathol 161:439–447

Zhou XP, Waite KA, Pilarski R, Hampel H, Fernandez MJ, Bos C, Dasouki M, Feldman GL, Greenberg LA, Ivanovich J, Matloff E, Patterson A, Pierpont ME, Russo D, Nassif NT, Eng C (2003) Germline PTEN promoter mutations and deletions in Cowden/Bannayan-Riley-Ruvalcaba syndrome result in aberrant PTEN protein and dysregulation of the phosphoinositol-3-kinase/Akt pathway. Am J Hum Genet 73:404–411

Zhou J, Brugarolas J, Parada LF (2009) Loss of Tsc1, but not Pten, in renal tubular cells causes polycystic kidney disease by activating mTORC1. Hum Mol Genet 18:4428–4441

Zhu D, Hattori H, Jo H, Jia Y, Subramanian KK, Loison F, You J, Le Y, Honczarenko M, Silberstein L, Luo HR (2006a) Deactivation of phosphatidylinositol 3,4,5-trisphosphate/Akt signaling mediates neutrophil spontaneous death. Proc Natl Acad Sci USA 103:14836–14841

Zhu Y, Hoell P, Ahlemeyer B, Krieglstein J (2006b) PTEN: a crucial mediator of mitochondria-dependent apoptosis. Apoptosis 11:197–207

Zhu Y, Hoell P, Ahlemeyer B, Sure U, Bertalanffy H, Krieglstein J (2007) Implication of PTEN in production of reactive oxygen species and neuronal death in in vitro models of stroke and Parkinson's disease. Neurochem Int 50:507–516

Zou J, Chang SC, Marjanovic J, Majerus PW (2009a) MTMR9 increases MTMR6 enzyme activity, stability, and role in apoptosis. J Biol Chem 284:2064–2071

Zou W, Lu Q, Zhao D, Li W, Mapes J, Xie Y, Wang X (2009b) Caenorhabditis elegans myotubularin MTM-1 negatively regulates the engulfment of apoptotic cells. PLoS Genet 5:e1000679

Glossary

A glossary of terms is provided for readers who are not experts of the inositol lipid field.

***myo*-Inositol:** This is one of nine possible stereoisomers of inositol which is a cyclohexanehexol. *My*o-inositol is the most commonly occurring stereoisomer in nature, therefore, the IUPAC-approved abbreviation "Ins" refers to *myo*-inositol (Nomenclature Committee of the International Union of Biochemistry, 1989, http://www.chem.qmul.ac.uk/iupac/cyclitol/myo.html). The conformation of *myo*-inositol is the so-called "chair" conformation with five equatorial and one axial hydroxyl groups. This conformation, and the numbering of the hydroxyls have been best visualized by Agranoff (1978) who compared the ring to a turtle and the hydroxyls to the appendages. Here, the numbering starts with the right front flipper going counterclockwise. The head of the turtle then corresponds to the axial hydroxyl at the 2nd position. It is notable that *myo*-inositol has an axis of symmetry going through the 2nd and 5th carbons. The numbering used refers to the D-enantiomers but it is important to remember that because of this symmetry D-Ins1*P* is the same as L-Ins3*P* and, therefore, isomers (such as Ins1*P* and Ins3*P*) that are enantiomeric twins cannot be separated with conventional HPLC methods.

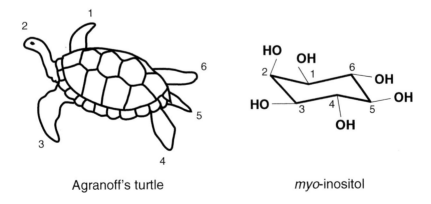

Agranoff's turtle   *myo*-inositol

Phosphatidylinositol: This is the base molecule for all phosphoinositides. The recommended abbreviation is PtdIns (http://www.chem.qmul.ac.uk/iupac/misc/phos2t8.html#t4) but the early literature often uses "PI" as the abbreviation. This short form is still in use in the context of kinases that phosphorylate PtdIns or its phosphorylated derivatives, such as in PI 3-kinases or PI 4-kinases. PtdIns consists of a diacylglycerol backbone in which the 1- and 2-positions of the glycerol are most often esterified with a stearoyl- and arachidonyl- fatty acid chains, respectively, and the *myo*-inositol ring is linked to the 3rd- position of the glycerol via a phosphodiester bond formed with the 1st hydroxyl of inositol. PtdIns can be phosphorylated in all but the 2nd and 6th positions of the inositol ring, giving rise to the seven known phosphoinositides.

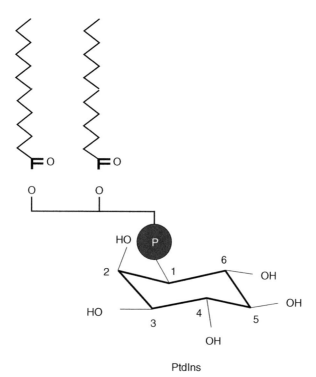

PtdIns

Polyphosphoinositides: This refers to any of the further phosphorylated PtdIns regardless of the number and positions of the phosphate groups. Sometimes they are abbreviated as PPIs but mostly in the 80's literature but this is still the recommended abbreviation (see in Michell et al. 2005).

Phosphoinositides: This is a term often used to designate collectively PtdIns and all of its phosphorylated derivatives regardless of their isomerism. "PI" is the abbreviation used lately for phosphoinositides but it often causes confusion so it should be avoided. There is no consensus abbreviation for this term that includes both PtdIns and the PPIs. The individual forms of PPIs are abbreviated specifying the positions phosphorylated on the inositol ring. For example, phosphatidylinositol 4,5-bisphosphate [PtdIns(4,5)P_2] is a double phosphorylated PtdIns with positions 4- and 5- phosphorylated.

It is worth pointing out here a few rules about the terminology: the numbers on the inositol ring are not "primed", since there is no other ring in the structure that would require the use of "prime" to discriminate the rings (as opposed to multiring structures such as the nucleosides). The number of phosphates are indicated by the prefixes "bis-" (latin for twice) "tris-" (greek for three times) to indicate that these molecules contain the indicated number of phosphates but all placed individually at various positions. (contrast this with ATP which is a "triphosphate" with the phosphates linked to one another). There are no higher numbers than trisphosphates in phosphoinositides but there are in the soluble inositol phosphates, for which the numbers continue as "tetrakis-", "pentakis-" and "hexakis–" 'kis' being a greek prefix for –times. In the abbreviations it is recommended to italicise the "*P*" if it designates a phosphomonoester (see Michell et al. (2005) in the list below for more details on nomenclature recommendations).

Old literature used terms such as "DPI" and "TPI" for diphosphoinositide and triphosphoinositide, respectively. These correspond to mono- and bis-phosphorylated PtdIns from a time when the exact configurations of the phosphates were not known.

Phosphoinositide kinases: Phosphoinositide kinases add a phosphate to a specific position onto the inositol ring of phosphoinositides using ATP. The kinases are named after the position they phosphorylate and hence we distinguish 3-, 4- and 5-kinases (no primes!!). There is an inherent inconsistency about the abbreviations used to designate these enzymes. For example, the term "PI 3-kinase" is used to refer to any of the 3-kinases, regardless of the substrates they phosphorylate. Since there are PI 3-kinases that can only phosphorylate PtdIns (and not further phosphorylated forms) (the Class III PI 3-kinases) they are also named PtdIns 3-kinases. However, the Class I PI 3-kinases that phosphorylate PtdIns(4,5)P_2 are rarely called PtdIns(4,5)P_2 3-kinases and in most articles unspecified "PI 3-kinase" refers to the Class I enzymes. In contrast, PI 4-kinases can only phosphorylate PtdIns (and not further phosphorylated forms) in which case it would be more correct to call them PtdIns 4-kinases. However, because of historical reasons, these inconsistencies are tolerated even if they defy logic based on current knowledge. The list of the various forms and classes of PI kinases are summarized in the respective chapters.

Phosphoinositide phosphatases: Phosphoinositide phosphatases remove one or more phosphates from PPIs. They can be specific to the position of the phosphate they remove and the substrate they can use. Some will dephosphorylate only PPIs while others can also use the water-soluble inositol phosphates as substrates. Phosphatases are usually named after the position of phosphate they attack such as 5-phosphatases or 3-phosphatases. Some PI phosphatases are not position specific, such as the monophosphatases (see Chapters 7 and 8 in Volume I for more details).

Phospholipase C: These enzymes (PLCs) hydrolyze PtdIns (or PPIs) by cleaving the phosphodiester group such that they leave diacylglycerol behind and release the inositol headgroup, which carries the phosphate still attached at the 1-position (or other phosphates if the substrate is any of the PPIs). To discriminate from other PLCs that use other phospholipids as substrate (such as PC-PLC), PLCs that hydrolyze phosphoinositides are called PI-PLCs. This, however, also causes some confusion, since mammalian PI-PLCs are believed to hydrolyze primarily PtdIns(4,5)P_2 *in vivo* (although they can also hydrolyze PtdIns and PtdIns4P *in vitro*). However, there are bacterial PLC enzymes that will use either PtdIns or phosphatidylinositol glycan (GPI) linkages but cannot hydrolyze polyphosphoinositides. The literature that deals with the bacterial enzymes uses the term PI-PLC to emphasize that the bacterial enzymes are specific for PtdIns or GPI. So the term "PI-PLC" means two different enzyme groups depending on whether used in mammalian or prokaryotic studies. However, in most cases PLC without any designation refers to the mammalian phosphoinositide-specific PLCs.

Inositol 1,4,5-trisphosphate: Ins(1,4,5)P_3 is the water soluble molecule liberated after PLC-mediated hydrolysis of PtdIns(4,5)P_2. This molecule has a receptor located in the ER membrane that also is a Ca^{2+} channel and which is gated by Ins(1,4,5)P_3 binding. Ins(1,4,5)P_3 is a bona fide second messenger liberated upon stimulation of cell surface receptors coupled to PLC activation.

PH domain: PH domains (for pleckstrin homology domains) are protein modules of roughly 150 amino acid length that were first recognized in pleckstrin (Tyers et al. 1988). These were the first protein modules that were shown to bind PPIs. Many PH domains can recognize and bind phosphoinositides with variable specificities earning these domains the reputation of being PPI binding modules. Although several PH domains can, indeed, recognize PIs with high affinity and specificity, many PH domains show promiscuous PPI recognition and many do not bind PIs at all. Moreover, PH domains also recognize proteins and often bind proteins and lipids simultaneously (Lemmon 2004).

FYVE domain: This was the second protein module identified with specific PPI recognition, namely to recognize PtdIns3P (Burd and Emr 1998). Its name originated from the four molecules (three from baker's yeast) in which this module was first described (Fab1, YOTB, Vac1 and EEA1). FYVE domains use two Zn^{2+} ions to stabilize their structure and they are also called FYVE zinc fingers. They show structural similarities to the C1 domains that recognize diacylglycerol (Misra and Hurley 1999; Kutateladze et al. 1999).

PX domain: Phox-homology domains were also recognized as capable of binding PtdIns3P. They were initially found in sorting nexins (Ponting 1996) and NADPH oxidase subunits (Bravo et al. 2001; Ellson et al. 2001; Kanai et al. 2001), but they are present in a large variety of signaling molecules. PX domains can also bind other phospholipids, such as PtdOH and PtdIns(3,4)P_2, and they also interact with proteins (Vollert and Uetz 2004).

Phosphoinositide binding protein domains: In addition to the above defined protein modules, several other modular protein domains have been identified as phosphoinositide effectors (Lemmon 2008). Because of their increasing number they will not be listed here but can be found in the individual Chapters.

References

Agranoff BW (1978) Trends Biochem Sci 3(12):N283–N285
Bravo J, Karathanassis D, Pacold CM, Pacold ME, Ellson CD, Anderson KE, Butler PJG, Lavenir I, Perisisc O, Hawkins PT, Stephens L, Williams RL (2001) Mol Cell 8:829–839
Burd CG, Emr SD (1998) Mol Cell 2:157–162
Ellson CD, Gobert-Gosse S, Anderson KE, Davidson K, Erdjument-Bromage H, Tempst P, Thuring JW, Cooper MA, Lim ZY, Holmes AB, Gaffney PRJ, Coadwell J, Chilvers ER, Hawkins PT, Stephens LR (2001) Nat Cell Biol 3:679–682
Kanai F, Liu H, Field SJ, Akbary H, Matsuo T, Brown GE, Cantley LC, Yaffe MB (2001) Nat Cell Biol 3:675–678
Kutateladze TG, Ogburn KD, Watson WT, deBeer T, Emr SD, Burd CG, Overduin M (1999) Mol Cell 3:805–811
Lemmon MA (2004) Biochem Soc Trans 32:707–711
Lemmon MA (2008) Nat Rev Mol Cell Biol 9(2):99–111
Michell RH, Heath VL, Lemmon MA, Dove SK (2005) Trends Biochem Sci 31(1):52–63
Misra S, Hurley JH (1999) Cell 97:657–666
Ponting, CP (1996) Prot Sci 5:2353–2357
Tyers M, Rachubinski RA, Stewart MI, Varrichio AM, Shorr RG, Haslam RJ, Harley CB (1988) Nature 333(6172):470–473
Vollert CS, Uetz P (2004) Mol Cell Proteomics 3(11):1053–1064

Index

A
A1 adenosine receptors, 127
Abl, 221
Actin cytoskeleton, 99, 104, 217, 226, 227, 238
Actin-regulatory protein, 68, 227
Acute lymphocytic leukemia, 226, 291
Acute myeloid leukemia, 226, 291
ADAMs, 241
α-Adaptin, 39
Adenocarcinoma cells, 235, 243, 292, 294, 296, 306
Adenylyl cyclase, 78, 79, 201, 202
3T3 L1 adipocytes, 228, 234
β-Adrenergic receptor, 79, 135, 198, 201, 202
Aedes aegypti, 75
age-1, 100
Airway hyper-responsiveness (AHR), 141, 189, 193
Akt, 100, 101, 103, 104, 134, 136, 139, 149, 196, 199, 202, 247, 251, 296
Akt1, 104, 130, 196, 199
Akt2, 104, 196
Akt3, 104
Akt activation, 99, 100, 131, 222, 223, 228–231, 241, 242, 247, 248, 250, 284, 297, 302
Alpha-actinin, 36
Alzheimer, 70, 302
Amoeboid migration, 99, 138
Amphiphysin 1, 231, 232
Amphiphysin 2, 231
Amyotrophic lateral sclerosis, 255
Angiotensin II, 11, 69, 139
ANTH domain, 135
Anti-nuclear autoantibodies, 142
AP-1, 13
AP-2, 38, 39, 41–44, 203, 231, 236, 237, 240
AP-3, 13, 14

AP-4, 39
AP180, 38, 43
Apoptosis signal-regulating kinase 1 (ASK1), 302
APPL1, 237–239
Arabidopsis thaliana, 252
Arf1, 13–15, 17
Arf6, 37
Arp2/3 complex, 68
ArPIKfyve protein, 254
AS160, 196
AS1949490, 230
ASH domain (ASPM/SPD2/hydin), 218, 236, 240
Aspergillus niger, 307
Asthma, 140, 141, 189, 193, 194, 225, 248
Astrocyte, 13, 78, 290
Ataxia, 74, 116, 232, 235, 247, 293, 295, 301
Ataxia telangiectasia-related (ATR), 116
Atg6, 130
Atg14L, 130
ATM (ataxia telangiectasia mutated), 8, 116
Atypical G-protein transglutaminase II, 69
Autoimmune disease, 126, 136, 138, 141–143, 192, 193, 224, 289, 297
Autoinhibition, 75, 81
Autophagosome, 312, 314, 315
Autophagy, 116, 130, 255, 306, 314, 315; *see also* PI3K complexes I and II
Autosomal-dominant cancers, 147, 255, 283, 289, 313
Auxilin, 102
Azospermia, 308, 313

B
B lymphocytes, 136, 185, 189, 191, 192, 222
BAD, 100, 134
BAF complex, 48
BaF3 cell, 146

Bannayan-Zonana syndrome, 183, 189, 301
Basal body, 235
BCR (B cell receptor), 126, 128, 137, 191, 222, 290
BCR/ABL (break point cluster region/Abelson kinase fusion protein), 126, 227
Beclin-1, 130
Bergmann glia, 13
Bilateral congenital cataracts, 236
Bi-polar disorder, 233
Bladder cancer, 243
Breast cancer, 102, 144, 248, 250, 251
Brefeldin A, 9, 14
Bronchoalveolar lavage (BAL), 189
Bruton's tyrosine kinase (Btk), 135, 140, 188, 226
Burkitt lymphoma, 243

C
C2 domain, 37, 65, 68, 128, 246, 284, 286, 287, 298
C3 botulinum toxin, 78
C5a, 128, 187
Ca^{2+} channel, 140
Ca^{2+} signaling, 80
Ca^{2+}/calmodulin kinase II, 239
CAAX motifs, 218, 234, 235, 239, 240
Caenorhabditis elegans, 4, 71, 75, 76, 100, 129, 232, 233, 307, 314, 315, 318
Calcineurin, 43, 232
Calpain, 38
cAMP, *see* Cyclic AMP
Capping protein/CapZ, 33
Cardiomyocyte, 101, 127, 139, 140, 198, 199, 201–203, 290, 291
Cardiotoxicity, 199
Caspase-3, 302
Caspase 9, 134
β-catenin, 10
CBRIII motif, 287
CD2-associated protein, 228
$CD4^+$ T cells, 143, 190, 193, 305, 316
$CD8^+$ T cells, 136, 190, 194
CD19, 126
CD40, 191
CD62L, 191
CD63, 13
CD81, 10
Cdc42 (cell division cycle 42), 35, 135, 136, 237, 240, 287
Cdc42/N-WASP, 35
Cdc42-GAP, 35, 135, 136, 240, 287

Cell death, 71, 150, 199, 225, 231, 247, 293, 302, 316
Cell motility, 70, 99, 251, 298
β-Cells, 16, 194, 195, 294
Cell senescence, 297, 300, 301
Cell wall abnormalities, 252
Cellular cilia, 235
Ceramide transfer protein (CERT), 16
Cerebellum, 71, 245, 250, 290
Charcot-Marie-Tooth (CMT) disorder, 112, 255, 313
Chemotaxis, 73, 74, 143, 187, 189, 190, 192, 222, 245, 249, 293, 298, 305
Chloride channel, 246
Chromaffin cells, 37
Chromosome 10q, 239
Chromosome 10q23, 101
Chromosome 22q12, 241
Chromosome 3p22, 70
Chromosome 17p13.3, 243
Chronic lymphocytic leukaemia, 243
Chronic myelogenous leukaemia (CML), 126, 151, 222, 227
Chronic obstructive pulmonary disease (COPD), 141, 194
Chylous ascites, 101
Ciliopathy syndrome, 220, 235
c-kit receptor, 141
Clathrin-coated vesicles, 39, 41, 43, 227, 232, 238
Clathrin-mediated endocytosis, 38, 39, 42, 43, 116, 232, 233
CMT4J patients, 255
Coated pits, 39, 43, 203, 228, 231, 232, 234, 237
Coatomer protein complex I (COP-I), 253
Coatomer protein complex II (COP-II), 253
Cochlea, 233
Cofilin, 33, 68
Coiled-coil domain, 71, 72, 307, 314
Compartmentalization, 202
Congenital cataracts, 236
COP-I-vesicle, *see* Coatomer protein complex I
COP-II vesicle, *see* Coatomer protein complex II
COS-7 cells, 14, 16, 80
Cowden disease, 283, 284, 286, 288, 289, 301
Craniofacial/limb abnormalities, 242
Cubulin, 239
Cul3-SPOP (cullin 3-speckle-type POZ domain protein), 47, 48
Cutaneous melanoma, 243

Index

CX$_5$R catalytic motif, 216, 282, 284
Cyclic AMP (cAMP), 78, 79, 101, 140, 201, 202
Cytochrome c, 302, 303

D
DAAX (death domain-associated protein), 288
daf16, 100
Danio rerio, 233, 315, 318
DAP12, 226
DAP160 (dynamin-associated protein 160), 231
Dbl, 235
dbl-homology (DH), 249
Degenerative neuropathy, 255
Degranulation, 141, 142, 188, 189, 225
Demyelinating neuropathy, 313
Dendritic spines, 293
Dent-2 disease, 236–240
DEP domain, 249
Desmin, 316
DHR-1 domain, 135, 136
Diabetic nephropathy, 228
Diacylglycerol (DAG), 16, 45, 62, 96
Diacylglycerol kinase, 45
Diacylglycerol-activated serine/threonine protein kinase (PKC), 96
Dictyostelium discoideum, 67, 99, 113, 138, 287, 298
Disc-large1 (Dlg1), 314
DJ-1, 302, 303
DNA repair, 47, 289, 307
DNA-dependent protein kinase (DNA-PK$_{cs}$), 116, 117, 130
DOCK2, 136, 138, 191
Dok 1, 221
Dok 2, 221, 223
Dominant-negative, 199, 230
Dorsal ruffles, 33
Down's syndrome, 233
Dpm1 (Dolichol phosphate mannosyltransferase), 253
Drosophila melanogaster, 113, 232, 233, 252, 305, 310, 315, 318
Drug target, 143
Dynamin-2, 38, 43
Dynamin oligomer, 38, 43, 129, 231; *see also* Clathrin-coated vesicles *and* Clathrin-mediated endocytosis

E
E-cadherin, 37, 40, 41
EEA1, 14, 129, 311; *see also* FYVE domain
EF-hand, 64, 80

EGFR, 10, 13, 36, 44, 76, 116, 144, 147, 300
EGFR degradation, 230, 244, 315, 316
Embryonic development, 134, 217, 234, 235, 242, 283, 287
Embryonic lethality, 70, 76, 101, 150, 220, 221, 235, 243, 289
Endocytic recycling compartment, 15, 36
Endometrioid carcinoma, 296
Endophilin 1, 232
Entamoeba histolytica, 307
ENTH domain, 135
ENTH/ANTH domain, 135
ENU mutagenesis screen, 233
Eosinophils, 185, 188, 189, 193
Eotaxin, 189
Epac, 79
Epinephrine, 201
Eps15, 231
Epsin, 38, 43
ERK, 78, 97, 103
Erlotinib, 151
Erythroleukemia, 45, 227
Erythropoiesis, 76
Erythropoietin (Epo), 227
Estrogen receptor, 102, 241, 248
Eukaryotic elongation factor 1A, 13
Everolimus, 120, 148
Excised patch recording, 19, 27, 83
Exo70, 37
Exocyst complex, 36, 37, 316
Extracellular matrix, 299

F
Fab1/PIKfyve, 135, 254; *see also* FYVE domain
Fab1p, 26, 27
FAK, 38, 286
FAPP1, 12, 14, 16
FcγR-mediated phagocytosis, 224, 225, 234, 305
FERM domain, 135
Fibroblast growth factor (FGF), 17, 74
Fibronectin, 37
Fig4, 254, 255
Filamin, 227
Filopodia, 33, 314
FK506-binding protein 12 (FKBP12), 147
Fluorescence resonance energy transfer (FRET), 104
FMLP (formylmethionyl leucyl phenylalanine), 97, 98, 187, 188
Focal adhesions, 36–38
Foxn1, 69

FoxO1, 131, 137, 196
FOXO3a, 100, 104
Francisella tularensis, 225
Frequenin (Frq1p), 15
Friend murine leukemia virus, 227
Frq1p, 16
FYVE domain, 27, 100, 135, 196, 310, 312

G

G protein-coupled receptor (GPCR), 62, 72, 96, 127, 202, 249
G_1/S boundary, 70
G_2/M phase, 70
G_2/M transition, 70
G6P (glucose-6-phosphatase), 230
Gastrointestinal stromal tumors, 144
Gefitinib, 151
Gelsolin, 33, 68
Gemcitabine, 236, 243
GGA protein, 13
GGA1, 13
GGA2, 13
GGA3, 13, 14
GIPC (GAIP-interacting protein, C terminus), 237–239
GK (glucokinase), 230
Gleevec, 144, 151
Glioblastoma, 102, 146, 230, 234, 287, 296, 300, 304
GLTSCR2, 286, 300
Gluconeogenesis, 194, 196, 198, 230
Glucose homeostasis, 196, 198, 217, 229, 230, 242
Glucose intolerance, 196, 198
Glucose uptake, 194–196, 228, 234, 242
GLUT4, 196, 197, 228, 234, 242, 254
Glutamate, 29, 247, 302
Glycogen synthase, 100, 134, 286
Glycogenolysis, 194
Glycosphingolipids, 12, 17, 18
Glycosylphosphatidylinositol (GPI), 340
Golgi resident proteins, 16
Golgi-ER shuttling, 253
Golgi-localised, γ-ear-containing, Arf-binding protein (GGA), 13
Granulocyte/macrophage colony-stimulating factor (GM-CSF), 139
Grb2 (growth-factor-receptor-bound protein 2), 231
Gsk3β (glycogen synthase kinase 3β), 11, 134, 199, 241, 254, 286
GTPase activating proteins (GAPs), 72, 103, 135

Guanine nucleotide exchange factors (GEFs), 35, 77, 103, 112, 135, 138, 191, 248, 249
Guanosine 3', 5'-monophosphate (GTP), 37, 65, 72, 77, 135
Gα_q, 63–65, 69, 71–74, 82

H

Hamartin, 131; *see also* TSC1
Hamartomas, 147, 283
HAUSP (herpes virus-associated ubiquitin-specific protease), 288
HBV core protein, 243
Hck, 221
Heart failure, 140, 198, 199, 201–203, 289
Hedgehog signalling, 17, 18
HEK cells, 228
HeLa cells, 230, 311, 314
Hemidesmosomes, 36
Hemostasis, 111, 146, 226
Heparin, 79, 188
Hepatitis C virus (HCV), 14
Hepatocellular carcinoma, 10, 146, 243, 291
Hepatocyte necrosis, 101
Heterodimerisation, 4
High fat diet, 220, 229, 242, 294, 295
Histone H1, 48, 70
Histone H3, 48
hJumpy, 307
Human neuropathies, 254, 255
hVps15, 130, 314
hVps34, 130, 314
Hyperactivation of Akt, 102
Hypergammaglobulinemia, 193
Hyperglycaemia, 230
Hyperinsulinaemia, 230
Hyperlipidemia, 198
Hypoglycemia, 101, 197, 294
Hypothalamus, 234

I

IC87114, 118, 141, 150, 187, 188, 193
IgE binding, 188
IgE high affinity receptor (FCeRI), 188
IgE switch, 191
IL-1β, 69, 70
IL-2, 142, 189, 190
IL-4, 126, 189, 191, 224, 305
IL-5, 189, 226
IL-6, 69, 70, 127, 142, 223, 224, 226
IL-8, 187
Imatinib, 144, 151

Immunoreceptor tyrosine-based activation motifs (ITAMs), 75, 126, 140, 188, 222, 225
Immunoreceptor tyrosine-based inhibitory motifs (ITIMs), 222, 225
Infertility, 71, 241
Inflammation, 69, 70, 112, 136, 138, 140, 142, 143, 185–189, 193, 194, 203, 223, 225, 250
ING2, 49, 244
Inhibitor of growth protein 2, *see* ING2
Inner and outer hair cells, 233
Inositide 5-phosphatase, 216–219
Inositol 1,3,4,5-tetrakisphosphate, 5, 97, 217–219, 231, 236, 239, 240, 283
Inositol 1,4,5-trisphosphate, 62, 63, 67, 68, 217, 218, 231, 236, 239
Inositol-phosphoceramides, 17, 19
Inositol polyphosphate 5-phosphatase, 216, 217, 241, 252
Inp51, 252
Inp52, 218, 252
Inp53, 252
INPP4A, 245–248, 257
INPP4B, 215, 245–248, 257
INPP5A, 217, 220, 239, 256
INPP5B, 220, 236–238, 240, 256
INPP5E, 220, 224, 225, 234–236, 256
INPP5F, 220, 236, 251, 252, 256, 257
INPP5J, 241, 242, 256
INPP5K, 257
INPPL-1, 220, 227, 228, 231, 256
Ins(1,3,4)P_3, 218, 239, 246
Ins(1,3,4,5)P_4, *see* Inositol 1,3,4,5-tetrakisphosphate
Ins(1,3,4,5)P_4 receptor, 217–219, 231, 240, 283
Ins(1,4,5)P_3, *see* Inositol 1,4,5-trisphosphate
Insulin granules, 129
Insulin mediated endocytosis, 196
Insulin resistance, 150, 196, 228–230, 255, 295
Insulin responsive cells, 100
Insulinaemia, 229
Insulin-like growth factor, 230
Integrin-linked kinase (ILK), 130
Intersectin 1, 231, 232
Invadopodia, 36
Invasion plasmid gene D (IpgD), 244
Ion selectivity, 12, 232, 233, 306, 316
IP$_3$ receptor, 49
IP$_3$ receptor mobility, 49
IpgD, 244, 246

Iressa, *see* Gefitinib
IκB kinase (IκBK), 134

J
Joubert syndrome, 235
Jun N-terminal kinase (JNK), 302

K
Keratinocytes, 70, 231
Kif13b, 316
Knock-in mice, 101, 187, 190, 193
Knock-out mouse, 71, 140, 196, 201, 218, 223, 232

L
L6 myotubes, 242
LAB/Lat2, 126, 140
Lamellipodia, 33, 35, 229, 234, 240, 251
Lamin B, 45
lamp-1, 13
LARG, 77, 78
LAT, 75, 126, 140, 223
Leishmania major, 190
Leptin receptor, 230
Leucine zipper, 253
Leukotriene B$_4$, 187
Lhermitte-Duclos disease, 283, 289, 301
Linker for activation of T cells (LAT), 75, 126, 140, 223
Lipid signaling pathway, 101
Lipid transfer activity, 17
Lissencephaly, 242
LL5β, 227
Long term potentiation (LTP), 250
Loss of heterozygosity (LOH), 101, 102, 241, 243, 248
Low-density lipoprotein, 138, 139
Lowe oculocerebrorenal (OCRL) syndrome, *see* Lowe syndrome
Lowe protein, 236; *see also* OCRL protein
Lowe syndrome, 220, 236–240, 256; *see also* OCRL protein
LPS, 69, 126, 127, 141, 191, 194
L-selectin, *see* CD62L
Lung adenocarcinoma, 243, 292
LY294002, 9, 18, 98, 116, 117, 149, 188–190, 193
Lymph node-homing receptor, 191
Lymphocytes, *see* B lymphocytes *and* T lymphocytes
Lyn, 75, 188, 221, 222
Lysophosphatidic acid, 77, 78, 127, 129

M

M1-mtm-1, 307
M1-muscarinic receptor, 73
Macroencephaly, 289, 293
Macrophages, 101, 136, 138, 139, 142, 185, 188, 221–226, 234
Mal (MyD88-adaptor-like), 126
Mannose-6-phosphate receptor (M6PR), 238
Mast cells, 16, 138, 140–142, 188, 193, 221, 222, 225, 226
Mastoparan, 7
MDCK cells, 15, 16
Megalin, 238, 239
MEL cells, 48
Membrane ruffles, 33, 35, 68, 238, 312
Memory formation, 230
Mental retardation, 235, 236, 240, 242
Mesenchymal migration, 287, 294
Metastatic adenocarcinoma, 235
Methacoline, 193
Mg^{2+}-dependent phosphoesterase, 217
Microspikes, 33
Microvilli, 33
Miller-Dieker syndrome (MDS), 242
MIP, 307, 308
Mitogen-activated protein kinase (MAPK), 17, 129, 147, 148, 151, 203, 222, 253
Mitogen-activated protein kinase-activated kinase 2 (MAPKAP-2), 130
MORM (mental retardation, obesity, congenital retinal dystrophy and micropenis), 235
Motile actin comets, 33
Mouse erythroleukemia (MEL), 45
Mozart, 233
mRNA export, 47
Mss4p, 27, 29, 31, 46
mtmr-9, 307, 314, 315
mTOR (mammalian target of rapamycin), 103, 104, 116–119, 130, 131, 144, 148–151, 301
mTORC1 (mTOR complex 1), 119–121
mTORC2 (mTOR complex 2), 100, 119, 249
Multiple myeloma, 120, 243
Multivesicular bodies, 9
Murine erthythroleukamia (MEL) cells, 45, 48
MVP (major vault protein), 288
myc translocation, 300
Myeloproliferative disease, 74, 222
Myocardial infarction, 138, 140, 199, 203
Myosin 1E, 231
Myosin light chain kinase, 37
Myotonic dystrophies, 316
Myotubular myopathy, 112, 308, 312, 315
Myotubularin, 112, 249, 282, 283, 306, 307, 310–318

N

NADPH oxidase, 128, 135
NCA (Na^+/Ca^{2+} antiporter), 232, 233
NCS-1, 8, 15
Nedd4-1, 286
Nephrin, 228
Neuroexcitatory cell death, 247
Neurofibrillary tangles, 70
Neuronal cell migration, 242
Neutrophils, 37, 73, 97–99, 128, 137, 138, 142, 185, 187, 192–194, 299, 305
N-formylmethionyl-leucyl-phenylalanine (fMLP), 97, 98, 187, 188
NGF (nerve growth factor), 238
NGF-differentiated PC12 cells, 241
NK cells, 222
nla (nebula), 233
NLS (nuclear localization signal), 46, 288
N-methyl-D-aspartate (NMDA) receptor
Non-Hodgkin's lymphoma, 118–121, 235
Non-kinase scaffolding function, 101
Non-steroidal anti-inflammatory drugs (NSAIDs), 143
NPXY motif, 221, 227
NS5A, 14
NTAL, 126, 140
Nuclear envelope, 44, 45, 80
Nuclear phosphoinositide cycle, 104
Nuclear speckles, 45, 47, 48
NVP-BEZ235, 103
N-WASP, 35

O

OCRL gene, 217, 237, 238, 256
OCRL protein, 220, 227, 236–240
μ-Opioid receptor, 74
OSBP1, 17
Osteoclast activation, 226
Ovalbumin (OVA), 141, 189, 193, 194
Oxysterol-binding protein (OSBP), 7, 8, 13, 17

P

p50, 197
p53, 49, 145, 244, 289, 299–301, 303
p55, 197
$p70^{S6K}$, 131, 147, 148, 199, 229, 242
p85, 97, 98, 123, 145, 149, 184, 193, 197, 198, 246
p110, 91, 101, 102, 104, 116
p115RhoGEF, 77, 78
p130Cas, 227

Palmitoylation, 5, 6
Pancreatic β-cells, 16, 129, 194, 195, 294
Pancreatitis, 143
Parkin, 302, 303
Parkinson disease, 302, 303
PCAF (p300/CBP-associated factor), 285
PDGF (platelet-derived growth factor), 7, 10, 11, 35, 97–99, 116, 129, 236
PDK-1 (phosphoinositide-dependent kinase-1), 100, 103, 126
PDZ domain, 80, 249
Penicillium funiculosum, 98
PEPCK (phosphoenolpyruvate carboxykinase), 230
Perinatal lethality, 101, 197
Pertussis toxin, 127
PEST sequence, 286
Peutz-Jeghers syndrome, 147
PH, *see* Pleckstrin homology
Phagocytosis, 224, 225, 234, 305
Pharbin, 234
PH-GRAM, 307, 310
5-phosphatase, 11, 217–219, 222, 223, 225–228, 231–243
43 kDa 5-phosphatase, 217, 239
75 kDa 5-phosphatase, 218, 240
Phosphatidylinositol 3,5-bisphosphate [PtdIns(3,5)P_2], 217, 218, 234, 254, 282, 283, 306, 310, 313
Phosphatidylinositol 4,5-bisphosphate [PtdIns(4,5)P_2], 10, 12, 67, 68, 81, 95, 117, 123, 218, 232, 236, 282, 284
Phosphatidylinositol kinase activity, 96
Phosphatidylserine (PtdSer), 69
Phosphodiesterase (PDE), 63, 101, 140, 201, 202
Phosphoinositide phosphatase, 102, 131, 216, 222, 248–251
Phospholamban, 201
Phospholipase C (PLC), 26, 33, 62
Phospholipase D (PLD), 138
Phospholipid binding pocket, 30
Phox2a, 229
PI3K/Akt signaling, 228, 230, 241, 242, 251, 284, 301
PI 3-K-C2, 98
PI 3-K effectors, 100, 102, 104
PI3K complex I, 130
PI3K complex II, 130
PI4KII, 2
PI4KIII, 2, 116, 117
PIAS1, 221
PICT-1, 286

PIK domain, 202, 203
Pik1, 4, 8, 9, 15
Pik1p, 8, 15
PIK3CA, 102, 103, 145, 146, 184, 226
PIK3CB, 102
PIK3R1, 101, 102, 146, 184
PIK3R2, 184
PIK3R3, 184
PIK93, 9, 12
PIKfyve kinase, 282
PINK1 (PTEN-induced putative kinase 1), 302, 303
PIP$_2$, *see* Phosphatidylinositol 4,5-bisphosphate
PIP$_3$, 26, 49, 187, 188, 195–199, 201, 202
PIPKI, 27, 29–31, 37, 46
PIPKII, 27, 29–31, 48
PIX, 136
PKA (protein kinase A), 27, 28, 31, 99, 201, 202, 249
PKC (protein kinase C), 8, 11, 31, 45, 99, 196
Platelet-activating factor (PAF), 189
PLCγ1, 221
Pleckstrin-homology (PH), 99, 126, 135, 249, 282
PML (promyelocytic leukemia protein), 288
Podosomes, 36, 226
Polycystic kidneys, 235, 296
Polydactyly, 235
Polymorphism, 129, 228
Polyoma tumor DNA virus, 96
Presynaptic nerve terminals, 231
P-REX1 (PtdIns(3,4,5)P_3-dependent Rac exchanger), 103, 249–251
P-REX2a, 249–251
P-REX2b, 249
Pro-apoptotic, 100, 302
Profilin, 26, 33, 36, 68
Progesterone receptor (PR), 248
Proline-rich inositol polyphosphate 5-phosphatase, 241
Proline-rich domains, 218, 227, 231
Prostate cancer, 145, 150, 185, 248, 251, 288, 296, 297, 300
Protein kinase D (PKD), 8, 17, 31
Protein phosphatase, 131, 283, 286–288, 310
Proteinuria, 236, 238, 239
Proteus, 283
Proteus-like syndrome, 283
PTB domains, 221
PtdIns3P, 96, 97, 99, 103, 112, 116, 117, 129
PtdIns4P, 96, 116, 117

PtdIns(3,4,5)P_3, 63, 75, 117, 126, 135, 217, 218, 223–225, 227, 231, 236, 242, 283, 302, 310
PtdIns(3,4,5)P_3 phosphatase, 101, 102
PtdIns(4,5)P_2, 10, 11, 67, 81, 117, 123, 217, 218, 231, 236, 240, 284, 310
PTEN (phosphatase and tensin homolog deleted on chromosome 10), 102
PTEN hamartoma tumor syndromes (PHTS), 301
Punctate opacities in the lens, 236
Purkinje cell, 13, 18, 247, 250
PX domain, 62, 103, 135
PxxP motifs, 221

Q
Quercetin, 149

R
R3-mtm-3, 307
R5-mtm-5, 307
R6-mtm-6, 307
Rab1, 237, 238, 240
Rab2, 240
Rab5, 128, 130, 237, 238, 240, 247, 311
Rab6, 237, 238, 240
Rab9, 240
Rab11, 8, 9, 37
Rac, 8, 11, 73, 76, 99, 135, 225, 237, 238, 240, 249, 251
Rac1, 35, 82, 234, 238, 239, 250, 251, 287
Raf, 99, 151
RalA, 69
RalGDS, 99
RAN (Ras-related nuclear protein), 288
RANTES, 187
Rap1, 77, 78
Rapalogue, 311
Raptor, 131, 146
Ras, 63, 77, 78, 84, 99, 127, 128, 144, 145, 149, 151, 301
Reactive oxygen species (ROS), 185, 188, 189, 225, 250, 301, 302, 305
Readily-releasable pool, 37
Respiratory burst, 101, 188, 189
Retinal degradation, 235
Retinoblastoma protein RB (pRB), 48
Retrograde trafficking, 238, 240, 253
Rheumatoid arthritis (RA), 141
Rho superfamily, 103
Rhodopsin, 71
Rictor, 131, 146
RID (Rac induced recruitment domain), 307
RNA interference, 286

RNA polymerase II, 47, 70
ROCK, 37, 38
ROS, *see* Reactive oxygen species
RTK (receptor tyrosine kinase), 74, 76, 96, 98, 249
Rubicon, 130
Ryanodine receptor, 201, 315

S
S phase, 70, 98, 104
S6K1 (p70 ribosomal protein S6 kinase-1), 98, 104
Sac domain, 218, 231, 251–253
Sac domain-containing inositol phosphatases (SCIPs), 251, 252
Sac domain phosphoinositide phosphatase, 251
Sac1, 12, 17, 252, 253, 283
Sac1 phosphatase, 18; *see also* Sac domain
SAC2/INPP5F, 251
Sac3, 254, 255
SAC3/FIG4, 251, 254
Salmonella, 246
Salmonella-containing vacuole (SCV), 246
SAM domain, 218, 227
Sarcopenia, 316
Schizophrenia, 73, 233
Schizosaccharomyces pombe, 67
Schwann cells, 313, 316
Ses1/2, 237, 238
Sec3, 37
Sec8, 316
Second messenger, 12, 44, 45, 49, 62, 69, 97, 201, 217, 239, 282
Sertoli cell, 241, 313
SGK3 (serum and glucocorticoid-regulated kinase-3), 103
SH2 (Src homology 2), 98
SH3 domain, 74, 75, 83
SH3-containing protein, 221
Shc, 221, 227, 286
Shigella flexneri, 244
SHIP (SH2-domain inositol phosphatase), 131, 137
SHIP1α, 219, 220
SHIP1δ, 219, 220
Ship2, 228–230
SID (set interacting domain), 307, 310, 314
SigD/SopB, 246
Single nucleotide polymorphisms (SNPs), 228, 316
SKICH (SKIP carboxyl homology domains), 218, 241, 242

SKIP (skeletal muscle and kidney inositol phosphatase), 236, 242, 243
SLP-76, 75
Smoothened, 17, 18
SopB, 246
Speckle targeted PIPKIalpha regulated-poly(A) polymerase (Star-PAP), 46
Sperm function, 241
Spermatids, 79, 241
Sphingomyelin, 12, 16
Sphingosine-1-phosphate receptor, 127
Splenomegaly, 222, 297
Squamous cell carcinomas (SCC), 231, 239
Src, 26, 36, 43, 75, 96, 97, 221, 246, 286
Staurosporine, 8, 149, 302
Steatohepatitis, 289
Sterile alpha motif (SAM), 218, 227
Stomach cancer, 235
Stress responses, 116
Stt4p, 11
Superoxide formation, 250
Suppressor of morphogenesis in genitalia-1 (SMG-1), 116
Swiss 3T3 cells, 45
Syk, 75, 126, 140, 188
Synaptic plasticity, 230, 232
Synaptic vesicles, 13, 16, 217, 232, 233
Synaptojanin 1, 218, 231–233, 252
Synaptojanin 2, 231, 233, 234
Synaptotagmin, 37
Syndapin, 231
SYNJ1, 231, 233
SYNJ2, 231, 233
Syntaxin 8, 13
Systemic lupus erythematosus (SLE), 142, 143, 193

T
T cell receptor interacting molecule (TRIM), 126
T loop kinase, 100
T lymphocytes, 136, 143, 189–191, 193, 222
Talin, 32, 36, 38
Tarceva, 151
Telomere, 297
Temsirolimus, 121, 148
TG100-115, 118, 140, 141, 194
Thr308, 100, 130
Thrombin, 78, 127, 226
Thrombin receptor, 77
Thymocytes, 190
TIRF (total internal reflection fluorescence) microscopy, 227

TMEM55A, 244, 245, 257
TMEM55B, 244, 245, 257
Toll—IL-1 receptor domain-containing adaptor protein (TIRAP), 126
Toll-like receptor (TLR), 126, 138
TORC1, 103, 131, 147, 148, 150
TORC2, 131, 146, 148
Transcription, 47–49, 131, 134, 248, 250, 284
Transferrin receptor, 238
Transformation/transcription domain-associated protein (TRRAP), 116
Transforming retroviral oncogene, 99
Trans-membrane phosphatase with tensin homology (TPTE), 283
Triose phosphate isomerase (TIM) barrel, 63, 65, 73, 74, 82, 83
TrkA (tropomyosin receptor kinase A), 238
Troponin I, 201
TSC1, 131
TSC2, 131, 147
Tuberin, 131; *see also* TSC2
Tumor suppressor, 48, 49, 70, 79, 102, 104, 145, 230, 248, 289, 305
Type 2 diabetic patient, 197, 228
Type III secretion machinery, 244
Type IV 5-phosphatase, see *INPP5E*

U
unc-80, 233
Uncoating, 232
3'-untranslated region (UTR), 228
UVRAG, 130

V
Vac14, 254
Vacuole, 26, 240, 246, 252
v-Akt, 99
Vascular endothelial growth factor (VEGF), 76, 144, 147
Vasculogenesis, 76
Vav, 136
Villin, 35
Vinculin, 36, 38
Vinexin, 227
Vps15p, 129
Vps30p, 130
Vps34, 116, 117, 129, 130, 282
VSVG, 16, 17

W
WASP, *see* Wiskott Aldrich Syndrome protein
Weight gain, 229
Winged helix/forkhead family, 69

Wingless, 10
Wiskott Aldrich Syndrome protein (WASP), 68
Wnt ligand, 10
Wnt signalling, 11
Wortmannin, 9, 11, 98, 101, 116, 117, 149, 150, 187–189, 193, 196, 197, 301
WW, 218, 227

X
Xenopus laevis, 75

X-linked centronuclear (myotubular) myopathy, 312
X-linked disorder, 236
X-linked myotubular myopathy, 112; *see also* Myotubular myopathy

Y
YOTB, 135; *see also* FYVE domain

Z
ZAP-70, 75, 126